Étale Cohomology

Princeton Mathematical Series

1. The Classical Groups *by Hermann Weyl*
3. An Introduction to Differential Geometry *by Luther Pfahler Eisenhart*
4. Dimension Theory *by W. Hurewicz and H. Wallman*
8. Theory of Lie Groups: I *by C. Chevalley*
9. Mathematical Methods of Statistics *by Harald Cramér*
10. Several Complex Variables *by S. Bochner and W. T. Martin*
11. Introduction to Topology *by S. Lefschetz*
12. Algebraic Geometry and Topology *edited by R. H. Fox, D. C. Spencer, and A. W. Tucker*
14. The Topology of Fibre Bundles *by Norman Steenrod*
15. Foundations of Algebraic Topology *by Samuel Eilenberg and Norman Steenrod*
16. Functionals of Finite Riemann Surfaces *by Menahem Schiffer and Donald C. Spencer*
17. Introduction to Mathematical Logic, Vol. I *by Alonzo Church*
19. Homological Algebra *by H. Cartan and S. Eilenberg*
20. The Convolution Transform *by I. I. Hirschman and D. V. Widder*
21. Geometric Integration Theory *by H. Whitney*
22. Qualitative Theory of Differential Equations *by V. V. Nemytskii and V. V. Stepanov*
23. Topological Analysis *by Gordon T. Whyburn* (revised 1964)
24. Analytic Functions *by Ahlfors, Behnke, Bers, Grauert et al.*
25. Continuous Geometry *by John von Neumann*
26. Riemann Surfaces *by L. Ahlfors and L. Sario*
27. Differential and Combinatorial Topology *edited by S. S. Cairns*
28. Convex Analysis *by R. T. Rockafellar*
29. Global Analysis *edited by D. C. Spencer and S. Iyanaga*
30. Singular Integrals and Differentiability Properties of Functions *by E. M. Stein*
31. Problems in Analysis *edited by R. C. Gunning*
32. Introduction to Fourier Analysis on Euclidean Spaces *by E. M. Stein and G. Weiss*
33. Étale Cohomology *by J. S. Milne*
34. Pseudodifferential Operators *by Michael E. Taylor*
35. Three-Dimensional Geometry and Topology, Volume 1 *by William P. Thurston, Edited by Silvio Levy*
36. Representation Theory of Semisimple Groups: An Overview Based on Examples *by Anthony W. Knapp*
37. Foundations of Algebraic Analysis *by Masaki Kashiwara, Takahiro Kawai, and Tatsuo Kimura. Translated by Goro Kato*
38. Spin Geometry *by H. Blaine Lawson, Jr., and Marie-Louise Michelsohn*
39. Topology of 4-Manifolds *by Michael H. Freedman and Frank Quinn*
40. Hypo-Analytic Structures: Local Theory *by François Treves*
41. The Global Nonlinear Stability of the Minkowski Space *by Demetrios Christodoulou and Sergiu Klainerman*
42. Essays on Fourier Analysis in Honor of Elias M. Stein *edited by C. Fefferman, R. Fefferman, and S. Wainger*
43. Harmonic Analysis: Real-Variable Methods, Orthogonality, and Oscillatory Integrals *by Elias M. Stein*
44. Topics in Ergodic Theory *by Ya. G. Sinai*
45. Cohomological Induction and Unitary Representations *by Anthony W. Knapp and David A. Vogan, Jr.*

Étale Cohomology

J. S. Milne

PRINCETON UNIVERSITY PRESS
PRINCETON AND OXFORD

Published by Princeton University Press
41 William Street, Princeton, New Jersey 08540
99 Banbury Road, Oxford OX2 6JX

press.princeton.edu

First published in 1980
New cloth and paperback printing, 2025
Cloth ISBN 9780691273792
Paperback ISBN 9780691273785
ISBN (e-book) 9780691273778

The Library of Congress has cataloged a prior edition of this book as follows:

Milne, James S 1942–
Étale cohomology.
(Princeton mathematical series : 33)
Bibliography: p.
Includes index.
I. Geometry, Algebraic. 2. Homology theory.
3. Sheaves, Theory of. I. Title. II. Series.
QA564.M52 514'23 79-84003
ISBN-13: 978-0-691-08238-7 (cloth)
ISBN-10: 0-691-08238-3 (cloth)

This book has been composed in Monophoto Times Roman

La mer était étale, mais le reflux commençait à se faire sentir.

Hugo, *Les travailleurs de la mer.*

Contents

Preface

The purpose of this book is to provide a comprehensive introduction to the étale topology, sheaf theory, and cohomology. When a variety is defined over the complex numbers, the complex topology may be used to define cohomology groups that reflect the structure of the variety much more strongly than do those defined, for example, by the Zariski topology. For an arbitrary scheme the complex topology is not available, but the étale topology, whose definition is purely algebraic, may be regarded as a replacement. It gives a sheaf theory and cohomology theory with properties very close to those arising from the complex topology. When both are defined for a variety over the complex numbers, the étale and complex cohomology groups are closely related. On the other hand, when the scheme is the spectrum of a field and hence has only one point, the étale topology need not be trivial, and in fact the étale cohomology of the scheme is precisely equivalent to the Galois cohomology of the field. Étale cohomology has achieved an importance for the study of schemes comparable to that of complex cohomology for the study of the geometry of complex manifolds or of Galois cohomology for the study of the arithmetic of fields.

The étale topology was initially defined by A. Grothendieck and developed by him with the aid of M. Artin and J.-L. Verdier in order to explain Weil's insight (Weil [1]) that, for polynomial equations with integer coefficients, the complex topology of the set of complex solutions of the equations should profoundly influence the number of solutions of the equations modulo a prime number. In this, the étale topology has been brilliantly successful. We give a sketch of the explanation it provides. It must be assumed the equations define a scheme proper and smooth over some ring of integers. The complex topology on the complex points of the scheme determines the complex cohomology groups. The comparison theorem says that these groups are essentially the same as the étale cohomology groups of the scheme regarded as a variety over the complex numbers. The proper and smooth base change theorems now show that these last groups are canonically isomorphic to the étale cohomology groups of the scheme regarded as a variety over the algebraic closure of a finite residue field. But the points of the scheme with coordinates in a finite field are the fixed points of the Frobenius operator acting on the set of points of the scheme with coordinates in the algebraic closure of the finite field. The Lefschetz trace formula now shows that the number of points in the finite field may be computed from the trace of the Frobenius operator acting on étale cohomology groups that are essentially equal to the original complex cohomology groups. A large part of this book may be regarded as a justification of this sketch.

To give the reader some idea of the similarities and differences to be expected between the étale and complex theories, we consider the case of a projective nonsingular curve X of genus g over an algebraically closed field k. If k is the complex numbers, then X may be regarded as a one-dimensional compact complex manifold $X(\mathbb{C})$, and its fundamental group $\pi_1(X, x)$ has $2g$ generators and a single, well-known relation. The most interesting cohomology group is H^1, and $H^1(X(\mathbb{C}), \Lambda) = \text{Hom}(\pi_1(X, x), \Lambda)$ for a constant abelian sheaf Λ; for example, $H^1(X(\mathbb{C}), \mathbb{Z}) = \mathbb{Z}^{2g}$. If k is arbitrary, then it is possible to define in a purely algebraic way a fundamental group $\pi_1^{\text{alg}}(X, x)$ that, when k is of characteristic zero, is the profinite completion of $\pi_1(X, x)$. The étale cohomology group $H^1(X_{\text{et}}, \Lambda) = \text{Hom}(\pi_1^{\text{alg}}(X, x), \Lambda)$ for any constant abelian sheaf Λ, but now Hom refers to continuous homomorphisms. Thus $H^1(X_{\text{et}}, \Lambda) = \Lambda^{2g}$, if Λ is finite or is the l-adic integers $\mathbb{Z}_l$. But $H^1(X_{\text{et}}, \mathbb{Z}) = 0$, for $\mathbb{Z}$ must be given the discrete topology, and the image of any continuous map $\pi_1^{\text{alg}}(X, x) \to \mathbb{Z}$ is finite. Therefore the étale cohomology is as expected in the first two cases but is anomolous in the last.

It may seem that the étale topology should be superfluous when k is the complex numbers, but this is not so; the étale groups have one important advantage over the complex groups, namely, that if X is defined over a subfield k_0 of k, then any automorphism of k/k_0 acts on $H^r(X_{\text{et}}, \Lambda)$.

The book's first chapter is concerned with the properties of étale morphisms, Henselian rings, and the algebraic fundamental group. It had been my original intention to state these without proof, but this would have been unsatisfactory since one of the essential differences between étale sheaf theory and the usual sheaf theory is that trivial facts from point set topology must frequently be replaced by subtle facts from algebraic geometry. On the other hand to give a complete treatment of these topics would require a book in itself. Thus Chapter I is a compromise; almost everything about étale morphisms and Henselian rings is proved and almost nothing about the fundamental group. The prerequisites for this chapter are a solid knowledge of basic commutative algebra, for example, the contents of Atiyah and Macdonald [1], plus a reasonable understanding of the language of schemes.

The next two chapters are concerned with the basic theory of étale sheaves and with elementary étale cohomology. The prerequisite for these chapters is some knowledge of homological algebra and Galois cohomology.

The fourth chapter treats Azumaya algebras over schemes and the Brauer groups of schemes. Here it is assumed that the reader is familiar with the corresponding objects over fields. This chapter may be skipped.

The fifth chapter contains a detailed analysis of the cohomology of curves and of surfaces. The section on curves assumes a knowledge of the representation theory of finite groups and that on surfaces assumes a more detailed knowledge of algebraic geometry than required earlier in the book.

The sixth chapter proves the fundamental theorems in étale cohomology and applies them to show the rationality of some very general classes of zeta functions and L-series.

The appendixes list definitions and results concerning limits, spectral sequences, and hypercohomology that the reader may find useful.

The most striking application of étale cohomology, that of Deligne to proving the Weil-Riemann hypothesis, is not included, but anyone who reads this book will find little difficulty with Deligne's original paper. Essentially the only results he uses that are not included here concern Lefschetz pencils of odd fiber dimension. However, we do treat Lefschetz pencils of fiber dimension one, and the general case is very similar and only slightly more difficult.

I have tried to keep things as concrete as possible. Only enough foundational material is included to treat the étale site and similar sites, such as the flat and Zariski sites. In particular, the word *topos* does not occur. Derived categories are not used although their spirit pervades the last part of Chapter VI.

For an account of the origins of étale cohomology and its results up to the mid 1960s, I recommend Artin's talk at the International Congress in Moscow, 1966 [3]; for a "popular" account of the history of the Weil conjectures (which is intimately related to the history of étale cohomology) and of Deligne's solution, I recommend Katz's article [2], and for a survey of the main ideas and results in étale cohomology and their relations to their classical analogues, I recommend Deligne's Arcata lectures [SGA. $4\frac{1}{2}$, Arcata]. The best introduction to the material from algebraic geometry required for reading this book is provided by Hartshorne [2].

It is a pleasure to thank M. Artin for explaining a number of points to me, R. Hoobler for his comments on Chapter IV, and the Institute for Advanced Study and l'Institut des Hautes Etudes Scientifiques where parts of the book were written.

Terminology and Conventions

All rings are Noetherian and all schemes are locally Noetherian. A variety is a geometrically reduced and irreducible scheme of finite-type over a field, and a curve or surface is a variety of dimension one or two.

For a field k, k_s or k_{sep} is the separable algebraic closure of k and k_{al} the algebraic closure. If K is Galois over k, then $\text{Gal}(K/k)$ or $G(K/k)$ is the corresponding Galois group; G_k denotes $\text{Gal}(k_s/k)$.

For a ring A, A^* denotes the group of units of A and $k(\mathfrak{p})$ the field of fractions of $A/\mathfrak{p}$, where $\mathfrak{p}$ is a prime ideal in A.

For a scheme X, $R(X)$ is the ring of rational functions on X, X_i the set of points x of codimension i (that is, such that $\dim O_{X,x} = i$), and X^i the set of points x of dimension i (that is, such that $\overline{\{x\}}$ has dimension i). A geometric point of X is a map $z \to X$ where z is the spectrum of a separably closed field.

Sets is the category of sets, **Ab** the category of abelian groups, **Gp** the category of groups, G-**sets** the category of finite sets on which G acts (continuously on the left), G-**mod** the category of (discrete) G-modules, **Sch**/X the category of schemes over X, **FEt**/X the category of schemes finite and étale over X, **LFT**/X the category of schemes locally of finite-type over X, and **Fun**(**C**, **A**) the category of functors from **C** to **A**.

The symbols $\mathbb{N}$, $\mathbb{Z}$, $\mathbb{Q}$, $\mathbb{R}$, $\mathbb{C}$, $\mathbb{F}_q$ denote respectively, the natural numbers, the ring of integers, the field of rational numbers, the field of real numbers, the field of complex numbers, and the finite field of q elements.

The symbols α_p, μ_n, $\mathbb{G}_m$, $\mathbb{G}_a$ denote certain group schemes (II.2.18).

An injection is denoted by $\hookrightarrow$, a surjection by $\twoheadrightarrow$, an isomorphism by $\approx$, a quasi-isomorphism (or homotopy) by $\sim$, and a canonical isomorphism by $=$. The symbol $X \overset{df}{=} Y$ means X is defined to be Y, or that X equals Y by definition.

The kernel and cokernel of multiplication by n, $M \xrightarrow{n} M$, are denoted respectively by M_n and $M^{(n)}$.

The empty set and empty scheme are both denoted by $\emptyset$.

The symbol $b \gg a$ means b is sufficiently greater than a.

Étale Cohomology

CHAPTER I

Étale Morphisms

A flat morphism is the algebraic analogue of a map whose fibers form a continuously varying family. For example, a surjective morphism of smooth varieties is flat if and only if all fibers have the same dimension. A finite morphism to a reduced scheme is flat if and only if, over any connected component, all fibers have the same number of points (counting multiplicities). The image of a flat morphism of finite-type is open, and flat morphisms that are surjective on the underlying spaces are epimorphisms in a very strong sense.

An étale morphism is a flat quasi-finite morphism $Y \to X$ with no ramification (that is, branch) points. Locally Y is then defined by an equation $T^m + a_1 T^{m-1} + \cdots + a_m = 0$, where $a_1, \ldots, a_m$ are functions on an open subset U of X and all roots of the equation over a point of U are simple. An étale morphism induces isomorphisms on the tangent spaces and so might be expected to be a local isomorphism. This is true over the complex numbers if local is meant in the sense of the complex topology, but the Zariski topology is too coarse for this to hold algebraically. However, an étale morphism induces an isomorphism on the completions of the local rings at a point where there is no residue field extension. Moreover, it has all the uniqueness properties of a local isomorphism.

A local scheme is Henselian if, for any scheme étale over it, any section of the closed fiber extends to a section of the scheme. It is strictly Henselian, or strictly local, if any scheme étale and faithfully flat over it has a section. The strictly local rings play the same role for the étale topology as local rings play for the Zariski topology.

The fundamental group of a scheme classifies finite étale coverings of it. For a smooth variety over the complex numbers, the algebraic fundamental group is simply the profinite completion of the topological fundamental group. There are algebraic analogues for many of the results on the topological fundamental group.

§1. Finite and Quasi-Finite Morphisms

Recall that a morphism of schemes $f: Y \to X$ is *affine* if the inverse image of any open affine subset U of X is an open affine subset of Y. If, moreover, $\Gamma(f^{-1}(U), \mathcal{O}_Y)$ is a finite $\Gamma(U, \mathcal{O}_X)$-algebra for every such U, then f is said to be *finite*. These conditions need only be checked for all U in some open affine covering of X (Mumford [3, III.1, Prop. 5]).

Examples of finite morphisms abound. Let X be an integral scheme with field of rational functions $R(X)$, and let L be a finite field extension of $R(X)$. The *normalization* of X in L is a pair (X', f) where X' is an integral scheme with $R(X') = L$ and $f: X' \to X$ is an affine morphism such that, for all open affines U of X, $\Gamma(f^{-1}(U), \mathcal{O}_{X'})$ is the integral closure of $\Gamma(U, \mathcal{O}_X)$ in L.

PROPOSITION 1.1. *If X is normal and $f: X' \to X$ is the normalization of X in some finite separable extension of $R(X)$, then f is finite.*

Proof. One has only to show that $\Gamma(f^{-1}(U), \mathcal{O}_{X'})$ is a finite $\Gamma(U, \mathcal{O}_X)$-algebra for U an open affine in X, but this is done in Atiyah-Macdonald [1, 5.17].

Remark 1.2. The above proposition holds for many schemes X without the separability assumption, for example, for reduced excellent schemes and so for varieties ([EGA. IV.7.8] and Bourbaki [2, V.3.2]). (A field is excellent and a Dedekind domain A is excellent if the completion $\hat{K}$ of its field of fractions K at any maximal ideal of A is separable over K; any scheme of finite type over an excellent scheme is excellent.)

PROPOSITION 1.3. (a) *A closed immersion is finite.*

(b) *The composite of two finite morphisms is finite.*

(c) *Any base change of a finite morphism is finite, that is, if $f: Y \to X$ is finite, then so also is $f_{(X')}: Y_{(X')} \to X'$ for any morphism $X' \to X$.*

Proof. These reduce easily to statements about rings, all of which are obvious.

The "going up" theorem of Cohen-Seidenberg has the following geometric interpretation.

PROPOSITION 1.4. *Any finite morphism $f: Y \to X$ is proper, that is, it is separated, of finite-type, and universally closed.*

Proof. For any open affine covering (U_i) of X, f restricted to $f^{-1}(U_i) \to U_i$ is separated for all i, and so f is separated. (Hartshorne [2, II.4.6]). To show that finite morphisms are universally closed it suffices, according to (1.3c), to show that they are closed, and for this it suffices, according to (1.3a,b), to show that they map the whole space onto a closed set. Thus we must show that $f(Y)$ is closed. This reduces easily to the affine case with, for example, $f = {}^a g$ where $g: A \to B$ is finite. Let $\mathfrak{I} = \ker(g)$. Then f factors into spec $B \to$ spec $A/\mathfrak{I} \to$ spec A. The first map is surjective (Atiyah-Macdonald [1, 5.10]), and the second is a closed immersion.

For morphisms $X \to$ spec k, with k a field, there is a topological characterization of finiteness.

PROPOSITION 1.5. *Let $f: X \to$ spec k be a morphism of finite-type with k a field. The following are equivalent:*

(a) *X is affine and $\Gamma(X, \mathcal{O}_X)$ is an Artin ring;*

(b) *X is finite and discrete (as a topological space);*

(c) *X is discrete;*

(d) *f is finite.*

Proof. See Atiyah-Macdonald [1, Chapter VIII, especially exercises 2, 3, 4].

A morphism $f: Y \to X$ is *quasi-finite* if it is of finite-type and has finite fibers, that is, $f^{-1}(x)$ is discrete (and hence finite) for all $x \in X$. Similarly an A-algebra B is *quasi-finite* if it is of finite-type and if $B \otimes_A k(\mathfrak{p})$ is a finite $k(\mathfrak{p})$-algebra for all prime ideals $\mathfrak{p} \subset A$.

Exercise 1.6. (a) Let A be a discrete valuation ring. Show that $A[T]/(P(T))$ is a quasi-finite A-algebra if and only if some coefficient of $P(T)$ is a unit, and that it is finite if and only if the leading coefficient of $P(T)$ is a unit.

(b) Let A be a Dedekind domain with field of fractions K. Show that spec $K \to$ spec A is never finite, that it is quasi-finite if it is of finite-type, and that it is of finite-type if and only if A has only finitely many prime ideals.

PROPOSITION 1.7. (a) *Any immersion is quasi-finite.*

(b) *The composite of two quasi-finite morphisms is quasi-finite.*

(c) *Any base change of a quasi-finite morphism is quasi-finite.*

Proof. (a) Let $f: Y \to X$ be an immersion. Clearly f has finite fibers, and to show that it is of finite-type it suffices to show that $f^{-1}(U)$ is quasi-compact for any open affine U in X. But U is a Noetherian topological space (recall that all schemes are locally Noetherian), and $f^{-1}(U) = U \cap Y$ is a subset of U.

(b) This is obvious.

(c) Let $f: Y \to X$ be quasi-finite and $X' \to X$ arbitrary. If $x' \mapsto x$ under $X' \to X$, then the fiber

$$f_{(X')}^{-1}(x') = f^{-1}(x) \otimes_{k(x)} k(x')$$

and hence is discrete.

If $f: Y \to X$ is finite and U is an open subscheme of Y, then it follows from the above proposition that $U \to X$ is quasi-finite. The remarkable thing is that essentially every quasi-finite morphism comes in this way.

THEOREM 1.8. (Zariski's Main Theorem). *If X is a quasi-compact, then any separated, quasi-finite morphism $f: Y \to X$ factors as $Y \xrightarrow{f'} Y' \xrightarrow{g} X$ where f' is an open immersion and g is finite.*

Proof. The most elementary proof may be found in Raynaud [3, Chapter IV]. We sketch the deduction of (1.8) from the following affine form of it, proved in Raynaud [3, p. 42]: let B be an A-algebra that is quasi-finite, and let A' be the integral closure of A in B; then the map spec $B \to$ spec A' is an open immersion.

Consider a scheme X. Associated with any quasi-coherent sheaf A of $\mathcal{O}_X$-algebras, there is a pair (X', g) where X' is a scheme and $g: X' \to X$ is an affine morphism such that $g_*\mathcal{O}_{X'} = A$ (Hartshorne [2, II. Ex. 5.17] and [EGA. I.9.1.4]). One writes $X' = \mathbf{spec}\, A$. For any X-scheme $Y \xrightarrow{f} X$, to give an X-morphism $Y \to X'$ is the same as to give a homomorphism $A \to f_*\mathcal{O}_Y$ of $\mathcal{O}_X$-algebras.

Now let $f: Y \to X$ be separated and of finite-type. Then $f_*\mathcal{O}_Y$ is a quasi-coherent $\mathcal{O}_X$-algebra [EGA. I.6.7.1], and the $\mathcal{O}_X$-algebra A' such that $\Gamma(U, A')$ is the integral closure of $\Gamma(U, \mathcal{O}_X)$ in $\Gamma(U, f_*\mathcal{O}_Y)$ for all open affines $U \subset X$ is also quasi-coherent [EGA. II.6.3.4]. The associated X-scheme $X' = \mathbf{spec}\, A'$ is called the normalization of X in Y.

Assume further that f is quasi-finite. It follows easily from the affine result quoted above, that the morphism $Y \to X'$ induced by the inclusion $A' \subset f_*\mathcal{O}_Y$ is an open immersion. Now let (A_i) be the family of all coherent $\mathcal{O}_X$-subalgebras of A'. One checks easily that the morphism $Y \to \mathbf{spec}\, A_i$, induced by the inclusion $A_i \subset f_*\mathcal{O}_Y$, is an open immersion for all sufficiently large A_i (using the fact that $A' = \bigcup A_i$; compare the proof of Raynaud [3, p. 42, Cor. 2(2)]). Since $\mathbf{spec}\, A_i$ is finite over X, this proves (1.8).

Remark 1.9. Zariski's main theorem is, more correctly, the main theorem of Zariski [2]. There he was interested in the behavior of a singularity on a normal variety under a birational map. The original statement is essentially that if $f: Y \to X$ is a birational morphism of varieties and $\mathcal{O}_{X,x}$ is integrally closed, then either $f^{-1}(x)$ consists of one point and the inverse morphism f^{-1} is defined in a neighborhood of x or all components of $f^{-1}(x)$ have dimension ≥ 1. To relate this to Grothendieck's version, note that if in (1.8) X and Y are varieties, f is birational and X is normal, then g is an isomorphism. For a more complete discussion of the theorem, see Mumford [3, III.9].

COROLLARY 1.10. *Any proper, quasi-finite morphism $f: Y \to X$ is finite.*

Proof. Let $f = gf'$ be the factorization as in (1.8). As g is separated and f is proper, f' is proper. (Use the factorization

$$f' = f_{(Y')} \circ \Gamma_{f'} : Y \to Y \times_X Y' \to Y'.)$$

Thus f' is an immersion with closed image, that is, a closed immersion. Now both f' and g are finite.

Remark 1.11. The separatedness is necessary in both of the above results; for if X is the affine line with the "origin doubled" (Hartshorne [2, II.2.3.6]), and $f:X \to \mathbb{A}^1$ is the natural map, then f is universally closed and quasi-finite, but is not finite. (It is even flat and étale; see the next two sections.)

Exercise 1.12. Let $f: Y \to X$ be separated and of finite-type with X irreducible. Show that if the fiber over the generic point η is finite, then there is an open neighborhood U of η in X such that $f^{-1}(U) \to U$ is finite.

§2. Flat Morphisms

A homomorphism $f:A \to B$ of rings is *flat* if B is flat when regarded as an A-module via f. Thus, f is flat if and only if the functor $-\otimes_A B$ from A-modules to B-modules is exact. In particular, if $\mathfrak{I}$ is any ideal of A and f is flat, then $\mathfrak{I} \otimes_A B \to A \otimes_A B = B$ is injective. The converse to this statement is also true.

PROPOSITION 2.1. *A homomorphism $f:A \to B$ is flat if $(a \otimes b \mapsto f(a)b)$: $\mathfrak{I} \otimes_A B \to B$ is injective for all ideals $\mathfrak{I}$ in A.*

Proof. Let $g:M' \to M$ be an injective map of A-modules where, following Atiyah-Macdonald [1, 2, 19], we may assume M to be finitely generated.

Case (a) M is free. We prove this case by induction on the rank r of M. If $r = 1$, then we may identify M with A and M' with an ideal in A; then the statement to be proved is the statement given. If $r > 1$, then $M = M_1 \oplus M_2$ with M_1 and M_2 free of rank $<r$. Consider the exact commutative diagram:

$$\begin{array}{ccccccccc}
0 & \longrightarrow & M_1 & \longrightarrow & M & \longrightarrow & M_2 & \longrightarrow & 0 \\
 & & \uparrow{\scriptstyle g_1} & & \uparrow{\scriptstyle g} & & \uparrow{\scriptstyle g_2} & & \\
0 & \longrightarrow & g^{-1}(M_1) & \longrightarrow & M' & \longrightarrow & pg(M') & \longrightarrow & 0
\end{array}$$

When tensored with B, the top row remains exact, and g_1 and g_2 remain injective. This implies that $g \otimes 1$ is injective.

Case (b) M arbitrary (finitely generated). Let $x_1, \ldots, x_r$ generate M, let M^* be the free A-module on $x_1, \ldots, x_r$, and consider the exact commutative diagram:

$$\begin{array}{ccccccccc} 0 & \longrightarrow & N & \xrightarrow{j} & M^* & \xrightarrow{h} & M & \longrightarrow & 0 \\ & & \| & & \uparrow{\scriptstyle i} & & \uparrow{\scriptstyle g} & & \\ 0 & \longrightarrow & N & \longrightarrow & h^{-1}g(M') & \longrightarrow & M' & \longrightarrow & 0. \end{array}$$

By case (a) $i \otimes 1$ is injective, and it follows that $g \otimes 1$ is injective.

PROPOSITION 2.2. *If $f: A \to B$ is flat, then so also is $S^{-1}A \to T^{-1}B$ for all multiplicative subsets $S \subset A$ and $T \subset B$ such that $f(S) \subset T$. Conversely, if $A_{f^{-1}(\mathfrak{n})} \to B_{\mathfrak{n}}$ is flat for all maximal ideals $\mathfrak{n}$ of B, then $A \to B$ is flat.*

Proof. $S^{-1}A \to S^{-1}B$ is flat according to Atiyah-Macdonald [1, 2.20], and $S^{-1}B \to T^{-1}B$ is flat according to Atiyah-Macdonald [1, 3.6]. For the converse, let $M' \to M$ be an injective map of A-modules. To show that $B \otimes_A M' \to B \otimes_A M$ is injective, it suffices to show that

$$B_{\mathfrak{n}} \otimes_B (B \otimes_A M') \to B_{\mathfrak{n}} \otimes_B (B \otimes_A M)$$

is injective for all $\mathfrak{n}$, but this follows from the flatness of $A_{\mathfrak{p}} \to B_{\mathfrak{n}}$ with $\mathfrak{p} = f^{-1}(\mathfrak{n})$ and the isomorphism $B_{\mathfrak{n}} \otimes_B (B \otimes_A N) \approx B_{\mathfrak{n}} \otimes_{A_{\mathfrak{p}}} (A_{\mathfrak{p}} \otimes_A N)$, which exists for any A-module N.

Remark 2.3. If $a \in A$ is not a zero-divisor and $f: A \to B$ is flat, then $f(a)$ is not a zero-divisor in B because the injectivity of $(x \mapsto ax): A \to A$ implies that of $(x \mapsto f(a)x): B \to B = A \otimes_A B$. Thus, if A is an integral domain and $B \neq 0$, then f is injective. Conversely, any injective homomorphism $f: A \to B$, where A is a Dedekind domain and B is an integral domain, is flat, In proving this, we may localize and hence assume that A is principal. According to (2.1), it suffices to prove that for any ideal $\mathfrak{I} \neq 0$ of A, $\mathfrak{I} \otimes_A B \to B$ is injective, but $\mathfrak{I} \otimes_A B$ is a free B-module of rank one, and we know that the generator of $\mathfrak{I}$ is not mapped to zero in B.

A morphism $f: Y \to X$ of schemes is *flat* if, for all points y of Y, the induced map $\mathcal{O}_{X,f(y)} \to \mathcal{O}_{Y,y}$ is flat. Equivalently f is flat if for any pair V and U of open affines of Y and X such that $f(V) \subset U$, the map $\Gamma(U, \mathcal{O}_X) \to \Gamma(V, \mathcal{O}_Y)$ is flat. From (2.2) it follows that the first condition needs only to be checked for closed points y of Y.

PROPOSITION 2.4. (a) *An open immersion is flat.*

(b) *The composite of two flat morphisms is flat.*

(c) *Any base extension of a flat morphism is flat.*

Proof. (a) and (b) are obvious.

(c) If $f: A \to B$ is flat and $A \to A'$ is arbitrary, then to see that $A' \to B \otimes_A A'$ is flat, one may use the canonical isomorphism $(B \otimes_A A') \otimes_{A'} M \approx B \otimes_A M$, which exists for any A'-module M.

In order to get less trivial examples of flat morphisms we shall need the following criterion.

PROPOSITION 2.5. *Let B be a flat A-algebra, and consider $b \in B$. If the image of b in $B/\mathfrak{m}B$ is not a zero-divisor for any maximal ideal $\mathfrak{m}$ of A, then $B/(b)$ is a flat A-algebra.*

Proof. After applying (2.2), we may assume that $A \to B$ is a local homomorphism of local rings. By assumption, if $c \in B$ and $bc = 0$, then $c \in \mathfrak{m}B$. We shall show by induction that in fact $c \in \mathfrak{m}^r B$ for all r, and hence $c \in \bigcap \mathfrak{m}^r B \subset \bigcap \mathfrak{n}^r = (0)$, where $\mathfrak{n}$ is the maximal ideal of B. Assume that $c \in \mathfrak{m}^r B$, and write

$$c = \sum a_i b_i$$

where the a_i form a minimal generating set for $\mathfrak{m}^r$. Then

$$0 = bc = \sum_i a_i b_i b,$$

and so, by one of the standard flatness criteria (proved in (2.10b′) below), there are equations

$$b_i b = \sum_j a_{ij} b'_j$$

with the $b'_j \in B$, $a_{ij} \in A$ such that

$$\sum_i a_i a_{ij} = 0$$

for all j. From the choice of the a_i, all $a_{ij} \in \mathfrak{m}$. Thus $b_i b \in \mathfrak{m}B$, and since b is not a zero-divisor in $B/\mathfrak{m}B$, this implies that $b_i \in \mathfrak{m}B$. Thus $c \in \mathfrak{m}^{r+1}B$, which completes the induction. We have shown that b is not a zero-divisor in B, and the same argument, with A replaced by $A/\mathfrak{I}$ and B by $B/\mathfrak{I}B$, shows that b is not a zero-divisor in $B/\mathfrak{I}B$ for any ideal $\mathfrak{I}$ of A.

Fix such an ideal, and consider the exact commutative diagram:

$$\begin{array}{ccccccccc}
 & & 0 & & 0 & & & & \\
 & & \downarrow & & \downarrow & & & & \\
 & & \mathfrak{I} \otimes B & \longrightarrow & \mathfrak{I} \otimes B & \longrightarrow & \mathfrak{I} \otimes (B/(b)) & \longrightarrow & 0 \\
 & & \downarrow & & \downarrow & & \downarrow & & \\
0 & \longrightarrow & B & \xrightarrow{b} & B & \longrightarrow & B/(b) & \longrightarrow & 0 \\
 & & \downarrow & & \downarrow & & \downarrow & & \\
0 & \longrightarrow & B/\mathfrak{I}B & \xrightarrow{b} & B/\mathfrak{I}B & \longrightarrow & (B/(b))/\mathfrak{I}(B/(b)) & \longrightarrow & 0 \\
 & & \downarrow & & \downarrow & & \downarrow & & \\
 & & 0 & & 0 & & 0 & &
\end{array}$$

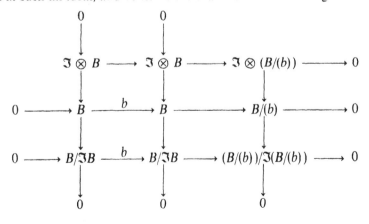

in which b means multiplication by b. An application of the snake lemma shows that $\mathfrak{J} \otimes B/(b) \to B/(b)$ is injective, which shows that $B/(b)$ is flat over A, according to (2.1).

Remarks 2.6. (a) For any ring A, $A[X_1, \ldots, X_n]$ is a free A-module, and so $\mathbb{A}^n_A$ is flat over A. Let Z be a hypersurface in $\mathbb{A}^n_A$, that is, a scheme of the form spec $(A[X_1, \ldots, X_n]/(P))$, $P \neq 0$. Then (2.5) shows that Z is flat over spec $A \Leftrightarrow$ for all maximal ideals $\mathfrak{m}$ of A, $Z \otimes_A k(\mathfrak{m}) \neq \mathbb{A}^n_{k(\mathfrak{m})} \Leftrightarrow$ the ideal generated by the coefficients of P is A (assuming that spec A is connected). Similar statements hold for hypersurfaces in $\mathbb{P}^n_A$.

(b) We may restate (a) as follows: a hypersurface Z is flat if and only if its closed fibers over spec A all have the same dimension. This generalizes. Firstly, if $f: Y \to X$ is flat, then

$$\dim(\mathcal{O}_{Y_x,y}) = \dim(\mathcal{O}_{Y,y}) - \dim(\mathcal{O}_{X,x}),$$

where $x = f(y)$. For varieties this means that $\dim(Y_x) = \dim(Y) - \dim(X)$ for any closed point x of X with Y_x nonempty. The proof, which is quite elementary, may be found in [EGA. IV.6.1] or Hartshorne [2, III.9.5]. Secondly, if X and Y are regular schemes and $f: Y \to X$ is such that

$$\dim(\mathcal{O}_{Y_x,y}) = \dim(\mathcal{O}_{Y,y}) - \dim(\mathcal{O}_{X,x})$$

for all closed points y of Y, where $x = f(y)$, then f is flat. The proof again may be found in [EGA. IV.6.1]. (See also Hartshorne [2, III. Ex. 10.9].)

(c) There is another criterion for flatness that is frequently very useful. It is easy to construct examples $Z \xrightarrow{f} Y \xrightarrow{g} X$ in which g and gf are flat, but f is not flat. However, if one also knows that the maps on fibers $f_x: Z_x \to Y_x$ are flat for all closed $x \in X$, then f is flat ([SGA. 1, IV.5.9], or Bourbaki [2, III.5.4 Prop. 2,3]).

(d) If B is flat over A and $b_1, \ldots, b_n$ is a sequence of elements of B whose image in $B/\mathfrak{m}B$ is regular for each maximal ideal $\mathfrak{m}$ of A, that is, b_i is not a zero-divisor in $B/\mathfrak{m} + (b_1, b_2, \ldots, b_{i-1})$ for any i, then $B/(b_1, \ldots, b_n)$ is flat over A. This follows by induction from (2.5).

(e) There is a second generalization of (a). Let X be an integral scheme and Z a closed subscheme of $\mathbb{P}^n_X$; for each $x \in X$, let $p_x \in \mathbb{Q}[T]$ be the Hilbert polynomial of the fiber $Z_x \subset \mathbb{P}^n_{k(x)}$; then Z is flat over X if and only if p_x is independent of x (Hartshorne [2, III.9.9]).

A flat morphism $f: A \to B$ is *faithfully flat* if $B \otimes_A M$ is nonzero for any nonzero A-module M. On taking M to be a principal ideal in A, we see that such a morphism is injective.

PROPOSITION 2.7. *Let $f: A \to B$ be a flat morphism with $A \neq 0$. The following are equivalent:*

(a) *f is faithfully flat;*

(b) *a sequence $M' \to M \to M''$ of A-modules is exact whenever $B \otimes_A M' \to B \otimes_A M \to B \otimes_A M''$ is exact;*

(c) *${}^a f$:spec $B \to$ spec A is surjective;*

(d) *for every maximal ideal $\mathfrak{m}$ of A, $f(\mathfrak{m})B \neq B$. In particular, a flat local homomorphism of local rings is automatically faithfully flat.*

Proof. (a) ⇒ (b). Suppose that $M' \overset{g_1}{\to} M \overset{g_2}{\to} M''$ becomes exact after tensoring with B. Then im $(g_2 g_1) = 0$ because

$$B \otimes_A \text{im}\,(g_2 g_1) = \text{im}\,((1 \otimes g_2)(1 \otimes g_1)) = 0,$$

and im (g_1) = ker (g_2) because

$$B \otimes (\ker g_2/\text{im}\, g_1) = \ker (1 \otimes g_2)/\text{im}\,(1 \otimes g_1) = 0.$$

(b) ⇒ (a). $M \overset{0}{\to} M \to 0$ is exact if and only if $M = 0$.

(a) ⇒ (c). For any prime ideal $\mathfrak{p}$ of A, $B \otimes_A k(\mathfrak{p}) \neq 0$, and so ${}^a f^{-1}(\mathfrak{p}) =$ spec $(B \otimes_A k(\mathfrak{p}))$ is nonempty.

(c) ⇒ (d). This is trivial.

(d) ⇒ (a). Let $x \in M$, $x \neq 0$. Because f is flat, it suffices to show that $B \otimes_A N \neq 0$, where $N = Ax \subset M$. But $N \approx A/\mathfrak{I}$ for some ideal $\mathfrak{I}$ of A, and hence $B \otimes N \approx B/\mathfrak{I}B$. If $\mathfrak{m}$ is a maximal ideal of A containing $\mathfrak{I}$, then $\mathfrak{I}B \subset f(\mathfrak{m})B \neq B$, and so $B/\mathfrak{I}B \neq 0$.

COROLLARY 2.8. *Let $f: Y \to X$ be flat; let $y \in Y$, and let x' be such that $x = f(y)$ is in the closure $\overline{\{x'\}}$ of $\{x'\}$. Then there is a y' such that $y \in \overline{\{y'\}}$ and $f(y') = x'$.*

Proof. The x' such that $x \in \overline{\{x'\}}$ are exactly the points in the image of the canonical map spec $\mathcal{O}_x \to X$. The corollary therefore follows from the fact that the map spec $\mathcal{O}_y \to$ spec $\mathcal{O}_x$ induced by f is surjective.

A morphism $f: Y \to X$ is *faithfully flat* if it is flat and surjective. According to (2.7c), this agrees with the previous definition for rings.

We now consider the question of flatness for finite morphisms. The next theorem shows that, for such a morphism $f: Y \to X$, flatness has a very explicit interpretation in terms of the properties of $f_* \mathcal{O}_Y$ as an $\mathcal{O}_X$-module.

THEOREM 2.9. *Let M be a finitely generated A-module. The following are equivalent:*

(a) *M is flat;*

(b) *$M_\mathfrak{m}$ is a free $A_\mathfrak{m}$-module for all maximal ideals $\mathfrak{m}$ of A;*

(c) *$\tilde{M}$ is a locally free sheaf on* spec A;

(d) *M is a projective A-module.*

Moreover, if A is an integral domain, they are equivalent to:

(e) dim$_{k(\mathfrak{p})}$ $(M \otimes_A k(\mathfrak{p}))$ *is the same for all prime ideals $\mathfrak{p}$ of A.*

Proof. (d) ⇒ (a). This implication does not use the finite generation of M. As tensor products commute with direct sums, any free module is flat, and any direct summand of a flat module is flat.

(b) ⇒ (c). Let $\mathfrak{m}$ be a maximal ideal of A, and let $x_1, \ldots, x_r$ be elements of M whose images in $M_{\mathfrak{m}}$ form a basis for $M_{\mathfrak{m}}$ over $A_{\mathfrak{m}}$. Then the homomorphism

$$g: A^r \to M, \qquad g(a_1, \ldots, a_r) = \sum a_i x_i,$$

induces an isomorphism $A_a^r \to M_a$ for some $a \in A$, $a \notin \mathfrak{m}$, because the kernel and cokernel of g are zero at $\mathfrak{m}$ and, being finitely generated, have closed support in spec A.

(c) ⇒ (a). Let $a_1, \ldots, a_r$ be elements of A such that the ideal $(a_1, \ldots, a_r) = A$ and M_{a_i} is a free A_{a_i}-module for all i. Let $B = \prod A_{a_i}$. Then B is faithfully flat over A, and $B \otimes_A M = \prod M_{a_i}$ is clearly a flat B-module. It follows that M is a flat A-module (apply (2.7b)).

To prove the remaining implications, (a) ⇒ (d), (a) ⇒ (b), we shall need the following lemma.

LEMMA 2.10. *Let $0 \to N \to F \xrightarrow{g} M \to 0$ be an exact sequence of A-modules with N a submodule of F.*

(a) *If M and F are flat over A, then $N \cap \mathfrak{J}F = \mathfrak{J}N$ for all ideals $\mathfrak{J}$ of A.*

(b) *Let M be flat and F free, with basis (y_i) over A. If*

$$n = \sum a_i y_i \in N,$$

then there exist $n_i \in N$ such that

$$n = \sum a_i n_i.$$

(b′) *Let M be any flat A-module. If*

$$\sum_i a_i x_i = 0,$$

$a_i \in A$, $x_i \in M$, then there are equations

$$x_i = \sum_j a_{ij} x'_j$$

with $x'_j \in M$, $a_{ij} \in A$, such that

$$\sum_i a_i a_{ij} = 0$$

for all j.

(c) *Let M be flat and F free. For any finite set $\{n_1, \ldots, n_r\}$ of elements of N, there exists an A-linear map $f: F \to N$ with $f(n_j) = n_j$, $j = 1, \ldots, r$.*

Proof. a. From the given exact sequence, we obtain exact sequences,

$$0 \longrightarrow N \cap \mathfrak{I}F \longrightarrow \mathfrak{I}F \longrightarrow \mathfrak{I}M \longrightarrow 0,$$
$$\mathfrak{I} \otimes N \longrightarrow \mathfrak{I} \otimes F \longrightarrow \mathfrak{I} \otimes M \longrightarrow 0.$$

As M and F are flat, $\mathfrak{I} \otimes F$ and $\mathfrak{I} \otimes M$ may be identified with $\mathfrak{I}F$ and $\mathfrak{I}M$, and then the image of $\mathfrak{I} \otimes N$ in $\mathfrak{I} \otimes F = \mathfrak{I}F$ becomes identified with $\mathfrak{I}N$. But from the first sequence, this is also $N \cap \mathfrak{I}F$.

(b) Let $\mathfrak{I}$ be the ideal generated by the a_i occurring in

$$n = \sum a_i y_i.$$

Then $n \in N \cap \mathfrak{I}F = \mathfrak{I}N$, and so there are $n_i \in N$ such that

$$n = \sum a_i n_i.$$

(b′) Write M as a quotient of a free module, as in (b), and let $a_1 x_1 + \cdots + a_r x_r = 0$. It is possible to choose F so that it has a basis (y_i) with $g(y_i) = x_i$, $i = 1, \ldots, r$. Then

$$n = \sum a_i y_i \in N,$$

and so it may be written

$$n = \sum a_i n_i,$$

$n_i \in N$. Write

$$n_i = y_i - \sum a_{ij} y_j,$$

some a_{ij}. Then

$$n = \sum a_i n_i = n - \sum_j \sum_i (a_i a_{ij}) y_j,$$

and so

$$\sum_i a_i a_{ij} = 0$$

each j. Also

$$x_i = \sum a_{ij} g(y_j),$$

and so x'_j may be taken to be $g(y_j)$.

(c) We use induction on r. Assume first that $r = 1$, and write

$$n_1 = \sum_{j=1}^{s} a_j y_{i_j}$$

where (y_i) is a basis for F. Then

$$n_1 = \sum_{j=1}^{s} a_j n'_j$$

for some $n'_j \in N$, and f may be taken to be the map such that $f(y_{i_j}) = n'_j$, $j = 1, \ldots, s$, and $f(y_i) = 0$ otherwise. Now suppose that $r > 1$, and there are maps $f_1, f_2: F \to N$ such that $f_1(n_1) = n_1$ and

$$f_2(n_i - f_1(n_i)) = n_i - f_1(n_i), \qquad i = 2, \ldots, r.$$

Then

$$f: F \to N, \qquad f(y) = f_1(y) + f_2(y) - f_2 f_1(y)$$

has the required property.

We now complete the proof of (2.9).

(a) $\Rightarrow$ (d). Embed M into an exact sequence

$$0 \to N \to F \to M \to 0$$

in which F is free and N and F are both finitely generated. According to (2.10c), this sequence splits, and so M is projective.

(a) $\Rightarrow$ (b). We may assume that M is a finitely generated flat module over a local ring A. Let $x_1, \ldots, x_r \in M$ be such that their images in $M/\mathfrak{m}M$ form a basis for this over the field $A/\mathfrak{m}$. Embed M in a sequence

$$0 \to N \to F \xrightarrow{g} M \to 0$$

where F is free with basis $\{y_1, \ldots, y_r\}$ and $g(y_i) = x_i$. As $N \subset \mathfrak{m}F$, $\mathfrak{m}N = N \cap (\mathfrak{m}F) = N$, and N is zero according to Nakayama's lemma.

(c) $\Rightarrow$ (e). This is obvious.

(e) $\Rightarrow$ (c). Fix a prime ideal $\mathfrak{p}$ of A, and choose elements $x_1, \ldots, x_r$ of M_a, some $a \notin \mathfrak{p}$, whose images in $M \otimes_A k(\mathfrak{p})$ form a basis. According to Nakayama's lemma the map

$$g: A_a^r \to M_a, \qquad g(a_1, \ldots, a_r) = \sum a_i x_i$$

defines a surjection $A_\mathfrak{p}^r \to M_\mathfrak{p}$. On changing a, we may assume that g itself is surjective. For any prime ideal $\mathfrak{q}$ of A_a, the map $k(\mathfrak{q})^r \to M \otimes_A k(\mathfrak{q})$ is surjective, and hence is an isomorphism because $\dim(M \otimes_A k(\mathfrak{q})) = r$. Thus $\ker(g) \subset \mathfrak{q}A_a^r$ for any $\mathfrak{q}$, which implies that it is zero as A_a is reduced. Thus M_a is free.

Remark 2.11. Let $f: Y \to X$ be finite and flat. I claim that f is open. Following (2.9), we may assume that $X = \operatorname{spec} A$, $Y = \operatorname{spec} B$, and $B \approx A^r$ as an A-module. Let $T^r + a_1 T^{r-1} + \cdots + a_r$ be the characteristic polynomial over A of an element $b \in B$. A prime ideal $\mathfrak{p}$ of A is in the image of $\operatorname{spec}(B_b) \to \operatorname{spec}(A)$ exactly when $B_b/\mathfrak{p}B_b$ is nonzero. But $B_b/\mathfrak{p}B_b \approx (B/\mathfrak{p}B)_{\bar{b}}$ and so this ring is nonzero exactly when $\bar{b}$ is not nilpotent in $B/\mathfrak{p}B$ or, equivalently, when some coefficient of $T^r + a_1 T^{r-1} + \cdots + a_r$ is nonzero in $A/\mathfrak{p}$. Thus the image of $\operatorname{spec} B_b$ in $\operatorname{spec} A$ is $\bigcup \operatorname{spec} A_{a_i}$, which is open. A much more general statement holds.

THEOREM 2.12. *Any flat morphism that is locally of finite-type is open.*

LEMMA 2.13. *Let $f: Y \to X$ be of finite-type. For all pairs (Z, U) where Z is a closed irreducible subset of Y and U is an open subset such that*

$U \cap Z \neq \emptyset$, there is an open subset V of X such that $f(U \cap Z) \supset V \cap \overline{f(Z)} \neq \emptyset$. (Here, $\overline{f(Z)}$ denotes the closure of the set $f(Z)$).

Proof. First note the following statements.

(a) The lemma is true for closed immersions.

(b) The lemma is true for f if it is true for

$$f_{\text{red}}: Y_{\text{red}} \to X_{\text{red}}.$$

(c) The lemma is true for gf if it is true for f and g.

For, if V' satisfies the conclusion of the lemma for the pair $(\overline{f(Z)}, V)$ and the map g, then it also satisfies the conclusion for the pair (Z, U) and the map gf.

(d) It suffices to check the lemma locally on Y and X.

(e) In checking the lemma for a given Z, we note that X may be replaced by $\overline{f(Z)}$, and hence may be assumed to be irreducible.

Using (a), (c), and (d), we may reduce the proof to the case that f is the projection $\mathbb{A}^n \times X \to X$ where X is affine. Using (b) and (e), we reduce the proof further to the case that $X = \operatorname{spec} A$, A an integral domain. Finally, using (c) again, we reduce the proof to the case that f is the projection $\mathbb{A}^1 \times X \to X$.

Let Z be a closed irreducible subset of $\mathbb{A}^1_X$, say $Z = \operatorname{spec} B$ where $B = A[T]/\mathfrak{q}$. We may assume that $\mathfrak{q} \neq 0$, for otherwise the lemma is easy. We may also assume, according to (e), that $\mathfrak{q} \cap A = (0)$, that is, that $f(Z) = X$. Let K be the field of fractions of A, and let $t = T \pmod{\mathfrak{q}}$. Since $\mathfrak{q}$ contains a nonconstant polynomial, t is algebraic over K, and so there is an $a \in A$, $a \neq 0$, such that at is integral over A. Then B_a is finite over A_a, and so $\operatorname{spec} B_a \to \operatorname{spec} A_a$ is surjective (Atiyah-Macdonald [1, 5.10]). Thus we are reduced to showing that the image of a nonempty open subset U of $\operatorname{spec} B_a$ contains a nonempty open subset of spec A. But if U contains $(\operatorname{spec} B_a)_b$, and b satisfies the polynomial $T^m + a_1 T^{m-1} + \cdots + a_m = 0$, $a_i \in A_a$, then $f(U) \supset \bigcup (\operatorname{spec} A_a)_{a_i}$.

Proof of (2.12). (Compare Hartshorne [2, III. Ex. 9.1].) Let $f: Y \to X$ be as in the proposition. It suffices to show that $f(Y)$ is open. We may assume that X is quasi-compact. Let $W = X - f(Y)$ and let $Z_1, \ldots, Z_n$ be the irreducible components of $\bar{W}$. Let z_j be the generic point of Z_j. If $z_j \in f(Y)$, say $z_j = f(y)$, then (2.13) applied to $(\overline{\{y\}}, Y)$ shows that there exists an open U in X such that $f(Y) \supset U \cap Z_j \supset \{z_j\}$. But then

$$f(Y) \supset U \cap \left(X - \bigcup_{i \neq j} Z_i\right) \supset \{z_j\},$$

and, as U and $(X - \bigcup_{i \neq j} Z_i)$ are open, this implies that $z_j \notin \bar{W}$, which is a contradiction. Thus $z_j \in W$, and, according to (2.8), all specializations of z_j belong to W. Thus $W \supset Z_j$, $W \supset \bigcup Z_j = \bar{W}$, and $f(Y)$ is open.

Remark 2.14. If $f: Y \to X$ is finite and flat, then it is both open and closed. Thus, if X is connected, then f is surjective and hence faithfully flat (provided $Y \neq \varnothing$).

Exercise 2.15. Give an example to show that (2.12) is false without the finiteness condition, even if f is surjective. (Start with the example in (1.6*b*)).

If $f: Y \to X$ is finite, and for some $y \in Y$, $\mathcal{O}_y$ is free as an $\mathcal{O}_{f(y)}$-module, then clearly $\Gamma(f^{-1}(U), \mathcal{O}_Y)$ is free over $\Gamma(U, \mathcal{O}_X)$ for some open affine U in X containing $f(y)$. (See the proof of (b) $\Rightarrow$ (c) in (2.9.) Thus the set of points $y \in Y$ such that $\mathcal{O}_y$ is flat over $\mathcal{O}_x$ is open in Y and is even nonempty if X is integral and $\overline{f(Y)} = X$. Again this holds more generally.

THEOREM 2.16. *Let $f: Y \to X$ be locally of finite-type. The set of points $y \in Y$ such that $\mathcal{O}_y$ is flat over $\mathcal{O}_{f(y)}$ is open in Y; it is nonempty if X is integral.*

Proof. A reasonably self-contained proof of this may be found in Matsumura [1, Chapter VIII]. See also [EGA, IV.11.1.1].

Recall that, in any category with fiber products, a morphism $Y \to X$ is a *strict epimorphism* if the sequence

$$Y \times_X Y \underset{p_2}{\overset{p_1}{\rightrightarrows}} Y \to X$$

is exact, that is, if the sequence of sets

$$\mathrm{Hom}\,(X, Z) \to \mathrm{Hom}\,(Y, Z) \underset{p_2^*}{\overset{p_1^*}{\rightrightarrows}} \mathrm{Hom}\,(Y \times Y, Z)$$

is exact for all Z, that is, if the first arrow maps Hom (X, Z) bijectively onto the subset of Hom (Y, Z) on which p_1^* and p_2^* agree.

Clearly the condition that a morphism of schemes be surjective is not sufficient to imply that it is a (strict) epimorphism (consider spec $k \to$ spec A, where A is a local Artin ring with residue field k), but for flat morphisms it is (almost).

THEOREM 2.17. *Any faithfully flat morphism $f: Y \to X$ of finite-type is a strict epimorphism.*

It is convenient to prove the following result first.

PROPOSITION 2.18. *If $f: A \to B$ is faithfully flat, then the sequence*

$$0 \to A \xrightarrow{f} B \xrightarrow{d^0} B^{\otimes 2} \to \cdots \to B^{\otimes r} \xrightarrow{d^{r-1}} B^{\otimes r+1} \to \cdots$$

is exact, where

$$B^{\otimes r} = B \otimes_A B \otimes \cdots \otimes_A B \qquad (r \text{ times})$$

$$d^{r-1} = \sum (-1)^i e_i$$

$$e_i(b_0 \otimes \cdots \otimes b_{r-1}) = b_0 \otimes \cdots \otimes b_{i-1} \otimes 1 \otimes b_i \otimes \cdots \otimes b_{r-1}.$$

Proof. The usual argument shows that $d^r d^{r-1} = 0$. We assume first that f admits a section, that is, that there exists a homomorphism $g: B \to A$ such that $gf = 1$, and we construct a contracting homotopy $k_r: B^{\otimes r+2} \to B^{\otimes r+1}$. Define

$$k_r(b_0 \otimes \cdots \otimes b_{r+1}) = g(b_0)b_1 \otimes b_2 \otimes \cdots \otimes b_{r+1}, \qquad r \geq -1.$$

It is easily checked that $k_{r+1}d^{r+1} + d^r k_r = 1$, $r \geq -1$, and this shows that the sequence is exact.

Now let A' be an A-algebra, let $B' = A' \otimes_A B$, and let $f' = 1 \otimes f$: $A' \to B'$. The sequence corresponding to f' is obtained from the sequence for f by tensoring with A' (because $B^{\otimes r} \otimes_A A' \approx B'^{\otimes r}$). Thus, if A' is a faithfully flat A-algebra, it suffices to prove the theorem for f'. Take $A' = B$, and then $f' = (b \mapsto b \otimes 1): B \to B \otimes_A B$ has a section, namely, $g(b \otimes b') = bb'$, and so the sequence is exact.

Remark 2.19. A similar argument to the above shows that if $f: A \to B$ is faithfully flat and M is an A-module, then the sequence

$$0 \to M \to M \otimes_A B \xrightarrow{1 \otimes d^0} M \otimes_A B^{\otimes 2} \to \cdots$$
$$\to M \otimes B^{\otimes r} \xrightarrow{1 \otimes d^{r-1}} M \otimes B^{\otimes r+1} \to \cdots$$

is exact. Indeed, one may assume again that f has a section and construct a contracting homotopy as before.

Proof of 2.17. We have to show that for any scheme Z and any morphism $h: Y \to Z$ such that $hp_1 = hp_2$, there exists a unique morphism $g: X \to Z$ such that $gf = h$.

Case (*a*) $X = \operatorname{spec} A$, $Y = \operatorname{spec} B$, and $Z = \operatorname{spec} C$ are all affine. In this case the theorem follows from the exactness of

$$0 \to A \to B \xrightarrow{e_0 - e_1} B \otimes_A B$$

(since ${}^a e_0 = p_2$, ${}^a e_1 = p_1$).

Case (*b*) $X = \operatorname{spec} A$ and $Y = \operatorname{spec} B$ affine, Z arbitrary. We first show the uniqueness of g. If $g_1, g_2: X \to Z$ are such that $g_1 f = g_2 f$, then g_1 and g_2 must agree on the underlying topological space of X because f is surjective. Let $x \in X$; let U be an open affine neighborhood of $g_1(x) (= g_2(x))$ in Z, and let $a \in A$ be such that $x \in X_a$ and $g_1(X_a) = g_2(X_a) \subset U$. Then B_b, where b is the image of a in B is faithfully flat over A_a, and it therefore follows from case (a) that $g_1 | X_a = g_2 | X_a$.

Now let $h: Y \to Z$ have $hp_1 = hp_2$. Because of the uniqueness just proved, it suffices to define g locally. Let $x \in X$, $y \in f^{-1}(x)$, and let U be an open affine neighborhood of $h(y)$ in Z. Then $f(h^{-1}(U))$ is open in X (2.12), and so it is possible to find an $a \in A$ such that $x \in X_a \subset f(h^{-1}(U))$. I claim $f^{-1}(X_a)$ is contained in $h^{-1}(U)$. Indeed, if $f(y_1) = f(y_2)$, there is a $y' \in Y \times Y$ such that $p_1(y') = y_1$ and $p_2(y') = y_2$; if $y_2 \in h^{-1}(U)$, then

$$h(y_1) = hp_1(y') = hp_2(y') = h(y_2) \in U,$$

which proves the claim. If now b is the image of a in B, then $h(Y_b) = h(f^{-1}(X_a)) \subset U$, and B_b is faithfully flat over A_a. Thus the problem is reduced to case (a).

Case (c) General case. It is easy to reduce to the case where X is affine. Since f is quasi-compact, Y is a finite union, $Y = Y_1 \cup \cdots \cup Y_n$, of open affines. Let Y^* be the disjoint union $Y_1 \amalg \cdots \amalg Y_n$. Then Y^* is affine and the obvious map $Y^* \to X$ is faithfully flat. In the commutative diagram,

$$\begin{array}{ccccc}
\mathrm{Hom}\,(X, Z) & \longrightarrow & \mathrm{Hom}\,(Y, Z) & \rightrightarrows & \mathrm{Hom}\,(Y \times Y, Z) \\
\| & & \downarrow & & \downarrow \\
\mathrm{Hom}\,(X, Z) & \longrightarrow & \mathrm{Hom}\,(Y^*, Z) & \rightrightarrows & \mathrm{Hom}\,(Y^* \times Y^*, Z),
\end{array}$$

the lower row is exact by case (b) and the middle vertical arrow is obviously injective. An easy diagram chase now shows that the top row is exact.

Exercise 2.20. Show that $\operatorname{spec} k[T] \to \operatorname{spec} k[T^3, T^5]$ is an epimorphism, but is not a strict epimorphism.

Remark 2.21. Let $f: A \to B$ be a faithfully flat homomorphism, and let M be an A-module. Write M' for the B-module $f_*M = B \otimes_A M$. The module $e_{0*}M' = (B \otimes_A B) \otimes_B M'$ may be identified with $B \otimes_A M'$ where $B \otimes_A B$ acts by $(b_1 \otimes b_2)(b \otimes m) = b_1 b \otimes b_2 m$, and $e_{1*}M'$ may be identified with $M' \otimes_A B$ where $B \otimes_A B$ acts by $(b_1 \otimes b_2)(m \otimes b) = b_1 m \otimes b_2 b$. There is a canonical isomorphism $\phi: e_{1*}M' \to e_{0*}M'$ arising from

$$e_{1*}M' = (e_1 f)_* M = (e_0 f)_* M = e_{0*}M';$$

explicitly it is the map

$$M' \otimes_A B \to B \otimes_A M', \qquad (b \otimes m) \otimes b' \mapsto b \otimes (b' \otimes m), \qquad m \in M.$$

Moreover, M can be recovered from the pair (M', ϕ) because

$$M = \{m \in M' \mid 1 \otimes m = \phi(m \otimes 1)\}$$

according to (2.19).

Conversely, every pair (M', ϕ) satisfying certain conditions does arise in this way from an A-module. Given $\phi: M' \otimes_A B \to B \otimes_A M'$ define

$$\phi_1: B \otimes_A M' \otimes_A B \to B \otimes_A B \otimes_A M',$$
$$\phi_2: M' \otimes_A B \otimes_A B \to B \otimes_A B \otimes_A M',$$
$$\phi_3: M' \otimes_A B \otimes_A B \to B \otimes_A M' \otimes_A B$$

by tensoring ϕ with id_B in the first, second, and third positions respectively. Then a pair (M', ϕ) arises from an A-module M as above if and only if $\phi_2 = \phi_1\phi_3$. The necessity is easy to check. For the sufficiency, define $M = \{m \in M' \mid 1 \otimes m = \phi(m \otimes 1)\}$. There is a canonical map $(b \otimes m \mapsto bm): B \otimes_A M \to M'$, and it suffices to show that this is an isomorphism (and that the map arising from M is ϕ). Consider the diagram

$$\begin{array}{ccc}
M' \otimes_A B & \underset{\beta\otimes 1}{\overset{\alpha\otimes 1}{\rightrightarrows}} & B \otimes_A M' \otimes_A B \\
\Big\downarrow{\scriptstyle \phi} & & \Big\downarrow{\scriptstyle \phi_1} \\
B \otimes_A M' & \underset{e_1\otimes 1}{\overset{e_0\otimes 1}{\rightrightarrows}} & B \otimes_A B \otimes_A M'
\end{array}$$

in which $\alpha(m) = 1 \otimes m$ and $\beta(m) = \phi(m \otimes 1)$. As the diagram commutes with either the upper or the lower horizontal maps (for the lower maps, this uses the relation $\phi_2 = \phi_1\phi_3$), ϕ induces an isomorphism on the kernels. But, by definition of M, the kernel of the pair $(\alpha \otimes 1, \beta \otimes 1)$ is $M \otimes_A B$, and, according to (2.19), the kernel of the pair $(e_0 \otimes 1, e_1 \otimes 1)$ is M'. This essentially completes the proof.

More details on this, and the following two results may be found in Murre [1, Chapter VII] and Knus-Ojanguren [1, Chapter II].

PROPOSITION 2.22. *Let $f: Y \to X$ be faithfully flat and quasi-compact. To give a quasi-coherent $\mathcal{O}_X$-module M is the same as to give a quasi-coherent $\mathcal{O}_Y$-module M' plus an isomorphism $\phi: p_1^*M' \to p_2^*M'$ satisfying*

$$p_{31}^*(\phi) = p_{32}^*(\phi)p_{21}^*(\phi).$$

(Here the p_{ij} are the various projections $Y \times Y \times Y \to Y \times Y$, that is $p_{ji}(y_1, y_2, y_3) = (y_j, y_i), j > i$).

Proof. In the case that Y and X are affine, this is a restatement of (2.21).

By using the relation between schemes affine over a scheme and quasi-coherent sheaves of algebras (Hartshorne [2, II. Ex. 5.17]), one can deduce from (2.22) the following result.

THEOREM 2.23. *Let $f: Y \to X$ be faithfully flat and quasi-compact. To give a scheme Z affine over X is the same as to give a scheme Z' affine over Y plus an isomorphism $\phi: p_1^*Z' \to p_2^*Z'$ satisfying*

$$p_{31}^*(\phi) = p_{32}^*(\phi)p_{21}^*(\phi).$$

Remark 2.24. The above is a sketch of part of descent theory. Another part describes which properties of morphisms descend. Consider a

Cartesian square

$$\begin{array}{ccc} Y & \longleftarrow & Y' \\ \big\downarrow{\scriptstyle f} & & \big\downarrow{\scriptstyle f'} \\ X & \longleftarrow & X' \end{array}$$

in which the map $X' \to X$ is faithfully flat and quasi-compact. If f' is quasi-compact (respectively separated, of finite-type, proper, an open immersion, affine, finite, quasi-finite, flat, smooth, étale), then f is also [EGA. IV.2.6,2.7]. The reader may check that this statement implies the same statement for faithfully flat morphisms $X' \to X$ that are locally of finite-type. (Use (2.12)).

Of a similar nature is the result that if $f: Y \to X$ is faithfully flat and Y is integral (respectively normal, regular), then so also is X [EGA. 0_{IV}, 17.3.3].

Finally, we quote a result that may be regarded as a vast generalization of the Hilbert Nullstellensatz. The Nullstellensatz says that any morphism of finite-type $f: X \to \operatorname{spec} k$, k a field, has a *quasi-section*, that is, that there exists a k-morphism $g: \operatorname{spec} k' \to X$ with k' a finite field extension of k.

PROPOSITION 2.25. *Let X be quasi-compact, and let $f: Y \to X$ be a faithfully flat morphism that is locally of finite-type. Then there exists an affine scheme X', a faithfully flat quasi-finite morphism $h: X' \to X$, and an X-morphism $g: X' \to Y$.*

Proof. One has to show that, locally, there exist sequences satisfying the conditions of (2.6d) and of length equal to the relative dimension of Y/X. (See [EGA. IV.17.16.2] for the details.)

§3. Étale Morphisms

Let k be a field and $\bar{k}$ its algebraic closure. A k-algebra A is *separable* if $\bar{A} = A \otimes_k \bar{k}$ has zero Jacobson radical, that is, if the maximal ideals of $\bar{A}$ have intersection zero.

PROPOSITION 3.1. *Let A be a finite algebra over a field k. The following are equivalent:*

(a) *A is separable over k;*

(b) *$\bar{A}$ is isomorphic to a finite product of copies of $\bar{k}$;*

(c) *A is isomorphic to a finite product of separable field extensions of k;*

(d) *the discriminant of any basis of A over k is nonzero (that is, the trace pairing $A \times A \to k$ is nondegenerate).*

Proof. (a) $\Rightarrow$ (b). From (1.5) we know that $\bar{A}$ has only finitely many prime ideals and that they are all maximal. Now (a) implies that their intersection is zero and (b) follows from the Chinese remainder theorem (Atiyah-Macdonald [1, 1.10]).

(b) $\Rightarrow$ (c). The Chinese remainder theorem implies that A/I_r, where I_r is the Jacobson radical of A, is isomorphic to a finite product $\prod k_i$ of finite field extensions of k. Write $[K:k]_s$ for the separable degree of a field extension K/k. Then $\mathrm{Hom}_{k\text{-alg}}(A, \bar{k})$ has

$$\sum [k_i:k]_s$$

elements. But

$$\mathrm{Hom}_{k\text{-alg}}(A, \bar{k}) \approx \mathrm{Hom}_{\bar{k}\text{-alg}}(\bar{A}, \bar{k}),$$

and this set has $[\bar{A}:\bar{k}]$ elements by (b). Thus

$$[\bar{A}:\bar{k}] = \sum [k_i:k]_s \leq \sum [k_i:k] = [A/I_r:k] \leq [A:k].$$

Since $[\bar{A}:\bar{k}] = [A:k]$, equality must hold throughout and we have (c).

(c) $\Rightarrow$ (d). If $A = \prod k_i$, where the k_i are separable field extensions of k, then $\mathrm{disc}(A) = \prod \mathrm{disc}(k_i)$, and this is nonzero by one of the standard criteria for a field extension to be separable.

(d) $\Rightarrow$ (a). The discriminants of A and $\bar{A}$ are the same. If x is in the radical of $\bar{A}$, then xa is nilpotent for any $a \in \bar{A}$, and so $Tr_{\bar{A}/\bar{k}}(xa) = 0$ all a. Thus $x = 0$.

A morphism $f: Y \to X$ that is locally of finite-type is said to be *unramified* at $y \in Y$ if $\mathcal{O}_{Y,y}/\mathfrak{m}_x\mathcal{O}_{Y,y}$ is a finite separable field extension of $k(x)$, where $x = f(y)$. In terms of rings, this indicates that a homomorphism $f: A \to B$ of finite-type is unramified at $\mathfrak{q} \in \mathrm{spec}\, B$ if and only if $\mathfrak{p} = f^{-1}(\mathfrak{q})$ generates the maximal ideal in $B_\mathfrak{q}$ and $k(\mathfrak{q})$ is a finite separable field extension of $k(\mathfrak{p})$. Thus this terminology agrees with that in number theory.

A morphism $f: Y \to X$ is *unramified* if it is unramified at all $y \in Y$.

PROPOSITION 3.2. *Let $f: Y \to X$ be locally of finite-type. The following are equivalent:*

(a) *f is unramified;*

(b) *for all $x \in X$, the fiber $Y_x \to$ spec $k(x)$ over x is unramified;*

(c) *all geometric fibers of f are unramified (that is, for all morphisms* spec $k \to X$, *with k separably closed,* $Y \times_X$ spec $k \to$ spec k *is unramified);*

(d) *for all $x \in X$, Y_x has an open covering by spectra of finite separable $k(x)$-algebras;*

(e) *for all $x \in X$, Y_x is a sum $\coprod$ spec k_i, where the k_i are finite separable field extensions of $k(x)$;*

(If f is of finite-type, then Y_x itself is the spectrum of a finite separable $k(x)$-algebra in (d), *and Y_x is a finite sum in* (e)*; in particular f is quasi-finite).*

Proof. (a) $\Leftrightarrow$ (b). This follows from the isomorphism $\mathcal{O}_{Y,y}/\mathfrak{m}_x\mathcal{O}_{Y,y} \approx \mathcal{O}_{Y_x,y}$.

(b) $\Rightarrow$ (d). Let U be an open affine subset of Y_x, and let $\mathfrak{q}$ be a prime ideal in $B = \Gamma(U, \mathcal{O}_{Y_x})$. According to (b), $B_\mathfrak{q}$ is a finite separable field extension of $k(x)$. Also

$$k(x) \subset B/\mathfrak{q} \subset B_\mathfrak{q}/\mathfrak{q}B_\mathfrak{q} = B_\mathfrak{q},$$

and so $B/\mathfrak{q}$ is also a field. Thus $\mathfrak{q}$ is maximal, B is an Artin ring (Atiyah-Macdonald [1, 8.5]), and $B = \prod B_\mathfrak{q}$, where $\mathfrak{q}$ runs through the finite set spec B. This proves (d).

A similar argument shows that (c) $\Rightarrow$ (d), and (d) $\Rightarrow$ (e) $\Rightarrow$ (c) and (d) $\Rightarrow$ (b) are trivial consequences of (3.1).

Notice that according to the above definition, any closed immersion $Z \hookrightarrow X$ is unramified. Since this does not agree with our intuitive idea of an unramified covering, for example, of Riemann surfaces, we need a more restricted notion. A morphism of schemes (or rings) is defined to be *étale* if it is flat and unramified (hence also locally of finite-type).

PROPOSITION 3.3. (a) *Any open immersion is étale.*

(b) *The composite of two étale morphisms is étale.*

(c) *Any base change of an étale morphism is étale.*

Proof. After applying (2.4), we only have to check that the three statements hold for unramified morphisms. Both (a) and (b) are obvious (any immersion is unramified). Also, (c) is obviously true according to (3.1) if the base change is of the form $k \to k'$, where k and k' are fields, but, according to (3.2), this is all that has to be checked.

Example 3.4. Let k be a field and $P(T)$ a monic polynomial over k. Then the monogenic extension $k[T]/(P)$ is separable (or unramified or étale) if and only if P is separable, that is, has no multiple roots in $\bar{k}$.

This generalizes to rings. A monic polynomial $P(T) \in A[T]$ is *separable* if $(P, P') = A[T]$, that is, if $P'(T)$ is a unit in $A[T]/(P)$ where $P'(T)$ is the formal derivative of $P(T)$. It is easy to see that P is separable if and only if its image in $k(\mathfrak{p})[T]$ is separable for all prime ideals $\mathfrak{p}$ in A.

Let $B = A[T]/(P)$, where P is any monic polynomial in $A[T]$. As an A-module, B is free of finite rank equal to the degree of P. Moreover, $B \otimes_A k(\mathfrak{p}) = k(\mathfrak{p})[T]/(\bar{P})$ where $\bar{P}$ is the image of P in $k(\mathfrak{p})[T]$. It follows from (3.2b) that B is unramified and so étale over A if and only if P is separable. More generally, for any $b \in B$, B_b is étale over A if and only if P' is a unit in B_b.

For example, $B = A[T]/(T^r - a)$ is étale over A if and only if ra is invertible in A (for $ra \in A^* \Leftrightarrow r\bar{a} \in k(\mathfrak{p})^*$, all $\mathfrak{p} \Leftrightarrow T^r - \bar{a}$ is separable in $k(\mathfrak{p})[T]$ all $\mathfrak{p}$).

For algebras generated by more than one element, there is the following Jacobian criterion: let $C = A[T_1, \ldots, T_n]$, let $P_1, \ldots, P_n \in C$, and let $B = C/(P_1, \ldots, P_n)$; then B is étale over A if and only if the image of $\det(\partial P_i/\partial T_j)$ in B is a unit. That B is unramified over A if and only if the condition holds follows directly from (3.5b) below. (The B-module $\Omega^1_{B/A}$ has generators $dT_1, \ldots, dT_n$ and relations $\sum$ "$(\partial P_i/\partial T_j)$" $dT_j = 0$.) That B is flat over A may be proved by repeated applications of (2.5). (See Mumford [3, III. §10. Thm. 3′] for the details.)

Note that if $Y = \operatorname{spec} B$ and $X = \operatorname{spec} A$ were analytic manifolds, then this criterion would indicate that the induced maps on the tangent spaces were all isomorphisms, and hence $Y \to X$ would be a local isomorphism at every point of Y by the inverse function theorem. It is clearly not true in the geometric case that $\operatorname{spec} B \to \operatorname{spec} A$ is a local isomorphism (unless local is meant in the sense of the étale topology: see later). For example, consider $\operatorname{spec} \mathbb{Z}[T]/(T^2 - 2) \to \operatorname{spec} \mathbb{Z}$, which is étale on the complement of $\{(2)\}$.

PROPOSITION 3.5. *Let $f: Y \to X$ be locally of finite-type. The following are equivalent:*

(a) *f is unramified;*

(b) *the sheaf $\Omega^1_{Y/X}$ is zero;*

(c) *the diagonal morphism $\Delta_{Y/X}: Y \to Y \times_X Y$ is an open immersion.*

Proof. (a) ⇒ (b). Since $\Omega^1_{Y/X}$ behaves well with respect to base change, one can reduce the proof to the case that $Y = \operatorname{spec} B$ and $X = \operatorname{spec} A$ are affine, then to the case that $A \to B$ is a local homomorphism of local rings and using Nakayama's lemma to the case where A and B are fields. Then B is a separable field extension of A, and it is a standard fact that this implies that $\Omega^1_{B/A} = 0$.

(b) ⇒ (c). Since the diagonal is always at least locally closed, we may choose an open subscheme U of $Y \times_X Y$ such that $\Delta_{Y/X}: Y \to U$ is a closed immersion and regard Y as a subscheme of U. Let I be the sheaf of ideals on U defining Y. Then I/I^2, regarded as a sheaf on Y, is isomorphic to $\Omega^1_{Y/X}$ and hence is zero. Using Nakayama's lemma, one sees this implies that $I_y = 0$ for all $y \in Y$, and it follows that $I = 0$ on some open subset V of U containing Y. Then $(Y, \mathcal{O}_Y) = (V, \mathcal{O}_V)$ is an open subscheme of $Y \times Y$.

(c) ⇒ (a). By passing to the geometric fiber over a point of X, we may reduce the problem to the case of a morphism $f: Y \to \operatorname{spec} k$ where k

is an algebraically closed field. Let y be a closed point of Y. Because k is algebraically closed, there exists a section g: spec $k \to Y$ whose image is $\{y\}$. The following square is Cartesian:

$$\begin{array}{ccc} Y & \xrightarrow{\Delta} & Y \times_X Y \\ \uparrow{\scriptstyle g} & & \uparrow{\scriptstyle (gf,\,1)} \\ \{y\} & \xrightarrow{g} & Y. \end{array}$$

Since Δ is an open immersion, this implies that $\{y\}$ is open in Y. Moreover, spec $\mathcal{O}_y = \{y\} \to$ spec k still has the property that spec $\mathcal{O}_y \xrightarrow{\Delta}$ spec $(\mathcal{O}_y \otimes_k \mathcal{O}_y)$ is an open immersion. But $\mathcal{O}_y$ is a local Artin ring with residue field k, and so spec $\mathcal{O}_y \otimes_k \mathcal{O}_y$ has only one point, and $\mathcal{O}_y \otimes_k \mathcal{O}_y \to \mathcal{O}_y$ must be an isomorphism. By counting dimensions over k, one sees then that $\mathcal{O}_y = k$. Thus, by applying (3.1) and (3.2), we have (a).

COROLLARY 3.6. *Consider morphisms $f: X \to S$, $g: Y \to X$. If fg is étale and f is unramified, then g is étale.*

Proof. Write $g = p_2\Gamma_g$ where $\Gamma_g: Y \to Y \times_S X$ is the graph of g and $p_2: Y \times_S X \to X$ is the projection on the second factor. Γ_g is the pull-back of the open immersion $\Delta_{X/S}: X \to X \times_S X$ by $g \times 1: Y \times_S X \to X \times_S X$, and p_2 is the pull-back of the étale map $fg: Y \to S$ by $f: X \to S$. Thus, by using (3.3), we see that g is étale.

Remark 3.7. Let $f: Y \to X$ be locally of finite-type. The annihilator of $\Omega^1_{Y/X}$ (an ideal in $\mathcal{O}_Y$) is called the *different* $\mathfrak{d}_{Y/X}$ of Y over X. That this definition agrees with the one in number theory is proved in Serre [7, III.7].

The closed subscheme of Y defined by $\mathfrak{d}_{Y/X}$ is called the *branch locus* of Y over X. The open complement of the branch locus is precisely the set on which $\Omega^1_{Y/X} = 0$, that is, on which $f: Y \to X$ is unramified. The theorem of the purity of branch locus states that the branch locus (if nonempty) has pure codimension one in Y in each of the two cases: (a) when f is faithfully flat and finite over X; or (b) when f is quasi-finite and dominating, Y is regular and X is normal. (See Altman and Kleiman [1, VI.6.8], [SGA. 1, X.3.1], and [SGA. 2, X.3.4].)

PROPOSITION 3.8. *If $f: Y \to X$ is locally of finite-type, then the set of points y of Y, such that $\mathcal{O}_{Y,y}$ is flat over $\mathcal{O}_{X,f(y)}$ and $\Omega^1_{Y/X,y} = 0$, is open in Y. Thus there is a unique largest open set U in Y on which f is étale.*

Proof. This follows immediately from (2.16).

Exercise 3.9. Let $f: Y \to X$ be finite and flat, and assume that X is connected. Then $f_*\mathcal{O}_Y$ is locally free, of constant rank r say. Show that there is a sheaf of ideals $\mathfrak{D}_{Y/X}$ on X, called the *discriminant* of Y over

X, with the property that if U is an open affine in X such that $B = \Gamma(f^{-1}(U), \mathcal{O}_Y)$ is free with basis $\{b_1, \ldots, b_r\}$ over $A = \Gamma(U, \mathcal{O}_X)$, then $\Gamma(U, \mathfrak{D}_{Y/X})$ is the principal ideal generated by det $(Tr_{B/A}(b_i b_j))$. Show that f is unramified, hence étale, at all $y \in f^{-1}(x)$ if and only if $(\mathfrak{D}_{Y/X})_x = \mathcal{O}_{X,x}$ (use (3.1d)). Use this to show that if f is unramified at all $y \in f^{-1}(x)$ for some $x \in X$, then there exists an open subset $U \subset X$ containing x such that $f : f^{-1}(U) \to U$ is étale. Show that if $B = A[T]/(P(T))$ with P monic, then the discriminant $\mathfrak{D}_{B/A} = (D(P))$, where $D(P)$ is the discriminant of P, that is, the resultant, res (P, P'), of P and P'. Show also that the different $\mathfrak{d}_{B/A} = (P'(t))$ where $t = T\ (\mathrm{mod}\ P)$. (See Serre [7, III.§6].)

The next proposition and its corollaries show that étale morphisms have the uniqueness properties of local isomorphisms.

PROPOSITION 3.10. *Any closed immersion $f : Y \to X$ that is flat (hence étale) is an open immersion.*

Proof. According to (2.12), $f(Y)$ is open in X and so, after replacing X by $f(Y)$, we may assume f to be surjective. As f is finite, $f_*\mathcal{O}_Y$ is locally free as an $\mathcal{O}_X$-module (2.9). Since f is a closed immersion, this implies that $\mathcal{O}_X \approx f_*\mathcal{O}_Y$, that is, that f is an isomorphism.

Remark 3.11. By using Zariski's main theorem, we may prove a stronger result, namely, that any étale, universally injective, separated morphism $f : Y \to X$ is an open immersion. (*Universally injective* is equivalent to *injective and all maps $k(f(y)) \to k(y)$ radicial* [EGA. I.3.7.1].) In fact, by proceeding as above, one can assume that f is universally bijective, hence a homeomorphism (2.12), hence proper, and hence finite (1.10). Now f being étale and radicial implies that $f_*\mathcal{O}_Y$ must be free of rank one.

COROLLARY 3.12. *If X is connected and $f : Y \to X$ is étale (respectively étale and separated), then any section s of f is an open immersion (respectively an isomorphism onto an open connected component). Thus there is a one-to-one correspondence between the set of such sections and the set of those open (respectively open and closed) subschemes Y_i of Y such that f induces an isomorphism $Y_i \to X$. In particular, a section is known when its value at one point is known if f is separated.*

Proof. Only the first assertion requires proof. Assume first that f is separated. Then s is a closed immersion because $fs = 1$ is a closed immersion, and f is separated (compare with the proof of (3.6)). According to (3.6) s is étale, and hence it is an open immersion. Thus s is an isomorphism onto its image, which is both open and closed in Y. If f is only assumed to be étale, then it is separated in a neighborhood of y and $x = f(y)$, and hence the above argument shows that s is a local isomorphism at x.

COROLLARY 3.13. *Let $f, g: Y' \to Y$ be X-morphisms where Y' is a connected X-scheme and Y is étale and separated over X. If there exists a point $y' \in Y'$ such that $f(y') = g(y') = y$ and the maps $k(y) \to k(y')$ induced by f and g coincide, then $f = g$.*

Proof. The graphs of f and g, $\Gamma_f, \Gamma_g: Y' \to Y' \times_X Y$, are sections to $p_1: Y' \times_X Y \to Y'$ The conditions imply that Γ_f and Γ_g agree at a point, and so Γ_f and Γ_g are equal (3.12). Thus $f = p_2\Gamma_f = p_2\Gamma_g = g$.

We saw in (3.4) above that given a monic polynomial $P(T)$ over A, it is possible to construct an étale morphism spec $C \to$ spec A by taking $C = B_b$ where $B = A[T]/(P)$ and b is such that $P'(T)$ is a unit in B_b. We shall call such an étale morphism *standard.* The interesting fact is that locally every étale morphism $Y \to X$ is standard. Geometrically this means that in a neighborhood of any point x of X, there are functions $a_1, \ldots, a_r$ such that Y is locally described by the equation $T^r + a_1T^{r-1} + \cdots + a_r = 0$, and the roots of the equation are all simple (at any geometric point).

THEOREM 3.14. *Assume that $f: Y \to X$ is étale in some open neighborhood of $y \in Y$. Then there are open affine neighborhoods V and U of y and $f(y)$, respectively, such that $f \mid V: V \to U$ is a standard étale morphism.*

Proof. Clearly, we may assume that $Y = $ spec C and $X = $ spec A are affine. Also, by Zariski's main theorem (1.8), we may assume that C is a finite A-algebra. Let $\mathfrak{q}$ be the prime ideal of C corresponding to y. We have to show that there is a standard étale A-algebra B_b such that $B_b \approx C_c$ for some $c \notin \mathfrak{q}$. It is easy to see (because everything is finite over A) that it suffices to do this with A replaced by $A_\mathfrak{p}$, where $\mathfrak{p} = f^{-1}(\mathfrak{q})$, that is, that we may assume that A is local and that $\mathfrak{q}$ lies over the maximal ideal $\mathfrak{p}$ of A.

Choose an element $t \in C$ whose image $\bar{t}$ in $C/\mathfrak{p}C$ generates $k(\mathfrak{q})$ over $k(\mathfrak{p})$, that is, $\bar{t}$ is such that $k(\mathfrak{p})[\bar{t}] = k(\mathfrak{q}) \subset C/\mathfrak{p}C$. Such an element exists because $C/\mathfrak{p}C$ is a product $k(\mathfrak{q}) \times C'$, and $k(\mathfrak{q})/k(\mathfrak{p})$ is separable. Let $\mathfrak{q}' = \mathfrak{q} \cap A[t]$. I claim that $A[t]_{\mathfrak{q}'} \to C_\mathfrak{q}$ is an isomorphism. Note first that $\mathfrak{q}$ is the only prime ideal of C lying over $\mathfrak{q}'$ (in checking this, one may tensor with $k(\mathfrak{p})$). Thus the semilocal ring $C \otimes_{A[t]} A[t]_{\mathfrak{q}'}$ is actually local and so equals $C_\mathfrak{q}$. As $A[t] \to C$ is injective and finite, it follows that

$$A[t]_{\mathfrak{q}'} \to C \otimes_{A[t]} A[t]_{\mathfrak{q}'} = C_\mathfrak{q}$$

is injective and finite. It is surjective because $k(\mathfrak{q}') \to k(\mathfrak{q})$ is surjective, and Nakayama's lemma may be applied.

$A[t]$ is finite over A (it is a submodule of a Noetherian A-module), and the isomorphism $A[t]_{\mathfrak{q}'} \to C_\mathfrak{q}$ extends to an isomorphism $A[t]_{c'} \xrightarrow{\approx} C_c$ for some $c \notin \mathfrak{q}$, $c' \notin \mathfrak{q}'$. Thus C may be replaced by $A[t]$, that is, we may assume that t generates C over A.

Let $n = [k(\mathfrak{q}):k(\mathfrak{p})]$, so that $1, \bar{t}, \ldots, \bar{t}^{n-1}$ generate $k(\mathfrak{q})$ as a vector space over $k(\mathfrak{p})$. Then $1, t, \ldots, t^{n-1}$ generate $C = A[t]$ over A (according to Nakayama's lemma), and so there is a monic polynomial $P(T)$ of degree n and a surjection $h: B = A[T]/(P) \to C$. Clearly $\bar{P}(T)$ is the characteristic polynomial of $\bar{t}$ in $k(\mathfrak{q})$ over $k(\mathfrak{p})$ and so is separable. Thus B_b is a standard étale A-algebra for some $b \notin h^{-1}(\mathfrak{q})$. With a suitable choice of b and c we get a surjection $h': B_b \to C_c$ with both B_b and C_c étale A-algebras. According to (3.6), h' is étale, and ${}^a h'$:spec $C_c \to$ spec B_b is a closed immersion. Hence, according to (3.10), ${}^a h'$ is an open immersion, which completes the proof.

Remark 3.15. The fact that f was flat was used only in the last step of the above proof. Thus the argument shows that locally every unramified morphism is a composite of a closed immersion with a standard étale morphism.

COROLLARY 3.16. *A morphism $f: Y \to X$ is étale if and only if for every $y \in Y$, there exist open affine neighborhoods $V =$ spec C of y and $U =$ spec A of $x = f(y)$ such that*

$$C = A[T_1, \ldots, T_n]/(P_1, \ldots, P_n)$$

and det $(\partial P_i/\partial T_j)$ *is a unit in C.*

Proof. Because of (3.4), we only have to prove the necessity. From the theorem, we may assume that $Y \to X$ is standard étale, for example $X =$ spec A, $Y =$ spec C, $C = B_b$, $B = A[T]/(P)$. Then $C \approx A[T, U]/(P(T), bU - 1)$, and the determinant corresponding to this is $P'(T)b$. Since the image of $P'(T)b$ is a unit in C, this proves the corollary with the added information that n may be taken to be two.

With this structure theorem, it is relatively easy to prove that if $Y \to X$ is étale, then Y inherits many of the good properties of X. (For the opposite inheritance, see (2.24).)

PROPOSITION 3.17. *Let $f: Y \to X$ be étale.*

(a) Dim $(\mathcal{O}_{Y,y}) =$ dim $(\mathcal{O}_{X,f(y)})$ *for all $y \in Y$.*

(b) *If X is normal, then Y is normal.*

(c) *If X is regular, then Y is regular.*

Proof. (a) We may assume that $X =$ spec A where A $(= \mathcal{O}_x)$ is local and that $Y =$ spec B. The proof uses only the assumption that B is quasi-finite and flat over A. Let $\mathfrak{q}$ be the prime ideal of B corresponding to y (so $\mathfrak{q}$ lies over $\mathfrak{p}$, the maximal ideal of A). Then spec $B_\mathfrak{q} \to$ spec A is surjective (2.7), so dim $(B_\mathfrak{q}) \geq$ dim (A). Conversely we may assume $B = B'_b$, where B' is finite over A (1.8). Then dim $(A) \geq$ dim (B') $(\geq$ dim $B_\mathfrak{q})$ (Atiyah-Macdonald [1, 5.9]).

(b) We may assume that $X =$ spec A where A is local (hence normal) and that $T =$ spec C where $C = B_b$ is a standard étale A-algebra with

$B = A[T]/(P(T))$. Let K be the field of fractions of A, let $L = C \otimes_A K = K[T]/(P(T))$, and let A' be the integral closure of A in L. Note that L is a product of separable field extensions of K. Then we have the inclusions

$$\begin{array}{ccccc} C & \subset & A'_b & \subset & L \\ \cup & & \cup & & \\ A \subset B & \subset & A' & & \end{array}$$

Write $t = T \pmod{P(T)}$. Choose an $a \in A'$. We have to show that a/b^s, or equivalently just a, is in C.

Let $\bar{K}$ be the algebraic closure of K and $\phi_1, \ldots, \phi_r$ the homomorphisms $L \to \bar{K}$ over K such that $\phi_1(t), \ldots, \phi_r(t)$ are the roots of $P(T)$ (so $r =$ degree P). Write

$$a = a_0 + a_1 t + \cdots + a_{r-1} t^{r-1}, \qquad a_i \in K.$$

Then we have r equations,

$$\phi_j(a) = a_0 + a_1 t_j + \cdots + a_{r-1} t_j^{r-1}$$

where $t_j = \phi_j(t)$. Let D be the determinant of these equations, regarding the a_i as unknowns, so that $D = \pm \prod_{i<j} (t_i - t_j)$, that is, $D^2 =$ discriminant of $P(T) = \mathfrak{D}_{B/A}$ (compare (3.9)). Since the $\phi_j(a)$ and t_j^i are integral over A, it follows from Cramer's rule that the Da_i, $i = 1, \ldots, r$, are also integral over A. Since the $Da_i \in K$ and A is normal, they belong to A, and this implies that $Da \in B \subset C$. Since D is a unit in C, it follows that $a \in C$.

(c) Let $y \in Y$. Then $\dim (\mathcal{O}_{Y,y}) = \dim (\mathcal{O}_{X,f(y)})$, and $m_y = m_x \mathcal{O}_{Y,y}$ can be generated by $\dim (\mathcal{O}_{X,f(y)})$ elements.

Remark 3.18. An argument, similar to that in (b), shows that if X is reduced, then Y is reduced (Raynaud [3, p. 74]).

We now determine the structure of étale morphisms $Y \to X$ when X is normal.

PROPOSITION 3.19. *Let $f: Y \to X$ be étale, where X is normal. Then locally f is a standard étale morphism of the form* spec $C \to$ spec A *where A is an integral domain, $C = B_b$, $B = A[T]/(P(T))$, and $P(T)$ is irreducible over the field of fractions of A.*

Proof. The only new fact to be shown is that $P(T)$ may be chosen to be irreducible over the field of fractions K of A. Clearly we may reduce the problem to the case that $X =$ spec A where A is a local ring and assume that $Y =$ spec C with C a standard étale A-algebra, say $C = B_b$, $B = A[T]/(P(T))$ with $P(T)$ possibly reducible. Fix a prime ideal $\mathfrak{q}$ in C such that $\mathfrak{p} = \mathfrak{q} \cap A$ is the maximal ideal of A.

Note that any monic factor $Q(T)$ of $P(T)$ in $K[T]$ automatically has coefficients in A. (Let K' be a splitting field for $Q(T)$; the roots of $Q(T)$ in K' are also roots of $P(T)$ and hence are integral over A; it follows that the coefficients of $Q(T)$ are also integral over A since they can be expressed in terms of the roots.) Choose $P_1(T)$ to be a monic irreducible factor of $P(T)$ whose image in $k(\mathfrak{q})$ is zero, and write $P(T) = P_1(T)Q(T)$ with $P_1, Q \in A[T]$. Then the images $\bar{P}_1$ and $\bar{Q}$ of P_1 and Q in $k(\mathfrak{p})[T]$ are coprime since $\bar{P}(T)$ is separable and so has no multiple roots. It follows that $(P_1, Q) = A[T]$ (compare (4.1a) below), and the Chinese remainder theorem shows that $B \approx A[T]/(P_1) \times A[T]/(Q)$. Let b_1 be the image of b in $B_1 = A[T]/(P_1)$. Obviously $C_1 = (B_1)_{b_1}$ is the standard A-algebra sought.

THEOREM 3.20. *Let X be a normal scheme and $f: Y \to X$ an unramified morphism. Then f is étale if and only if, for any $y \in Y$, $\mathcal{O}_{X,f(y)} \to \mathcal{O}_{Y,y}$ is injective.*

Proof. If f is flat, then $\mathcal{O}_{f(y)} \to \mathcal{O}_y$ is injective according to (2.3). For the converse, note that locally f factors into $Y \xrightarrow{f'} Y' \xrightarrow{g} X$ with f' a closed immersion and g étale (3.15). Write $A = \mathcal{O}_{X,f(y)}$; following (3.19), we may write $\mathcal{O}_{Y',f'(y)} = C_\mathfrak{q}$ where $C = A[T]/(P(T))$ with $P(T)$ irreducible over the field of fractions K of A. We have $A \to C_\mathfrak{q} \to \mathcal{O}_{Y,y}$, which, when tensored with K, becomes $K \to C_\mathfrak{q} \otimes_A K \to \mathcal{O}_{Y,y} \otimes_A K$. As $A \to \mathcal{O}_{Y,y}$ is injective, $K \to \mathcal{O}_{Y,y} \otimes_A K$ is injective, which shows that $C_\mathfrak{q} \otimes_A K \to \mathcal{O}_{Y,y} \otimes_A K$ is not the zero map. But $C_\mathfrak{q} \otimes_A K = K[T]/(P)$ is a field, and so this last map is injective. Hence $C_\mathfrak{q} \to \mathcal{O}_{Y,y}$ is injective, and we already know that it is surjective because f' is a closed immersion. Thus $\mathcal{O}_{Y,y} = C_\mathfrak{q}$ is flat over A.

THEOREM 3.21. *Let X be a connected normal scheme, and let $K = R(X)$. Let L be a finite separable field extension of K, let X' be the normalization of X in L, and let U be any open subscheme of X' that is disjoint from the support of $\Omega^1_{X'/X}$. Then $U \to X$ is étale, and conversely any separated étale morphism $Y \to X$ of finite-type can be written $Y = \coprod U_i \to X$ where each $U_i \to X$ is of this form.*

Proof. $\Omega^1_{U/X} = \Omega^1_{X'/X}|U = 0$, and so $U \to X$ is unramified according to (3.5). It is étale according to (3.20).

Conversely, let $Y \to X$ be separated, étale, and of finite-type. The connected components Y_i of Y are irreducible (because the irreducible components of Y containing y are in one-to-one correspondence with the minimal prime ideals of $\mathcal{O}_{Y,y}$ and Y is normal). If $\operatorname{spec} L_i \to \operatorname{spec} K$ is the generic fiber of $Y_i \to X$ and X_i is the normalization of X in L_i, then Zariski's main theorem implies that $Y_i \to X_i$ is an open immersion (see (1.8), especially the proof).

Remark 3.22. In [EGA. IV.17] the following functorial definitions are made. Let X be a scheme and F a contravariant functor $\mathbf{Sch}/X \to \mathbf{Sets}$. Then F is said to be *formally smooth* (*lisse*) (respectively, *formally unramified* (*net*), *formally étale*) if for any affine X-scheme X' and any subscheme X'_0 of X' defined by a nilpotent ideal $\mathfrak{I}$, $F(X') \to F(X'_0)$ is surjective (respectively, injective, bijective).

A scheme Y over X is said to be *formally smooth, formally unramified,* or *formally étale* over X when the functor $h_Y = \mathrm{Hom}_X(-, Y)$ it defines has the corresponding property. If, in addition, Y is locally of finite presentation over X, then one says simply that Y is smooth, unramified, or étale over X.

We show that a morphism $f: Y \to X$ that is étale in our sense is also étale in the above sense. (The converse, which is more difficult, may be found, for example, in Artin [9, I.1.1].) Thus, given an X-morphism $g_0: X'_0 \to Y$, we must show that there is a unique X-morphism $g: X' \to Y$ lifting it:

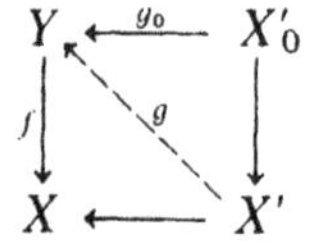

The uniqueness implies that it sufficies to do this locally. Thus we may assume that f is standard, for example, $X = \operatorname{spec} A$, $Y = \operatorname{spec} C$, $C = B_b$, $B = A[T]/(P) = A[t]$. Let $X' = \operatorname{spec} R$, $X'_0 = \operatorname{spec} R_0$ and $R_0 = R/\mathfrak{I}$. Then we are given an A-homomorphism $g_0: C \to R_0$, and we want to find a unique $g: C \to R$ lifting it:

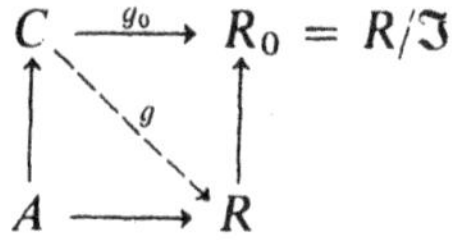

By using induction on the length of $\mathfrak{I}$, we may reduce the question to the case that $\mathfrak{I}^2 = 0$. Let $r \in R$ be such that $g_0(t) = r \pmod{\mathfrak{I}}$. We have to find an $r' \in R$ such that $r' \equiv r \pmod{\mathfrak{I}}$ and $P(r') = 0$. Write $r' = r + h$, $h \in \mathfrak{I}$. Then h must satisfy the equation $P(r + h) = 0$. But $P(r + h) = P(r) + hP'(r)$, where $P(r) \in \mathfrak{I}$ and $P'(r)$ is a unit (since $P'(t) \in C^* \Rightarrow P'(r) \in R_0^*$), and so there is a unique h.

Alternatively, this may be proved by applying (3.12) to $Y \times_X X'/X'$.

THEOREM 3.23. (*Topological invariance of étale morphisms.*) *Let X_0 be the closed subscheme of a scheme X defined by a nilpotent ideal. The functor $Y \mapsto Y_0 = Y \times_X X_0$ gives an equivalence between the categories of étale X-schemes and étale X_0-schemes.*

Proof. To give an X-morphism $Y \to Z$ of étale X-schemes is the same as to give its graph, that is, a section to $Y \times_X Z \to Y$. According to (3.12), such sections are in one-to-one correspondence with the open subschemes of $Y \times_X Z$ that map isomorphically onto Y. Since the same is true for X_0-morphisms $Y_0 \to Z_0$, it is easy to see using (3.10) or (3.11) that our functor is faithfully full. Thus it remains to show that it is essentially surjective on objects. Because of the uniqueness assertion for morphisms, it suffices to locally lift an étale X_0-scheme Y_0 to an X-scheme Y. But then we may assume that $Y_0 \to X_0$ is standard, and the assertion is obvious.

For completeness, we list some conditions equivalent to smoothness.

PROPOSITION 3.24. *Let $f: Y \to X$ be locally of finite-type. The following are equivalent:*

(a) *f is smooth* (in the sense of (3.22));

(b) *for any $y \in Y$, there exist open affine neighborhoods V of y and U of $f(y)$ such that $f|V$ factors into $V \to V' \to U \hookrightarrow X$ where $V \to V'$ is étale and V' is affine n-space over U;*

(c) *for any $y \in Y$, there exist open affine neighborhoods $V =$ spec C of y and $U =$ spec A of $x = f(y)$ such that*

$$C = A[T_1, \ldots, T_n]/(P_1, \ldots, P_m), \qquad m \leq n,$$

and the ideal generated by the $m \times m$ minors of $(\partial P_i/\partial T_j)$ is C;

(d) *f is flat and for any algebraically closed geometric point $\bar{x}$ of X, the fiber $Y_{\bar{x}} \to \bar{x}$ is smooth;*

(e) *f is flat and for any algebraically closed geometric point $\bar{x}$ of X, $Y_{\bar{x}}$ is regular;*

(f) *f is flat and $\Omega^1_{Y/X}$ is locally free of rank equal to the relative dimension of Y/X.*

Proof. See [SGA. 1, II] or Demazure-Gabriel: [1, I. §4.4].

Remark 3.25. (a) In the case that f is of finite-type, conditions (d) and (e) may be paraphrased by saying that Y is a flat family of nonsingular varieties over X.

(b) Condition (b) shows that for a morphism of finite-type *étale* is equivalent to *smooth and quasi-finite.*

Finally we note that (2.25) has an analogue for smooth morphisms.

PROPOSITION 3.26. *Let $f: Y \to X$ be smooth and surjective, and assume that X is quasi-compact. Then there exists an affine scheme X', a surjective étale morphism $h: X' \to X$, and an X-morphism $g: X' \to Y$.*

Proof. See [EGA. IV.17.16.3].

Exercise 3.27. (Hochster). Let A be the ring $k[T^2, T^3]$ localized at its maximal ideal (T^2, T^3) (that is, A is the local ring at a cusp on a curve); let $B = A[S]/(S^3Y^2 + S + T^2)$, and let C be the integral closure of A in B. Show that B is étale over A, but that C is not flat over A. (Hint: show that TS and T^2S are in C; hence $TS \in (T^2:T^3)_C$. If C were flat over A, then

$$(T^2:T^3)_C = (T^2:T^3)_A C = (T^2, T^3);$$

but $TS \in (T^2, T^3)$ would imply $S \in C$.)

Exercise 3.28. Let Y and X be smooth varieties over a field k; show that a morphism $Y \to X$ is étale if and only if it induces an isomorphism on tangent spaces for any closed point of Y.

Exercise 3.29. Do Hartshorne [2, III. Ex. 10.6].

§4. Henselian Rings

Throughout this section, A will be a local ring with maximal ideal $\mathfrak{m}$ and residue field k. The homomorphisms $A \to k$ and $A[T] \to k[T]$ will be written as $(a \mapsto \bar{a})$ and $(f \mapsto \bar{f})$.

Two polynomials $f(T)$, $g(T)$ with coefficients in a ring B are *strictly coprime* if the ideals (f) and (g) are coprime in $B[T]$, that is, if $(f, g) = B[T]$. For example, $f(T)$ and $T - a$ are coprime if and only if $f(a) \neq 0$ and are strictly coprime if and only if $f(a)$ is a unit in B.

If A is a complete discrete valuation ring, then Hensel's lemma (in number theory) states the following: if f is a monic polynomial with coefficients in A such that $\bar{f}$ factors as $\bar{f} = g_0 h_0$ with g_0 and h_0 monic and coprime, then f itself factors as $f = gh$ with g and h monic and such that $\bar{g} = g_0$, $\bar{h} = h_0$. In general, any local ring A for which the conclusion of Hensel's lemma holds is said to be *Henselian*.

Remark 4.1. (a) The g and h in the above factorization are strictly coprime. More generally, if $f, g \in A[T]$ are such that $\bar{f}, \bar{g}$ are coprime in $k[T]$ and f is monic, then f and g are strictly coprime in $A[T]$. Indeed, let $M = A[T]/(f, g)$. As f is monic, this is a finitely generated A-module; as $(\bar{f}, \bar{g}) = k[T]$, $(f, g) + \mathfrak{m}A[T] = A[T]$ and $\mathfrak{m}M = M$, and so Nakayama's lemma implies that $M = 0$.

(b) The factorization $f = gh$ is unique, for let $f = gh = g'h'$ with g, h, g', h' all monic, $\bar{g} = \bar{g}'$, $\bar{h} = \bar{h}'$, and $\bar{g}$ and $\bar{h}$ coprime. Then g and h' are strictly coprime in $A[T]$, and so there exist $r, s \in A[T]$ such that $gr + h's = 1$. Now

$$g' = g'gr + g'h's = g'gr + ghs,$$

and so g divides g'. As both are monic and have the same degree, they must be equal.

THEOREM 4.2. *Let x be the closed point of $X =$ spec A. The following are equivalent:*

(a) *A is Henselian;*

(b) *any finite A-algebra B is a direct product of local rings* $B = \prod B_i$ (*the* B_i *are then necessarily isomorphic to the rings* $B_{\mathfrak{m}_i}$, *where the* $\mathfrak{m}_i$ *are the maximal ideals of B*);

(c) *if* $f: Y \to X$ *is quasi-finite and separated, then* $Y = Y_0 \amalg Y_1 \amalg \cdots Y_n$ *where* $f(Y_0)$ *does not contain x and* Y_i *is finite over X and is the spec of a local ring,* $i \geq 1$;

(d) *if* $f: Y \to X$ *is étale and there is a point* $y \in Y$ *such that* $f(y) = x$ *and* $k(y) = k(x)$, *then f has a section* $s: X \to Y$;

(d′) *let* $f_1, \ldots, f_n \in A[T_1, \ldots, T_n]$; *if there exists an* $a = (a_1, \ldots, a_n) \in k^n$ *such that* $\bar{f}_i(a) = 0$, $i = 1, \ldots, n$, *and* $\det((\partial \bar{f}_i/\partial T_j)(a)) \neq 0$, *then there exists a* $b \in A^n$ *such that* $\bar{b} = a$ *and* $f_i(b) = 0$, $i = 1, \ldots, n$;

(e) *let* $f(T) \in A[T]$; *if* $\bar{f}$ *factors as* $\bar{f} = g_0 h_0$ *with* g_0 *monic and* g_0 *and* h_0 *coprime, then f factors as* $f = gh$ *with g monic and* $\bar{g} = g_0$, $\bar{h} = h_0$.

Proof. (a) ⇒ (b). According to the going-up theorem, any maximal ideal of B lies over $\mathfrak{m}$. Thus B is local if and only if $B/\mathfrak{m}B$ is local.

Assume first that B is of the form $B = A[T]/(f)$ with $f(T)$ monic. If $\bar{f}$ is a power of an irreducible polynomial, then $B/\mathfrak{m}B = k[T]/(\bar{f})$ is local and B is local. If not, then (a) implies that $f = gh$ where g and h are monic, strictly coprime, and of degree ≥ 1. Then $B \approx A[T]/(g) \times A[T]/(h)$ (Atiyah-Macdonald [1, 1.10]), and this process may be continued to get the required splitting.

Now let B be an arbitrary finite A-algebra. If B is not local, then there is a $b \in B$ such that $\bar{b}$ is a nontrivial idempotent in $B/\mathfrak{m}B$. Let f be a monic polynomial such that $f(b) = 0$; let $C = A[T]/(f)$, and let $\phi: C \to B$ be the map that sends T to b. Since C is monogenic over A, the first part implies that there is an idempotent $c \in C$ such that $\overline{\phi(c)} = \bar{b}$. Now $\phi(c) = e$ is a nontrivial idempotent in B; $B = Be \times B(1 - e)$ is a nontrivial splitting, and the process may be continued.

(b) ⇒ (c). According to (1.8), f factors into $Y \xrightarrow{f'} Y' \xrightarrow{g} X$ with f' an open immersion and g finite. Then (b) implies that $Y' = \coprod \operatorname{spec}(\mathcal{O}_{Y',y})$ where the y run through the (finitely many) closed points of Y'. Let $Y_* = \coprod \operatorname{spec}(\mathcal{O}_{Y',y})$, where the y runs through the closed points of Y' that are in Y. Then Y_* is contained in Y and is both open and closed in Y because it is so in Y'. Let $Y = Y_* \coprod Y_0$. Then clearly $f(Y_0)$ does not contain x.

(c) ⇒ (d). Using (c), we may reduce the question to the case of a finite étale local homomorphism $A \to B$ such that $k(\mathfrak{m}) = k(\mathfrak{n})$ where $\mathfrak{n}$ is the maximal ideal of B. According to (2.9b), B is a free A-module, and since $k(\mathfrak{n}) = B \otimes_A k(\mathfrak{m}) = k(\mathfrak{m})$ it must have rank 1, that is, $A \approx B$.

(d) ⇒ (d′). Let

$$B = A[T_1, \ldots, T_n]/(f_1, \ldots, f_n) = A[t_1, \ldots, t_n],$$

and let $J(T_1, \ldots, T_n) = \det(\partial f_i/\partial T_j)$. The conditions imply that there exists a prime ideal $\mathfrak{q}$ in B lying over $\mathfrak{m}$ such that $J(t_1, \ldots, t_n)$ is a unit in $B_\mathfrak{q}$. It follows that $J(t_1, \ldots, t_n)$ is a unit in B_b for some $b \in B$, $b \notin \mathfrak{q}$ and thus that B_b is étale over A (compare (3.4); to convert B_b to an algebra of the form considered there, use the trick of the proof of (3.16). Now apply (d) to lift the solution in k^n to one in A^n.

(d′) ⇒ (e). Write

$$f(T) = a_n T^n + a_{n-1} T^{n-1} + \cdots + a_0,$$

and consider the equations,

$$\begin{aligned} X_0 Y_0 &= a_0, \\ X_0 Y_1 + X_1 Y_0 &= a_1, \\ X_0 Y_2 + X_1 Y_1 + X_2 Y_0 &= a_2, \\ &\cdots\cdots \\ X_{r-1} Y_s + Y_{s-1} &= a_{n-1}, \\ Y_s &= a_n \end{aligned}$$

where $r = \deg(g_0)$ and $s = n - r$. Clearly $(b_0, \ldots, b_{r-1}; c_0, \ldots, c_s)$ is a solution to this system of equations if and only if

$$f(T) = (T^r + b_{r-1}T^{r-1} + \cdots + b_0)(c_s T^s + \cdots + c_0).$$

The Jacobian of the equations is

$$\det\begin{pmatrix} Y_0 & & X_0 & & \\ Y_1 \; Y_0 & & X_1 X_0 & & \\ \vdots \;\; Y_1 & & X_2 X_1 X_0 & & \\ Y_s \;\; \vdots & & \vdots \;\; \vdots \;\; \vdots & & \\ & & 1 & & \end{pmatrix} = \operatorname{res}(g, h), \text{ the resultant of } g \text{ and } h,$$

where $g = T^r + X_{r-1}T^{r-1} + \cdots + X_0$ and $h = Y_s T^s + \cdots + Y_0$. To prove (e) we have only to show that $\operatorname{res}(g_0, h_0) \neq 0$. But $\operatorname{res}(g_0, h_0)$ can be zero only if both $\deg(g_0) < r$ and $\deg(h_0) < s$, or g_0 and h_0 have a common factor, and neither of these occurs.

(e) ⇒ (a). This is trivial.

COROLLARY 4.3. *If A is Henselian, then so also is any finite local A-algebra B and any quotient ring $A/\mathfrak{I}$.*

Proof. Obviously B satisfies condition (b) of the theorem, and $A/\mathfrak{I}$ satisfies the definition of a Henselian ring.

PROPOSITION 4.4. *If A is Henselian, then the functor $B \mapsto B \otimes_A k$ induces an equivalence between the category of finite étale A-algebras and the category of finite étale k-algebras.*

Proof. After applying (4.2b), we need only consider local A-algebras B. The canonical map

$$\mathrm{Hom}_A(B, B') \to \mathrm{Hom}_k(B \otimes k, B' \otimes k)$$

is injective according to (3.13). To see the surjectivity, note that a k-homomorphism $B \otimes k \to B' \otimes k$ induces an A-homomorphism $g: B \to B' \otimes_A k$ by composition with $B \to B \otimes k$ and hence an A-homomorphism

$$(b' \otimes b \mapsto b'g(b)): B' \otimes_A B \to B' \otimes_A k.$$

Now apply (4.2d) to the map $\mathrm{spec}(B' \otimes B) \to \mathrm{spec}(B')$ to get an A-homomorphism $B' \otimes_A B \to B'$ that induces the required map $B \to B'$. Thus the functor is fully faithful. To complete the proof, one only has to observe that any local étale k-algebra k' can be written in the form $k[T]/(f_0(T))$, where $f_0(T)$ is monic and irreducible and then that $B = A[T]/(f(T))$, where $\bar{f}(T) = f_0(T)$ and f is monic, has the property that $B \otimes_A k = k'$.

So far, we have had no examples of Henselian rings. The following is a generalization of Hensel's lemma in number theory.

PROPOSITION 4.5. *Any complete local ring A is Henselian.*

Proof. Let B be an étale A-algebra, and suppose that there exists a section $s_0: B \to k$. We have to show (4.2d) that this lifts to a section $s: B \to A$. Write $A_r = A/\mathfrak{m}^{r+1}$; if we can prove that there exist compatible sections $s_r: B \to A_r$, then these maps will induce a section $s: B \to \varprojlim A_r = A$. For $r = 0$ the existence of s_r is given. For $r > 0$ the existence of s_r follows from that of s_{r-1} because of the property of the functor defined by an étale morphism (3.22).

Remark 4.6. (a) The last two propositions show that the functor $B \mapsto B \otimes_A \hat{A}$ gives an equivalence between the categories of finite étale algebras over A and over its completion when A is Henselian. Under certain circumstances, notably when X is proper over a Henselian ring A, this result extends to the categories of schemes finite and étale over X and over $\hat{X} = X \otimes_A \hat{A}$. See Artin [2] and [5].

(b) The result in (4.4) has the following generalization. Let X be a scheme proper over a Henselian local ring A, and let X_0 be the closed fiber of X. The functor $Y \mapsto Y \times_X X_0$ induces an equivalence between the category of schemes Y finite and étale over X and those over X_0. When A is complete, proofs of this may be found in Artin [9, VII.11.7] and Murre [1, 8.1.3]. The Henselian case is deduced from the complete case by means of the approximation theorem (Artin [9, II]; see also Artin [5, Theorem 3.1]).

(c) Part (c) of (4.2) also has a generalization. Let $f: Y \to X$ be separated and of finite-type, where $X = \operatorname{spec} A$ with A Henselian local. If y is an isolated point in the closed fiber Y_0 of f, so that $Y_0 = \{y\} \amalg Y'_0$ (as schemes), then $Y = Y'' \amalg Y'$ with Y'' finite over X and Y'' and Y' having closed fibers $\{y\}$ and Y'_0 respectively. For a proof, see Artin [9, I.1.10].

(d) If X is an analytic manifold over $\mathbb{C}$, then the local ring at a point x of X is Henselian. (For this, and similar examples, see Raynaud [3, VII.4].)

Remark 4.7. Let $f: Y \to X$ be étale, and suppose $k(x) = k(y)$ for some $y \in Y$, $x = f(y)$. Then the map on the completions $\hat{\mathcal{O}}_{X,x} \to \hat{\mathcal{O}}_{Y,y}$ is étale and, according to (4.5) and (4.2), has a section, which implies (3.12) it is an isomorphism $\hat{\mathcal{O}}_{X,x} \xrightarrow{\approx} \hat{\mathcal{O}}_{Y,y}$. (See Hartshorne [2, III. Ex. 10.4] for a converse statement.)

This may be used to give an example of an injective unramified map of rings that is not étale. Let X be a curve over a field, which has a node at x_0, and let $f: Y \to X$ be the normalization of X in $k(X)$. It is obvious this map is unramified, and $\mathcal{O}_{X,f(y)} \hookrightarrow \mathcal{O}_{Y,y}$ is injective for all y, but if y lies over x_0, then $\hat{\mathcal{O}}_{X,x_0} \to \hat{\mathcal{O}}_{Y,y}$ is not an isomorphism because $\hat{\mathcal{O}}_{X,x_0}$ is not an integral domain. (It has two minimal prime ideals; see Hartshorne [2, I.5.6.3].)

A is a subring of its completion $\hat{A}$ (since we require rings to be Noetherian). Thus any local ring A is a subring of a Henselian ring, and the smallest such ring will be called the Henselization of A. More precisely, let $i: A \to A^{\mathrm{h}}$ be a local homomorphism of local rings; A^{h} is the *Henselization* of A if it is Henselian and if any other local homomorphism from A into a Henselian local ring factors uniquely through i. Clearly (A^{h}, i) is unique, up to a unique isomorphism, if it exists.

Before proving the existence of A^{h}, we introduce the notion of an *étale neighborhood* of a local ring A. It is a pair $(B, \mathfrak{q})$ where B is an étale A-algebra and $\mathfrak{q}$ is a prime ideal of B lying over $\mathfrak{m}$ such that the induced map $k \to k(\mathfrak{q})$ is an isomorphism.

LEMMA 4.8. (a) *If $(B, \mathfrak{q})$ and $(B', \mathfrak{q}')$ are étale neighborhoods of A such that* spec B' *is connected, then there is at most one A-homomorphism $f: B \to B'$ such that $f^{-1}(\mathfrak{q}') = \mathfrak{q}$.*

(b) *Let $(B, \mathfrak{q})$ and $(B', \mathfrak{q}')$ be étale neighborhoods of A; there is an étale neighborhood $(B'', \mathfrak{q}'')$ of A with* spec B'' *connected and A-homomorphisms $f: B \to B''$, $f': B' \to B''$ such that $f^{-1}(\mathfrak{q}'') = \mathfrak{q}$, $f'^{-1}(\mathfrak{q}'') = \mathfrak{q}'$.*

Proof. (a) This is an immediate consequence of (3.13).

(b) Let $C = B \otimes_A B'$. The maps $B \to k(\mathfrak{q}) = k$, $B' \to k(\mathfrak{q}') = k$ induce a map $C \to k$. Let $\mathfrak{q}''$ be the kernel of this map. Then $(B'', \mathfrak{q}''B'')$, where $B'' = C_c$ with some $c \notin \mathfrak{q}''$ such that spec B'' is connected, is the required étale neighborhood.

It follows from the lemma that the étale neighborhoods of A with connected spectra form a filtered direct system. Define $(A^h, \mathfrak{m}^h)$ to be its direct limit, $(A^h, \mathfrak{m}^h) = \varinjlim (B, \mathfrak{q})$. It is easy to check that A^h is a local A-algebra with maximal ideal $\mathfrak{m}^h$, $A^h/\mathfrak{m}^h = k$ and A^h is the Henselization of A. Also A^h is obviously flat over A. (Slightly less trivial is the fact that A^h is Noetherian, that is, we have not passed outside our category. This may be found in Artin [1, III.4.2].)

Exercise 4.9. Instead of building A^h from below, it is possible to descend on it from above. Let $\tilde{A}$ be the intersection of all local Henselian subrings H of $\hat{A}$, containing A, which have the property that $\hat{\mathfrak{m}} \cap H = \mathfrak{m}_H$. Show that $(\tilde{A}, i)$, where $i: A \hookrightarrow \tilde{A}$ is the inclusion map, is a Henselization of A. (Hint: to show that $\tilde{A}$ satisfies the definition of Henselian ring, note that the factorization $\bar{f} = g_0 h_0$ lifts to a factorization $f = gh$ in any H; use the uniqueness of the factorization to show that g and h are in $\bigcap H = \tilde{A}$.)

Examples 4.10. (a) Let A be normal; let K be the field of fractions of A, and let K_s be a separable closure of K. The Galois group G of K_s over K acts on the integral closure B of A in K_s. Let $\mathfrak{n}$ be a maximal ideal of B lying over $\mathfrak{m}$, and let $D \subset G$ be the decomposition group of $\mathfrak{n}$, that is, $D = \{\sigma \in G \mid \sigma(\mathfrak{n}) = \mathfrak{n}\}$. Let A^h be the localization at $\mathfrak{n}^D$ of the integral closure B^D of A in K_s^D. (Here

$$B^D = \{b \in B \mid \sigma(b) = b \text{ all } \sigma \in D\}$$

etc.) I claim that A^h is the Henselization of A.

Indeed, if A^h were not Henselian, there would exist a monic polynomial $f(T)$ that is irreducible over A^h but whose reduction $\bar{f}(T)$ factors into relatively prime factors. But from such an f one can construct a finite Galois extension L of K_s^D such that the integral closure A' of A^h in L is not local. This is a contradiction since the Galois group of L over K_s^D permutes the prime ideals of A' lying over $\mathfrak{n}^D$ and hence cannot be a quotient of D. To see that A^h is the Henselization, one only has to show that it is a union of étale neighborhoods of A, but this is easy using (3.21).

(b) Let k be a field, and let A be the localization of $k[T_1, \ldots, T_n]$ at $(T_1, \ldots, T_n)$. The Henselization of A is the set of power series $P \in k[[T_1, \ldots, T_n]]$ that are algebraic over A. (For a good discussion of why this should be so, see Artin [8]; for a proof, see Artin [9, II.2.9].)

(c) The Henselization of $A/\mathfrak{J}$ is $A^h/\mathfrak{J}A^h$. This is immediate from the definition of the Henselization and (4.3).

Every ring is a quotient of a normal ring, and so it would have sufficed to construct A^h for A normal. This is the approach adopted by Nagata [1].

Remark 4.11. We have seen that if A is normal, then so also is A^{h}. It is also true that if A is reduced or regular, then A^{h} is reduced or regular and $\dim A^{\mathrm{h}} = \dim A$. These statements follow from (3.18) or (3.17).

Let X be a scheme and let $x \in X$. An *étale neighborhood* of x is a pair (Y, y) where Y is an étale X-scheme and y is a point of Y mapping to x such that $k(x) = k(y)$. The connected étale neighborhoods of x form a filtered system and clearly the limit $\varinjlim \Gamma(Y, \mathcal{O}_Y) = \mathcal{O}_{X,x}^h$.

By definition, A being Henselian means that it has no finite étale extensions with trivial residue field extension (4.2d), except those of the form $A \to A^{\mathrm{r}}$. Thus if the residue field of A is separably algebraically closed, then A has no finite étale extensions at all. Such a Henselian ring is called *strictly Henselian* or *strictly local*. Most of the above theory can be rewritten for strictly Henselian rings. In particular, the *strict Henselization* of A is a pair (A^{sh}, i) where A^{sh} is a strictly Henselian ring and $i: A \to A^{\mathrm{sh}}$ is a local homomorphism such that any other local homomorphism $f: A \to H$ with H strictly Henselian extends to a local homomorphism $f': A^{\mathrm{sh}} \to H$; moreover, f' is to be uniquely determined once the induced map $A^{\mathrm{sh}}/\mathfrak{m}^{\mathrm{sh}} \to H/\mathfrak{m}_H$ on residue fields is given.

Fix a separable closure k_s of k. Then $A^{\mathrm{sh}} = \varinjlim B$, where the limit runs over all commutative diagrams

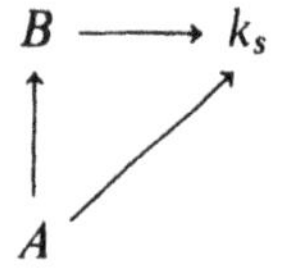

in which $A \to B$ is étale. If $A = k$ is a field, then A^{sh} is any separable closure of k; if A is normal, then A^{sh} can be constructed the same way as A^{h} except that the decomposition group must be replaced by the inertia group; if A is normal and Henselian, then A^{sh} is the maximal unramified extension of A in the sense of the number theorists. Finally, $(A/\mathfrak{I})^{\mathrm{sh}} = A^{\mathrm{sh}}/\mathfrak{I}A^{\mathrm{sh}}$.

Let X be a scheme and $\bar{x} \to X$ a geometric point of X. An *étale neighborhood* of $\bar{x}$ is a commutative diagram:

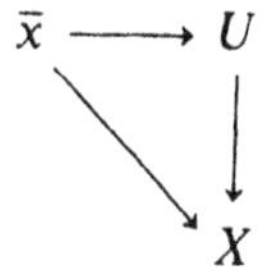

with $U \to X$ étale. Clearly $\mathcal{O}_{X,x}^{\mathrm{sh}} = \varinjlim \Gamma(U, \mathcal{O}_U)$ where the limit is taken over all étale neighborhoods of $\bar{x}$. We write $\mathcal{O}_{X,\bar{x}}^{\mathrm{sh}}$, or simply $\mathcal{O}_{X,\bar{x}}$ for this limit. As we shall see, $\mathcal{O}_{X,\bar{x}}$ is the analogue for the étale topology of the local ring for the Zariski topology, that is, it is the local ring relative to a stronger notion of localization. Note that its definition is formally the same, for $\mathcal{O}_{X,x} = \varinjlim \Gamma(U, \mathcal{O}_U)$ where the limit is taken over all Zariski (open) neighborhoods of x.

Exercise 4.12. Study the properties of the ring $A^{\mathrm{qh}} = \varinjlim B$, where the limit is taken over all diagrams

$$\begin{array}{ccc} B & & \\ \uparrow & \searrow & \\ A & \longrightarrow & k \end{array}$$

in which spec B is connected and spec $B \to$ spec A is finite and étale or an open immersion or a composite of such morphisms.

Exercise 4.13. Let X be a smooth scheme over spec A, where A is Henselian with residue field k. Show that $X(A) \to X(k)$ is surjective. (Use 3.24b).

§5. The Fundamental Group: Galois Coverings

In this section we summarize some of the basic properties of the fundamental group of a scheme. Proofs may be found in Murre [1].

The fundamental group $\pi_1(X, x_0)$ of an arcwise connected, locally connected, and locally simply connected topological space with base point x_0 may be defined in two ways: either as the group of closed paths through x_0 modulo homotopy equivalence or as the automorphism group of the universal covering space of X. The first definition does not generalize well to schemes—there are simply too few algebraically defined closed paths—but the second does. Thus the important, defining property of the fundamental group of a scheme is that it classifies in a natural way the étale coverings of X, étale being the most natural analogue of local homeomorphism.

Thus let X be a connected scheme, and let $\bar{x} \to X$ be a geometric point of X. Define a functor F:**FEt**/$X \to$ **Sets**, where **FEt**/X is the category of X-schemes finite and étale over X, by setting $F(Y) = \mathrm{Hom}_X(\bar{x}, Y)$. Thus to give an element of $F(Y)$ is to give a point $y \in Y$ lying over x and a $k(x)$-homomorphism $k(y) \to k(\bar{x})$. It may be shown that this functor is strictly prorepresentable, that is, that there exists a directed set I, a projective system $(X_i, \phi_{ij})_{i \in I}$ in **FEt**/X in which the transition morphisms $\phi_{ij}: X_j \to X_i$ $(i \leq j)$ are epimorphisms and elements $f_i \in F(X_i)$ such that

(a) $f_i = \phi_{ij} \circ f_j$, and

(b) the natural map $\varinjlim \operatorname{Hom}(X_i, Z) \to F(Z)$ induced by the f_i is an isomorphism for any Z in $\mathbf{FEt}/X$.

The projective system $\tilde{X} = (X_i, \phi_{ij})$ will play the role of the universal covering space of a topological space, and we want to define π_1 to be its automorphism group.

For an X-scheme Y, we write $\operatorname{Aut}_X(Y)$ for the group of X-automorphisms of Y acting on the right. For any $Y \in \mathbf{FEt}/X$, $\operatorname{Aut}_X(Y)$ acts on $F(Y)$ (on the right), and if Y is connected, then this action is faithful, that is, for any $g \in F(Y)$, $\sigma \mapsto \sigma \circ g : \operatorname{Aut}_X(Y) \to F(Y)$ is injective (this follows from (3.13)). If Y is connected and $\operatorname{Aut}_X(Y)$ acts transitively on $F(Y)$ so that the above map $\operatorname{Aut}_X(Y) \to F(Y)$ is bijective, then Y is said to be *Galois* over X. For any $Y \in \mathbf{FEt}/X$ there is a $Y' \in \mathbf{FEt}/X$ that is Galois and an X-morphism $Y' \to Y$ (see 5.4 below). It follows that the objects X_i in $\tilde{X}$ may be assumed to be Galois over X. Now, given $j \geq i$, we can define a map $\psi_{ij}: \operatorname{Aut}_X(X_j) \to \operatorname{Aut}_X(X_i)$ by requiring that $\psi_{ij}(\sigma) f_i = \phi_{ij} \circ \sigma \circ f_j$. We define $\pi_1(X, \bar{x})$ to be the profinite group $\varprojlim \operatorname{Aut}_X(X_i)$.

Remark 5.1. (a) If $\bar{x}'$ is any other geometric point of X, then $\pi_1(X, \bar{x}')$ is isomorphic to $\pi_1(X, \bar{x})$ and the isomorphism is canonically determined up to an inner automorphism of $\pi_1(X, \bar{x})$.

(b) If the above process is carried through for an arcwise connected, locally connected, and locally simply connected topological space X, a point x on it, and the category of covering spaces of X, then one finds that F is representable, that is, not merely prorepresentable, by the universal covering space $\tilde{X}$ of X. Thus $\pi_1(X, x)$ can be directly defined as the automorphism group of $\tilde{X}$ over X.

(c) Let X be a smooth projective variety over $\mathbb{C}$, and let X^{an} be the associated analytic manifold. The Riemann existence theorem states that the functor, which associates with any finite étale map $Y \to X$ the local isomorphism of analytic manifolds $Y^{an} \to X^{an}$, is an equivalence of categories. Thus the étale fundamental group $\pi_1(X)$ and the analytic fundamental group $\pi_1(X^{an})$ have the same finite quotients. It follows that their completions with respect to the topology defined by the subgroups of finite index are equal. But $\pi_1(X)$ by definition, is already complete, and hence $\pi_1(X) \approx \widehat{\pi_1(X^{an})}$.

The reason that no algebraically defined fundamental group can equal $\pi_1(X^{an})$ is that the covering space does not exist algebraically (at least, not as an element of $\mathbf{FEt}/X$).

(d) A theorem of Grauert and Remmert implies that (c) also holds for nonprojective varieties. This fact is the basis of the proof of the theorem comparing étale and complex cohomology. (See III.3.)

Examples 5.2. (a) Let $X = \operatorname{spec} k$, k a field. The X_i may be taken to be the spectra of all finite Galois extensions K_i of k contained in $k(\bar{x})$.

Thus $\pi_1(X, \bar{x})$ is the Galois group over k of the separable closure of $k(x)$ in $k(\bar{x})$. Changing $\bar{x}$ corresponds therefore to choosing a different separable algebraic closure.

(b) Let X be a normal scheme. Let $\bar{x} = \operatorname{spec} k(x)_{\text{sep}}$ where x is the generic point of X. Then the X_i may be taken to be the normalizations of X in K_i, where the K_i run through the finite Galois extensions of $k(x)$ contained in $k(\bar{x})$ such that the normalization of X in K_i is unramified. Thus $\pi_1(X, \bar{x})$ is the Galois group of $k(x)_{\text{un}}$ over $k(x)$, where $k(x)_{\text{un}} = \bigcup K_i$.

(c) Let $X = \operatorname{spec} A$, where A is a strictly Henselian local ring. Then $\pi_1(X, \bar{x}) = \{1\}$ since **FEt**$/X$ consists only of direct sums of copies of X.

(If X is a scheme and $\bar{x}$ a geometric point of X, then $\operatorname{spec} \mathcal{O}_{X,\bar{x}}$ is the algebraic analogue of a sufficiently small ball about a point x on a manifold; thus $\pi_1 = \{1\}$ agrees with the ball being contractible.)

(d) Let $X = \operatorname{spec} A$, where A is Henselian. Let $\bar{x}$ be a geometric point over the closed point x of X. The equivalence of categories **FEt**$/X \leftrightarrow$ **FEt**$/\operatorname{spec} k(x)$ (4.4) induces an isomorphism $\pi_1(X, \bar{x}) \approx \pi_1(\operatorname{spec} k(x), \bar{x})$.

(e) Let $X = \operatorname{spec} K$ where K is the field of fractions of a strictly Henselian discrete valuation ring A. Then X is the algebraic analogue of a punctured disc in the plane (compare with (c) above) and so one might hope that $\pi_1(X, \bar{x}) \approx \hat{\mathbb{Z}}$. This is true if the residue field $A/\mathfrak{m}$ has characteristic zero because (Serre [7, IV. Proposition 8]) the Galois extensions of K are exactly the Kummer extensions K_n/K, where $K_n = K[t^{1/n}]$ with t a uniformizing parameter. The map

$$(\sigma \mapsto \sigma(t^{1/n})/t^{1/n}) \colon \operatorname{Gal}(K_n/K) \to \mu_n(K)$$

is an isomorphism from the Galois group onto the nth roots of unity. Thus

$$\pi_1(X, \bar{x}) = \varprojlim \operatorname{Gal}(K_n/K) = \varprojlim \mu_n(K) \approx \hat{\mathbb{Z}}.$$

If the residue field has characteristic p, then this is no longer true because of wild ramification. However, any tamely ramified extension of K is still Kummer (Serre [7, IV]) and the tame fundamental group

$$\pi_1^t(X, \bar{x}) = \varprojlim_{p \nmid n} \mu_n(K) \approx \varprojlim_{p \nmid n} \mathbb{Z}/n\mathbb{Z}.$$

(In general, if K is a field and A is a discrete valuation ring with field of fractions K, then a finite separable field extension L of K is *tamely ramified* with respect to A if, for each valuation ring B of L lying over A, the residue field extension $B/\mathfrak{n} \supset A/\mathfrak{m}$ is separable and the ramification index of B/A is not divisible by p (the characteristic of $A/\mathfrak{m}$). Let X be a connected normal scheme; let D be a finite union $D = \bigcup D_i$ of irreducible divisors on X, and let x_i be the generic point of D_i. Then a map $f\colon Y \to X$

is a *tamely ramified* covering if it is finite and étale over $X - D$, Y is connected and normal, and $R(Y)/R(X)$ is tamely ramified with respect to the rings $\mathcal{O}_{X,x_i}$. The tame fundamental group π_1^t is defined so as to classify such coverings. (See Grothendieck-Murre [1].) A basic theorem of Abhyankar generalizes the statement that, in the above example, tame coverings are all Kummer. It says (Grothendieck-Murre [1, 2.3]) that if the D_i have only normal crossings, then for any tamely ramified covering $f: Y \to X$ there is an étale surjective map $X' \to X$ such that $Y \times_X X' \to X'$ is a Kummer covering.)

(f) Let $X = \mathbb{P}^1_k$, where k is separably closed. If $k = \mathbb{C}$, then X is topologically a sphere, and so $\pi_1(X, \bar{x}) = \{1\}$. To see that this is true in general we have to show that **FEt**/X is trivial or, more precisely, that any finite étale map $Y \to X$ with Y connected is an isomorphism. But if ω is the differential dt on $\mathbb{P}^1$, then ω has a double pole at infinity and no other poles or zeros. Thus $f^*(\omega)$ has $2n$ poles, where n is the degree of f, and no zeros. But $-2n = 2g - 2 \geq -2$, where g is the genus of Y, and so $n = 1$ and f is an isomorphism.

(The same argument shows that there is no nontrivial map $Y \to \mathbb{P}^1$ that is étale over $\mathbb{A}^1 \subset \mathbb{P}^1$ and tamely ramified at infinity. In this case $f^*(\omega)$ has $\leq 2n - (n - 1) = n + 1$ poles and no zeros, and $-(n + 1) \geq 2g - 2 \geq -2$ shows again that $n = 1$.)

(g) Let X be a proper scheme over a Henselian local ring A such that the closed fiber X_0 of X is geometrically connected. Then it follows from (4.6b) that $\pi_1(X, \bar{x})$ is canonically isomorphic to $\pi_1(X_0, \bar{x}_0)$ if $\bar{x}$ is the composite $\bar{x}_0 \to X_0 \hookrightarrow X$.

(h) Let U be an open subscheme of a regular scheme X, where the closed complement $Z = X - U$ has codimension ≥ 2. It follows immediately from the theorem of the purity of branch locus (3.7) and the description of the fundamental group of a normal scheme given in (b) above, that $\pi_1(U, \bar{x}) \approx \pi_1(X, \bar{x})$ for any geometric point $\bar{x}$ of U.

It follows that the fundamental group is a birational invariant for varieties complete and regular over a field k (because any dominating rational map of such varieties is defined on the complement of a closed subset of codimension ≥ 2; see Hartshorne [2, V.5.1]).

(i) Let X be a scheme proper and smooth over spec A where A is a complete discrete valuation ring with algebraically closed residue field of characteristic p. Assume that $X_{\bar{K}}$ is connected, where $\bar{K}$ is the algebraic closure of the field of fractions of A, and that the special fiber of X is connected. Then the kernel of the surjective homomorphism $\pi_1(X_{\bar{K}}, \bar{x}) \to \pi_1(X, \bar{x})$, for any geometric point $\bar{x}$ of $X_{\bar{K}}$, is contained in the kernel of any homomorphism of $\pi_1(X_{\bar{K}}, \bar{x})$ into a finite group of order prime to p (see [SGA. 1, X.] or Murre [1]).

(j) Let X_0 be a smooth projective curve of genus g over an algebraically closed field k of characteristic p. Then there is a complete discrete valuation ring A with residue field k and a smooth projective curve X over A such that $X_0 = X \otimes_A k$. (The obstructions to lifting a smooth projective variety lie in second (Zariski) cohomology groups and hence vanish for a curve ([SGA. 1, III.7])). It follows from (g) above that $\pi_1(X_0, \bar{x}) \approx \pi_1(X, \bar{x})$ and from (i) that there is a surjection $\pi_1(X_{\bar{K}}, \bar{x}) \to \pi_1(X, \bar{x})$ with small kernel. The comparison theorem (5.1c) shows that $\pi_1(X_{\bar{K}}, \bar{x})$ is the profinite completion of the topological fundamental group of a curve of genus g over $\mathbb{C}$ and hence is well-known. On putting these facts together, one finds that $\pi_1(X_0, \bar{x}_0)^{(p)} \approx G^{(p)}$ where G is the free group on $2g$ generators u_i, v_i ($i = 1, \ldots, g$) with the single relation

$$(u_1 v_1 u_1^{-1} v_1^{-1}) \cdots (u_g v_g u_g^{-1} v_g^{-1}) = 1$$

and where the superscript (p) on a group H means replace H by $\varprojlim H_i$ with the H_i running through all finite quotients of H of order prime to p.

This computation of the prime-to-p part of the fundamental group of a curve in characteristic p was one of the first major successes of Grothendieck's approach to algebraic geometry.

Once $\pi_1(X, \bar{x})$ has been constructed, the important result is that it really does classify finite étale maps $Y \to X$.

THEOREM 5.3. *Let $\bar{x}$ be a geometric point of the connected scheme X. The functor F defines an equivalence between the category* **FEt**/X *and the category* $\pi_1(X, \bar{x})$**-sets** *of finite sets on which $\pi_1(X, \bar{x})$ acts continuously (on the left).*

Proof. See Murre [1].

Remark 5.4. Recall that if $Y \to X$ is finite and étale and both Y and X are connected, then we defined Y to be Galois over X if $\mathrm{Aut}_X(Y)$ has as many elements as the degree of Y over X. We wish to extend this notion slightly.

If G is a finite group, then G_X (for any scheme X) denotes the scheme $\coprod_{\sigma \in G} X_\sigma$, where $X_\sigma = X$ for each σ. Note that we may define an action of G on G_X (on the right) by requiring $\sigma | X_\tau$ to be the identity map $X_\tau \to X_{\tau\sigma}$.

Let G act on Y over X; then $Y \to X$ is *Galois*, with *Galois group* G, if it is faithfully flat and the map

$$\psi: G_Y \to Y \times Y, \psi | Y_\sigma = (y \mapsto (y, y\sigma))$$

is an isomorphism. Equivalently, $Y \to X$ is Galois with Galois group G if there is a faithfully flat morphism $U \to X$, locally of finite-type, such that Y_U is isomorphic (with its G-action) to G_U. In terms of rings, a ring $B \supset A$ on which G acts by A-automorphisms (on the left) is a Galois extension of

A with Galois group G if B is a finite flat A-algebra and $\mathrm{End}_A(B)$ has a basis $\{\sigma \mid \sigma \in G\}$ as a left B-module with multiplication table

$$(b\sigma)(c\tau) = b\sigma(c)\sigma\tau, \qquad b, c \in B, \qquad \sigma, \tau \in G.$$

We leave it to the reader to check that these are all equivalent and that if $Y \to X$ is Galois with Galois group G, then so also is $Y_{(X')} \to X'$ for any $X' \to X$. Also the reader may check that any finite étale morphism $Y \to X$ can be embedded in a Galois extension, that is, that there exists a Galois morphism $Y' \to X$ that factors through $Y \to X$. (For X normal, this is obvious from the Galois theory of fields; the general case is not much more difficult; see, for example, Murre [1, 4.4.1.8].)

Let X be connected, with geometric point $\bar{x}$. According to (5.3), to give an $\bar{x}$-pointed Galois morphism $Y \to X$ with a given action of G on Y over X is the same as to give a continuous morphism $\pi_1(X, \bar{x}) \to G$. We shall sometimes write $\pi^1(X, \bar{x}; G) \overset{\mathrm{df}}{=} \mathrm{Hom}_{\mathrm{conts}}(\pi_1(X, \bar{x}), G) \approx$ set of $\bar{x}$-pointed Galois coverings of X with Galois group G (modulo isomorphism). As an application of étale cohomology we shall show how to compute $\pi^1(X, \bar{x}; G)$ for some schemes X when G is commutative. Note that in this case it is not necessary to give the Galois coverings $\bar{x}$-points.

Remark 5.5. Let A be a ring with no idempotents other than 0 and 1, that is, such that $X = \mathrm{spec}\, A$ is connected, and let $\tilde{X} = (X_i, \phi_{ij})$ be the universal covering scheme (as above) relative to some geometric point $\bar{x}$ of X. Each X_i is an affine scheme Spec A_i and $\tilde{A} = \varinjlim A_i$ has the following properties: spec $\tilde{A} = \varprojlim X_i$; there is no nontrivial finite étale map $\tilde{A} \to \mathbf{B}$; $\pi_1(X, \bar{x}) = \mathrm{Aut}_A(\tilde{A})$. Thus $\tilde{A}$ may reasonably be called the *étale closure* of A. If A has only a finite number of idempotents, then $A = \prod A_j$ (finite product), $X = \coprod X_j$ ($X_j = \mathrm{spec}\, A_j$), and $\mathbf{FEt}/X \approx \prod \mathbf{FEt}/X_j$. If A has an infinite number of idempotents, then $A = \bigcup A_j$ where each A_j has only a finite number of idempotents, $X = \varprojlim X_j$, and $\mathbf{FEt}/X =$ "$\varprojlim$" $\mathbf{FEt}/X_j$ [EGA. IV.17]. Thus the study of $\mathbf{FEt}/X$, X affine, essentially can be reduced to the case with X connected.

Remark 5.6. We have seen that $\pi_1(X, \bar{x})$ prorepresents the functor that takes a finite group N to the set of isomorphism classes of $\bar{x}$-pointed Galois coverings of X with a given action of N. Clearly this property determines $\pi_1(X, \bar{x})$, and, since the category of finite groups is Artinian, a standard theorem (Grothendieck [3]) shows that the existence of $\pi_1(X, \bar{x})$ is equivalent to the left exactness of the functor.

It is natural to ask whether there exists a larger fundamental group that, in addition, classifies coverings whose structure groups are finite group schemes. More precisely, consider a variety X over a field k; fix a spec (k_{al})-point $\bar{x}$ on X, and consider the functor that takes a finite group scheme N over k to the set of isomorphism classes of pairs $(Y, \bar{y})$ where Y is a principal homogeneous space for N over X (III.4) and $\bar{y}$ is a spec (k_{al})-point of Y lying over $\bar{x}$. Unless X is complete and k is algebraically closed, this functor will not be left exact, as one may see by looking at the cohomology sequence of a short exact sequence of commutative finite group schemes. However, when these conditions hold, then the functor is left exact and is represented by a profinite group scheme $\pi_1(X_{\mathrm{fl}}, \bar{x})$, (Nori [1]). The fundamental group $\pi_1(X, \bar{x})$ is the maximal proétale quotient of this true fundamental group $\pi_1(X_{\mathrm{fl}}, \bar{x})$. See also [SGA. 1, p. 271, 289, 309].

Comments on the Literature

The first source for much of the material in this chapter is Grothendieck's Bourbaki talks [3] and [SGA. 1], and the ultimate source is Chapter IV of [EGA.]. Some of the same material can be found in Raynaud [3] and Iverson [2] in the affine case and in the notes of Altman-Kleiman [1]. The first chapter of Artin [9] also contains an elegant, if brief, treatment of étale maps and Henselian rings. See also Kurke-Pfister-Roczen [1]. The most useful introduction to the fundamental group is Murre [1]. The theory of the higher étale homotopy groups is developed in Artin-Mazur [1].

CHAPTER II

Sheaf Theory

In order to obtain a sufficiently fine topology on a scheme X, it is necessary to relax the requirement that a covering consists of subsets of X. Thus a covering for the étale topology is a surjective family of étale morphisms $(U_i \to X)$. Most of the definitions and results from sheaf theory generalize quite easily to the new situation. For example, it is possible to define the sheaf associated with a presheaf, the stalk of a sheaf at a geometric point, and the direct and inverse images of sheaves relative to a morphism of schemes. As a first indication that the theory we obtain is interesting, we show that the category of sheaves over the spectrum of a field is equivalent to the category of discrete modules over the Galois group of the field. Any group scheme or quasi-coherent module defines a sheaf for the étale topology. There are many examples of sequences of sheaves that are not exact relative to the Zariski topology, but become exact when considered relative to the finer étale topology. Sometimes, particularly when considering p-torsion sheaves in characteristic p, it is necessary to use an even finer topology than the étale topology: the flat topology.

§1. Presheaves and Sheaves

We shall be concerned with classes E of morphisms of schemes satisfying the following conditions:

(e_1) all isomorphisms are in E;

(e_2) the composite of two morphisms in E is in E;

(e_3) any base change of a morphism in E is in E.

A morphism in such a class E will be referred to as an *E-morphism*. The full subcategory of **Sch**/X of X-schemes whose structure morphism is an E-morphism will be written E/X.

The following examples of such classes will be particularly important:

(a) the class E = (Zar) of all open immersions;

(b) the class E = (ét) of all étale morphisms of finite-type;

(c) the class E = (fl) of all flat morphisms that are locally of finite-type.

In each of these examples, the E-morphisms are open (I.2.12) and any open immersion is an E-morphism. This will be so in almost all examples. The E-morphisms are to play the role of the open subsets in an E-topology.

Now fix a base scheme X, a class E as above, and a full subcategory $\mathbf{C}/X$ of $\mathbf{Sch}/X$ that is closed under fiber products and is such that, for any $Y \to X$ in $\mathbf{C}/X$ and any E-morphism $U \to Y$, the composite $U \to X$ is in $\mathbf{C}/X$. An *E-covering* of an object Y of $\mathbf{C}/X$ is a family $(U_i \xrightarrow{g_i} Y)_{i \in I}$ of E-morphisms such that $Y = \bigcup g_i(X_i)$. The class of all such coverings of all such objects is the *E-topology* on $\mathbf{C}/X$. The category $\mathbf{C}/X$ together with the E-topology is the *E-site* over X, and will be written $(\mathbf{C}/X)_E$ or simply as X_E. The *small E-site* on X is $(E/X)_E$ and, in the case that all E-morphisms are locally of finite-type, the *big E-site* on X is $(\mathbf{LFT}/X)_E$ where $\mathbf{LFT}/X$ is the full subcategory of $\mathbf{Sch}/X$ of X-schemes whose structure morphism is locally of finite-type. By the *Zariski site* X_{Zar} on X we shall always mean the small (Zar)-site $((\mathrm{Zar})/X)_{\mathrm{Zar}}$; by the *étale site* X_{et} we shall always mean the small (ét)-site $((\text{ét})/X)_{\mathrm{et}}$, and by the *flat site* X_{fl}, we shall always mean the big (fl)-site $(\mathbf{LFT}/X)_{\mathrm{fl}}$. Note that in the first two of these examples, but not in the third, any morphism in $\mathbf{C}/X$ is an E-morphism (see (I.3.6) for the étale site).

A small site is to be thought of as being analogous to a topological space in the usual sense and a big site to the category of all topological spaces and continuous maps over a given topological space. Note that the (Zar)-topology on a scheme in the above sense is the same as the Zariski topology in the usual sense, up to the identification of any open immersion with its image.

Remark 1.1. The category $\mathbf{C}/X$, together with the family of E-coverings, satisfies the following conditions:

1. if $\phi: U \to U$ is an isomorphism in $\mathbf{C}/X$, then it is a covering;
2. if $(U_i \to U)_i$ is a covering, and, for each i, $(V_{ij} \to U_i)_j$ is a covering, then $(V_{ij} \to U)_{i,j}$ is a covering;
3. if $(U_i \to U)$ is a covering, then, for any morphism $V \to U$ in $\mathbf{C}/X$, $(U_i \times_U V \to V)$ is a covering.

Thus $\mathbf{C}/X$, together with the family of E-coverings, is a *Grothendieck topology* in the sense of Artin [1].

A *presheaf* P on a site $(\mathbf{C}/X)_E$ is a contravariant functor $(\mathbf{C}/X)^\circ \to \mathbf{Ab}$. Thus P associates with each U in $\mathbf{C}/X$ an abelian group $P(U)$, which we shall sometimes write as $\Gamma(U, P)$, and whose elements we shall sometimes refer to as *sections* of P over U. With each morphism $f: U' \to U$ of $\mathbf{C}/X$, P associates a morphism $P(f): P(U) \to P(U')$. Since $P(f)$ corresponds to the usual notion of restricting a function to an open subset, we shall sometimes write it as res_f or simply as $\mathrm{res}_{U',U}$ or $(s \mapsto s \mid U')$. Note, however, the confusion in these last notations; unlike the case of topological spaces, there will in general be many maps $U' \to U$ and their restriction morphisms need not agree.

A *morphism* $\phi: P \to P'$ of presheaves on $(\mathbf{C}/X)_E$ is simply a morphism of functors $P \to P'$. Thus ϕ associates with each U in $\mathbf{C}/X$ a homomorphism $\phi(U): P(U) \to P'(U)$ and these homomorphisms commute with restriction maps. The presheaves and presheaf-morphisms over $(\mathbf{C}/X)_E$ form a category $\mathbf{P}(X_E) = \mathbf{P}((\mathbf{C}/X)_E)$, which inherits most of the properties of **Ab**. If P and P' are presheaves on X_E, then the presheaf Q with $Q(U) = P(U) \oplus P'(U)$ and $Q(f) = P(f) \oplus P(f')$ is the direct sum of P and P' in $\mathbf{P}(X_E)$. Thus $\mathbf{P}(X_E)$ is an additive category. The kernel (respectively cokernel) of a morphism $\phi: P \to P'$ is the presheaf Q with $Q(U) = \ker(\phi(U))$ (respectively coker $(\phi(U))$) for all U and the obvious restriction maps. The direct sum (respectively product) of a family of presheaves $(P_i)_{i \in I}$ is the presheaf Q with $Q(U) = \oplus P_i(U)$ (respectively $\prod P_i(U)$). A similar statement holds for direct and inverse limits of presheaves; a sequence $P' \xrightarrow{\phi} P \xrightarrow{\phi'} P''$ of presheaves is exact if and only if all of the sequences $P'(U) \xrightarrow{\phi(U)} P(U) \xrightarrow{\phi'(U)} P''(U)$ are exact. If one puts all of these obvious remarks together, then one sees that $\mathbf{P}(X_E)$ is an abelian category satisfying AB5 and AB4* (Appendix A).

Examples 1.2. (a) For any abelian group M, the constant presheaf P_M on X_E is defined to have $P_M(U) = M$, all U, and $P_M(f) = 1_M$, all f, except that we require $P_M(\varnothing) = 0$ if $\varnothing$ is the empty scheme.

(b) The presheaf $\mathbb{G}_a$ has $\mathbb{G}_a(U) = \Gamma(U, \mathcal{O}_U)$ regarded as an additive group for all U; for any morphism $f: U \to U'$, $\mathbb{G}_a(f)$ is the map $\Gamma(U', \mathcal{O}_{U'}) \to \Gamma(U, \mathcal{O}_U)$ induced by f.

(c) The presheaf $\mathbb{G}_m$ has $\mathbb{G}_m(U) = \Gamma(U, \mathcal{O}_U)^*$ for all U, and the obvious restriction maps.

(d) Let F be a sheaf of $\mathcal{O}_X$-modules in the usual sense of the Zariski topology. Define $W(F)$ as the presheaf with $W(F)(U) = \Gamma(U, F \otimes_{\mathcal{O}_X} \mathcal{O}_U)$ for all U and with the obvious restriction maps. For example, $W(\mathcal{O}_X) = \mathbb{G}_a$. If X is an S-scheme and $F = \Omega^1_{X/S}$, then $W(\Omega^1_{X/S})$ should not be confused with $U \mapsto \Gamma(U, \Omega^1_{U/S})$. There is a canonical map from $W(\Omega^1_{X/S})$ to this second presheaf, but it is an isomorphism only for the small étale (or coarser) site.

PROPOSITION 1.3. *For any U/X étale, $W(\Omega^1_{X/S})(U) = \Gamma(U, \Omega^1_{U/S})$.*

Proof. For any sequence of morphisms $U \to X \to S$, there is a canonical map $(g \otimes df \mapsto g\,df): \mathcal{O}_U \otimes \Omega^1_{X/S} \to \Omega^1_{U/S}$. In the case that $S = \operatorname{spec} R$, $X = \operatorname{spec} A$, $U = \operatorname{spec} C$, and the map $U \to X$ is a standard étale morphism (I.3.14), this map is an isomorphism because any R-derivation $D: A \to M$, where M is a C-module, extends uniquely to an

R-derivation $D'\colon C \to M$. As any étale morphism is locally of this form (I.3.14), we see that $\mathcal{O}_U \otimes \Omega^1_{X/S} \approx \Omega^1_{U/S}$ as sheaves (in the usual sense) on U and that

$$\Gamma(U, \mathcal{O}_U \otimes \Omega^1_{X/S}) \xrightarrow{\approx} \Gamma(U, \Omega^1_{U/S}).$$

The usual criterion for a presheaf to be a sheaf has an obvious translation in terms of the present set-up once it is realized that the intersection of two open sets should be replaced by the fiber product of two objects of **Sch**$/X$. Indeed, the fiber product, in the category of topological spaces, of two open subsets of a space is their intersection.

A presheaf P on X_E is a *sheaf* if it satisfies:

(S_1) if $s \in P(U)$ and there is a covering $(U_i \to U)_{i\in I}$ of U such that $\mathrm{res}_{U_i,U}(s) = 0$ for all i, then $s = 0$;

(S_2) if $(U_i \to U)_{i\in I}$ is a covering and the family $(s_i)_{i\in I}$, $s_i \in P(U_i)$ is such that

$$\mathrm{res}_{U_i\times_U U_j,U_i}(s_i) = \mathrm{res}_{U_i\times_U U_j,U_j}(s_j)$$

for all i and j, then there exists an $s \in P(U)$ such that $\mathrm{res}_{U_i,U}(s) = s_i$ for all i.

In other words, P is a sheaf if a section is determined by its restriction to a covering, and a compatible family of sections on a covering always arises from a global section. In other symbols, P is a sheaf if the sequence

$$\text{(S)} \qquad P(U) \to \prod_i P(U_i) \rightrightarrows \prod_{i,j} P(U_i \times_U U_j)$$

is exact for all coverings $(U_i \to U)$.

A presheaf P that satisfies (S_1) is called a *separated presheaf*. Note that the empty scheme $\varnothing$ has an empty covering, that is, the empty set of morphisms is a covering of $\varnothing$. Since a product of abelian groups over an empty indexing set is zero, $P(\varnothing) = 0$ for any separated presheaf. (Those who find this reasoning too bizarre may prefer to take this as part of the definition of a separated presheaf and sheaf).

In the case that E contains all open immersions, any open covering $U = \bigcup U_i$ in the usual sense of the Zariski topology is an open covering in the sense of the E-topology, and so a sheaf F on X_E defines by restriction sheaves in the usual sense on all schemes U in $\mathbf{C}/X$. The condition (S) for coverings by open subsets is just the usual sheaf condition. To get some idea of what (S) means for other coverings, we look at a Galois covering.

PROPOSITION 1.4. *Let Y be Galois over X with Galois group G, and let P be a presheaf for the étale (or finer) topology on X that takes disjoint sums of schemes to direct products. Then G acts on $P(Y)$ (on the left), and the sequence (S) for the covering $(Y \to X)$ may be canonically identified with* *the sequence,*

$$P(X) \to P(Y) \underset{(\sigma_1, \ldots, \sigma_n)}{\overset{(1,1,\ldots,1)}{\rightrightarrows}} P(Y)^n, \text{ where } G = \{\sigma_1, \ldots, \sigma_n\}.$$

Thus P satisfies (S) for the covering $(Y \to X)$ if and only if $P(X) \to P(Y)$ identifies $P(X)$ with the subset $P(Y)^G = \{a \in P(Y) | \sigma(a) = a, \text{ all } \sigma \in G\}$ of $P(Y)$.

Proof. From the definition of a Galois covering (I.5.4) we know that $Y \times_X Y \approx G_Y = \coprod Y_{\sigma_i}$, and once $Y \times Y$ has been identified via this map with $\coprod Y_{\sigma_i}$, the diagram

$$Y \times_X Y \underset{p_2}{\overset{p_1}{\rightrightarrows}} Y$$

becomes identified with

$$\coprod Y_{\sigma_i} \underset{(\sigma_1, \ldots, \sigma_n)}{\overset{(1,1,\ldots,1)}{\rightrightarrows}} Y.$$

On applying P, we obtain

$$P(Y) \underset{(\sigma_1, \ldots, \sigma_n)}{\overset{(1,\ldots,1)}{\rightrightarrows}} P(Y)^n,$$

and this gives the required statement.

We make the sheaves on $(\mathbf{C}/X)_E$ into a category $\mathbf{S}((\mathbf{C}/X)_E) = \mathbf{S}(X_E)$ by defining a morphism of sheaves to be the same as a morphism of presheaves. Thus $\mathbf{S}(X_E)$ is a full subcategory of $\mathbf{P}(X_E)$. We shall prove in the next section that $\mathbf{S}(X_E)$ inherits most, but not all, of the good properties of $\mathbf{P}(X_E)$.

Before giving some examples of sheaves, we prove a result that makes it easier to check that a presheaf is a sheaf.

PROPOSITION 1.5. *Let P be a presheaf for the étale or flat site on X. Then P is a sheaf if and only if it satisfies the following two conditions:*

(a) *for any U in $\mathbf{C}/X$, the restriction of P to the usual Zariski topology on U is a sheaf;*

(b) *for any covering $(U' \to U)$ with U and U' both affine, $P(U) \to P(U') \rightrightarrows P(U' \times_U U')$ is exact.*

Proof. The necessity of the two conditions is obvious. We prove the sufficiency in the case of the flat site and leave the reader to make the minor changes required for the étale site.

Condition (a) implies that if a scheme V is a sum $V = \coprod V_i$ of subschemes V_i, then $P(V) = \prod P(V_i)$. From this it follows that the sequence (S) for a covering $(U_i \to U)_i$ is isomorphic to the sequence (S) arising from the single morphism $(U' \to U)$, $U' = \coprod U_i$ because

$$(\coprod U_i) \times_U (\coprod U_i) = \coprod (U_i \times_U U_j)$$

[EGA. I.3.2.4]. Thus (b) implies that (S) is exact for coverings $(U_i \to U)_{i \in I}$ in which the indexing set I is finite and each U_i is affine, for then $\coprod U_i$ is affine.

Let $f: U' \to U$ be the morphism $\coprod U'_j \to U$, where $(U'_j \to U)_j$ is a covering, and write U as a union of open affines $U = \bigcup U_i$. Then $f^{-1}(U_i)$ is a union of open affines $f^{-1}(U_i) = \bigcup U'_{ik}$. Each $f(U'_{ik})$ is open in U_i (I.2.12), and U_i is quasi-compact, and therefore there is a finite set K_i of k's such that $(U'_{ik} \to U_i)_{k \in K_i}$ is a covering. By proceeding in this way and possibly introducing infinite redundances into the union $U = \bigcup U_i$, one may write $U = \bigcup U_i$ and $U' = \bigcup U'_{ik}$ as unions of open affines in such a way that for any i, $(U'_{ik} \to U_i)_{k \in K_i}$ is a finite covering of U_i. Consider the diagram:

$$
\begin{array}{ccccc}
P(U) & \longrightarrow & P(U') & \rightrightarrows & P(U' \times_U U') \\
\downarrow & & \downarrow & & \downarrow \\
\prod_i P(U_i) & \longrightarrow & \prod_i \prod_k P(U'_{ik}) & \rightrightarrows & \prod_i \prod_{k,l} P(U'_{ik} \times_U U'_{il}) \\
\downarrow\downarrow & & \downarrow\downarrow & & \\
\prod_{i,j} P(U_i \cap U_j) & \longrightarrow & \prod_{i,j} \prod_{k,l} P(U'_{ik} \cap U'_{jl}) & &
\end{array}
$$

From (a) the two columns are exact, and from the extension of (b) the middle row is a product of exact sequences and so is exact. It follows that $P(U) \to P(U')$ is injective and that P is a separated presheaf. Hence the bottom arrow is injective and an easy diagram chase now shows that the top row is exact.

COROLLARY 1.6. *For any quasi-coherent $\mathcal{O}_X$-module F, $W(F)$ is a sheaf on X_{fl} and, a fortiori, on X_{et}.*

Proof. Obviously $W(F)$ satisfies condition (a), and (1.2.19) shows that it satisfies condition (b).

In particular, the presheaf $\mathbb{G}_a$ is a sheaf for the flat and étale topologies. We give a second proof of this fact. Let $\mathbb{G}_{a,X}$ be the scheme $X \times_{\operatorname{spec} \mathbb{Z}} \operatorname{spec} \mathbb{Z}[T]$. Then, for any scheme U over X,

$$
\begin{aligned}
\operatorname{Hom}_X(U, \mathbb{G}_{a,X}) &= \operatorname{Hom}_{\operatorname{spec}(\mathbb{Z})}(U, \operatorname{spec} \mathbb{Z}[T]) \\
&= \operatorname{Hom}(\mathbb{Z}[T], \Gamma(U, \mathcal{O}_U)) = \Gamma(U, \mathcal{O}_U).
\end{aligned}
$$

Hence $\mathbb{G}_a(U) = \operatorname{Hom}_X(U, \mathbb{G}_{a,X})$. Now (I.2.17) implies that condition (b) of (1.5) is satisfied.

Similarly, $\mathbb{G}_m(U) = \operatorname{Hom}_X(U, \mathbb{G}_{m,X})$ where $\mathbb{G}_{m,X} = X \otimes_{\mathbb{Z}} \mathbb{Z}[T, T^{-1}]$, and this description of $\mathbb{G}_m(U)$ shows, as for $\mathbb{G}_a$, that both conditions of

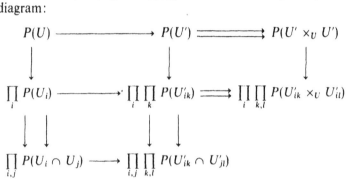

(1.5) are satisfied. Thus $\mathbb{G}_m$ is a sheaf for the flat or étale sites on any scheme X.

Both $\mathbb{G}_a$ and $\mathbb{G}_m$ are examples of sheaves defined by commutative group schemes. Recall that an X-scheme G is a commutative *group scheme* if there is a given factorization

$$\begin{array}{ccc} \mathbf{Sch}/X & \longrightarrow & \mathbf{Ab} \\ & {\scriptstyle G}\searrow \quad \swarrow{\scriptstyle f} & \\ & \mathbf{Sets} & \end{array}$$

in which f is the forgetful functor. This simply says that $G(Y)$ is an abelian group for all Y over X and that all the maps $G(Y) \to G(Y')$ corresponding to X-morphisms $Y' \to Y$ are homomorphisms. Alternatively one can say that G is a commutative group in the category $\mathbf{Sch}/X$. From this last description it is easy to see how to make

$$G_X = \coprod_{\sigma \in G} X_\sigma, \; X_\sigma = X$$

into a group scheme, where G is any commutative group (namely, $G_X \times G_X = \coprod (X_\sigma \times_X X_\tau)$, and $X_\sigma \times_X X_\tau \approx X$ (canonically), and so $G_X \times G_X \to G_X$ may be defined to be the map that, on $X_\sigma \times_X X_\tau$, is simply the identity map $X_\sigma \times_X X_\tau \xrightarrow{\approx} X_{\sigma\tau}$). Any commutative group scheme obviously defines a presheaf for any site on X.

COROLLARY 1.7. *The presheaf defined by a commutative group scheme on X is a sheaf for the flat, étale, and Zariski sites on X.*

Proof. (1.5b) follows from (I.2.17), and (1.5a) is easy to check.

If G is a commutative group scheme over X (or spec $\mathbb{Z}$), then we shall write G for both the group scheme and the sheaf defined by it (or by $G_{(X)}$).

The next example is very important. It shows that giving an étale sheaf over the spectrum of a field is the same as giving a discrete module over the Galois group of the separable closure of the field.

Fix a field K and consider the étale site on $X = \text{spec } K$. Recall (I.3) that any scheme U that is étale and of finite-type over X is finite over X and is a disjoint union $U = \coprod U_i$ of spectra of finite separable field extensions of K. Fix a separable closure K_s of K and let $G = \text{Gal}(K_s/K)$, that is, choose a geometric point $\bar{x} \to X$ of X and write $G = \pi_1(X, \bar{x})$. Note that G acts on K_s on the left and on spec$(K_s) = \bar{x}$ on the right.

Let P be a presheaf on X_{et}. If K' is a finite separable field extension of K, then we write $P(K')$ instead of $P(\text{spec } K')$. Define $M_P = \varinjlim P(K')$ where the limit is taken over all subfields K' of K_s that are finite over K. Then G acts on $P(K')$ on the left through its action on K' whenever K'/K

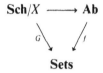

is Galois, and it follows that G acts on the limit M_P. Clearly $M_P = \bigcup M_P^H$, where H runs through the open subgroups of G, and so M_P is a discrete G-module (in the sense of Serre [8, I.2]).

Conversely, given a discrete G-module M, we can define a presheaf F_M such that:

(a) $F_M(K') = M^H$, if $H = \mathrm{Gal}\,(K_s/K')$;

(b) $F_M(\prod K_i) = \coprod F_M(K_i)$.

One only has to take

$$F_M(U) = \mathrm{Hom}_G\,(F(U), M)$$

where F is the functor $\mathbf{FEt}/X \to G\text{-}\mathbf{sets}$ defined by $F(U) = \mathrm{Hom}_X\,(\bar{x}, U)$ (compare I.5.3) and note that (a) $F(K') = G/H$; (b) $F(\prod K_i) = \prod F(K_i)$.

LEMMA 1.8. *F_M is a sheaf.*

Proof. It suffices to check conditions (a) and (b) of (1.5). Part (b) of the definition of F_M implies condition (a) of (1.5). It is clear from the proof of (1.5) that in checking (b) we may take U' and U to be "arbitrarily Zariski-small" affines and, in this case, to be spec L' and spec L respectively where $L' \supset L$ are finite separable field extensions of K. Let L'' be a finite Galois extension of L containing L', and consider the diagram,

$$\begin{array}{ccccc} F_M(L) & \longrightarrow & F_M(L') & \rightrightarrows & F_M(L' \otimes_L L') \\ \| & & \downarrow & & \downarrow \\ F_M(L) & \longrightarrow & F_M(L'') & \rightrightarrows & F_M(L'' \otimes_L L'') \end{array}$$

The bottom row is exact according to (1.4) since, by definition, $F_M(L) = F_M(L'')^{\mathrm{Gal}\,(L''/L)}$. Also $F_M(L) \to F_M(L')$ and $F_M(L') \to F_M(L'')$ are injective. An easy diagram chase now shows that the top row is exact.

THEOREM 1.9. *The correspondences $F \leftrightarrow M_F$, $M \leftrightarrow F_M$ induce an equivalence between the category $\mathbf{S}(X_{\mathrm{et}})$ and the category G-$\mathbf{mod}$ of discrete G-modules.*

Proof. Clearly a G-homomorphism $M \to M'$ induces a morphism $F_M \to F_{M'}$. Conversely, let H be an open normal subgroup of G, and let $K' = K_s^H$. If $\phi: F \to F'$ is a morphism of sheaves, then $\phi(K'): F(K') \to F'(K')$ commutes with the action of G because of the functoriality of ϕ. Thus $\varinjlim \phi(K')$ is a G-homomorphism $M_F \to M_{F'}$.

It is trivial to check that $\mathrm{Hom}_G\,(M, M') \to \mathrm{Hom}\,(F_M, F_{M'})$ is an isomorphism and that the canonical map $F \to F_{M_F}$ is an isomorphism, and this suffices to show that the two categories are equivalent.

Remark 1.10. With obvious modifications, the above theory (and some of what follows) is also valid for presheaves and sheaves of sets, groups, rings, and modules over a sheaf of rings. There is a canonical presheaf of rings on the site $(\mathbf{C}/X)_E$ for any E, X and $\mathbf{C}$, namely, $U \mapsto \Gamma(U, \mathcal{O}_U)$. This presheaf of rings will be denoted by $\mathcal{O}_{X_E}$ or simply $\mathcal{O}_X$. Note that $\mathcal{O}_{X_{\text{et}}}$ and $\mathcal{O}_{X_{\text{fl}}}$ are actually sheaves of rings and that any quasi-coherent $\mathcal{O}_X$-module F (in the usual sense) defines a sheaf of modules $W(F)$ on X_{et} and X_{fl}.

Remark 1.11. The proof of (1.9) actually shows more than we have stated. For any profinite group G, let G-**sets** be the category of finite sets with a continuous left G-action. Call a covering of a G-set S any surjective family of morphisms $(S_i \to S)$. Then, with a slightly more general notion of site, these coverings define a structure of a site on the category G-**sets**, and consequently we get presheaves and sheaves on G-**sets**. The same proof as in (1.9) shows that the category of sheaves is equivalent to the category G-**mod** under the functor that sends a sheaf F to the G-module $\varinjlim F(G/H)$, the limit being taken over all open subgroups H of G.

In the situation of (1.9), the functor $U \mapsto F(U) = \text{Hom}_X(\bar{x}, U)$ defines an equivalence of categories (ét)/$X \to G$-**sets** $(G = \pi_1(X, \bar{x}))$ under which coverings correspond to coverings. Thus there are equivalences of categories {sheaves on G-**sets**} $\approx$ {sheaves on X_{et}} $\approx$ {G-modules}. Every profinite group arises as a Galois group of fields (Douady [1]; see also Waterhouse [1]) and so the theory of modules over profinite groups is equivalent to the theory of étale sheaves over fields.

Remark 1.12. Corollary (1.7) shows that any presheaf P on the étale or flat site representable by a group scheme is a sheaf. More generally, the same argument shows that any presheaf of sets on the étale or flat sites representable by a scheme is a sheaf of sets.

One can show that on any category there is a finest topology relative to which all representable presheaves of sets are sheaves. This is called the *canonical topology* (Artin [1, Chapter I]). The above statement shows that the étale and flat topologies are coarser than the canonical topology. Whenever the E-topology on $\mathbf{C}/X$ is coarser than the canonical topology, $\mathbf{C}/X$ can be embedded as a full subcategory of the category of sheaves of sets on $(\mathbf{C}/X)_E$.

Occasionally it may happen that all sheaves are representable or ind-representable. This is the case for étale sheaves (of sets or groups) over a field K. If F is a sheaf of sets, let $S = \varinjlim F(K_i)$ where the K_i run through finite extensions of K contained in K_s. Then S is a continuous G-set and can be written $S = \coprod_{i \in I} S_i$ where each S_i is finite and $S_i \approx G/H_i$ for some open subgroup H_i of G. Now F is represented by $U = \coprod U_i$ where $U_i = \text{spec } K_i$, $K_i = K_s^{H_i}$. As $U = \varinjlim (\coprod_J U_i)$, where the limit is taken over all finite subsets J of I, U may be regarded as an ind-object of (ét)/spec (K).

Remark 1.13. Finite inverse limits exists in E/X if $E =$ (Zar), (ét), or **LFT**, that is, finite inverse limits exist in the underlying categories of what we have called the Zariski, étale, and flat sites. To prove this one only has to show that finite products exist and that the difference kernel of any two morphisms exists (Appendix A6). The product of a family $U_1, U_2, \ldots, U_n$ in E/X is $U_1 \times_X \cdots \times_X U_n$. Difference kernels obviously exist in (Zar)/X. For $E =$ (ét), first recall that any morphism $f: U' \to U$ in (ét)/X is étale (I.3.6). Thus $\Gamma_f: U' \to U' \times_X U$ is étale (it is even an open immersion, (I.3.5)). The difference kernel, in **Sch**/X, of two morphisms $f_1, f_2: U' \rightrightarrows U$ is $f: U_0 \to U'$ where f is such that

$$\begin{array}{ccc} U_0 & \xrightarrow{f} & U' \\ {\scriptstyle f}\downarrow & & \downarrow{\scriptstyle \Gamma_{f_1}} \\ U' & \xrightarrow{\Gamma_{f_2}} & U' \times_X U' \end{array}$$

is Cartesian. The statement now follows from (I.3.3). The proof for $E =$ **LFT** is similar.

Remark 1.14. Let **C** be any category. A *sieve* in **C** is a full subcategory such that if B is an object in the sieve and there is a morphism from A to B, then A also is in the sieve. Now let $(\mathbf{C}/X)_E$ be a site. To any covering $\mathcal{U} = (U_i \to U)$ in the site there is an associated sieve $s(\mathcal{U})$ in $\mathbf{C}/U$ $(=(\mathbf{C}/X)/U)$ whose objects are all U-schemes $Y \to U$ in $\mathbf{C}/X$ such that $Y \to U$ factors through some $U_i \to U$. For example, if $\mathcal{U} = (U \overset{id}{\to} U)$, then $s(\mathcal{U}) = \mathbf{C}/U$.

Let P be a presheaf on $(\mathbf{C}/X)_E$. The sheaf condition (S) for the covering $\mathcal{U}$ is equivalent to the natural map $P(U) \to \varprojlim_{\mathcal{U}} P(U_i)$ being an isomorphism (Appendix A5). Clearly

$$\varprojlim_{\mathcal{U}} P(U_i) = \varprojlim_{\mathcal{U}} \varprojlim_{\mathbf{C}/U_i} P(V) = \varprojlim_{s(\mathcal{U})} P(V),$$

and so the condition that P be a sheaf is that

$$P(U) \xrightarrow{\approx} \varprojlim_{s(\mathcal{U})} P(V)$$

for all U and all sieves $s(\mathcal{U})$ arising from coverings.

In [SGA. 4] a site is defined to be a category together with a family of sieves rather than coverings. What we have called a topology is there called a "pre-topology." Starting from a site $(\mathbf{C}/X)_E$ in our sense, we can obtain a site in the sense of [SGA. 4] as follows: the underlying category is $\mathbf{C}/X$; the sieves in $\mathbf{C}/U$ are all sieves that contain a sieve of the form $s(\mathcal{U})$ where $\mathcal{U}$ is a covering of U. Since the notion of presheaf and sheaf on a site (in our sense) is identical to the notion of presheaf and sheaf on the corresponding site (in the sense of [SGA. 4]), this difference in the definition of site is unimportant [SGA. 4, II.1].

§2. The Category of Sheaves

Consider sites $(\mathbf{C}'/X')_{E'}$ and $(\mathbf{C}/X)_E$. A morphism $\pi: X' \to X$ of schemes defines a *morphism of sites* $(\mathbf{C}'/X')_{E'} \to (\mathbf{C}/X)_E$ if:

(a) for any Y in $\mathbf{C}/X$, $Y_{(X')}$ is in $\mathbf{C}'/X'$;

(b) for any E-morphism $U \to Y$ in $\mathbf{C}/X$, $U_{(X')} \to Y_{(X')}$ is an E'-morphism.

Since the base change of a surjective family of morphisms is again surjective, we see that π defines a functor

$$\pi^{\cdot} = (Y \mapsto Y_{(X')}): \mathbf{C}/X \to \mathbf{C}'/X'$$

that takes coverings to coverings. Frequently we shall simply refer to π as a *continuous morphism* $\pi: X'_{E'} \to X_E$.

Examples 2.1. (a) The identity map of X defines a morphism of sites $(\mathbf{C}/X)_{E'} \to (\mathbf{C}/X)_E$ if and only if the E'-topology on X is as fine as the E-topology.

(b) The identity map on X defines continuous morphisms, $X_{\mathrm{fl}} \to X_{\mathrm{et}} \to X_{\mathrm{Zar}}$.

(c) Any morphism $\pi: X' \to X$ defines a continuous morphism $\pi: X'_E \to X_E$ provided it respects the underlying categories.

Let $\pi: X'_{E'} \to X_E$ be continuous. With any presheaf P' on $X'_{E'}$ we may associate the presheaf $\pi_p(P') = P' \circ \pi^{\cdot}$ on X_E. Explicitly, $\pi_p(P')$ is the presheaf on X_E such that $\Gamma(U, \pi_p(P')) = \Gamma(U_{(X')}, P')$. The presheaf $\pi_p(P')$ is called the *direct image* of P'.

Clearly π_p defines a functor $\mathbf{P}(X'_{E'}) \to \mathbf{P}(X_E)$, and we define the *inverse image* functor $\pi^p: \mathbf{P}(X_E) \to \mathbf{P}(X'_{E'})$ to be the left adjoint of π_p, that is, π^p is such that

$$\mathrm{Hom}_{\mathbf{P}(X')}(\pi^p P, P') \approx \mathrm{Hom}_{\mathbf{P}(X)}(P, \pi_p P').$$

The existence of π^p is given by a standard result in category theory.

PROPOSITION 2.2. *Let* $\mathbf{C}$ *and* $\mathbf{C}'$ *be small categories, and let* p *be a functor* $\mathbf{C} \to \mathbf{C}'$. *Let* $\mathbf{A}$ *be a category with direct limits, and write* $\mathbf{Fun}(\mathbf{C}, \mathbf{A})$ *and* $\mathbf{Fun}(\mathbf{C}', \mathbf{A})$ *for the categories of functors* $\mathbf{C} \to \mathbf{A}$ *and* $\mathbf{C}' \to \mathbf{A}$. *The functor*

$$(f \mapsto f \circ p): \mathbf{Fun}(\mathbf{C}', \mathbf{A}) \to \mathbf{Fun}(\mathbf{C}, \mathbf{A})$$

has a left adjoint.

Proof. See, for example, Hilton-Stammbach [1, p. 321].

We apply this to $\mathbf{C} = \mathbf{C}/X$, $\mathbf{C}' = \mathbf{C}'/X'$, $\mathbf{A} = \mathbf{Ab}$ and $p = \pi$: The smallness condition on $\mathbf{C}$ and $\mathbf{C}'$ is no particular problem if each is a subcategory of a category of schemes of finite-type over some scheme X, for such a category has a small subcategory containing at least one object in every isomorphism class. (Cover X by open affines $U_i = \operatorname{spec} A_i$, and consider all schemes that can be built up of schemes of the form spec B, where B is a quotient of some $A_i[X_1, X_2, \ldots]$.) In any specific situation such set-theoretic questions will cause no special difficulty, but to be both rigorous and general one should use universes [SGA. 4, I. Appendix] or the following method of Waterhouse. Intuition suggests that the only presheaves and sheaves we need consider are determined by a small amount of information, and this prevents difficulties even though we may be considering functors on $\mathbf{Sch}/X$. In Waterhouse [2] a presheaf P is defined to be *basically bounded* if there is an infinite cardinal m such that $P(\operatorname{spec} A) = \varinjlim P(\operatorname{spec} B)$ for all affine schemes spec A/X, where the limit is over subrings $B \subset A$ of cardinality $\leq m$. By considering only basically bounded presheaves and sheaves, one (apparently) eliminates all set-theoretic problems because in any particular construction one essentially has only to use schemes made up of rings that are subsets of some fixed set of cardinality m. That these problems are not entirely trivial is shown by an example (Waterhouse [2]) of a basically unbounded presheaf that has no associated sheaf. (The second proof of (2.11) below fails in his situation because the limit, $\varinjlim \check{H}^0(\mathcal{U}, P_1)$, is over a class and cannot be replaced by a limit over a set because of the unboundedness of P.)

We shall need to know the inverse image presheaf $\pi^p P$ explicitly, and so we sketch the proof of (2.2) in our particular situation.

Roughly, $\Gamma(U', \pi^p P)$ is the union of the groups of sections of P over all open sets containing the image of U'. The precise definition is that $(\pi^p P)(U') = \varinjlim P(U)$ where the limit runs over all commutative squares

$$(g, U) = \begin{array}{ccc} U' & \xrightarrow{g} & U \\ \downarrow & & \downarrow \\ X' & \xrightarrow{\pi} & X \end{array}$$

with $U \to X$ in $\mathbf{C}/X$. A morphism from one such square (g, U) to a second (g_1, U_1) is an X-morphism $h: U \to U_1$ such that $hg = g_1$.

If $h': V' \to U'$ is a morphism in $\mathbf{C}'/X'$, then there is an obvious restriction map $\Gamma(U', \pi^p P) \to \Gamma(V', \pi^p P)$ because the first group is the direct limit over a smaller category than the second. Thus $\pi^p P$ is a presheaf.

To give a morphism $P \to \pi_p P'$ is the same as to give maps $P(U) \to P'(U_{(X')})$ for all U in $\mathbf{C}/X$, which are compatible with the restriction maps. To give a morphism $\pi^p P \to P'$ is the same as to give maps $P(U) \to P'(U')$ for all commutative squares

$$\begin{array}{ccc} U' & \longrightarrow & U \\ \downarrow & & \downarrow \\ X' & \longrightarrow & X \end{array}$$

which are compatible with the restriction maps. But a morphism $U' \to U$ factors uniquely as $U' \to U_{(X')} \to U$, and so if we are given a map $P(U) \to P'(U_{(X')})$, then we get maps $P(U) \to P'(U')$ automatically. Thus a morphism $P \to \pi_p P'$ defines a morphism $\pi^p P \to P'$ and conversely.

PROPOSITION 2.3. *The limit defining $\Gamma(U', \pi^p P)$ is cofiltered if finite inverse limits exist in* $\mathbf{C}/X$.

Proof. One has only to construct the appropriate fibered product or difference kernel to show that the category satisfies the duals of (f_1) and (f_2) and is connected (Appendix A).

Note that (2.3) holds for the Zariski, étale, and flat sites (1.13). When the limits are cofiltered, it is possible to give a slightly more explicit description of $\Gamma(U', \pi^p P)$. A section of $\pi^p P$ over U' is represented by a pair (s, g) where $g: U' \to U$ is a morphism over π and $s \in P(U)$; two such pairs (s_1, g_1) and (s_2, g_2) represent the same element in $\Gamma(U', \pi^p P)$ if there is a commutative diagram

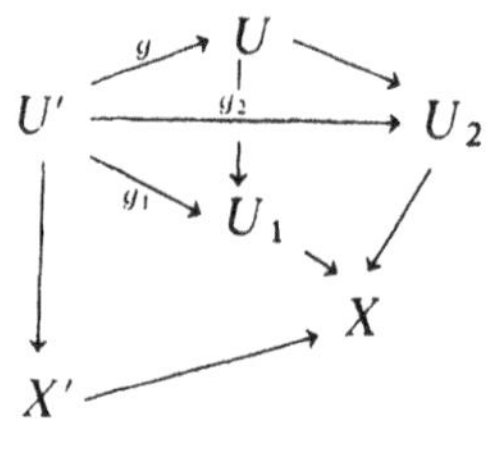

such that $\mathrm{res}_{U,U_1}(s_1) = \mathrm{res}_{U,U_2}(s_2)$. The element "$(s, g)$" in $\Gamma(U', \pi^p P)$ restricts to the element "(s, gh)" in $\Gamma(V', \pi^p P)$ under $V' \xrightarrow{h} U'$.

Examples 2.4. (a) If P is a constant presheaf, then $\pi^p P$ is a constant presheaf and is defined by the same group.

(b) If $\pi: X' \to X$ is in $\mathbf{C}/X$ and $\mathbf{C}'/X' = (\mathbf{C}/X)/X'$, then there is an initial object in the category over which the limit is being taken, namely,

$$\begin{array}{ccc} U' & \xrightarrow{id} & U' \\ \downarrow{\scriptstyle f'} & & \downarrow{\scriptstyle \pi f'} \\ X' & \xrightarrow{\pi} & X \end{array}$$

Thus $\Gamma(U', \pi^p P) = \Gamma(U', P)$, and so π^p simply restricts the functor P to the category $\mathbf{C}'/X'$. In this case we often write $P|X'$ for $\pi^p P$.

(c) If $\pi: X \to X$ is the identity map, then $\pi_p \pi^p(P) = P$ for all presheaves P on X_E (we assume $\mathbf{C}'/X \supset \mathbf{C}/X$).

Exercise 2.5. Consider π^p when π is the identity map $X_{fl} \to X_{et}$ or $X_{et} \to X_{Zar}$; for example, does $\pi^p \mathbb{G}_a = \mathbb{G}_a$?

PROPOSITION 2.6. *The functor π_p is exact and π^p is right exact. Moreover, π^p is left exact in each of the following two cases:*

(a) *finite inverse limits exist in* $\mathbf{C}/X$ (*for example, when* $\mathbf{C}/X = \mathbf{LFT}/X$, (ét)$/X$, *or* (Zar$/X$), (1.13));

(b) π *is in* $\mathbf{C}/X$ *and* $\mathbf{C}'/X' = (\mathbf{C}/X)/X'$.

Proof. The exactness of π_p is obvious from its definition, and π^p is right exact because arbitrary direct limits are right exact in **Ab**. The left exactness of π^p follows from (2.3) and the fact that filtered direct limits are exact in **Ab** (Appendix A3) in case (a), and is obvious from (2.4b) in case (b).

PROPOSITION 2.7. *If F is a sheaf, then $\pi_p F$ is a sheaf.*

Proof. Let π be a morphism $(\mathbf{C}'/X')_E \to (\mathbf{C}/X)_E$. For any U in $\mathbf{C}/X$ we write $U' = U \times_X X'$. If $(U_i \to U)$ is a covering, then so also is $(U_i' \to U')$, and so

$$F(U') \to \prod_i F(U_i') \rightrightarrows \prod_{i,j} F(U_i' \times_{U'} U_j')$$

is exact. But $U_i' \times_{U'} U_j' \approx (U_i \times_U U_j)'$, and so this sequence is isomorphic to the sequence

$$(\pi_p F)(U) \to \prod_i \pi_p F(U_i) \rightrightarrows \prod_{i,j} \pi_p F(U_i \times_U U_j),$$

which proves $\pi_p F$ is a sheaf.

Remark 2.8. It is not true in general that $\pi^p F$ is a sheaf when F is. However, it is obviously true in the example described in (2.4b).

It is possible to define stalks of sheaves and presheaves at a point of a site once one has decided what a point is. Although there is a theory of points and stalks for almost any site, we shall only study it in the case of the étale site because there the points may be described very explicitly.

The important property of a point of a topological space for the usual theory of sheaves is that to give a sheaf on a one-point space is simply to give a set (or abelian group). This is not true for étale sheaves on a one-point scheme unless the scheme is the spectrum of a separably closed field. Thus a point for the étale topology on X should be a geometric point.

Let x be a point (of the underlying topological space) of X. We shall always use $\bar{x}$ to denote the spectrum of some separably closed field $k(\bar{x})$ containing $k(x)$ and $u_x:\bar{x} \to X$ to denote the map induced by the inclusion of $k(x)$ in $k(\bar{x})$. According to (1.9), the functor $F \mapsto F(\bar{x})$ gives an equivalence of categories between $\mathbf{S}(\bar{x}_{\text{et}})$ and $\mathbf{Ab}$.

Let P be a presheaf on X_{et}. The *stalk* of P at $\bar{x}$, $P_{\bar{x}}$, is the abelian group $(u_x^p P)(\bar{x})$. More explicitly, $P_{\bar{x}} = \varinjlim P(U)$ where the limit runs over all commutative triangles

$$\begin{array}{ccc} U & \longleftarrow & \bar{x} \\ \downarrow & \swarrow {\scriptstyle u_x} & \\ X & & \end{array}$$

with U étale over X, that is, over all étale neighborhoods of $\bar{x}$ in X. Clearly $P_{\bar{x}}$ is independent of the field $k(\bar{x})$ chosen.

Remarks 2.9. (a) The functor $(P \mapsto P_{\bar{x}}): \mathbf{P}(X_{\text{et}}) \to \mathbf{Ab}$ is exact, because u^p is exact (2.6).

(b) The abelian group $P_{\bar{x}}$ is acted on by Gal $(k(x)_{\text{sep}}/k(x))$. Indeed the morphism u_x factors into $\bar{x} \xrightarrow{\pi} x \xrightarrow{i_x} X$ and $P_{\bar{x}}$ is the underlying abelian group of the Gal $(k(x)_{\text{sep}}/k(x))$-module associated with the presheaf $i_x^p(P)$ on spec $k(x)$.

(c) Let $U \to X$ be an étale morphism whose image contains x. In general U can be given an $\bar{x}$-point, that is, the structure of an étale neighborhood, in many different ways, and so there is no canonical map $P(U) \to P_{\bar{x}}$. Of course, once U has been given an $\bar{x}$-point, there will be such a canonical map $P(U) \to P_{\bar{x}}$, which we often write $s \mapsto s_{\bar{x}}$.

(d) Let Y be a scheme locally of finite-type over X. If (U_i) is a filtered inverse system of X-schemes with each U_i affine, then it is easy to see that the canonical map $Y(\varprojlim U_i) \to \varinjlim Y(U_i)$ is an isomorphism ((3.3) below). Thus if P is the sheaf defined by a group scheme G that is locally of finite-type over X, then

$$P_{\bar{x}} = \varinjlim G(U) = G(\varprojlim U) = G(\mathcal{O}_{X,\bar{x}}) \quad (\text{I}.4).$$

For example, $(\mathbb{G}_a)_{\bar{x}} = \mathcal{O}_{X,\bar{x}}$ and $(\mathbb{G}_m)_{\bar{x}} = \mathcal{O}_{X,\bar{x}}^*$.

PROPOSITION 2.10. *Let F be a sheaf on X_{et}. If $s \in F(U)$ is nonzero, then there is an $x \in X$ and an $\bar{x}$-point of U such that $s_{\bar{x}}$ is nonzero.*

Proof. Suppose that s maps to zero under all such maps $F(U) \to F_{\bar{x}}$. Clearly any $u \in U$ is the image of an $\bar{x}$-point of U, where x is the image of u in X. Thus, from the definition of $F_{\bar{x}}$, for any $u \in U$, there is a scheme V_u, étale over U, whose image in U contains u and which is such that $s|V_u = 0$. This family $(V_u \to U)_{u \in U}$ is a covering of U, and so, by (S_1), $s = 0$.

The next theorem shows that every presheaf has an associated sheaf.

THEOREM 2.11. *For any presheaf P on X_E there is a sheaf aP on X_E and a morphism $\phi : P \to aP$ such that any other morphism ϕ' from P into a sheaf F factors uniquely as*

$$\begin{array}{ccc} P & \xrightarrow{\phi} & aP \\ & \searrow^{\phi'} \quad \swarrow_{\psi} & \\ & F & \end{array}$$

Proof. We give one proof for the étale and Zariski sites and sketch a second that works in general.

LEMMA 2.12. *Let X_E be a site.*

(a) *A product of sheaves on X_E is again a sheaf.*

(b) *If $(F_i)_{i \in I}$ is a family of subsheaves of a sheaf F, then $\bigcap F_i$ (where $(\bigcap F_i)(U) = \bigcap (F_i(U))$ is again a sheaf.*

(c) *If $\phi : F \to F'$ is a morphism of sheaves and F_0 is a subsheaf of F', then $\phi^{-1}F_0$ (where $(\phi^{-1}F_0)(U) = \phi^{-1}(F_0(U))$ is again a sheaf; in particular,* ker (ϕ) *is a sheaf.*

Proof. It is easy to check these directly. Alternatively they follow from the description of the sheaf condition in (1.14) and the fact that inverse limits commute with inverse limits.

We now prove (2.11) for the étale site (the proof for the Zariski site is similar). Note first that if X is the spectrum of a separably closed field, then any scheme U étale of finite-type over X is uniquely a finite disjoint union $U = \coprod X_i$ with $X_i = X$. The sheaf aP associated with a presheaf P is

$$U \mapsto \prod_{i=1}^{r} P(X_i) = P(X)^r,$$

and $\phi(U)$ is the map $P(U) \to \prod P(X_i)$ defined by the restriction maps.

Now consider an arbitrary X. For each $x \in X$ we choose an $\bar{x}$ and write $P_{\bar{x}}^* = a(u_x^p P)$ with the above definition of a. Let P^* be the sheaf $\prod_{x \in X} u_{xp} P_{\bar{x}}^*$, and let ϕ be the map $P \to P^*$ induced by the maps $P \to u_{xp}(u_x^p P) \to u_{xp}(P_{\bar{x}}^*)$, where the first map is that corresponding (by adjointness) to the identity map $u_x^p P \to u_x^p P$ and the second is induced by $u_x^p P \to a(u_x^p P) = P_{\bar{x}}^*$. Let aP be the intersection of all subsheaves of P^* containing $\phi(P)$. Let $\phi' : P \to F$ be another morphism from P into a sheaf F, and consider the diagram

$$\begin{array}{ccccc} P & \xrightarrow{\phi} & aP & \hookrightarrow & P^* \\ & \searrow^{\phi'} & & & \downarrow{\psi} \\ & & F & \hookrightarrow & F^* \end{array}$$

where ψ is induced by the maps $\phi'_{\bar{x}}: P^*_{\bar{x}} \to F_{\bar{x}}$. The injectivity of $F \to F^*$ is a restatement of (2.10). The diagram commutes because of the functoriality of the maps $P \to u_{xp}P^*_{\bar{x}}$. As $\psi^{-1}(F)$ is a subsheaf of P^* containing $\phi(P)$ and hence aP, ψ induces a morphism $\psi_0: aP \to F$ such that $\psi_0\phi = \phi'$. If $\psi_1: aP \to F$ is a second such morphism, then ker $(\psi_0 - \psi_1)$ is a subsheaf of P^* containing $\phi(P)$ and so must contain aP, that is, $\psi_0 = \psi_1$. This completes the first proof.

In the case of an arbitrary site $(\mathbf{C}/X)_E$ the theorem may be proved as follows. For any U in $\mathbf{C}/X$ define $P_0(U)$ to be the set of all $s \in P(U)$ such that $\text{res}_{U_i,U}(s) = 0$ for all U_i in some covering of U. It is easily seen that P_0 is a subpresheaf of P and that the quotient presheaf $P_1 = P/P_0$ (with $P_1(U) = P(U)/P_0(U)$) satisfies (S_1).

Now, with the notations of (III.2) below, define $(aP)(U) = \varinjlim \check{H}^0(\mathcal{U}, P_1)$ where the limit runs over all coverings $\mathcal{U}$ of U. Thus, a section $s \in (aP)(U)$ is represented by a pair of families $((s_i), (U_i))$ where $(U_i \to U)$ is a covering of U, $s_i \in P_1(U_i)$, and

$$\text{res}_{U_i \times U_j, U_i}(s_i) = \text{res}_{U_i \times U_j, U_j}(s_j)$$

for all i, j. Two such families $((s_i), (U_i))$ and $((s'_j), (U'_j))$ represent the same element of $(aP)(U)$ if (U_i) and (U'_j) have a common refinement (V_k) such that the families (t_k), $t_k \in P_1(V_k)$ arising by restriction from (s_i) and (s'_j) are equal.

Define restriction maps as follows: if $V \to U$ is in $\mathbf{C}/X$ and $s \in (aP)(U)$ is represented by $((s_i), (U_i))$, then $\text{res}_{V,U}(s)$ is represented by

$$((\text{res}_{V \times_U U_i, U_i}(s_i)), (V \times_U U_i)).$$

This clearly makes aP into a presheaf.

Also, any morphism $\phi': P \to F$ from P into a sheaf factors uniquely through the canonical map $\phi: P \to aP$. For, according to (S_1), $\phi'(P_0) = 0$ and so ϕ' factors uniquely through $P \to P_1$. Also according to (S_2), any morphism $P_1 \to F$ extends to a morphism $aP \to F$ ($\check{H}^0(\mathcal{U}, -)$ is a functor of presheaves, and $\check{H}^0(\mathcal{U}, F) = F(U)$), and according to (S_1), this extension is unique.

The only real complication in the proof occurs in showing that aP is a sheaf. For this we refer the reader to Artin [1, II.1.4(ii)]. (The condition (+) there is our (S_1), and we have denoted P_1^+ by aP.)

Exercise 2.13. Make explicit the map $P(U) \to P^*(U)$, where U is étale over X, that is defined in the first proof above.

Remark 2.14. (a) The above theorem may be restated as follows: the natural inclusion functor $\mathbf{S}(X_E) \to \mathbf{P}(X_E)$ has a left adjoint $a: \mathbf{P}(X_E) \to \mathbf{S}(X_E)$.

(b) Let $\pi: X'_{E'} \to X_E$ be a continuous morphism such that π^p takes sheaves to sheaves. Then for any presheaf P on X_E and sheaf F on $X'_{E'}$,

$$\begin{aligned}\mathrm{Hom}_{\mathbf{S}'}(\pi^p aP, F) \approx \mathrm{Hom}_{\mathbf{S}}(aP, \pi_p F) &\approx \mathrm{Hom}_{\mathbf{P}}(P, \pi_p F)\\ &\approx \mathrm{Hom}_{\mathbf{P}'}(\pi^p P, F)\\ &\approx \mathrm{Hom}_{\mathbf{S}'}(a\pi^p P, F),\end{aligned}$$

which implies there is a canonical isomorphism $\pi^p aP \approx a\pi^p P$.

(c) By using the fact that $P \in P(\bar{x}_{\mathrm{et}})$ is a sheaf if and only if it takes finite sums of schemes to products of abelian groups, one may verify that u_x^p takes sheaves to sheaves. Thus $u_x^p a \approx au_x^p$, and the canonical map

$$P_{\bar{x}}\ (=(au_x^p P)(\bar{x})) \to (aP)_{\bar{x}}\ (=(u_x^p aP)(\bar{x}))$$

is an isomorphism, that is, P and aP have the same stalks.

THEOREM 2.15. (a) *The inclusion functor* $\mathbf{S}(X_E) \to \mathbf{P}(X_E)$ *is left exact and preserves inverse limits;* $a: \mathbf{P}(X_E) \to \mathbf{S}(X_E)$ *is exact and preserves direct limits.*

(b) *The following statements are equivalent:* $0 \to F' \to F \to F''$ *is exact in* $\mathbf{S}(X_E)$; $0 \to F' \to F \to F''$ *is exact in* $\mathbf{P}(X_E)$; $0 \to F'(U) \to F(U) \to F''(U)$ *is exact for all* U *in* $\mathbf{C}/X$. *For the étale topology, they are also equivalent to the sequence*

$$0 \to F'_{\bar{x}} \to F_{\bar{x}} \to F''_{\bar{x}}$$

being exact for all geometric points $\bar{x}$.

(c) $\phi: F \to F'$ *is surjective in* $\mathbf{S}(X_E)$ *if and only if for any* $s \in F'(U)$ *there exists a covering* $(U_i \to U)$ *of* U *and elements* $s_i \in F(U_i)$ *such that* $\phi(s_i) = \mathrm{res}_{U_i,U}(s)$ *for all* i. *For the étale topology, these statements are also equivalent to* $F_{\bar{x}} \to F'_{\bar{x}}$ *being surjective for all* $\bar{x}$.

(d) *To form arbitrary inverse limits (for example, kernels, products) in* $\mathbf{S}(X_E)$ *form the inverse limit in* $\mathbf{P}(X_E)$, *and then the resulting presheaf is a sheaf and is the inverse limit in* $\mathbf{S}(X_E)$. *To form arbitrary direct limits (for example, cokernels, sums) in* $\mathbf{S}(X_E)$ *form the direct limit in* $\mathbf{P}(X_E)$, *and then the associated sheaf is the direct limit in* $\mathbf{S}(X_E)$.

(e) *The category* $\mathbf{S}(X_E)$ *is abelian and satisfies* AB5 *and* AB3* *(but not in general* AB4*).

Proof. Note that $\mathbf{S}(X_E)$ is a subcategory of an abelian category and so, in particular, is additive.

(a) All statements except the left exactness of a are formal consequences of the adjointness properties of the two functors (Appendix A). The proof of the left exactness of a will make use of the construction of aP.

Consider first the étale site. If $P \to P'$ is injective, then, with the notation of the proof of (2.11), $P^* \to P'^*$ is injective (since u_x^p is exact) and

aP and aP' are subsheaves of P^* and P'^*. Thus $aP \to aP'$ is injective in $\mathbf{P}(X_{et})$. Let F be the kernel of $aP \to aP'$ in $\mathbf{S}(X_{et})$. Then $F = 0$ because the inclusion functor $\mathbf{S}(X_{et}) \to \mathbf{P}(X_{et})$ is left exact.

Now consider an arbitrary site. We assume that $P \to P'$ is injective and show first that $P_1 \to P'_1$ is injective. Suppose that $s \in P(U)$ maps to $s' \in P'_0(U)$; then there exists a covering (U_i) of U such that $\mathrm{res}_{U_i,U}(s') = 0$ for all i. The injectivity of $P(U_i) \to P'(U_i)$ shows that $\mathrm{res}_{U_i,U}(s) = 0$ for all i, which implies that $s \in P_0(U)$.

The commutativity of

$$\begin{array}{ccc} \check{H}^0(\mathcal{U}, P_1) & \hookrightarrow & \prod_i P_1(U_i) \\ \downarrow & & \downarrow \\ \check{H}^0(\mathcal{U}, P'_1) & \hookrightarrow & \prod_i P'_1(U_i) \end{array}$$

for a covering $\mathcal{U}$ of U shows that $\check{H}^0(\mathcal{U}, P_1) \to \check{H}^0(\mathcal{U}, P'_1)$ is injective. By passing to the limit over all coverings $\mathcal{U}$ of U, one finds that $aP(U) \to aP'(U)$ is injective. The proof may now be completed as before.

(b) That the first two statements are equivalent follows from (a); that the next two are equivalent has already been noted; that the fourth statement follows from the third is obvious. To see that the fourth implies the third, let $s' \in F'(U)$ map to zero in $F(U)$. For any $\bar{x}$-point of U, $s'_{\bar{x}}$ maps to zero under $F'_{\bar{x}} \to F_{\bar{x}}$, which implies that it is zero as this map is injective, which in turn implies that $s' = 0$ (2.10). Now let $s \in F(U)$ map to zero in $F''(U)$. This implies that $s_{\bar{x}}$ lies in the subgroup $F'_{\bar{x}}$ of $F_{\bar{x}}$ for all $\bar{x}$-points of U. It follows that, for any $u \in U$, there is an étale map $V_u \to U$ whose image contains u and that is such that $s|V_u$ lies in the subgroup $F'(V_u)$ of $F(V_u)$. As F' is a sheaf and (V_u) is a covering of U, this shows that $s \in F'(U)$.

(c) Let $F \to F'$ be a morphism of sheaves, and let P be its cokernel in $\mathbf{P}(X_E)$. The sequence $F \to F' \to aP \to 0$ is exact because a is exact. Thus $F \to F'$ is surjective $\Leftrightarrow aP = 0 \Leftrightarrow P = P_0$ (in the notation of the proof of (2.11)) $\Leftrightarrow$ the statement in (c) holds. On the étale site, $aP = 0$ if and only if $(aP)_{\bar{x}} = 0$ for all x in X, that is, if and only if $F_{\bar{x}} \to F'_{\bar{x}}$ is surjective for all x.

(d) Both statements follow immediately from (a). (See Appendix A4.)

(e) To show that $\mathbf{S}(X_E)$ is abelian, it only remains to show that for any morphism $\phi: F \to F'$ in $\mathbf{S}$, the induced morphism $\bar{\phi}: \mathrm{coim}(\phi) \to \mathrm{im}(\phi)$ is an isomorphism. But $\mathbf{P}(X_E)$ is an abelian category, a takes isomorphisms to isomorphisms, and $\bar{\phi}: \mathrm{coim}(\phi) \to \mathrm{im}(\phi)$ in $\mathbf{S}$ is obtained by applying a to the same map in $\mathbf{P}(X_E)$.

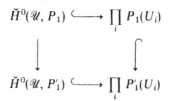

The fact that $\mathbf{S}(X_E)$ satisfies AB5 and AB3* follows from (a) and (d). (For AB5 only the left exactness of direct limits has to be proved.)

Exercise 2.16. Let $(M_i)_{i \in I}$ be a family of discrete G-modules where G is a profinite group. Let $M_* = \prod M_i$ in the usual sense, and let M be the submodule $\bigcup M_*^H$ of M_*, where the union runs through all open subgroups H of G. Show that M is the product of the M_i in the category of discrete G-modules. Deduce that products are not exact, that is, that AB4* does not hold in the category of discrete G-modules if G is infinite. Deduce the same result for $\mathbf{S}(X_{\text{et}})$, with X the spectrum of a nonseparably closed field. Examine the situation in $\mathbf{S}(X_{\text{Zar}})$, with X any scheme.

Remark 2.17. (a) Let (F_i) be a pseudofiltered direct system of sheaves on some site, and let F be the presheaf direct limit of the system. Then F satisfies the sheaf condition (S) for finite coverings $(U_i \to U)$. Indeed, each sequence

$$F_i(U) \to \prod_j F_i(U_j) \rightrightarrows \prod_{j,k} F_i(U_j \times U_k)$$

is exact. As each product is finite, it may be replaced by a direct sum. As direct limits commute with direct sums, on passing to the direct limit, we obtain an exact sequence,

$$F(U) \to \prod_j F(U_j) \rightrightarrows \prod_{j,k} F(U_j \times U_k).$$

On some sites, namely, the Noetherian sites, this is enough to show that F is a sheaf (III.3).

(b) If X is a Jacobson scheme, then in (2.10), the proof of (2.11), (2.15) and similar situations only geometric points $\bar{x}$ lying over closed points of X need be considered. (Recall [EGA. I.6.4] that a scheme X is Jacobson if every closed subset of X is the closure of its set of closed points; spec $\mathbb{Z}$ and spec of a field are Jacobson, and any scheme that is locally of finite-type over a Jacobson scheme is Jacobson.) More generally, any set S of geometric points such that $\{x \mid \bar{x} \in S\}$ is very dense in X suffices.

(c) Theorem (2.15) implies that the functors $(F \mapsto F_{\bar{x}}): \mathbf{S}(X_{\text{et}}) \to \mathbf{Ab}$, $x \in X$, form a faithful and conservative family: two morphisms ϕ, ϕ' are equal if the maps on stalks $\phi_{\bar{x}}$, $\phi'_{\bar{x}}$ are equal for all $x \in X$ and ϕ is an isomorphism if ϕ_x is an isomorphism for all $x \in X$. Also a sequence $F' \to F \to F''$ of sheaves is exact if and only if $F'_{\bar{x}} \to F_{\bar{x}} \to F''_{\bar{x}}$ is exact for all x.

Examples 2.18. (a) The *constant sheaf* on X defined by an abelian group M is the sheaf $F_M = a(P_M)$. I claim that for the Zariski, étale, flat, ... topologies, F_M is the sheaf defined by the constant group scheme M_X (compare 1.7). It suffices to show that the two take equal values on quasi-compact schemes, and so, for the rest of this paragraph, we assume our schemes to be quasi-compact. Let π_0: **Sch** $\to$ **Sets** be the functor that maps a quasi-compact scheme to its set of connected components. Then π_0 is left adjoint to the functor **Sets** $\to$ **Sch** that sends a set T to $T_{(\mathrm{spec}\,\mathbb{Z})} = \coprod_{t \in T} (\mathrm{spec}\,\mathbb{Z})_t$ (disjoint union of copies of spec $\mathbb{Z}$). Thus

$$\mathrm{Hom}_X(Y, M_X) = \mathrm{Hom}_{\mathrm{spec}\,\mathbb{Z}}(Y, M_{(\mathrm{spec}\,\mathbb{Z})}) = \mathrm{Hom}(\pi_0(Y), M).$$

The maps $M \to \mathrm{Hom}(\pi_0(Y), M)$ that send $m \in M$ to the constant function with value m define a morphism from the presheaf P_M into the sheaf M_X, and it is easy to check that this has the correct universal mapping property. Note that $F_M(Y) = M^r$ if Y has r connected components.

In the remaining examples, the site will always be the Zariski, étale or flat site.

(b) Define a subsheaf μ_n of $\mathbb{G}_m$ by setting $\mu_n(U) =$ group of nth roots of 1 in $\Gamma(U, \mathcal{O}_U)$. This is a sheaf; in fact it is the sheaf defined by the group scheme $\mu_n = \mathrm{spec}\,\mathbb{Z}[T]/(T^n - 1)$. Consider the *Kummer sequence* $0 \to \mu_n \to \mathbb{G}_m \xrightarrow{n} \mathbb{G}_m \to 0$ where "n" is the map $(u \mapsto u^n)$: $\mathbb{G}_m(U) \to \mathbb{G}_m(U)$. The sequence $0 \to \mu_n \to \mathbb{G}_m \to \mathbb{G}_m$ is clearly exact in $\mathbf{P}(X_E)$ and so is exact in $\mathbf{S}(X_E)$. However, $\mathbb{G}_m \xrightarrow{n} \mathbb{G}_m$ is rarely an epimorphism in $\mathbf{P}(X_E)$, for it is unusual for every element of $\Gamma(U, \mathcal{O}_U)^*$ to be an nth power, for all U. Moreover, it is usually not an epimorphism in $\mathbf{S}(X_{\mathrm{Zar}})$, for this would mean that every element of $\Gamma(U, \mathcal{O}_U)^*$ is locally an nth power for the Zariski topology. On the other hand, if A is a strictly local ring such that n is a unit in A, then Hensel's lemma implies that

$$0 \to \mu_n(A) \to A^* \xrightarrow{n} A^* \to 0$$

is exact (because $A[T]/(T^n - a)$ is étale over A for all $a \in A^*$ (I.3.4)). Thus (2.15c) implies that

$$0 \to \mu_n \to \mathbb{G}_m \xrightarrow{n} \mathbb{G}_m \to 0$$

is an exact sequence in $\mathbf{S}(X_{\mathrm{et}})$ provided that the characteristic of $k(x)$ does not divide n for any $x \in X$.

The same sequence is exact in $\mathbf{S}(X_{\mathrm{fl}})$ without any restriction on the residue field characteristics. Let $U \to X$ be in **LFT**$/X$, and let $u \in \Gamma(U, \mathcal{O}_U)^*$. Let (U_i) be a covering of U by open affines, and let $U_i' \to U_i$ be the map defined by $A_i \to A_i[T]/(T^n - u_i)$ where $A_i = \Gamma(U_i, \mathcal{O}_{U_i})$ and u_i is the restriction of u to U_i. Then $(U_i' \to U)$ is a flat covering of U, and the restriction of u to U_i' is in the image of $\mathbb{G}_m(U_i') \xrightarrow{n} \mathbb{G}_m(U_i')$ for all i. Thus (2.15c) implies that $\mathbb{G}_m \xrightarrow{n} \mathbb{G}_m$ is an epimorphism in $\mathbf{S}(X_{\mathrm{fl}})$.

(c) Let $(\mathbb{Z}/p\mathbb{Z})_X$ be the constant sheaf defined by the abelian group $\mathbb{Z}/p\mathbb{Z}$ where X is a scheme of characteristic p (that is, such that the canonical map $X \to \operatorname{spec} \mathbb{Z}$ factors through $\operatorname{spec} \mathbb{F}_p$). As $T^p - T = \prod_{i=0}^{p-1} (T - i)$ (in characteristic p) the ring

$$(\mathbb{F}_p[T]/(T^p - T)) \approx \prod_{i=0}^{p-1} \mathbb{F}_p,$$

and so

$$(\operatorname{spec} (\mathbb{F}_p[T]/(T^p - T)))_X \approx (\mathbb{Z}/p\mathbb{Z})_X.$$

Thus

$$(\mathbb{Z}/p\mathbb{Z})_X(U) = \{a \in \Gamma(U, \mathcal{O}_U) \mid a^p - a = 0\}.$$

Consider the *Artin-Schreier sequence* of sheaves

$$0 \to \mathbb{Z}/p\mathbb{Z} \to \mathbb{G}_a \xrightarrow{F-1} \mathbb{G}_a \to 0$$

where F is the map $a \mapsto a^p$. This sequence is rarely exact relative to the Zariski topology because $F - 1: A \to A$ is rarely surjective for A a local ring. However, if A is strictly local, then $F - 1: A \to A$ is surjective (because $A[T]/(T^p - T)$ is étale over A), and so the sequence is exact relative to the étale (and hence also the flat) topology.

(d) Let X be a scheme of characteristic p, and let α_p be the subsheaf of $\mathbb{G}_a$ defined by

$$\alpha_p(U) = \{a \in \Gamma(U, \mathcal{O}_U) \mid a^p = 0\}.$$

(If X is not of characteristic p, this is not a group.) Then α_p is the sheaf defined by the group scheme $\mathbb{F}_p[T]/(T^p)$. The sequence

$$0 \to \alpha_p \to \mathbb{G}_a \xrightarrow{F} \mathbb{G}_a \to 0,$$

where F is the map $a \mapsto a^p$, is exact relative to the flat topology but not usually the étale or Zariski topologies.

Exercise 2.19. Let $\phi: G \to G''$ be a morphism of commutative group schemes over a scheme X, and assume that G and G'' are locally of finite-type and flat over X. Consider the following statements:

(a) ϕ is an étale morphism;

(a′) ϕ is a flat morphism;

(b) $\ker(\phi)$ is an étale group scheme;

(b′) $\ker(\phi)$ is a flat group scheme;

(c) ϕ defines a surjective map of sheaves on X_{et};

(c′) ϕ defines a surjective map of sheaves on X_{fl}.

Show that (a) $\Leftrightarrow$ (b) $\Rightarrow$ (c), (c) $\not\Rightarrow$ (b); (a′) $\Leftrightarrow$ (b′) $\Rightarrow$ (c′), (c′) $\not\Rightarrow$ (b′). Ker (ϕ) means the kernel of ϕ in the category of group schemes over X, that is,

$$\begin{array}{ccc} \ker(\phi) = G \times_{G''} X & \longrightarrow & X \\ \downarrow & & \downarrow e \\ G & \xrightarrow{\phi} & G''. \end{array}$$

§3. Direct and Inverse Images of Sheaves

Suppose that $\pi: X' \to X$ defines a morphism of sites $(\mathbf{C}'/X')_{E'} \to (\mathbf{C}/X)_E$. The *direct image* of a sheaf F' on $X'_{E'}$ is defined to be $\pi_* F' = \pi_p F'$ and the *inverse image* of a sheaf F on X_E is defined to be $\pi^* F = a(\pi^p F)$. Recall (2.7) that $\pi_p F'$ is a sheaf. There are canonical isomorphisms

$$\mathrm{Hom}_{\mathbf{S}(X_E)}(F, \pi_* F') \approx \mathrm{Hom}_{\mathbf{P}(X'_{E'})}(\pi^p F, F') \approx \mathrm{Hom}_{\mathbf{S}(X'_{E'})}(\pi^* F, F')$$

that show that π_* and π^* are adjoint functors $\mathbf{S}(X'_{E'}) \rightleftarrows \mathbf{S}(X_E)$. Thus π_* is left exact and commutes with inverse limits, and π^* is right exact and commutes with direct limits. If π^p is exact (2.6), then

$$\pi^* = \mathbf{S}(X'_{E'}) \hookrightarrow \mathbf{P}(X'_{E'}) \xrightarrow{\pi^p} \mathbf{P}(X_E) \xrightarrow{a} \mathbf{S}(X_E)$$

is left exact as well as right exact.

Remarks 3.1. (a) If $\pi: X' \to X$ is in $\mathbf{C}/X$, then

$$\pi^*: \mathbf{S}((\mathbf{C}/X)_E) \to \mathbf{S}((\mathbf{C}/X')_E)$$

is simply the restriction functor; we usually write it $F \mapsto F|X'$. (We sometimes also use this notation for $\pi^* F$ when π is not of this form.)

(b) If $E' \subset E$, $\mathbf{C}'/X \subset \mathbf{C}/X$, and $\pi: X \to X$ is the identity map, then it is not necessarily true that $\pi_* \pi^*(F) = F$. However, it is true for the identity map $X_{\mathrm{fl}} \to X_{\mathrm{et}}$ (compare (III.3.11b)).

(c) Note that π^* depends on the underlying categories of the sites and not just the topologies. For example, let $\pi: \operatorname{spec} \bar{k} \to \operatorname{spec} k$ be the morphism defined by the inclusion of a field k into its algebraic closure $\bar{k}$. If we give $\operatorname{spec} k$ and $\operatorname{spec} \bar{k}$ their big étale sites, then π^* is simply the restriction functor and so commutes with arbitrary products of sheaves. However, we leave it as an exercise to the reader to show that if $\operatorname{spec} k$ and $\operatorname{spec} \bar{k}$ are given their small étale sites, then π^* does not commute with products (use 2.16).

(d) Let $\pi: X' \to X$ be a morphism; let G be a group scheme on X, and assume that the E-topology is coarser than the canonical topology so that G defines sheaves G_X and $G_{X'}$ on X_E and X'_E. The map $\pi^p G_X \to G_{X'}$, which sends the element of $\Gamma(U', G)$ represented by (s, g) with $g: U' \to U$ and $s \in \Gamma(U, G)$ to

$$sg \in G_X(U') = G_{X'}(U') = \Gamma(U', G_{X'}),$$

factors uniquely through $\pi^* G_X$. Thus there is a canonical map $\phi_G: \pi^* G_X \to G_{X'}$.

This need not be an isomorphism. For example, let $X = \operatorname{spec} k$ with k a field of characteristic p, let $X' = \operatorname{spec} A$ with A a k-algebra with many nilpotents, let $G = \alpha_p$, and consider $\pi: X'_{\mathrm{et}} \to X_{\mathrm{et}}$. Then $(\alpha_p)_X = 0$, so $\pi^*(\alpha_p)_X = 0$, but $(\alpha_p)_{X'}$ need not be zero. Thus ϕ_G in this case is injective but not surjective. As a second example consider $\pi: X'_{\text{et}} \hookrightarrow X_{\text{et}}$ where $X' = \operatorname{spec} \mathbb{F}_p$ and $X = \operatorname{spec} \mathbb{Z}/(p^2)$. There is an exact sequence $0 \to \mathbb{G}_a \to \pi^*\mathbb{G}_{m,X} \xrightarrow{\phi} \mathbb{G}_{m,X'} \to 0$, and so ϕ_G is surjective but not injective in this case.

There are two important situations where ϕ_G is an isomorphism, namely, where π^* is a restriction map (that is, $X' \to X$ is in $\mathbf{C}/X$) and where G is in $\mathbf{C}/X$. To see the second case note that, by definition of π_*, $\Gamma(G_{X'}, F) \approx \Gamma(G_X, \pi_* F)$ for any sheaf F on X'_E. Since, under this isomorphism, homomorphisms correspond to homomorphisms,

$$\operatorname{Hom}_{\mathbf{S}(X')}(G_{X'}, F) \approx \operatorname{Hom}_{\mathbf{S}(X)}(G_X, \pi_* F).$$

This shows that $G_{X'} \approx \pi^* G_X$ by the uniqueness of adjoints.

(e) Let $\pi: \operatorname{spec} K' \to \operatorname{spec} K$ correspond to an inclusion of fields $K \hookrightarrow K'$. There is a noncanonical commutative diagram:

$$\begin{array}{ccc} K'_s & \hookleftarrow & K_s \\ \uparrow & & \uparrow \\ K' & \hookleftarrow & K \end{array}$$

Let $\psi: G_{K'} \to G_K$ be the homomorphism of Galois groups induced by making $G_{K'}$ act on K_s as a subfield of K'_s. If $\mathbf{S}(\operatorname{spec}(K)_{\text{et}})$ and $\mathbf{S}(\operatorname{spec}(K')_{\text{et}})$ are identified with G_K-**mod** and $G_{K'}$-**mod**, then π^* and π_* become identified with the following functors: for any $N \in G_K$-**mod**, $\pi^* N$ has the same underlying group as N and has the $G_{K'}$-action defined by $\sigma n = \psi(\sigma) n$, $\sigma \in G_{K'}$, $n \in N$; for any $N \in G_{K'}$-**mod**, $\pi_* N = M_{G_K}^{\psi(G_{K'})}(N^{\ker(\psi)})$ where M_G^H has the same meaning as in Serre [8, 1.2.5].

It is easy to see that this description of π^* is correct and that the two functors are adjoint, which implies that the description of π_* is correct.

(f) Consider continuous morphisms $X''_{E''} \xrightarrow{\pi} X'_{E'} \xrightarrow{\pi'} X_E$. Clearly $(\pi'\pi)_* = \pi'_*\pi_*$ and $\pi^*\pi'^*$ is adjoint to $\pi'_*\pi_*$, which implies that $\pi^*\pi'^* = (\pi'\pi)^*$.

For the remainder of this section all schemes will have their étale topology unless explicitly stated otherwise. We first investigate the effect of π_* and π^* on stalks.

THEOREM 3.2. *Let $\pi: X' \to X$ be a morphism.*

(a) *For any sheaf F on X_{et} and any $x' \in X'$, $(\pi^*F)_{\bar{x}'} \approx F_{\overline{\pi(x')}}$, that is, the stalk of π^*F at $\bar{x}'$ is isomorphic to the stalk of F at $\overline{\pi(x')}$. In particular, if π is the canonical morphism $\pi: \operatorname{spec} \mathcal{O}_{X,\bar{x}} \to X$, then*

$$F_{\bar{x}} = (\pi^*F)_{\bar{x}} = \Gamma(\operatorname{spec} \mathcal{O}_{X,\bar{x}}, \pi^*F).$$

(b) *Assume that π is quasi-compact. Let $x \in X$; let $\bar{x} = \operatorname{spec} k(x)_{\mathrm{sep}}$; let f be the canonical morphism $\tilde{X} = \operatorname{spec} \mathcal{O}^{\mathrm{sh}}_{X,\bar{x}} \to X$, and let $\tilde{X}' = X' \times_X \tilde{X}$:*

$$\begin{array}{ccc} X' & \xleftarrow{f'} & \tilde{X}' \\ \downarrow{\scriptstyle \pi} & & \downarrow{\scriptstyle \tilde{\pi}} \\ X & \xleftarrow{f} & \tilde{X} \end{array}$$

Then $(\pi_* F)_{\bar{x}} = \Gamma(\tilde{X}', \tilde{F})$ for any sheaf F on X' and its restriction $\tilde{F} = f'^* F$ to $\tilde{X}'$.

Proof. (a) Write $x = \pi(x')$. We may take $\bar{x} = \bar{x}'$ and so have a commutative diagram:

$$\begin{array}{ccc} X' & \xleftarrow{u_{x'}} & \bar{x}' = \bar{x} \\ \downarrow{\scriptstyle \pi} & \swarrow{\scriptstyle u_x} & \\ X & & \end{array}$$

Then

$$(\pi^* F)_{\bar{x}'} = (u_{x'}^* \pi^* F)(\bar{x}') = (u_x^* F)(\bar{x}) = F_{\bar{x}}.$$

The final equality uses the fact that if $\bar{x} \to U \to X$ is an étale neighborhood of $\bar{x} \to X$, then $\bar{x} \to U$ factors through $\bar{x} \to \operatorname{spec} \mathcal{O}_{X,\bar{x}}$ (uniquely if U is connected). (See (I.4).)

(b) By definition $(f'^* F)(\tilde{X}') = \varinjlim F(U')$ where the limit is taken over all commutative diagrams with $U' \to X'$ étale:

$$\begin{array}{ccc} & U' & \\ \swarrow & & \nwarrow \\ X' & \xleftarrow{f'} & \tilde{X}' \end{array}$$

On the other hand, from the definitions of π_* and stalks, we see that $(\pi_* F)_{\bar{x}} = \varinjlim F(U')$ where the limit is taken over all such diagrams that come by base extension from a commutative diagram with $U \to X$ étale:

$$\begin{array}{ccc} & U & \\ \swarrow & & \nwarrow \\ X & \xleftarrow{f} & \tilde{X} \end{array}$$

To prove that these two limits are isomorphic, we must show that the second set of diagrams is cofinal in the first, that is, that $\tilde{X}' \to U'$ factors through $\tilde{X}' \to U_{(X')}$ for some étale $U \to X$. But $\tilde{X}$ is a limit, $\tilde{X} = \varprojlim U$, in which each U is affine and is étale over X. As inverse limits commute with fiber products (in any category) and π is quasi-compact, $\tilde{X}'$ is a limit, $\tilde{X}' = X' \times (\varprojlim U) = \varprojlim U_{(X')}$, in which each $U_{(X')}$ is quasi-compact and the transition morphisms are all affine. Now U' being of finite-type over X' implies that $\tilde{X}' \to U'$ factors through some $U_{(X')}$, as follows from the next lemma.

LEMMA 3.3. *Let X be a scheme and let $Y = \varprojlim Y_i$, where (Y_i) is a filtered inverse system of X-schemes such that the transition morphisms $Y_i \leftarrow Y_j$ are affine. Assume that the Y_i are quasi-compact and let Z be a scheme that is locally of finite-type over X. Then any X-morphism $Y \to Z$ factors through $Y \to Y_i$ for some i. More precisely,* $\mathrm{Hom}_X(Y, Z) = \varinjlim \mathrm{Hom}_X(Y_i, Z)$.

Proof. In the case that $X = \operatorname{spec} A$, $Y_i = \operatorname{spec} B_i$, and $Z = \operatorname{spec} C$ all are affine, the lemma is obvious because it merely asserts that if $B = \varinjlim B_i$ and C is finitely generated over A, then any A-homomorphism $C \to B$ factors through some B_i. The general case may be deduced from this case by the usual patching argument [EGA. IV.8.13.1].

Remark 3.4. If F is the sheaf on X'_{et} defined by a group scheme G that is locally of finite-type over X', then $(f'^*F)(\tilde{X}') = G(\tilde{X}')$. This follows from (3.3) and the fact that $\tilde{X}' = \varprojlim U'$, where the limit is taken over all diagrams as at the start of the proof of (3.2b).

COROLLARY 3.5. (a) *Let $i: Z \to X$ be a closed immersion, and let F be a sheaf on Z_{et}. Let $x \in X$. Then,*

$$(i_*F)_{\bar{x}} = 0 \quad \text{if } x \notin i(Z), \text{ and}$$
$$(i_*F)_{\bar{x}} = F_{\bar{x}_0} \quad \text{if } x = i(x_0),\ x_0 \in Z.$$

(b) *Let $j: U \to X$ be an open immersion, and let F be a sheaf on U_{et}. If $x \in j(U)$, $x = j(x_0)$, then $(j_*F)_{\bar{x}} = F_{\bar{x}_0}$. (However, $(j_*F)_{\bar{x}}$ need not be zero if $x \notin j(U)$.)*

(c) *Let $\pi: X' \to X$ be a finite morphism and F a sheaf on X'_{et}. For any $x \in X$, $(\pi_*F)_{\bar{x}} = \prod F_{\bar{x}'}^{d(x')}$ where the product is taken over all x' such that $\pi(x') = x$, and $d(x')$ is the separable degree of $k(x')$ over $k(x)$. In particular, if π is étale of constant degree d, then $(\pi_*F)_{\bar{x}} = F_{\bar{x}'}^d$ where x' is any point of X' such that $\pi(x') = x$.*

Proof. (a) With the notation of the theorem $\tilde{Z}$ is empty if $x \notin i(Z)$ and $\tilde{Z} = \operatorname{spec}(\mathcal{O}_{X,\bar{x}}/\mathfrak{J}\mathcal{O}_{X,\bar{x}}) = \operatorname{spec} \mathcal{O}_{Z,\bar{z}}$ if $i(z) = x$.

(b) Obvious.

(c) As $\tilde{X}'$ is finite over $\tilde{X} = \operatorname{spec} \mathcal{O}_{X,\bar{x}}$ and $\mathcal{O}_{X,\bar{x}}$ is Henselian, the connected components of $\tilde{X}'$ are in one-to-one correspondence with the connected components of $\pi^{-1}(\bar{x})$ (I.4.2b). Thus

$$\tilde{X}' = \coprod_{\pi(x')=x} \operatorname{spec}(\mathcal{O}_{X',\bar{x}'}^{d(x')}),$$

and so $F(\tilde{X}') = \prod F_{\bar{x}'}^{d(x')}$.

COROLLARY 3.6. *If π is a finite morphism, for example, a closed immersion, then π_* is exact (for the étale topology).*

Proof. This follows immediately from (3.5) and (2.17c).

Exercise 3.7. Let X be an integral scheme with generic point $g:\eta \to X$. Show that if X is normal, then $g_*M_\eta = M_X$ for any constant sheaf M_η on η. Show that if X is a curve with a node $i:z \hookrightarrow X$, then there is an exact sequence,

$$0 \to M_X \to g_*M_\eta \to i_*M_z \to 0.$$

What is true in general? (Hint: write g as the composite $\eta \to X' \to X$ where X' is the normalization of X.)

Remark 3.8. For a proper morphism $\pi:X' \to X$ a much sharper result than (3.2b) holds, namely, $(\pi_*F)_{\bar{x}} = \Gamma(X'_{\bar{x}}, F|X'_{\bar{x}})$ where F is a sheaf on X' and $X'_{\bar{x}} \hookrightarrow X'$ is the geometric fiber of X' over $\bar{x} \hookrightarrow X$. Thus the stalks of π_*F can be computed on the geometric fibers of X'/X.

After (3.2b), it suffices to prove this with X a strictly local scheme with closed point $x = \bar{x}$. The stalk of π_*F, $(\pi_*F)_{\bar{x}}$, is then equal to $\Gamma(X, \pi_*F) = \Gamma(X', F)$. Thus to prove the above assertion we must verify that if $\pi:X' \to X$ is proper and X is strictly local, then $\Gamma(X', F) \xrightarrow{\approx} \Gamma(X'_x, F|X'_x)$ for any sheaf F on X' (where X'_x is the closed fiber of X'/X).

This statement is part of the proper base change theorem and will be considered in (VI.2). We only make some remarks here. Firstly, if $\pi:X' \to X$ is finite, then it may be proved by the same argument as in (3.5c) above. Secondly, if it is true for two morphisms, then it is true for their composite. Finally, if F is a constant sheaf $F = M_{X'}$, then it follows from the existence of a Stein factorization. Recall [EGA. III.4.3.1] or Hartshorne [2, III.11.5] that the Stein factorization of a proper morphism $\pi:X' \to X$ is a sequence $X' \xrightarrow{\pi_1} Y \xrightarrow{\pi_2} X$ with π_2 finite ($Y = \mathbf{spec}\,\pi_*\mathcal{O}_{X'}$) and π_1 proper with connected fibers. Thus we need only consider the case that X is strictly local and that X'_x (and X') are connected. But then $\Gamma(X'_x, M) = M = \Gamma(X', M)$.

Example 3.9. Let X be integral and quasi-compact. Let $g:\operatorname{spec} K \hookrightarrow X$ be the inclusion of the generic point, and let $\mathbb{G}_{m,K}$ denote the sheaf defined by $\mathbb{G}_m$ on spec K. From the definition of $g_*\mathbb{G}_{m,K}$,

$$\Gamma(U, g_*\mathbb{G}_{m,K}) = \Gamma(U \times_X \operatorname{spec} K, \mathbb{G}_m) = R(U)^*$$

for any $U \to X$ étale, where $R(U)$ is the ring of rational functions on U.

There is a canonical injection $\phi:\mathbb{G}_{m,X} \to g_*\mathbb{G}_{m,K}$ that on any U is simply the inclusion $\Gamma(U, \mathcal{O}_U^*) \hookrightarrow R(U)^*$. The cokernel of this map is defined to be the sheaf of *Cartier divisors*, Div_X, on X_{et}.

In the case that X is regular, Cartier divisors may be interpreted as Weil divisors. Let X_1 be the set of points x on X of codimension 1, that is, such that $\mathcal{O}_{X,x}$ has dimension 1 and so is a discrete valuation ring. The sheaf D_X of *Weil divisors* on X_{et} is defined to be $\bigoplus_{x \in X_1} i_{x*}\mathbb{Z}$, where $\mathbb{Z}$ denotes the constant sheaf on x defined by $\mathbb{Z}$ and $i_x : x \hookrightarrow X$. I claim that $D_X \approx \mathrm{Div}_X$. For any U étale over X,

$$\Gamma(U, D_X) = \bigoplus_{x \in U_1} \mathbb{Z},$$

and so we may define a map $\psi: g_*\mathbb{G}_{m,K} \to D_X$ by requiring that $f \in R(U)^*$ maps to the family $(\mathrm{ord}_x(f))_x$ where ord_x is the discrete valuation on $R(U)$ defined by $\mathcal{O}_{U,x}$. To prove the claim, it suffices to show that

$$0 \to \mathbb{G}_{m,X} \xrightarrow{\phi} g_*\mathbb{G}_{m,K} \xrightarrow{\psi} D_X \to 0$$

is exact. But, for any $x \in X$, the sequence of stalks at $\bar{x}$ is $0 \to A^* \to L^* \to \bigoplus \mathbb{Z} \to 0$ where $A = \mathcal{O}_{X,\bar{x}}$. L is the field of fractions of A, and the sum is over the primes of height 1. But A is regular (I.4), hence factorial, and so the prime ideals of height 1 are principal, which shows that ψ is surjective. The exactness at the other points is trivial. (Alternatively one may simply use the fact that the sequence is known to be exact when restricted to U_{Zar}, for any U étale over X. See Mumford [2, Lecture 9] or Hartshorne [2, II.6.11].)

We now consider the situation: X is a scheme; U is an open subscheme of X, and Z is a subscheme of X whose underlying set is the complementary closed subset $Z = X - U$. We denote the inclusion maps by i and j,

$$Z \xrightarrow{i} X \xleftarrow{j} U.$$

If F is a sheaf on X_{et}, we get sheaves $F_1 = i^*F$ and $F_2 = j^*F$ on Z and U. Moreover, since $\mathrm{Hom}(F, j_*j^*F) \approx \mathrm{Hom}(j^*F, j^*F)$, there is a canonical morphism $F \to j_*j^*F$ corresponding to the identity map of j^*F. By applying i^* to this, we get a canonical morphism $\phi_F : F_1 \to i^*j_*F_2$. Thus, there is associated with any sheaf F on X_{et} a triple (F_1, F_2, ϕ_F) where $F_1 \in \mathbf{S}(Z)$, $F_2 \in \mathbf{S}(U)$, and $\phi : F_1 \to i^*j_*F_2$. We shall make this into an equivalence of categories.

Define $\mathbf{T}(X)$ to be the category whose objects are triples (F_1, F_2, ϕ) with $F_1 \in \mathbf{S}(Z_{et})$, $F_2 \in \mathbf{S}(U_{et})$, and ϕ a morphism $F_1 \to i^*j_*F_2$. A morphism $(F_1, F_2, \phi) \to (F'_1, F'_2, \phi')$ is a pair (ψ_1, ψ_2) where ψ_1 is a morphism

$F_1 \to F'_1$, ψ_2 is a morphism $F_2 \to F'_2$, and the two are compatible with ϕ and ϕ' in the sense that the following diagram commutes:

$$\begin{array}{ccc} F_1 & \xrightarrow{\phi} & i^*j_*F_2 \\ \downarrow{\scriptstyle \psi_1} & & \downarrow{\scriptstyle i^*j_*(\psi_2)} \\ F'_1 & \xrightarrow{\phi'} & i^*j_*F'_2 \end{array}$$

THEOREM 3.10. *There is an equivalence between the categories* $\mathbf{S}(X_{\text{et}})$ *and* $\mathbf{T}(X)$ *under which* $F \in \mathbf{S}(X_{\text{et}})$ *corresponds to the triple* (i^*F, j^*F, ϕ_F) *defined above.*

Proof. Let $\psi \in \operatorname{Hom}_{\mathbf{S}}(F, F')$; then $(i^*(\psi), j^*(\psi))$ is a morphism

$$(i^*F, j^*F, \phi_F) \to (i^*F', j^*F', \phi_{F'}).$$

Thus there is a functor $t:\mathbf{S}(X_{\text{et}}) \to \mathbf{T}(X)$.

For $(F_1, F_2, \phi) \in \operatorname{Ob}(\mathbf{T}(X))$, define $s(F_1, F_2, \phi)$ to be the fiber product of i_*F_1 and j_*F_2 over $i_*i^*j_*F_2$, so that

$$\begin{array}{ccc} s(F_1, F_2, \phi) & \longrightarrow & j_*F_2 \\ \downarrow & & \downarrow{\scriptstyle (\text{canonical})} \\ i_*F_1 & \xrightarrow{i_*(\phi)} & i_*i^*j_*F_2 \end{array}$$

is Cartesian. From the universal property of fiber products, we find that any morphism

$$(\psi_1, \psi_2):(F_1, F_2, \phi) \to (F'_1, F'_2, \phi')$$

induces a canonical morphism

$$s(\psi_1, \psi_2):s(F_1, F_2, \phi) \to s(F'_1, F'_2, \phi').$$

Thus we have a functor $s:\mathbf{T}(X) \to \mathbf{S}(X_{\text{et}})$.

For any $F \in \mathbf{S}(X_{\text{et}})$ the canonical maps $F \to i_*i^*F$ and $F \to j_*j^*F$ induce a map $F \to st(F)$. To show that this is an isomorphism, it suffices to show that

$$\begin{array}{ccc} F & \longrightarrow & j_*j^*F \\ \downarrow & & \downarrow \\ i_*i^*F & \longrightarrow & i_*i^*j_*j^*F \end{array}$$

is Cartesian.

Since the functor that takes a sheaf to its stalk preserves fiber products and the family of such functors is conservative, we may replace the diagram by its stalk. If $x \in U$, then the diagram becomes

$$\begin{array}{ccc} F_{\bar{x}} & \longrightarrow & F_{\bar{x}} \\ \downarrow & & \downarrow \\ 0 & \longrightarrow & 0 \end{array}$$

(see (3.5)), which is clearly Cartesian. If $x \in Z$, then the diagram becomes

$$\begin{array}{ccc} F_{\bar{x}} & \longrightarrow & (j_*j^*F)_{\bar{x}} \\ \downarrow & & \downarrow \\ F_{\bar{x}} & \longrightarrow & (j_*j^*F)_{\bar{x}}, \end{array}$$

which is again Cartesian.

Since s and t induce inverse maps on morphisms (this again only has to be checked on the stalks (2.17c)), this proves that the categories are equivalent.

If Y is any subscheme of X and F is a sheaf on X_{et}, we say that F has its *support* on Y if $F_{\bar{x}} = 0$ for any $x \notin Y$.

COROLLARY 3.11. *If $i:Z \hookrightarrow X$ is a closed immersion, then the functor $i_*:\mathbf{S}(Z_{\text{et}}) \to \mathbf{S}(X_{\text{et}})$ induces an equivalence between $\mathbf{S}(Z_{\text{et}})$ and the full subcategory of $\mathbf{S}(X_{\text{et}})$ comprising those sheaves with support on $i(Z)$.*

Proof. This follows from (3.10) and the obvious fact that F has support in Z if and only if $t(F)$ has the form $(F_1, 0, 0)$.

Example 3.12. Let $X = \operatorname{spec} A$ where A is a discrete valuation ring; let K be the field of fractions of A, and let k be the residue field of A. Let G_K be the Galois group of K_s over K, and let $G_K \supset D \supset I \supset \{1\}$ be its usual, noncanonical filtration by the decomposition group D and the inertia group I. Then $D/I = G_k$, the Galois group of k; K_s^D is the field of fractions of A^{h}, and K_s^I is the field of fractions of A^{sh} (I.4). Let $U = \operatorname{spec} K$ and $Z = \operatorname{spec} k$. If $F \in \mathbf{S}(U_{\text{et}})$ corresponds to $N \in G_K$-**mod**, then one sees, by using (3.2), for example, that i^*j_*N corresponds to $N^I \in G_k$-**mod**. Thus, to give a sheaf on X_{et} is the same as to give a triple (N_1, N_2, ϕ) where $N_1 \in G_k$-**mod**, $N_2 \in G_K$-**mod**, and $\phi:N_1 \to N_2^I$ is a G_k-homomorphism. Note that ϕ may more conveniently be regarded as a homomorphism $N_1 \to N_2$ that is compatible with the actions of G_k and G_K.

Remark 3.13. A sequence in $\mathbf{T}(X)$

$$(F_1', F_2', \phi') \to (F_1, F_2, \phi) \to (F_1'', F_2'', \phi'')$$

is exact if and only if the sequences

$$F_1' \to F_1 \to F_1'', \qquad F_2' \to F_2 \to F_2''$$

are exact in $\mathbf{S}(Z_{\text{et}})$, $\mathbf{S}(U_{\text{et}})$ respectively. This follows from (3.10), (2.15), and the fact that

$$\begin{aligned} s(F_1, F_2, \phi)_{\bar{x}} &= (F_1)_{\bar{x}} & x \in Z \\ &= (F_2)_{\bar{x}} & x \in U \end{aligned}$$

for any (F_1, F_2, ϕ) in $\mathbf{T}(X)$. For example, for any sheaf F on X there is an exact sequence in $T(X)$,

$$0 \to (0, j^*F, 0) \to (i^*F, j^*F, \phi_F) \to (i^*F, 0, 0) \to 0.$$

In the notation introduced below, this corresponds to the sequence

$$0 \to j_!j^*F \to F \to i_*i^*F \to 0$$

in $\mathbf{S}(X)$.

It is possible to define six functors,

$$\begin{array}{ccccc} & \overset{i^*}{\leftarrow} & & \overset{j_!}{\leftarrow} & \\ \mathbf{S}(Z) & \overset{i_*}{\rightarrow} & \mathbf{S}(X) & \overset{j^*}{\rightarrow} & \mathbf{S}(U). \\ & \overset{i^!}{\leftarrow} & & \overset{j_*}{\leftarrow} & \end{array}$$

If $\mathbf{S}(X)$ is identified with $\mathbf{T}(X)$, then these are described as follows:

$$\begin{array}{ll} i^*: F_1 \leftarrow\!\shortmid (F_1, F_2, \phi) & j_!: (0, F_2, 0) \longleftarrow\!\shortmid F_2 \\ i_*: F_1 \mapsto (F_1, 0, 0) & j^*: (F_1, F_2, \phi) \longmapsto F_2 \\ i^!: \ker \phi \leftarrow\!\shortmid (F_1, F_2, \phi) & j_*: (i^*j_*F_2, F_2, 1) \leftarrow\!\shortmid F_2. \end{array}$$

($j_!$ is "extension by zero"; $i^!$ is "form subsheaf of sections with support on Z"; in j_*, 1 is the identity map on $i^*j_*F_2$.)

PROPOSITION 3.14. (a) *Each functor is left adjoint to the one listed below it.*

(b) *The functors i^*, i_*, j^*, $j_!$ are exact; j_*, $i^!$ are left exact.*

(c) *The composites $i^*j_!$, $i^!j_!$, $i^!j_*$, j^*i_* are zero.*

(d) *The functors i_*, j_* are fully faithful, and $F \in \mathbf{S}(X_{\text{et}})$ has support in Z if and only if $F \approx i_*F_1$ for some $F_1 \in \mathbf{S}(Z_{\text{et}})$.*

(e) *The functors j_*, j^*, $i^!$, i_* map injectives to injectives.*

Proof. Parts (a), (b), (c) and (d) all follow easily from what has already been proved, while (e) follows from (a) and (b) and the trivial fact that a functor with an exact left adjoint preserves injectives (see (III.1.2) below).

Example 3.15. Let $X = \operatorname{spec} A$, where A is a discrete valuation ring as in (3.12). In terms of the triples in (3.12), the functors are as follows:

$$\begin{array}{ll} i^*(N_1, N_2, \phi) = N_1 & j_!(N) = (0, N, 0) \\ i_*(N) = (N, 0, 0) & j^*(N_1, N_2, \phi) = N_2 \\ i^!(N_1, N_2, \phi) = \ker(\phi) & j_*(N) = (N^I, N, 1). \end{array}$$

Exercise 3.16. Let X be a normal connected scheme of dimension one, for example, a smooth curve, and let $g:\eta \hookrightarrow X$ be its generic point. Write $G = \mathrm{Gal}(k(\bar{\eta})/k(\eta))$ and $G(v) = \mathrm{Gal}(k(\bar{v})/k(v))$, $v \in X^0$, where $\bar{\eta}$ and $\bar{v}$ are chosen so that $k(\bar{\eta}) = k(\eta)_s$, $k(\bar{v}) = k(v)_s$. For each $v \in X^0$, choose an embedding $\mathcal{O}_{X,\bar{v}} \hookrightarrow k(\bar{\eta})$; this embedding determines a filtration $G \supset D_v \supset I_v \supset \{1\}$ and an isomorphism $D_v/I_v \xrightarrow{\approx} G(v)$. Consider families $(M, (M_v, \phi_v)_{v \in X^0})$ where M is a discrete G-module, M_v is a discrete $G(v)$-module, $\phi_v : M_v \to M^{I_v}$ is compatible with the actions of $G(v)$ and G, and there is a nonempty open subscheme $U \subset X$ such that $\phi_v : M_v \to M$ is an isomorphism for $v \in U$. Make the set of such families into a category $\mathbf{M}(X_{\mathrm{et}})$ by defining a morphism to be a family of maps $(\psi, (\psi_v)_{v \in X^0})$ preserving all the structure.

(a) If $F \in \mathbf{S}(X_{\mathrm{et}})$, define $m(F) = (F_{\bar{\eta}}, (F_{\bar{v}}, \phi_v))$ where $\phi_v : F_{\bar{v}} \to F_{\bar{\eta}}$ is the *cospecialization* map induced by the embedding $\mathcal{O}_{X,\bar{v}} \hookrightarrow k(\bar{\eta})$. ($\phi_v$ may loosely be regarded as a map $F(\mathcal{O}_{X,\bar{v}}) \to F(k(\bar{\eta}))$.) Show that $m(F) \in \mathbf{M}(X_{\mathrm{et}})$ if and only if F is *generically locally constant*, that is, there exists an étale map $U' \to X$, $U' \neq \emptyset$, such that $F|U'$ is constant.

(b) Show that m defines an equivalence between the category of generically constant sheaves and $\mathbf{M}(X_{\mathrm{et}})$.

(c) If $m(F) = (M, (M_v, \phi_v))$, show that

$$\Gamma(X, F) = \{(s, (s_v)_{v \in X^0}) \mid s \in M^G,\ s_v \in M_v^{G(v)},\ \phi_v(s_v) = s\}.$$

(d) Let M be a G-module such that $M = M^{I_v}$ for all v in a nonempty open subset of X, and let F_0 be the sheaf on η corresponding to M; show that $m(g_* F_0) = (M, (M^{I_v}, \phi_v))$ with ϕ_v the inclusion $M^{I_v} \hookrightarrow M$.

(e) Describe the six functors $i^*, j_!, \ldots$ in terms of $\mathbf{M}$.

Remark 3.17. (3.11) applies, in particular, when U is the empty scheme, that is, when $i:Z \to X$ is a surjective closed immersion. This occurs when Z is the closed subscheme of X defined by a nilpotent ideal. In this case $i_*:\mathbf{S}(Z_{\mathrm{et}}) \to \mathbf{S}(X_{\mathrm{et}})$ is an equivalence of categories with quasi-inverse i^*. In fact this statement follows directly from the statement (I.3.23) that the functor $Y \mapsto Y_{(Z)}$ is an equivalence of categories (ét)$/X \to$ (ét)$/Z$.

More generally [SGA. 4, VIII.1.1] if $\pi: Y \to X$ is any universal homeomorphism, then both of these last two statements still hold, that is, (ét)$/Y \hookrightarrow$ (ét)$/X$, and $\pi_*:\mathbf{S}(T_{\mathrm{et}}) \xrightarrow{\approx} \mathbf{S}(X_{\mathrm{et}})$. It is known [SGA. 4, VIII.1.3] that π is a universal homeomorphism if and only if it is integral, surjective, and radicial. Examples of such morphisms are a morphism $X \otimes_k k' \to X$ where X is a scheme over a field k and k' is a purely inseparable extension of k or a morphism $X' \to X$ where X is geometrically unibranch and X' is the normalization of X_{red}.

Remark 3.18. Let $j: U \to X$ be an object of $\mathbf{C}/X$ for some site $(\mathbf{C}/X)_E$ (not recessarily the étale site). We shall show that j^* has a left adjoint $j_!$

with many of the same properties as the functor $j_!$ defined in the special situation of (3.14). This functor will again be called the *extension by zero* functor.

Let p be the functor $\mathbf{C}/U \to \mathbf{C}/X$, $p(Y \xrightarrow{g} U) = (Y \xrightarrow{jg} X)$. The functor

$$(f \mapsto f \circ p): \mathbf{Fun}\,(\mathbf{C}/X, \mathbf{Ab}) \to \mathbf{Fun}\,(\mathbf{C}/U, \mathbf{Ab})$$

is the functor $j^p: \mathbf{P}(X) \to \mathbf{P}(U)$ of §2. According to (2.2), it has a left adjoint, which we write $j_!: \mathbf{P}(U) \to \mathbf{P}(X)$. Explicitly, for $P \in \mathbf{P}(U)$ and $V \in \mathbf{C}/X$, $(j_!P)(V) = \varinjlim P(V')$ where the limit is taken over all commutative squares,

$$\begin{array}{ccc} V' & \longleftarrow & V \\ \downarrow & & \downarrow \\ U & \longrightarrow & X \end{array}$$

in $\mathbf{C}/X$. The limit breaks up into

$$(j_!\,P)(V) = \bigoplus_{\phi \in \mathrm{Hom}_X(V,U)} \varinjlim_{S(\phi)} P(V')$$

where $S(\phi)$ is the set of squares with $(V \to V' \to U) = \phi$. Since $S(\phi)$ contains a final object (the square with $V' = V$), we see that

$$(j_!P)(V) = \bigoplus_{\phi \in \mathrm{Hom}_X(V,U)} P(V_\phi)$$

where V_ϕ is the object $V \xrightarrow{\phi} U$ of $\mathbf{C}/U$. Thus $j_!$ is exact. Note that if j is an open immersion, then $(j_!P)(V) = P(V)$ if $V \to X$ factors through U and is zero otherwise.

We define $j_!$ on sheaves to be the composite

$$\mathbf{S}(X) \hookrightarrow \mathbf{P}(X) \xrightarrow{j_!} \mathbf{P}(U) \xrightarrow{a} \mathbf{S}(U).$$

This clearly is adjoint to j^*, the restriction functor. Being a left adjoint, it is automatically right exact, and it is left exact because it is a composite of left exact functors. Of course, if $(\mathbf{C}/X)_E = X_{\mathrm{et}}$ and j is an open immersion, this agrees with the old definition.

Finally in this section, we generalize to sheaves on arbitrary sites some of the standard constructions for modules. Let A be a sheaf of commutative rings on a site $(\mathbf{C}/X)_E$, and write $\mathbf{S}(X_E, A)$ for the category of sheaves of A-modules. If A is the constant sheaf $\mathbb{Z}$, then $\mathbf{S}(X, A) = \mathbf{S}(X)$.

Let S be a presheaf of sets on X_E. The presheaf PS of A-modules with $PS(U) =$ free $A(U)$-module on $S(U)$ for all U, and the obvious restriction maps, is called the *free presheaf* of A-modules generated by S. The associated sheaf $FS = aPS$ is the *free sheaf* of A-modules generated by S. One has

$$\operatorname{Hom}(S, M) \approx \operatorname{Hom}_{\mathbf{P}(X,A)}(PS, M) \approx \operatorname{Hom}_{\mathbf{S}(X,A)}(FS, M)$$

for any sheaf M of A-modules.

For any pair F_1 and F_2 of sheaves of A-modules, $\underline{\operatorname{Hom}}_A(F_1, F_2)$ is defined to be the sheaf

$$U \mapsto \operatorname{Hom}_{\mathbf{S}(U,A|U)}(F_1|U, F_2|U).$$

It is trivial to check that this is in fact a sheaf.

For any pair F_1 and F_2 of sheaves of A-modules, $F_1 \otimes_A F_2$ is defined to be the sheaf associated with the presheaf $U \mapsto F_1(U) \otimes_{A(U)} F_2(U)$.

It is also possible to give a direct construction of tensor products. Regard $F_1 \times F_2$ as a sheaf of sets, and consider the following maps:

$$\begin{aligned}
&F_1 \times F_1 \times F_2 \to F_1 \times F_2,\\
&\qquad (m_1, m_1', m_2) \mapsto (m_1 + m_1', m_2) - (m_1, m_2) - (m_1', m_2)\\
&F_1 \times F_2 \times F_2 \to F_1 \times F_2,\\
&\qquad (m_1, m_2, m_2') \mapsto (m_1, m_2 + m_2') - (m_1, m_2) - (m_1, m_2')\\
&A \times F_1 \times F_2 \to F_1 \times F_2,\\
&\qquad (a, m_1, m_2) \mapsto (am_1, m_2) - a(m_1, m_2)\\
&A \times F_1 \times F_2 \to F_1 \times F_2,\\
&\qquad (a, m_1, m_2) \mapsto (m_1, am_2) - a(m_1, m_2).
\end{aligned}$$

Then $F_1 \otimes_A F_2$ is the largest quotient sheaf of $F(F_1 \times F_2)$ such that the images of the above maps go to zero under $F(F_1 \times F_2) \to F_1 \otimes_A F_2$. This follows from the fact that the presheaf $U \mapsto F_1(U) \otimes_{A(U)} F_2(U)$ is the largest quotient presheaf of $P(F_1 \times F_2)$ with the corresponding property.

There are the usual adjunction isomorphisms. If F_1, F_2, and F_3 are sheaves or presheaves of A-modules, then a map $F_1 \times F_2 \to F_3$ is said to be *A-bilinear* if the maps $F_1(U) \times F_2(U) \to F_3(U)$ are all $A(U)$-bilinear. We write $\operatorname{Bilin}_A(F_1, F_2; F_3)$ for the group of such maps.

PROPOSITION 3.19. *There are canonical isomorphisms*

$$\operatorname{Hom}_A(F_1 \otimes F_2, F_3) \approx \operatorname{Hom}_A(F_1, \underline{\operatorname{Hom}}_A(F_2, F_3)) \approx \operatorname{Bilin}_A(F_1, F_2; F_3).$$

Proof. The isomorphism

$$\operatorname{Hom}_A(F_1 \otimes_A F_2, F_3) \approx \operatorname{Bilin}_A(F_1, F_2; F_3)$$

is obvious from the second definition of $F_1 \otimes_A F_2$. The proof that

$$\text{Bilin}_A(F_1, F_2; F_3) \approx \text{Hom}_A(F_1, \underline{\text{Hom}}_A(F_2, F_3))$$

is the same as the usual proof for the case of modules over a ring.

A sheaf F of A-modules on X_{et} is *pseudocoherent* at a geometric point $\bar{x}$ of X if there exists an étale neighborhood $U \to X$ of $\bar{x}$ and an exact sequence $(A|U)^m \to (A|U)^n \to F|U \to 0$ of sheaves on U_{et} with m, n finite.

PROPOSITION 3.20. *Let $\bar{x}$ be a geometric point of x.*

(a) *For any sheaf S of sets on X_{et}, $(FS)_{\bar{x}} = F(S_{\bar{x}})$, the free $A_{\bar{x}}$-module associated with $S_{\bar{x}}$.*

(b) *If F_1 is pseudocoherent at $\bar{x}$, then*

$$\underline{\text{Hom}}_A(F_1, F_2)_{\bar{x}} = \text{Hom}_{A_{\bar{x}}}(F_{1\bar{x}}, F_{2\bar{x}}).$$

(c) *For any pair of sheaves F_1 and F_2,*

$$(F_1 \otimes_A F_2)_{\bar{x}} = F_{1\bar{x}} \otimes_{A_{\bar{x}}} F_{2\bar{x}}.$$

Proof. (a) The operation of associating a free module to a set commutes with the formation of direct limits (being a left adjoint). Thus

$$(FS)_{\bar{x}} = (PS)_{\bar{x}} = \varinjlim PS(U) = F(\varinjlim S(U)) = F(S_{\bar{x}}).$$

(b) The conclusion is clearly valid for $F_1 = A$, and both sides commute with finite direct sums in the first factor. The result now follows from the five-lemma.

(c) Tensor products commute with the formation of filtered direct limits.

For any sheaf F_0, the functor $F \mapsto F \otimes_A F_0$ is right exact. This is obvious from (3.20c) for the étale site and may be checked directly for an arbitrary site. The sheaf F_0 is said to be a *flat* sheaf of A-modules if the functor is also left exact. Clearly, on the étale site, F_0 is flat if and only if its stalks are flat.

PROPOSITION 3.21. *Let P be a presheaf and F a sheaf. Write $P * F$ for the functor $U \mapsto P(U) \otimes_{A(U)} F(U)$. Then $a(P * F) = aP \otimes F$.*

Proof. $\text{Hom}_A(a(P * F), F') \approx \text{Hom}_A(P * F, F') \approx \text{Bilin}_A(P, F; F') \approx \text{Hom}_A(F, \underline{\text{Hom}}_A(P, F')) \approx \text{Bilin}_A(aP, F; F') \approx \text{Hom}_A(aP \otimes F, F')$ for any sheaf F'.

Exercise 3.22. (a) Let $\pi: X'_{E'} \to X_E$ be a morphism of sites. Show that $\pi_* \underline{\text{Hom}}(\pi^*F, F') = \underline{\text{Hom}}(F, \pi_*F')$ for any sheaves of abelian groups F on X_E and F' on X'_E.

(b) Write **Set** $(\mathbf{C}/X)_E$ for the category of sheaves of sets on a site $(\mathbf{C}/X)_E$. If $\pi:(\mathbf{C}'/X')_{E'} \to (\mathbf{C}/X)_E$ is continuous, show that there exist adjoint functors

$$\pi_*, \pi^*: \mathbf{Set}\,(X'_{E'}) \rightleftarrows \mathbf{Set}\,(X_E)$$

with $(\pi_* F)(U) = F(U_{(X')})$ for all U in $\mathbf{C}/X$. Show that if finite inverse limits exist in $\mathbf{C}/X$, then these functors agree with the same functors defined for sheaves of abelian groups. If $j: U \to X$ is in $\mathbf{C}/X$, show that $j^*: \mathbf{Set}\,(\mathbf{C}/X)_E \to \mathbf{Set}\,(\mathbf{C}/U)_E$ has a left adjoint $j_!$. For a sheaf F represented by a scheme Z/U, show that $j_! F$ is represented by the X-scheme $Z \to U \to X$. Deduce that if G is a commutative group scheme over U, then $j_! G$ (regarding G as a sheaf of abelian groups) $\neq j_! G$ (regarding G as a sheaf of sets).

Comments on the Literature

Most of the material of this chapter may be found in Artin [1] and, in a slightly different language, in [SGA. 4]. A brief but useful summary in the language of [SGA. 4] may be found in Giraud [1]. See also Miyanishi [1].

CHAPTER III

Cohomology

Most of the formalism of derived functor cohomology, Čech cohomology, and higher direct images continues to hold for sheaves on the étale and flat sites. It is even true, under quite general conditions, that Čech étale cohomology agrees with derived functor étale cohomology. Étale cohomology agrees with Galois cohomology when the base scheme is the spectrum of a field, with cohomology relative to the complex topology when the scheme is a smooth variety over $\mathbb{C}$ and the sheaf is torsion, with Zariski cohomology when the sheaf is defined by a quasi-coherent module, and with flat cohomology when the sheaf is defined by a smooth group scheme. As usual, H^1 has an explicit interpretation in terms of principal homogeneous spaces, and thus the techniques of étale cohomology may be used in their classification.

§1. Cohomology

We first recall a few definitions and facts from the theory of abelian categories. (See Bucur-Deleanu [1, Chapter 6, 7], for example, or Grothendieck [1].) Functors between abelian categories will be assumed to be additive.

Let $\mathbf{A}$ be an abelian category. An object I of $\mathbf{A}$ is *injective* if the functor $M \mapsto \mathrm{Hom}_{\mathbf{A}}(M, I): \mathbf{A} \to \mathbf{Ab}$ is exact. $\mathbf{A}$ has *enough injectives* if, for every M in $\mathbf{A}$, there is a monomorphism from M into an injective object. If $\mathbf{A}$ has enough injectives and $f: \mathbf{A} \to \mathbf{B}$ is a left exact functor from $\mathbf{A}$ into a second abelian category $\mathbf{B}$, then there is an essentially unique sequence of functors $R^if: \mathbf{A} \to \mathbf{B}$, $i \geq 0$, called the *right derived functors* of f with the properties:

(a) $R^0f = f$;

(b) $R^if(I) = 0$ if I is injective and $i > 0$;

(c) for any exact sequence $0 \to M' \to M \to M'' \to 0$ in $\mathbf{A}$, there are morphisms $\partial^i: R^if(M'') \to R^{i+1}f(M')$, $i \geq 0$, such that the sequence

$$\cdots \to R^if(M) \to R^if(M'') \xrightarrow{\partial^i} R^{i+1}f(M') \to R^{i+1}f(M) \to \cdots$$

is exact;

(d) the association in (c) of the long exact sequence to the short exact sequence is functorial.

An object M in $\mathbf{A}$ is *f-acyclic* if $R^i f(M) = 0$ for all $i > 0$. If

$$0 \to M \to N^0 \to N^1 \to N^2 \to \cdots$$

is a resolution of M by f-acyclic objects N^i, then the objects $R^i f(M)$ are canonically isomorphic to the cohomology objects of the complex

$$0 \to fN^0 \to fN^1 \to fN^2 \to \cdots.$$

In order to apply these results to functors from $\mathbf{S}(X_E)$, we must show that it has enough injectives.

PROPOSITION 1.1. *The category* $\mathbf{S}(X_E)$ *has enough injectives.*

Proof. We give two proofs, one which works for the étale site and one which works in general.

LEMMA 1.2. (a) *A product of injectives is injective.*

(b) *Any functor* $f: \mathbf{A} \to \mathbf{B}$ *that has an exact left adjoint* $g: \mathbf{B} \to \mathbf{A}$ *preserves injectives; for example, if* π *is a continuous morphism of sites, then* π_* *preserves injectives when* π^* *is exact.*

Proof. (a) Easy.

(b) Let I be injective in $\mathbf{A}$. The functor $M \mapsto \operatorname{Hom}_{\mathbf{B}}(M, fI)$ is isomorphic to the functor $M \mapsto \operatorname{Hom}_{\mathbf{A}}(gM, I)$, and the latter is exact because it is the composite of two exact functors.

Now assume that we are on the étale site, and let $u_x: \bar{x} \to X$ be a geometric point of X. The category $\mathbf{S}(\bar{x}_{\mathrm{et}})$ is isomorphic to $\mathbf{Ab}$ and so has enough injectives. Let $F \in \mathbf{S}(X_{\mathrm{et}})$, and choose for each $x \in X$ an embedding $u_x^* F \to F'_x$ of $u_x^* F$ into an injective sheaf. Define $F^* = \prod_{x \in X} u_{x*} u_x^* F$ and $F^{**} = \prod_{x \in X} u_{x*} F'_x$. Then the canonical maps $F \to F^*$ and $F^* \to F^{**}$ are monomorphisms (II.2), and F^{**} is injective according to (1.2).

Now consider an arbitrary site. Recall that a family $(A_j)_{j \in J}$ of objects of a category $\mathbf{A}$ is a *family of generators* for $\mathbf{A}$ if, given a monomorphism $B \to A$ in $\mathbf{A}$ that is not an isomorphism, there is a j and a morphism $A_j \to A$ that does not factor through $B \to A$.

LEMMA 1.3. *Any abelian category that satisfies AB5, AB3* and possesses a family of generators* (A_i) *has enough injectives.*

Proof. Let $A = \prod A_i$. Then $M \mapsto \operatorname{Hom}(A, M)$ is an embedding of $\mathbf{A}$ into the category of left R-modules, where R is the ring $\operatorname{Hom}_{\mathbf{A}}(A, A)$. The proof proceeds by examining this functor (see Bucur-Deleanu [1, 6.32]).

To prove (1.1), it remains to show that $\mathbf{S}(X_E)$ has a family of generators. For any $f:U \to X$ in $\mathbf{C}/X$ define $\mathbb{Z}_U$ to be the sheaf $f_!\mathbb{Z}$, where $\mathbb{Z}$ is the constant sheaf on U_E defined by $\mathbb{Z}$ (II.3.18). Then

$$\mathrm{Hom}_X(\mathbb{Z}_U, F) \approx \mathrm{Hom}_U(\mathbb{Z}, F|U) \approx F(U),$$

and so a family of generators for $\mathbf{S}(X_E)$ can be formed by taking one sheaf $\mathbb{Z}_U$ for each of sufficiently many isomorphism classes of objects of $\mathbf{C}/X$. (Of course, it has to be shown that this may be taken to be a set; compare with the remarks following (II.2.2).)

Remark 1.4. (a) If we consider the étale site on $X = \operatorname{spec} K$, K a field, then the above construction gives $(\mathbb{Z}[G/H])_H$, where H runs through the open subgroups of $G = \mathrm{Gal}(K_s/K)$ as a set of generators for G-**mod** $= \mathbf{S}(X_{\mathrm{et}})$. The earlier construction gives an injective embedding for $N \in G$-**mod** of the form $N \to M_G(N')$ where $N \to N'$ is an injective embedding of N as an abelian group and $M_G(N') = \text{Conts Maps}(G, N')$ is as in Serre [8, p. I-12].

(b) If, in (a), G is infinite, then it is not difficult to show that the only projective object in G-**mod** $= \mathbf{S}(X_{\mathrm{et}})$ is 0. The categories $\mathbf{S}(X_E)$ will rarely have enough projectives.

With this, we are able to define the right derived functors of any left exact functor from $\mathbf{S}(X_E)$ into an abelian category.

DEFINITIONS 1.5. (a) The functor $\Gamma(X, -):\mathbf{S}(X_E) \to \mathbf{Ab}$ with $\Gamma(X, F) = F(X)$, is left exact and its right derived functors are written

$$R^i\Gamma(X, -) = H^i(X, -) = H^i(X_E, -).$$

The group $H^i(X_E, F)$ is called the ith *cohomology group* of X_E with values in F.

(b) For any $U \to X$ in $\mathbf{C}/X$, the right derived functors of $F \mapsto F(U)$: $\mathbf{S}(X_E) \to \mathbf{Ab}$ are written $H^i(U, F)$. They should be distinguished (temporarily) from the groups $H^i(U, F|U)$.

(c) The inclusion functor $i:\mathbf{S}(X_E) \to \mathbf{P}(X_E)$ is left exact. Its right derived functors are written $\underline{H}^i(X_E, F)$ or simply $\underline{H}^i(F)$.

(d) For any fixed sheaf F_0 on X_E, the functor $F \mapsto \mathrm{Hom}_{\mathbf{S}}(F_0, F)$ is left exact. Its right derived functors are written $R^i\,\mathrm{Hom}_{\mathbf{S}}(F_0, -) = \mathrm{Ext}^i_{\mathbf{S}}(F_0, -)$.

(e) If F_0 and F_1 are sheaves on X_E, then $\underline{\mathrm{Hom}}(F_0, F_1)$ is the sheaf $U \mapsto \mathrm{Hom}(F_0|U, F_1|U)$. Fix F_0 and consider the functor

$$F \mapsto \underline{\mathrm{Hom}}(F_0, F):\mathbf{S}(X_E) \to \mathbf{S}(X_E).$$

It is left exact, and its right derived functors are written $\underline{\mathrm{Ext}}^i(F_0, F)$.

(f) For any continuous morphism $\pi: X'_{E'} \to X_E$, the right derived functors $R^i\pi_*$ of the functor $\pi_*: \mathbf{S}(X'_{E'}) \to \mathbf{S}(X_E)$ are defined. The sheaves $R^i\pi_* F$ are called the *higher direct images* of F.

The rest of this section will be devoted to the study of these functors.

Remark 1.6. (a) By definition of Ext^i, a short exact sequence $0 \to F' \to F \to F'' \to 0$ of sheaves induces a long exact sequence of Exts in the second variable. It also induces such a long exact sequence in the first variable, that is, for a fixed F_0 there is a long exact sequence

$$\cdots \to \mathrm{Ext}^i\,(F', F_0) \to \mathrm{Ext}^{i+1}\,(F'', F_0) \to \mathrm{Ext}^{i+1}\,(F, F_0) \to \cdots.$$

Indeed, if $F_0 \to I^{\cdot}$ is an injective resolution of F_0, then

$$0 \to \mathrm{Hom}\,(F'', I^{\cdot}) \to \mathrm{Hom}\,(F, I^{\cdot}) \to \mathrm{Hom}\,(F', I^{\cdot}) \to 0$$

is an exact sequence of complexes, from which the required long exact sequence can be constructed by the usual process.

(b) As in any abelian category, $\mathrm{Ext}^i\,(F, F')$ may also be interpreted as the group of Yoneda extensions, that is, the group of all exact sequences

$$0 \to F' \to F_{i-1} \to \cdots \to F_0 \to F \to 0$$

modulo a certain equivalence relation; see Mitchell [1].

(c) $H^i(X_E, F)$ is a contravariant functor on X_E; if $\pi^*: \mathbf{S}(X_E) \to \mathbf{S}(X'_{E'})$ is exact then the maps $H^i(X_E, F) \to H^i(X'_{E'}, \pi^* F)$ are induced by the obvious map $H^0(X, F) \to H^0(X', \pi^* F)$ and the universal property of derived functors (Bucur-Deleanu [1, 7.12]).

(d) The functors in (1.5) and their right derived functors are, to some extent, dependent only on the category $\mathbf{S}(X_E)$ of sheaves. For example, if $\pi: X' \to X$ is a universal homeomorphism, then $\mathbf{S}(X'_{\mathrm{et}}) \approx \mathbf{S}(X_{\mathrm{et}})$ (II.3.17), and if an F on X_{et} corresponds to an F' on X'_{et}, then

$$H^i(X_{\mathrm{et}}, F) \approx H^i(X'_{\mathrm{et}}, F'),$$
$$\mathrm{Ext}^i_{\mathbf{S}(X_{\mathrm{et}})}\,(F_0, F) \approx \mathrm{Ext}^i_{\mathbf{S}(X'_{\mathrm{et}})}\,(F'_0, F'),$$

and

$$\pi_* \underline{H}^i(X', F') = \underline{H}^i(X, F).$$

(e) The functors in (1.5) are not unrelated. For example, there is an isomorphism of functors $\Gamma(X, -) \approx \mathrm{Hom}\,(\mathbb{Z}, -)$ where $\mathbb{Z}$ denotes the constant sheaf on X_E. Thus $H^i(X, -) \approx \mathrm{Ext}^i\,(\mathbb{Z}, -)$, that is, the cohomology groups are special cases of Ext groups. Also it is easy to show that $\underline{H}^i(F)$ is the presheaf $U \mapsto H^i(U, F)$ (take an injective resolution and compute the two groups $\underline{H}^i(F)(U)$ and $H^i(U, F)$). Deeper relations, involving spectral sequences, will be given later in this section and the next.

Example 1.7. We interpret some of these functors in the case of the étale site on $X = \operatorname{spec} K$, K a field. Recall $\mathbf{S}(X_{\text{et}}) \approx G\text{-}\mathbf{mod}$ where $G = \operatorname{Gal}(K_{\text{sep}}/K)$.

(a) If F corresponds to the module M, then $\Gamma(X, F) = M^G$, and so

$$H^i(X, F) = H^i(G, M) = H^i(K, M),$$

where the two right hand groups refer to Galois cohomology in the sense of Serre [8].

(b) If F_0 and F correspond to M_0 and M, then $\operatorname{Hom}(F_0, F) = \operatorname{Hom}_G(M_0, M)$ and $\operatorname{Ext}^i_{\mathbf{S}}(F_0, F) = \operatorname{Ext}^i_G(M_0, M)$, where Ext^i_G refers to Exts in the category G-**mod**.

(c) If F_0 and F_1 correspond to M_0 and M_1, then $\underline{\operatorname{Hom}}(F_0, F_1)$ corresponds to

$$\bigcup_H \operatorname{Hom}_{\mathbf{Ab}}(M_0, M_1)^H = \bigcup_H \operatorname{Hom}_H(M_0, M_1),$$

where H runs through the open (normal) subgroups of G. (The action of H on $\operatorname{Hom}_{\mathbf{Ab}}(M_0, M_1)$ is $\sigma(f) = \sigma f \sigma^{-1}$, $\sigma \in H$, $f: M_0 \to M_1$.) We denote this group by $\underline{\operatorname{Hom}}_G(M_0, M_1)$. If M_0 is finitely generated, then $\underline{\operatorname{Hom}}_G(M_0, M_1) = \operatorname{Hom}_{\mathbf{Ab}}(M_0, M_1)$, but not usually otherwise. The derived functors of $M \mapsto \underline{\operatorname{Hom}}_G(M_0, M)$ will be written $\underline{\operatorname{Ext}}^i_G(M_0, M)$.

The next lemma allows us to compute right-derived functors using classes other than the class of injective objects.

LEMMA 1.8. *Let $f: \mathbf{A} \to \mathbf{B}$ be a left exact functor of abelian categories, and assume that $\mathbf{A}$ has enough injectives. Let T be a class of objects such that:*

(a) *every object of $\mathbf{A}$ is a subobject of an object of T;*

(b) *if $A \oplus A' \in T$, then $A \in T$;*

(c) *if $0 \to A' \to A \to A'' \to 0$ is exact and A' and A are in T, then $A'' \in T$ and $0 \to fA' \to fA \to fA'' \to 0$ is exact.*

Then all injectives are in T and all elements of T are f-acyclic. Thus the functors R^if can be computed using resolutions by objects in T.

Proof. Let I be an injective object of $\mathbf{A}$, and embed it into an object A of T. Since I is injective, it is a direct summand of A and so, according to (b), is in T.

Let $A \in T$ and form an injective resolution

$$0 \to A \to I^0 \to I^1 \to \cdots \to I^{i-1} \to I^i \to \cdots.$$

Define Z^i inductively so that the sequences

$$0 \to A \to I^0 \to Z^1 \to 0,$$
$$0 \to Z^i \to I^i \to Z^{i+1} \to 0$$

are exact. From (c), Z^1 is in T, and by induction it follows that each Z^i is in T. Also the sequences

$$0 \to fZ^i \to fI^i \to fZ^{i+1} \to 0$$

are all exact, which implies that $R^if(A) = 0$, all $i > 0$.

Example 1.9. (a) The class of all injective objects in a category **A** with enough injectives satisfies the conditions of (1.8) for any f. We have assumed (a), and it is easy to see that a direct summand of an injective is injective. For (c), note that A' being injective implies that the sequence $0 \to A' \to A \to A'' \to 0$ splits, hence that A'', being a direct summand of A, is injective, and finally that any left exact functor preserves the exactness of the sequence because it preserves direct sums.

(b) A sheaf F on a topological space X (in the usual sense) is flabby if the restriction maps $F(X) \to F(U)$ are all surjective. The class of flabby sheaves on X satisfies the conditions of (1.8) with $f = \Gamma(X, -)$ etc. (see Godement [1, II.3]).

(c) A sheaf F on a site $(\mathbf{C}/X)_E$ will be said to be *flabby* if $H^i(U, F) = 0$ for all U in $\mathbf{C}/X$ and all $i > 0$. This agrees with the notion of a flask sheaf in Artin [1, p. 39] (compare (2.12) below), but disagrees with the notion of flasque sheaf in [SGA. 4, V.4.1]. (The definition in [SGA. 4] is that F is flasque if for any sheaf S of sets on X_E, the higher right derived functors of $F' \mapsto \text{Hom}(S, F')$ are zero on F; thus a sheaf is flabby if it is flasque.)

Let T be the class of all flabby sheaves on X_E. Clearly T contains all injective sheaves, which implies that it satisfies (a) of (1.8). Also, since cohomology commutes with finite direct sums (a finite direct sum of injective resolutions is again an injective resolution), it satisfies (b). The first part of (c) is obvious, and the second part also is obvious for $f = \Gamma(X, -)$ and $f = H^0(U, -)$ and hence for $\underline{H}^0(-)$. It also holds, but less obviously, for $f = \pi_*$ (1.14).

PROPOSITION 1.10. *For any sheaf F on X_E and any $U \to X$ in $\mathbf{C}/X$, the groups $H^i(U, F)$ and $H^i(U_E, F|U)$ are canonically isomorphic. (U_E means the site $(\mathbf{C}/U)_E$.)*

Proof. The functor $(F \mapsto F|U)$:$\mathbf{S}(X_E) \to \mathbf{S}(U_E)$ is exact, and so the proposition follows from the next lemma.

LEMMA 1.11. *For any $\pi: U \to X$ in $\mathbf{C}/X$, the functor $\pi^* = (F \mapsto F|U)$ takes injective sheaves to injective sheaves.*

Proof. The functor π^* has an exact left adjoint $\pi_!:\underline{S}(U) \to \underline{S}(X)$ (II.3.18), and hence preserves injectives (1.2).

Remark 1.12. (a) It follows from the proposition that the functor $F \mapsto F|U$ preserves flabby sheaves. An alternative proof of this fact and hence of the proposition will be given in the next section (2.13).

(b) It follows from the proposition that $\underline{H}^i(F)$ is the presheaf $U \mapsto H^i(U_E, F|U)$.

We now consider the problem of computing the higher direct images of a sheaf and their cohomology.

PROPOSITION 1.13. *Let $\pi: X'_{E'} \to X_E$ be continuous, and let $F \in \mathbf{S}(X'_{E'})$. Then $R^i\pi_*F = a\pi_p(\underline{H}^i(F))$, that is, $R^i\pi_*F$ is the sheaf associated with the presheaf $U \mapsto H^i(U', F|U')$, where we have written $U' = U \times_X X'$.*

Proof. By definition, $\pi_* = a\pi_p i$, where i is the inclusion $\mathbf{S}(X') \to \mathbf{P}(X')$. Let $F \to I^{\cdot}$ be an injective resolution of F in $\mathbf{S}(X')$. Then $R^i\pi_*F$ is the ith cohomology group of the complex of sheaves $a\pi_p(iI^{\cdot})$. But a and π_p are exact, (II.2.15) and (II.2.6), and so commute with the formation of cohomology. Thus

$$R^i\pi_*F = a\pi_p(H^i(iI^{\cdot})) = a\pi_p(\underline{H}^i(F)).$$

COROLLARY 1.14. *If F is flabby, then $R^i\pi_*F = 0$ for $i > 0$. Thus flabby resolutions may be used to compute the functors $R^i\pi_*$.*

Proof. By definition, $H^i(U', F|U') = 0$ for any flabby F and any U'. The second statement follows from (1.8), using only the fact that $R^1\pi_*F = 0$ for F flabby.

THEOREM 1.15. *Let $\pi: Y \to X$ be a quasi-compact morphism; let F be a sheaf on Y_{et}. Let $\bar{x}$ be a geometric point of X such that $k(\bar{x})$ is the separable closure of $k(x)$. Let $\tilde{X} = \operatorname{spec} \mathcal{O}_{X,\bar{x}}$, let $\tilde{Y} = Y \times_X \tilde{X}$, and let $\tilde{F}$ be the inverse image of F on $\tilde{Y}$:*

$$\begin{array}{ccc} Y & \longleftarrow & \tilde{Y} \\ \downarrow & & \downarrow \\ X & \longleftarrow & \tilde{X} \end{array}$$

Then $R^p\pi_*(F)_{\bar{x}} \xrightarrow{\approx} H^p(\tilde{Y}, \tilde{F})$.

Proof. According to (II.2.14c) and (1.13),

$$R^p\pi_*(F)_{\bar{x}} = \varinjlim H^p(U \times_X Y, F|U \times Y),$$

where the limit may be taken over affine U such that $\tilde{X} = \varprojlim U$. Note that

$$\tilde{Y} = (\varprojlim U) \times Y = \varprojlim (U \times Y),$$

and that the transition morphisms in this inverse limit are affine morphisms of quasi-compact schemes. The theorem now follows from the fact that étale cohomology commutes with inverse limits of schemes.

LEMMA 1.16. *Let I be a filtered category and $(i \mapsto X_i)$ a contravariant functor from I to schemes over X. Assume that all schemes X_i are quasi-*

compact and that the maps $X_i \leftarrow X_j$ are affine. Let $X_\infty = \varprojlim X_i$, and, for a sheaf F on X_{et}, let F_i and F_∞ be its inverse images on X_i and X_∞ respectively. Then

$$\varinjlim H^p((X_i)_{\text{et}}, F_i) \xrightarrow{\approx} H^p((X_\infty)_{\text{et}}, F_\infty).$$

Proof. The proof is highly technical. It is based on the fact [EGA. IV.17] that the category of étale schemes of finite-type over X_∞ is the direct limit of the categories of such schemes over the X_i and (III.3) that étale cohomology commutes with direct limits of sheaves. (See [SGA. 4, VII.5.8] or Artin [1, III.3] for the details.)

Remark 1.17. (a) If in (1.15), F is defined by a group scheme G that is locally of finite-type over Y, then $\tilde{F}$ is defined by $G_{(\tilde{Y})}$. Thus the isomorphism may be written: $(R^p\pi_*G)_{\bar{x}} \approx H^p(\tilde{Y}, G)$.

A similar remark holds for (1.16).

(b) For a proper morphism, a much sharper result than (1.15) again holds. This is the proper base change theorem: let $\pi: Y \to X$ be proper, and let $\bar{x}$ be a geometric point of X; for any torsion sheaf F on Y, there is an isomorphism $(R^p\pi_*F)_{\bar{x}} \xrightarrow{\approx} H^p(Y_{\bar{x}}, F \mid Y_{\bar{x}})$ (VI.2).

(c) If one is prepared to use Čech cohomology, then (1.16) is an easy exercise (see (3.17) below).

(d) The results of [SGA. 4, VII.5], and in particular Lemma 1.16, are equally valid for the flat site (compare Grothendieck [4, p. 172]).

THEOREM 1.18. (a) (Leray spectral sequence). *For any continuous morphism $\pi:(\mathbf{C}'/X')_{E'} \to (\mathbf{C}/X)_E$, there is a spectral sequence*

$$H^p(X_E, R^q\pi_*F) \Rightarrow H^{p+q}(X'_{E'}, F),$$

where F is a sheaf on $X'_{E'}$.

(b) *For any continuous morphisms $X''_{E''} \xrightarrow{\pi'} X'_{E'} \xrightarrow{\pi} X_E$ there is a spectral sequence*

$$(R^p\pi_*)(R^q\pi'_*)F \Rightarrow R^{p+q}(\pi\pi')_*F,$$

where F is a sheaf on $X''_{E''}$.

Proof. Since flabby sheaves are acyclic for both $\Gamma(X_E, -)$ and π_*, both parts are consequences of (Appendix B1) and the following lemma.

LEMMA 1.19. *Let $\pi: X'_{E'} \to X_E$ be continuous. Then π_* maps flabby sheaves to flabby sheaves.*

Proof. This is proved in the next section (2.13).

Remark 1.20. (a) If in (1.19), π^* is exact (this is usually so; see the first paragraph of (II.3)), then π_* has an exact left adjoint and preserves injectives, and it is not necessary to wait until the next section for the proof of (1.18).

(b) Let G be a profinite group. Recall that for any abelian group N, $M_G(N)$ is the G-module $\{f: G \to N \mid f \text{ continuous}\}$ with G acting by $(\sigma f)(\tau) = f(\tau\sigma)$. Such a module $M_G(N)$ is called an *induced* G-module. If $G = \text{Gal}(\bar{k}/k)$ for some field k and $\bar{k} = k_{\text{sep}}$, then the induced G-modules correspond exactly to the sheaves on X_{et} of the form $u_* F$ where $u: \text{spec } \bar{k} \to \text{spec } k = X$. Call such a sheaf $u_* F$ *induced.*

Now assume (1.19). As any sheaf on $(\text{spec } k)_{\text{et}}$ is flabby, any induced sheaf will be flabby and hence acyclic for $\Gamma(X, -)$, $\Gamma(U, -)$, π_*, and $\underline{H}^0(-)$. Any sheaf on X can be embedded in an induced sheaf because the canonical map $F \to u_* u^* F$ is injective, and it follows that $H^i(X, -)$, $H^i(U, -)$, $\underline{H}^i(-)$ and $R^i\pi_*$ may be computed using resolutions by induced sheaves. The dictionary (II.1.9) allows us to deduce that similar statements holds for induced G-modules. In particular, resolutions by induced modules may be used to compute $H^i(G, -)$. In the case that M_0 is projective as an abelian group this is also true for the functors $\text{Hom}_G(M_0, -)$ and $\underline{\text{Hom}}_G(M_0, -)$. (Exercise!)

(c) The above remark suggests the following definition: a sheaf F on X_{et} is *induced* if it is of the form $\prod_{x \in X} u_{x*}(F_x)$ where F_x is a sheaf on $\bar{x}$ (we have chosen, for each $x \in X$, a geometric point $u_x: \bar{x} \to X$). Write $X' = \coprod_{x \in X} \bar{x}$ (disjoint union of schemes). To give a sheaf on X'_{et} is the same as to give a family of abelian groups $(F_x)_{x \in X}$, where F_x is to be regarded as a constant sheaf on $\bar{x}$. If $u: X' \to X$ is the obvious map, then $u_*((F_x)_{x \in X}) = \prod u_{x*} F_x$, which shows that an induced sheaf is flabby (since every sheaf on X'_{et} is flabby). As $F \to u_* u^* F$ is injective, resolutions by induced sheaves may be used to compute cohomology and higher direct images. An important fact is that every sheaf F on X_{et} has a canonical resolution by induced sheaves called its *Godement resolution*, which is defined as follows:

(i) $C^0(F) = u_* u^*(F)$; there is a canonical map $\varepsilon: F \to C^0(F)$;

(ii) $C^1(F) = C^0(\text{coker}(\varepsilon))$; there is a canonical map $d^0: C^0(F) \to C^1(F)$;

(iii) (Inductively) $C^i(F) = C^0(\text{coker}(d^{i-2}))$; there is a canonical map $d^{i-1}: C^{i-1}(F) \to C^i(F)$.

Then $F \to C^{\cdot}(F)$ is a flabby resolution of F, functorial in F, and each functor $F \mapsto C^n(F)$ is exact.

Exercise 1.21. If X is Jacobson, it is more natural to define F to be induced if $F = \prod_{x \in X^0} u_{x*}(F_x)$ where X^0 is the set of closed points of X (compare (II.2.17b)). For each $x \in X^0$, let $i_x: x \hookrightarrow X$ denote the inclusion map. Show that F is induced in this sense if and only if:

(a) $F \xrightarrow{\approx} \prod_{x \in X^0} i_{x*} i_x^*(F)$;

(b) $i_x^*(F)$ is induced (in the sense of (1.20*b*)) for all $x \in X^0$.

We now investigate the relation between the local Exts, $\underline{\mathrm{Ext}}^i(F_1, F_2)$, and the global Exts, $\mathrm{Ext}^i(F_1, F_2)$.

THEOREM 1.22. (Local-global spectral sequence for Exts). *For any sheaves F_1 and F_2 on X_E, there is a spectral sequence*

$$H^p(X_E, \underline{\mathrm{Ext}}^q(F_1, F_2)) \Rightarrow \mathrm{Ext}^{p+q}(F_1, F_2).$$

Proof. This follows from (Appendix B1) and the next lemma.

LEMMA 1.23. *If F_2 is injective, then $\underline{\mathrm{Hom}}(F_1, F_2)$ is flabby.*

Proof. This will be shown in the next section (2.13).

Remark 1.24. The sheaf $\underline{\mathrm{Ext}}^p(F_1, F_2)$ is in fact the sheaf associated with the presheaf $U \mapsto \mathrm{Ext}^p_{\mathbf{S}(U_E)}(F_1|U, F_2|U)$. To see this, we only have to check the following statements.

(a) the sheaves agree for $p = 0$; but this is the definition of $\underline{\mathrm{Hom}}(F_1, F_2)$;

(b) $\underline{\mathrm{Ext}}^p(F_1, F_2) = 0$ and $\mathrm{Ext}^p_{\mathbf{S}(U)}(F_1|U, F_2|U) = 0$, $p > 0$, if F_2 is injective; this is true by definition in the first case and follows from (1.11) in the second;

(c) both functors associate long exact sequences to short exact sequences in the second variable. This is part of the definition of the first family of functors and follows from the fact that $a:\mathbf{P} \to \mathbf{S}$ is exact for the second family.

As a consequence of this and (1.6a) we find that associated with any exact sequence $0 \to F' \to F \to F'' \to 0$ of sheaves on X_E, there is a long exact sequence of sheaves,

$$0 \to \underline{\mathrm{Hom}}(F'', F_2) \to \cdots \to \underline{\mathrm{Ext}}^p(F, F_2) \to \underline{\mathrm{Ext}}^p(F', F_2)$$
$$\to \underline{\mathrm{Ext}}^{p+1}(F'', F_2) \to \cdots.$$

We return now to the situation studied in (II.3.10), namely, to

$$Z \xrightarrow{i} X \xleftarrow{j} U$$

where i is a closed immersion and j is an open immersion such that X is the disjoint union of $i(Z)$ and $j(U)$. For any sheaf F on X_{et}, $i_*i^!F$ is the largest subsheaf of F that is zero outside Z. The group

$$\Gamma(X, i_*i^!F) = \Gamma(Z, i^!F) = \mathrm{Ker}(F(X) \to F(U))$$

is called the group of sections of F with support on Z. The functor $F \mapsto \Gamma(Z, i^!F)$ is left exact, and its right derived functors, $H^p_Z(X, F)$, are called the *cohomology groups of F with support on Z*. The functors $H^p_Z(X, F)$ are contravariant in (X, U). (The topologists usually write $H^p(X, U, -)$ for $H^p_Z(X, -)$.)

PROPOSITION 1.25. *For any sheaf F on X_{et} there is a long exact sequence,*

$$0 \to (i^!F)(Z) \to F(X) \to F(U) \to \cdots \to H^p(X, F) \to H^p(U, F)$$
$$\to H_Z^{p+1}(X, F) \to \cdots.$$

Proof. For any sheaf F on X_{et}, there is an exact sequence

$$0 \to j_!j^*F \to F \to i_*i^*F \to 0$$

(II.3.13). In particular, if F is the constant sheaf $\mathbb{Z}$, there is an exact sequence

$$0 \to \mathbb{Z}_U \to \mathbb{Z} \to \mathbb{Z}_Z \to 0$$

where we have written $\mathbb{Z}_U$ and $\mathbb{Z}_Z$ for $j_!j^*\mathbb{Z}$ and $i_*i^*\mathbb{Z}$. Correspondingly there is a long exact sequence,

$$\cdots \to \mathrm{Ext}^p(\mathbb{Z}, F) \to \mathrm{Ext}^p(\mathbb{Z}_U, F) \to \mathrm{Ext}^{p+1}(\mathbb{Z}_Z, F) \to \cdots.$$

We have already noted that $\mathrm{Ext}^p(\mathbb{Z}, F) = H^p(X, F)$. The groups

$$\mathrm{Hom}_{\mathbf{S}(X)}(\mathbb{Z}_U, F) \quad \text{and} \quad \mathrm{Hom}_{\mathbf{S}(U)}(\mathbb{Z}, j^*F)$$

are canonically isomorphic (II.3.14), and so $\mathrm{Ext}^p(\mathbb{Z}_U, F)$ is the ith right derived functor of

$$F \mapsto \mathrm{Hom}_{\mathbf{S}(U)}(\mathbb{Z}, j^*F) = \Gamma(U, F|U).$$

Since j^* preserves injectives and is exact, this shows that $\mathrm{Ext}^p(\mathbb{Z}_U, F) = H^p(U, F|U)$. There are canonical isomorphisms

$$\mathrm{Hom}_{\mathbf{S}(X)}(\mathbb{Z}_Z, F) \approx \mathrm{Hom}_{\mathbf{S}(Z)}(\mathbb{Z}, i^!F) \approx H_Z^0(X, F).$$

Thus $\mathrm{Ext}^p(\mathbb{Z}_Z, F) \approx H_Z^p(X, F)$.

Remark 1.26. A slight refinement of the above argument shows that for any triple $X \supset U \supset V$, where U and V are open subschemes of X, and any sheaf F on X_{et}, the following sequence is exact:

$$\cdots \to H_{X-U}^p(X, F) \to H_{X-V}^p(X, F) \to H_{U-V}^p(U, F|U)$$
$$\to H_{X-U}^{p+1}(X, F) \to \cdots.$$

If V is empty, this is the same as the above sequence.

PROPOSITION 1.27. (Excision). *Let $Z \subset X$ and $Z' \subset X'$ be closed subschemes, and let $\pi: X' \to X$ be an étale morphism such that the restriction of π to Z' is an isomorphism $\pi|Z': Z' \xrightarrow{\approx} Z$ and $\pi(X' - Z') \subset X - Z$. Then $H_Z^p(X, F) \to H_{Z'}^p(X', \pi^*F)$ is an isomorphism for all $p \geq 0$ and all sheaves F on X_{et}.*

Proof. Since π^* is exact and preserves injectives, it suffices to prove this for $p = 0$. There is an exact commutative diagram:

$$\begin{array}{ccccccc} 0 \to & H_Z^0(X, F) & \to & \Gamma(X, F) & \to & \Gamma(U, F) & \qquad (U = X - Z) \\ & \downarrow & & \downarrow & & \downarrow & \\ 0 \to & H_{Z'}^0(X', F|X') & \to & \Gamma(X', F) & \to & \Gamma(U', F) & \qquad (U' = X' - Z'). \end{array}$$

Let $\gamma \in H_Z^0(X, F)$ map to zero in $H_{Z'}^0(X', F|X')$. If we regard γ as an element of $\Gamma(X, F)$, then this means that $\gamma|U = 0 = \gamma|X'$. As $(U \to X, X' \to X)$ is a covering of X, this implies that $\gamma = 0$. For the surjectivity, let $\gamma' \in H_{Z'}^0(X', F|X')$ and regard γ' as an element of $\Gamma(X', F)$. As the sections $\gamma' \in \Gamma(X', F)$ and $0 \in \Gamma(U, F)$ agree on $X' \times_X U \subset U'$, they arise from a section $\gamma \in \Gamma(X, F)$, which must in fact be in $H_Z^0(X, F)$.

COROLLARY 1.28. *Let z be a closed point of X. Then $H_z^p(X, F) \xrightarrow{\approx} H_z^p(\operatorname{spec} \mathcal{O}_{X,z}^h, F)$ for any sheaf F on X_{et}.*

Proof. By the proposition, $H_z^p(X, F) = H_y^p(Y, F)$ for any étale neighborhood (Y, y) of z such that only y maps to z. According to (1.16),

$$\varinjlim H_y^p(Y, F) = H_z^p(\operatorname{spec} \mathcal{O}_{X,z}^h, F).$$

Finally in this section we define the *cohomology groups with compact support* $H_c^p(X, F)$ of an F on a separated variety X. The group of sections of F with compact support is defined to be

$$\Gamma_c(X, F) = \bigcup \ker(\Gamma(X, F) \to \Gamma(X - Z, F))$$

where the Z run through the complete subvarieties of X. Since $\Gamma_c(X, -)$ is left exact, it is tempting to define its right derived functors to be the cohomology groups of F with compact support, but these are uninteresting. For example, if X is affine, then $\Gamma_c(X, F) = \bigoplus_{x \in X^0} H_x^0(X, F)$ and so $R^p\Gamma_c(F) = \bigoplus_{x \in X^0} H_x^p(X, F)$. Instead we assume that X can be embedded $j: X \hookrightarrow \bar{X}$ as an open subvariety of a complete variety $\bar{X}$ and define $H_c^p(X, F) = H^p(\bar{X}, j_! F)$. (Recall that $j_!$ is exact but does not preserve injectives.) We shall see later (VI.3, 11) that if F is torsion, $H_c^p(X, F)$ is independent of the choice of $\bar{X}$ and satisfies a Poincaré duality theorem.

PROPOSITION 1.29. *The assumptions on X, $\bar{X}$, and F are as above.*

(a) $H_c^0(X, F) = \Gamma_c(X, F)$.

(b) *The functors $H_c^p(X, -)$ form a δ-functor, that is, they functorially associate long exact sequences of abelian groups to short exact sequences of sheaves.*

(c) *For any complete subvariety Z of X, there is a canonical morphism of δ-functors $H_Z^p(X, -) \to H_c^p(X, -)$.*

Proof. (a) From the exact sequence of sheaves on $\bar{X}$

$$0 \to j_! F \to j_* F \to i_* i^* j_* F \to 0,$$

where $i: \bar{X} - X \hookrightarrow \bar{X}$, we see that

$$H_c^0(X, F) = \ker(\Gamma(X, F) \to \Gamma(\bar{X} - X, i^* j_* F)).$$

But

$$\Gamma(\bar{X} - X, i^* j_* F) = \varinjlim \Gamma(V \times_{\bar{X}} X, F)$$

where $V \to \bar{X}$ is étale and contains $\bar{X} - X$ in its image. Thus

$$H_c^0(X, F) = \bigcup \ker(\Gamma(X, F) \to \Gamma(V \times_{\bar{X}} X, F)).$$

Suppose that $s \in \Gamma_c(X, F)$, so that $s | X - Z = 0$ for some complete $Z \subset X$; then Z, being complete, is closed in $\bar{X}$ and $s | V \cap X = 0$ if $V = \bar{X} - Z$; thus $s \in H_c^0(X, F)$. Conversely, suppose that $s \in H_c^0(X, F)$, so that $s | V \times X = 0$ for some V; then the image V' of V in $\bar{X}$ is open and contains $\bar{X} - X$; thus $Z = \bar{X} - V'$ is a complete subvariety and $X - Z = V' \cap X$; as $V \times X \to V' \cap X$ is an étale covering, $s | V \times X = 0$ implies that $s | X - Z = 0$ and $s \in \Gamma_c(X, F)$.

(b) As $j_!$ is exact, an exact sequence $0 \to F' \to F \to F'' \to 0$ on X gives rise to an exact sequence $0 \to j_! F' \to j_! F \to j_! F'' \to 0$ on $\bar{X}$ and hence to an exact cohomology sequence on $\bar{X}$.

(c) There is a canonical map $H_Z^0(X, F) \to H_c^0(X, F)$ for any sheaf F, and so (c) follows from the universality of derived functors.

Remark 1.30. Let Z be a closed subscheme of X. For any sheaf F on X, there is an exact sequence

$$0 \to j'_!(F | X - Z) \to F \to i_*(F | Z) \to 0$$

where j' and i are the immersions $X - Z \hookrightarrow X$, $Z \hookrightarrow X$ (II.3.13). Thus there is a long sequence of cohomology groups,

$$\cdots \to H_c^p(X - Z, F) \to H_c^p(X, F) \to H_c^p(Z, F) \to \cdots.$$

More generally, if $X = X_0 \supset X_1 \supset \cdots \supset X_r \neq \varnothing$ is a sequence of closed subschemes of X, then there is a spectral sequence,

$$E_1^{q,p} = H_c^{p+q}(X_p - X_{p+1}, F) \Rightarrow H_c^{p+q}(X, F).$$

Exercise 1.31. Write A for a (Noetherian) ring and the sheaf it defines on X_{et}.

(a) Show that if F is an injective sheaf of A-modules on X_{et}, then $F_{\bar{x}}$ is an injective A-module. (Use Godement [1, I.1.4.1] and (1.11).)

(b) Show that if F_0 is pseudocoherent at $\bar{x}$, then

$$\underline{\mathrm{Ext}}^p_A(F_0, F)_{\bar{x}} = \mathrm{Ext}^p_A(F_{0\bar{x}}, F_{\bar{x}}).$$

(Take an injective resolution $F \to I^{\cdot}$ of F and apply (II.3.20b) and (a).)

(c) Deduce that $\underline{\mathrm{Ext}}^p_A(F, A) = 0$, $p > 0$, if F is a locally free sheaf of A-modules of finite rank (that is, $F|U_i \approx A^{r_i}$ for each U_i in some covering (U_i) of X) or if F is pseudocoherent at all geometric points $\bar{x}$ of X and A is an injective A-module, for example, $A = \mathbb{Z}/(n)$.

§2. Čech Cohomology

It is possible to define a Čech cohomology theory that is closely analogous to the usual theory over a topological space.

Let $\mathscr{U} = (U_i \xrightarrow{\phi_i} X)_{i \in I}$ be a covering of X in the E-topology on X. For any $(p+1)$-tuple $(i_0, \ldots, i_p)$ with the i_j in I we write $U_{i_0} \times_X \cdots \times_X U_{i_p} = U_{i_0 \cdots i_p}$. Let P be a presheaf on X_E. The canonical projection

$$U_{i_0 \cdots i_p} \to U_{i_0 \cdots \hat{i}_j \cdots i_p} = U_{i_0} \times \cdots \times U_{i_{j-1}} \times U_{i_{j+1}} \times \cdots \times U_{i_p}$$

induces a restriction morphism

$$P(U_{i_0 \cdots \hat{i}_j \cdots i_p}) \to P(U_{i_0 \cdots i_p})$$

which we write ambiguously as res_j. Define a complex

$$C^{\cdot}(\mathscr{U}, P) = (C^p(\mathscr{U}, P), d^p)_p$$

as follows:

$$C^p(\mathscr{U}, P) = \prod_{I^{p+1}} P(U_{i_0 \cdots i_p}); \qquad d^p: C^p(\mathscr{U}, P) \to C^{p+1}(\mathscr{U}, P)$$

is the homomorphism such that if $s = (s_{i_0 \cdots i_p}) \in C^p(\mathscr{U}, P)$, then

$$(d^p s)_{i_0 \cdots i_{p+1}} = \sum_{j=0}^{p+1} (-1)^j \,\mathrm{res}_j\,(s_{i_0 \cdots \hat{i}_j \cdots i_{p+1}}).$$

The usual argument shows that $d^{p+1}d^p = 0$, that is, that this is a complex. The cohomology groups of $(C^p(\mathscr{U}, P), d^p)$ are called the *Čech cohomology groups*, $\check{H}^p(\mathscr{U}, P)$, of P with respect to the covering $\mathscr{U}$ of X.

Note that

$$\check{H}^0(\mathscr{U}, P) = \ker\left(\prod P(U_i) \to \prod P(U_{ij})\right),$$

and so there is a canonical map $P(X) \to \check{H}^0(\mathscr{U}, P)$ which is an isomorphism whenever P is a sheaf.

A second covering $\mathscr{V} = (V_j \xrightarrow{\psi_j} X)_{j \in J}$ is called a *refinement* of $\mathscr{U}$ if there is a map $\tau: J \to I$ such that for each j, ψ_j factors through $\phi_{\tau j}$, that is,

$\psi_j = \phi_{\tau j}\eta_j$ for some $\eta_j: V_j \to U_{\tau j}$. The map τ, together with the family (η_j), induces maps $\tau^p: C^p(\mathcal{U}, P) \to C^p(\mathcal{V}, P)$ as follows: if $s = (s_{i_0 \cdots i_p}) \in C^p(\mathcal{U}, P)$, then

$$(\tau^p s)_{j_0 \cdots j_p} = \mathrm{res}_{\eta_{j_0} \times \eta_{j_1} \times \cdots \times \eta_{j_p}}(s_{\tau j_0 \cdots \tau j_p}).$$

These maps τ^p commute with d and hence induce maps on the cohomology,

$$\rho(\mathcal{V}, \mathcal{U}, \tau): \check{H}^p(\mathcal{U}, P) \to \check{H}^p(\mathcal{V}, P).$$

LEMMA 2.1. *The map $\rho(\mathcal{V}, \mathcal{U}, \tau)$ does not depend on τ or the η_j.*

Proof. Suppose that τ', with family (η'_j), is another such map $J \to I$. For $s \in C^p(\mathcal{U}, P)$ define

$$(k^p s)_{j_0 \cdots j_{p-1}} = \sum (-1)^r \mathrm{res}_{\eta_{j_0} \times \cdots \times (\eta_{j_r}, \eta'_{j_r}) \times \cdots \times \eta'_{j_{p-1}}}(s_{\tau j_0 \cdots \tau j_r \tau' j_r \cdots \tau' j_{p-1}}).$$

Then k^p is a homomorphism $C^p(\mathcal{U}, P) \to C^{p-1}(\mathcal{V}, P)$ with the property that

$$d^{p-1}k^p + k^{p+1}d^p = \tau'^p - \tau^p.$$

On passing to the cohomology, we find this equation becomes

$$\rho(\mathcal{V}, \mathcal{U}, \tau') - \rho(\mathcal{V}, \mathcal{U}, \tau) = 0.$$

Hence, if $\mathcal{V}$ is a refinement of $\mathcal{U}$, we get a homomorphism $\rho(\mathcal{V}, \mathcal{U})$: $\check{H}^p(\mathcal{U}, P) \to \check{H}^p(\mathcal{V}, P)$ depending only on $\mathcal{V}$ and $\mathcal{U}$. It follows that if $\mathcal{U}, \mathcal{V}, \mathcal{W}$ are three coverings of X such that $\mathcal{W}$ is a refinement of $\mathcal{V}$ and $\mathcal{V}$ is a refinement of $\mathcal{U}$, then $\rho(\mathcal{W}, \mathcal{U}) = \rho(\mathcal{W}, \mathcal{V})\rho(\mathcal{V}, \mathcal{U})$. Thus we may define the *Čech cohomology groups* of P over X to be $\check{H}^p(X_E, P) = \varinjlim \check{H}^p(\mathcal{U}, P)$, where the limit is taken over all coverings $\mathcal{U}$ of X.

Remark 2.2. (a) The category over which the above limit is taken is not cofiltered. However, let J_X be the set of coverings of X modulo the equivalence relation: $\mathcal{U} \equiv \mathcal{V}$ if each is a refinement of the other. The notion of refinement induces a partial ordering on J_X, which is even filtered because any two coverings $\mathcal{U} = (U_i)$ and $\mathcal{V} = (V_j)$ have a common refinement $(U_i \times V_j)$. According to (2.1), the functor $\mathcal{U} \mapsto \check{H}^p(\mathcal{U}, P)$ factors through J_X, and limit may be taken over J_X.

(b) If $U \to X$ is in $\mathbf{C}/X$ and P is a presheaf on $(\mathbf{C}/X)_E$, then, as above, we may define cohomology groups $\check{H}^p(\mathcal{U}/U, P)$ and

$$\check{H}^p(U, P) = \varinjlim \check{H}^p(\mathcal{U}/U, P)$$

where $\mathcal{U}$ now denotes a covering of U. Note that $\check{H}^p(U, P)$ is defined intrinsically in terms of P. Thus, for example, $\check{H}^p(U, P)$ is obviously the same as $\check{H}^p(U, P|U)$. The mapping $U \mapsto \check{H}^p(U, P)$ extends to give a functor $\mathbf{C}/X \to \mathbf{Ab}$, that is, a presheaf on $(\mathbf{C}/X)_E$, which we denote by $\underline{\check{H}}^p(X_E, P)$ or simply $\underline{\check{H}}^p(P)$.

(c) There is a canonical map $P \to \check{\underline{H}}^0(P)$ whose kernel is the presheaf P_0 defined in the proof of (II.2.11), that is, $P_0(U)$ is the group of $s \in P(U)$ that restrict to zero on some covering of U. Thus $P \to \check{\underline{H}}^0(P)$ is injective if and only if P is a separated presheaf. $\check{\underline{H}}^0(P)$ is separated, and $\check{\underline{H}}^0(P) = aP$ for any separated P. In general $\check{\underline{H}}^0(\check{\underline{H}}^0(P)) = aP$ (see the above-mentioned proof).

(d) It is not true that Čech cohomology can be computed using alternating cochains unless the maps $U_i \to X$ are universal monomorphisms, for example, open immersions. In particular, the cohomology groups arising from a covering $(V \to U)$ consisting of a single morphism are not necessarily zero for $p > 0$ (see (e) below, or (2.6)). The proof in Serre [1] fails in our situation because the restriction maps $\rho_s^{s'}$, s' a subsimplex of s, are not well-defined.

(e) In the case that $X = \operatorname{spec} A$ is affine, $\mathcal{U}$ consists of one faithfully flat morphism $\operatorname{spec} B \to \operatorname{spec} A$ of finite-type, and P is the sheaf defined by $\mathbb{G}_m$, the groups $\check{H}^p(U, P)$ admit a very explicit definition: they are the cohomology groups of the complex

$$0 \to B^* \to (B \otimes_A B)^* \to (B \otimes_A B \otimes_A B)^* \to \cdots$$

In this form these groups were first defined by Amitsur [1] in the case that both A and B are fields. In the general case when A and B are rings and P is defined by any commutative group scheme, the groups $H^p(\mathcal{U}, P)$ have been extensively studied under the name of Amitsur cohomology groups. This gives a very elementary approach to a fragment of flat cohomology but unfortunately without the full techniques many results, which should be easy exercises, become very cumbersome to prove.

If $0 \to P' \to P \to P'' \to 0$ is an exact sequence of presheaves and $\mathcal{U}$ is any covering, then

$$0 \to C^p(\mathcal{U}, P') \to C^p(\mathcal{U}, P) \to C^p(\mathcal{U}, P'') \to 0$$

is exact for all p because it is a product of exact sequences of abelian groups. Thus, we get a short exact sequence of complexes

$$0 \to C^{\cdot}(\mathcal{U}, P') \to C^{\cdot}(\mathcal{U}, P) \to C^{\cdot}(\mathcal{U}, P'') \to 0,$$

and the general theory shows that there is a naturally associated long exact sequence

$$0 \to \check{H}^0(\mathcal{U}, P') \to \cdots \to \check{H}^p(\mathcal{U}, P) \to \check{H}^p(\mathcal{U}, P'') \to \check{H}^{p+1}(\mathcal{U}, P') \to \cdots.$$

The exactness is preserved if we pass to the direct limit over all coverings of X (better, over J_X; see (2.2a) above). Thus to any exact sequence $0 \to P' \to P \to P'' \to 0$ of presheaves there is a naturally corresponding long exact sequence of Čech cohomology groups,

$$0 \to \check{H}^0(U, P') \to \cdots \to \check{H}^p(U, P) \to \check{H}^p(U, P'') \to \check{H}^{p+1}(U, P') \to \cdots.$$

Unfortunately, this is no longer true if we start with a short exact sequence of sheaves that is exact only in $\mathbf{S}(X_E)$, that is, not in $\mathbf{P}(X_E)$, because the maps $C^p(\mathscr{U}, P) \to C^p(\mathscr{U}, P'')$ need not be surjective. Thus, in general, there will be no long exact sequence of Čech cohomology groups associated with a short exact sequence of sheaves, and $\check{H}^p(X_E, F)$ cannot be isomorphic to $H^p(X_E, F)$. However, these groups are related by a spectral sequence and in some cases are known to be isomorphic (see (2.7), (2.16), (2.17) below).

Notwithstanding the above remark, the $\check{H}^p(X_E, -)$ are right derived functors, but they are derived functors from the category of presheaves, not sheaves.

PROPOSITION 2.3. *The functors $\check{H}^p(\mathscr{U}/U, -)$ (respectively $\check{H}^p(U, -)$) are the right derived functors of $\check{H}^0(\mathscr{U}/U, -): \mathbf{P}(X_E) \to \mathbf{Ab}$ (respectively $H^0(U, -): \mathbf{P}(X_E) \to \mathbf{Ab}$) for any U in $\mathbf{C}/X$ and any covering $\mathscr{U}$ of U.*

Proof. Since we know that $\check{H}^*(\mathscr{U}/U, -)$ and $\check{H}^*(U, -)$ associate long exact sequences to short exact sequences, we only have to show that $\mathbf{P}(X_E)$ has enough injectives and that $\check{H}^p(\mathscr{U}/U, -)$ is zero on injectives. The proof that $\mathbf{P}(X_E)$ has enough injectives is identical to the proof of the same fact for $\mathbf{S}(X_E)$ using (1.3). Thus the proposition is a consequence of the next lemma.

LEMMA 2.4. *$\check{H}^p(\mathscr{U}/U, P) = 0$, $p > 0$, if P is injective.*

Proof. We have to show that the cochain complex

$$\prod P(U_i) \xrightarrow{d^1} \prod P(U_{i_0 i_1}) \xrightarrow{d^2} \prod P(U_{i_0 i_1 i_2}) \to \cdots$$

is exact. Recall (II.3.18) that for any $W \to U$ in $\mathbf{C}/X$ there is a presheaf $\mathbb{Z}_W$ on X with the properties that $\operatorname{Hom}(\mathbb{Z}_W, P) = P(W)$ for any $P \in \mathbf{P}(X)$ and

$$\mathbb{Z}_W(V) = \bigoplus_{\operatorname{Hom}_X(V, W)} \mathbb{Z}.$$

The above complex may be written

$$\prod \operatorname{Hom}(\mathbb{Z}_{U_i}, P) \to \prod \operatorname{Hom}(\mathbb{Z}_{U_{i_0 i_1}}, P) \to \cdots$$

or,

$$\operatorname{Hom}(\bigoplus \mathbb{Z}_{U_i}, P) \to \operatorname{Hom}(\bigoplus \mathbb{Z}_{U_{i_0 i_1}}, P) \to \cdots$$

Since P is injective, it suffices to show that

$$\bigoplus \mathbb{Z}_{U_i} \leftarrow \bigoplus \mathbb{Z}_{U_{i_0 i_1}} \leftarrow \bigoplus \mathbb{Z}_{U_{i_0 i_1 i_2}} \leftarrow \cdots$$

is exact in $\mathbf{P}(X)$, that is, that for all V in $\mathbf{C}/X$,

$$\bigoplus \mathbb{Z}_{U_i}(V) \leftarrow \bigoplus \mathbb{Z}_{U_{i_0 i_1}}(V) \leftarrow \bigoplus \mathbb{Z}_{U_{i_0 i_1 i_2}}(V) \leftarrow \cdots$$

is exact. For any U-scheme W and $\phi \in \mathrm{Hom}_X(V, U)$, write $\mathrm{Hom}_\phi(V, W)$ for the set of morphisms $\psi: V \to W$ such that

$$\begin{array}{ccc} V & \xrightarrow{\psi} & W \\ & {\scriptstyle\phi}\searrow \quad \swarrow & \\ & U & \end{array}$$

commutes. Then

$$\begin{aligned} \mathrm{Hom}_X(V, U_{i_0 i_1 \cdots}) &= \bigcup_{\mathrm{Hom}_X(V,U)} \mathrm{Hom}_\phi(V, U_{i_0 i_1 \cdots}) \qquad \text{(disjoint union)} \\ &= \bigcup_{\mathrm{Hom}_X(V,U)} (\mathrm{Hom}_\phi(V, U_{i_0}) \times \mathrm{Hom}_\phi(V, U_{i_1}) \times \cdots) \end{aligned}$$

Write $S(\phi) = \bigcup_i \mathrm{Hom}_\phi(V, U_i)$ (disjoint union). Then

$$\bigcup_{i_0, \ldots, i_p} \mathrm{Hom}_X(V, U_{i_0 i_1 \cdots i_p}) = \bigcup_{\mathrm{Hom}_X(V,U)} (S(\phi) \times \cdots \times S(\phi))$$

($p + 1$ copies of $S(\phi)$) and $\bigoplus \mathbb{Z}_{U_{i_0 \cdots i_p}}(V)$ is the free abelian group on $\bigcup_{\mathrm{Hom}_X(V,U)} (S(\phi) \times \cdots \times S(\phi))$. Hence the complex may be written,

$$\bigoplus_{\mathrm{Hom}_X(V,U)} \left(\bigoplus_{S(\phi)} \mathbb{Z} \leftarrow \bigoplus_{S(\phi) \times S(\phi)} \mathbb{Z} \leftarrow \bigoplus_{S(\phi) \times S(\phi) \times S(\phi)} \mathbb{Z} \leftarrow \cdots \right)$$

The complex inside the brackets is the standard complex associated with $\bigoplus_{S(\phi)} \mathbb{Z}$, which is exact; a contracting homotopy for it is given by $k^p(m)_{i_0 \cdots i_{p-1}} = m_{1 i_0 \cdots i_{p-1}}$, where 1 is some fixed element of $S(\phi)$ and

$$m = (m_{i_0 \cdots i_p}) \in \bigoplus_{S(\phi)^{p+1}} \mathbb{Z}.$$

COROLLARY 2.5. *Čech cohomology $\check{H}^p(X, -)$ agrees with derived functor cohomology $H^p(X, -)$ on sheaves if and only if to every short exact sequence of sheaves there is a functorially associated long exact sequence of Čech cohomology groups. This will be true, for example, if for every surjection $F \to F''$ of sheaves, the map $\varinjlim(\prod F(U_{i_0 \ldots i_p}) \to \prod F''(U_{i_0 \ldots i_p}))$ is surjective, where the limit is over all coverings of X.*

Proof. The necessity is obvious. Since $H^0(X, F) = \check{H}^0(X, F)$, the sufficiency follows from (2.4) and the fact that an injective sheaf is injective as a presheaf (because a is an exact left adjoint for $i: \mathbf{S} \hookrightarrow \mathbf{P}$). The last statement is clear.

Example 2.6. Let $Y \to X$ be a finite Galois covering of X with Galois group G. Recall (I.5.4) that this means that for every X-scheme T there is an action $Y(T) \times G_X(T) \to Y(T)$ of $G_X(T)$ on $Y(T)$ and that the map $Y \times G_X \to Y \times Y$, which on points is $(y, \sigma) \to (y, y\sigma)$, is an isomorphism. I claim that in fact there exists a commutative diagram:

$$\begin{array}{ccccccccc}
Y & \leftleftarrows & Y \times G & \overset{\leftarrow}{\leftleftarrows} & Y \times G^2 & \cdots & Y \times G^n & \overset{d_n^0}{\underset{d_n^1}{\longleftarrow}} \vdots & Y \times G^{n+1} \cdots \\
\| & & \downarrow{\scriptstyle \approx \, \phi_1} & & \downarrow{\scriptstyle \approx \, \phi_2} & & \downarrow{\scriptstyle \approx \, \phi_n} & & \downarrow{\scriptstyle \approx \, \phi_{n+1}} \\
Y & \leftleftarrows & Y \times Y & \overset{\leftarrow}{\leftleftarrows} & Y \times Y \times Y & \cdots & Y^{n+1} & \overset{p_0}{\underset{p_1}{\longleftarrow}} \vdots & Y^{n+2} \cdots
\end{array}$$

in which (on points):

$$\begin{aligned}
\phi_n(y, \sigma_1, \ldots, \sigma_n) &= (y, y\sigma_1, y\sigma_1\sigma_2, \ldots, y\sigma_1 \cdots \sigma_n) \\
p_i(y_0, \ldots, y_{n+1}) &= (y_0, \ldots, y_{i-1}, y_{i+1}, \ldots, y_{n+1}) \\
d_n^0(y, \sigma_1, \ldots, \sigma_{n+1}) &= (y\sigma_1, \sigma_2, \ldots, \sigma_{n+1}) \\
d_n^i(y, \sigma_1, \ldots, \sigma_{n+1}) &= (y, \sigma_1, \ldots, \sigma_i\sigma_{i+1}, \ldots, \sigma_{n+1}) \qquad 0 < i \leq n \\
d_n^{n+1}(y, \sigma_1, \ldots, \sigma_{n+1}) &= (y, \sigma_1, \ldots, \sigma_n).
\end{aligned}$$

Indeed, the commutativity is trivial to check; ϕ_1 is an isomorphism by definition of Galois, and

$$\phi_n = ((Y \times G^{n-1}) \times G \xrightarrow{\phi_{n-1} \times 1} Y^n \times G \xrightarrow{1 \times \phi_1} Y^{n+1})$$

is an isomorphism by induction.

Now let P be a presheaf that takes finite disjoint unions of schemes to direct products, for example, a sheaf. As $Y \times G^n = \coprod_{\sigma \in G^n} Y$, we see that

$$P(Y \times G^n) = \prod_{\sigma \in G^n} P(Y) = \operatorname{Hom}(G^n, P(Y)).$$

If P is applied to the above diagram and we take alternating sums of the maps, then the top row becomes isomorphic to the complex of inhomogeneous cochains of G with values in $P(Y)$ (see Serre [7, VII. §3]) and the bottom row becomes isomorphic to the Čech complex $C^{\cdot}(Y/X, P)$ of the covering $(Y \to X)$. Thus $\check{H}^p(Y/X, P)$ is canonically isomorphic to the cohomology group $H^p(G, P(Y))$, where $P(Y)$ is a left G-module via the action of G on Y.

PROPOSITION 2.7. *Let $U \to X$ be in $\mathbf{C}/X$; let $\mathscr{U}$ be a covering of U, and let F be a sheaf on X_E. There are spectral sequences*

$$\check{H}^p(\mathscr{U}/U, \underline{H}^q(F)) \Rightarrow H^n(U, F),$$
$$\check{H}^p(U, \underline{H}^q(F)) \Rightarrow H^n(U, F).$$

Proof. In order to apply (Appendix B1), we must show that

$$\check{H}^0(\mathscr{U}/U, \underline{H}^0(F)) = H^0(U, F) = \check{H}^0(U, \underline{H}^0(F)),$$

for any F, and that

$$\check{H}^p(\mathscr{U}/U, \underline{H}^0(F)) = 0 = \check{H}^p(U, \underline{H}^0(F)), p > 0,$$

if F is injective. The first statement is obvious from the various definitions, and the second follows from (2.4) since $\underline{H}^0$, which is simply the inclusion functor $\mathbf{S}(X_E) \hookrightarrow \mathbf{P}(X_E)$, preserves injectives.

COROLLARY 2.8. *There is a spectral sequence*

$$\underline{H}^p(X, \underline{H}^q(F)) \Rightarrow \underline{H}^n(X, F).$$

Proof. Let U vary in the second of the above spectral sequences.

The next proposition shows that this spectral sequence has at least one column that consists, except for one term, entirely of zeros.

PROPOSITION 2.9. $\check{\underline{H}}^0(X, \underline{H}^q(F)) = 0$ *for* $q > 0$; *that is,* $\check{H}^0(U, \underline{H}^q(F)) = 0$ *for* $q > 0$ *and all* $U \to X$ *in* $\mathbf{C}/X$.

Proof. Let $F \to I^{\cdot}$ be an injective resolution of F. Then $\underline{H}^q(F)$ is the qth cohomology presheaf of the complex $i(I^{\cdot})$. Since a is exact, it commutes with taking cohomology, and so $a(\underline{H}^q(F))$ is the qth cohomology sheaf of the complex $ai(I^{\cdot}) = I^{\cdot}$. Thus $a(\underline{H}^q(F)) = 0$. But we know according to (2.2c) that $\check{\underline{H}}^0(\underline{H}^q(F))$ is a subpresheaf of $a\underline{H}^q(F)$, which shows that it is zero.

COROLLARY 2.10. *For any sheaf* F *and any* $U \to X$ *in* $\mathbf{C}/X$, *there are isomorphisms*

$$\check{H}^0(U, F) \approx H^0(U, F)$$
$$\check{H}^1(U, F) \approx H^1(U, F)$$

and an exact sequence

$$0 \to \check{H}^2(U, F) \to H^2(U, F) \to \check{H}^1(U, \underline{H}^1(F)) \to \check{H}^3(U, F) \to H^3(U, F).$$

Proof. This is immediate from (2.7) and (2.9) (Appendix B).

Remark 2.11. (a) Proposition (2.9) can be given a more intuitive statement as follows: for any cohomology class $s \in H^q(U, F)$, $q > 0$, there is a covering $(U_i \to U)_i$ of U such that s restricts to zero in $H^q(U_i, F)$, for all i.

(b) The isomorphism between the first cohomology groups in (2.10) is important because it allows us to interpret the elements of $H^1(U, F)$ as principal homogeneous spaces (or torsors); see §4 below.

We now return to the theory of flabby sheaves. The careful reader will check that we have not in Section 2 used any of the results of Section 1 after (1.19).

PROPOSITION 2.12. *Let F be a sheaf on X_E. The following are equivalent:*

(a) *F is flabby;*

(b) $\check{H}^q(\mathcal{U}/U, F) = 0, q > 0$, *for any $U \to X$ in $\mathbf{C}/X$ and any covering $\mathcal{U}$ of U (in some cofinal system of coverings);*

(c) $\check{H}^q(U, F) = 0, q > 0$, *for all $U \to X$ in $\mathbf{C}/X$.*

Proof. (a) ⇒ (b). As F is flabby, $\underline{H}^q(F) = 0$, $q > 0$. Thus (2.7) gives isomorphisms $\check{H}^p(\mathcal{U}/U, F) \xrightarrow{\approx} H^p(U, F) = 0$.

(b) ⇒ (c). Pass to the direct limit in (b).

(c) ⇒ (a). The hypothesis says that $\underline{\check{H}}^q(F) = 0$, $q > 0$. It follows from (2.10) that $\underline{H}^1(F) = 0$. We proceed by induction on q, using the spectral sequence $\underline{\check{H}}^p(\underline{H}^q(F)) \Rightarrow \underline{H}^n(F)$ (2.8). We know that $\underline{\check{H}}^2(\underline{H}^0(F)) = \underline{\check{H}}^2(F) = 0$ by hypothesis, $\underline{\check{H}}^1(\underline{H}^1(F)) = 0$ as $\underline{H}^1(F) = 0$, and $\underline{\check{H}}^0(\underline{H}^2(F)) = 0$ according to (2.9). It follows from the spectral sequence that $\underline{H}^2(F) = 0$. Assume now that $\underline{H}^i(F) = 0$ for $i < q$. Then the same arguments show that $\underline{\check{H}}^i(\underline{H}^j(F)) = 0$ for all i, j, $i + j \leq q$, and this again implies that $\underline{H}^q(F) = 0$.

COROLLARY 2.13. (a) *If F is flabby, then $F|U$ is flabby for any $U \to X$ in $\mathbf{C}/X$.*

(b) *If $\pi: X'_{E'} \to X_E$ is continuous and F' is flabby, then $\pi_* F'$ is flabby.*

(c) *If F is injective, then $\underline{\mathrm{Hom}}\,(F_0, F)$ is flabby for any sheaf F_0.*

Proof. (a) The sheaf $F|U$ obviously satisfies condition (c) of (2.12).

(b) Let $(U_i \to U)$ be a covering of U and write $V \times_X X' = V'$ for any X-scheme V. Then $(U'_i \to U')$ is a covering of U' and $(U'_{i_0 \cdots i_p}) = (U_{i_0 \cdots i_p})'$. It follows from the definition of $\pi_* F'$ that the complexes $C^{\cdot}(\mathcal{U}', F')$ and $C^{\cdot}(\mathcal{U}, \pi_* F')$ are isomorphic. Since the former is exact, so is the latter.

(c) First note that if F is flat (as a sheaf of $\mathbb{Z}$-modules) and I is an injective sheaf, then $\underline{\mathrm{Hom}}\,(F, I)$ is injective. Indeed, the functors

$$F_0 \mapsto \mathrm{Hom}\,(F \otimes_Z F_0, I))$$
$$F_0 \mapsto \mathrm{Hom}\,(F_0, \underline{\mathrm{Hom}}\,(F, I))$$

are isomorphic (II.3.19). As F is flat, the first functor is exact, which implies that the second functor is exact and that $\underline{\mathrm{Hom}}\,(F, I)$ is injective.

Secondly, note that for all $U \to X$ in $\mathbf{C}/X$, the sheaf $a\mathbb{Z}_U$ is flat ($\mathbb{Z}_U$ is as in the proof of (2.4)). To prove this it suffices to show that $F \mapsto F \otimes a\mathbb{Z}_U$ is a left exact functor $\mathbf{S}(X_E) \to \mathbf{S}(X_E)$. Let $0 \to F' \to F \to F''$ be an exact sequence of sheaves. This sequence is also exact as a sequence of presheaves, and, for each $V \to X$ in $\mathbf{C}/X$,

$$0 \to F'(V) \otimes \mathbb{Z}_U(V) \to F(V) \otimes \mathbb{Z}_U(V) \to F''(V) \otimes \mathbb{Z}_U(V)$$

is exact (because $\mathbb{Z}_U(V)$ is a free $\mathbb{Z}$-module). By letting V vary in this last sequence, we obtain an exact sequence of presheaves, and by applying a to this sequence of presheaves, we obtain an exact sequence of sheaves

$$0 \to F' \otimes a\mathbb{Z}_U \to F \otimes a\mathbb{Z}_U \to F'' \otimes a\mathbb{Z}_U$$

(II.3.21).

Now let F be an injective sheaf on X_E, and let $\mathscr{U} = (U_i \to U)_{i \in I}$ be a covering of $U \in Ob(\mathbf{C}/X)$. Write $\mathbb{Z}.$ for the exact sequence of sheaves

$$\bigoplus_{i \in I} a\mathbb{Z}_{U_i} \leftarrow \bigoplus_{I \times I} a\mathbb{Z}_{U_{i_0 i_1}} \leftarrow \bigoplus_{I \times I \times I} a\mathbb{Z}_{U_{i_0 i_1 i_2}} \leftarrow \cdots$$

(compare the proof of (2.4)). Then, for any sheaf F_0 on X_E, the complex $C^{\cdot}(\mathscr{U}, \underline{\mathrm{Hom}}(F_0, F)) \approx \underline{\mathrm{Hom}}(\mathbb{Z}., \mathrm{Hom}(F_0, F)) \approx \mathrm{Hom}(\mathbb{Z}. \otimes F_0, F) \approx \mathrm{Hom}(F_0, \underline{\mathrm{Hom}}(\mathbb{Z}., F))$. This last complex is exact because $\underline{\mathrm{Hom}}(\mathbb{Z}., F)$ is a complex of sheaves that is exact (because F is injective) and in which every term is an injective sheaf (see above). Thus $H^p(\mathscr{U}/U, \underline{\mathrm{Hom}}(F_0, F)) = 0$ for $p > 0$, and $\underline{\mathrm{Hom}}(F_0, F)$ is flabby.

Next we consider more carefully the relation between the Čech cohomology groups and the derived functor cohomology groups. In the case of the Zariski topology there is the following result.

Proposition 2.14. *Let F be a quasi-coherent sheaf of $\mathcal{O}_X$-modules (for the Zariski topology on X) and assume that X is separated. Then there are canonical isomorphisms $\check{H}^p(X_{\mathrm{Zar}}, F) \xrightarrow{\approx} H^p(X_{\mathrm{Zar}}, F)$ for all p.*

Proof. The main step is the following lemma.

Lemma 2.15. *If U is affine and F is a quasi-coherent sheaf of $\mathcal{O}_U$-modules, then $H^p(U_{\mathrm{Zar}}, F) = 0$ for all $p > 0$.*

Proof. This is standard. One first shows that $\check{H}^p(\mathscr{U}/U, F) = 0$ for any covering of U by basic open affines, which implies that $\check{H}^p(U, F) = 0$, and then one uses Cartan's criterion to show that $H^p(U, F) = \check{H}^p(U, F)$ [EGA. III.1.3.1]. Cartan's criterion, which is a slight refinement of (c) $\Rightarrow$ (a) of (2.12), may be found in Godement [1, II.5.9.2].

Proof of (2.14). Let $\mathscr{U} = (U_i)$ be an open affine covering of X and consider the spectral sequence

$$\check{H}^p(\mathscr{U}/X, \underline{H}^q(F)) \Rightarrow H^{p+q}(X, F).$$

As $U_{i_0 \cdots i_p}$ is affine,

$$\Gamma(U_{i_0 \cdots i_p}, \underline{H}^q(F)) = \underline{H}^q(U_{i_0 \cdots i_p}, F) = 0$$

for $q > 0$, and so $\check{H}^p(\mathscr{U}/X, \underline{H}^q(F)) = 0$ for $q > 0$. Thus $H^p(X, F) = \check{H}^p(\mathscr{U}/X, F)$, $p \geq 0$. (See also Hartshorne [2, III.4.5].)

Remark 2.16. In (3.7), below we show that

$$H^p(X_{\text{fl}}, W(F)) = 0 = H^p(X_{\text{et}}, W(F))$$

if X is affine and F is quasi-coherent. Thus the same argument shows that for separated schemes X, the Čech and derived functor cohomologies of $W(F)$ agree for both the étale and flat topologies.

For the étale topology there is the following result of Artin [7].

THEOREM 2.17. *Let X be a quasi-compact scheme such that every finite subset of X is contained in an open affine set (for example, X quasi-projective over an affine scheme), and let F be a sheaf on X_{et}. Then there are canonical isomorphisms $\check{H}^p(X_{\text{et}}, F) \xrightarrow{\approx} H^p(X_{\text{et}}, F)$ for all p.*

Proof. One first needs the following result from commutative algebra.

LEMMA 2.18. *Let A be a ring, $\mathfrak{p}_1, \ldots, \mathfrak{p}_r$ prime ideals of A, and $A^{\text{sh}}_{\mathfrak{p}_1}, \ldots, A^{\text{sh}}_{\mathfrak{p}_r}$ strict Henselizations of the local rings $A_{\mathfrak{p}_1}, \ldots, A_{\mathfrak{p}_r}$. Then $A' = A^{\text{sh}}_{\mathfrak{p}_1} \otimes_A \cdots \otimes_A A^{\text{sh}}_{\mathfrak{p}_r}$ has the property that any faithfully flat étale map $A' \to B$ has a section $B \to A'$.*

Proof. This is (3.4(iii)) of Artin [7].

For any morphism $U \to X$ we write $U^0 = X$ and $U^n = U \times_X \cdots \times_X U$ (n copies). Also if $(\bar{x}) = (\bar{x}_1, \ldots, \bar{x}_r)$ is an r-tuple of geometric points, then $X_{(\bar{x})} = X$ if $r = 0$ and

$$X_{(\bar{x})} = \operatorname{spec}(\mathcal{O}_{\bar{x}_1}) \times_X \cdots \times_X \operatorname{spec}(\mathcal{O}_{\bar{x}_r})$$

if $r \neq 0$.

LEMMA 2.19. *Let X be as in the statement of the theorem, let $U \to X$ be étale of finite-type, and let $(\bar{x}) = (\bar{x}_1, \ldots, \bar{x}_r)$ be a family of geometric points of X. Let $W \to U^n \times X_{(\bar{x})}$ be an étale surjective morphism. Then there is an étale surjective morphism $U' \to U$ such that the induced X-morphism $U'^n \times X_{(\bar{x})} \to U^n \times X_{(\bar{x})}$ factors through W (provided either $n > 0$ or $r > 0$).*

Proof. First assume $n = 0$. Then $x_1, \ldots, x_r$ are contained in some open affine spec A of X, and it follows that $X_{(\bar{x})} = \operatorname{spec}(A^{\text{sh}}_{\mathfrak{p}_1} \otimes \cdots \otimes A^{\text{sh}}_{\mathfrak{p}_r})$ for some prime ideals $\mathfrak{p}_i$ of A. Thus $W \to X_{(\bar{x})}$ has a section by the first lemma.

The case $n = 1$, $r = 0$ is obvious (take $U' = W$).

We proceed by induction on n. Suppose that $V \to U$ is étale and of finite-type but not surjective and that the map $V^n \times X_{(\bar{x})} \to U^n \times X_{(\bar{x})}$ factors through W (for example, take V to be empty). Let $\bar{y}$ be a geometric point of U not covered by V, and let $X_{\bar{y}} \approx U_{\bar{y}}$ be the spectra of the strictly local rings at $\bar{y}$. The disjoint union $V \amalg X_{\bar{y}}$ has as nth power a sum of schemes of the form $V^i \times X^j_{\bar{y}} (i + j = n)$. The induction hypothesis applied to the pull-back of W to $V^i \times X^j_{\bar{y}} \times X_{(\bar{x})}$ shows that each of the maps $V^i \times (X^j_{\bar{y}} \times X_{(\bar{x})}) \to U^n \times X_{(\bar{x})}$ for $i < n$ factors through W after V has been replaced by a suitable étale covering V'. For $i = n$ the map factors by assumption. Thus we may assume that

$$(V \amalg X_{\bar{y}})^n \times X_{(\bar{x})} \to U^n \times X_{(\bar{x})}$$

factors through W. Now $X_{\bar{y}}$ is a limit of schemes Y étale over U, and so for some such Y, the map $(V \amalg Y)^n \times X_{(\bar{x})} \to U^n \times X_{(\bar{x})}$ also factors through W. Replace V by $V \amalg Y$. By proceeding in this way, we obtain a sequence of maps $V_1 \to U$, $V_2 \to U, \ldots$ whose images form a strictly increasing sequence of open subsets of U. As U is a Noetherian topological space, we arrive at a surjective map after only finitely many steps. Thus the lemma is proven.

To complete the proof of the theorem, we use (2.5). Let $F \to F''$ be a surjective map of sheaves, and let V be in (ét)/X. Clearly the coverings $U \to V$ consisting of a single morphism are cofinal in the set of all coverings of V (as X, and V, are quasi-compact). Let $s'' \in F''(U^n)$. There exists an étale covering $W \to U^n$ and an $s \in F(W)$ such that $s \mapsto s''|W$. According to the lemma, there exists a covering $U' \to U$ such that $U'^n \to U^n$ factors through W. Thus $s|U'^n \mapsto s''|U'^n$, which proves the theorem.

Little seems to be known about the relation between flat Čech cohomology and flat derived functor cohomology (but see Shatz [2, Thm 42, p. 208]).

THEOREM 2.20. (Hochschild-Serre spectral sequence). *Let $\pi: X' \to X$ be a finite Galois covering with Galois group G, and let F be a sheaf for the étale topology on X. There is a spectral sequence,*

$$H^p(G, H^q(X'_{\text{et}}, F)) \Rightarrow H^{p+q}(X_{\text{et}}, F).$$

Proof. Since G acts on X' on the right, it acts on $F(X')$ on the left. The composite of the functors

$$F \mapsto F(X'): \mathbf{S}(X_{\text{et}}) \to G\text{-}\mathbf{mod} \text{ and } N \mapsto N^G: G\text{-}\mathbf{mod} \to \mathbf{Ab}$$

is the section functor $\Gamma(X, -): \mathbf{S}(X_{\text{et}}) \to \mathbf{Ab}$ (II.1.4). Thus the proposition will follow from (Appendix B1) once we have shown that $H^p(G, I(X')) = 0$, $p > 0$, for any injective sheaf I on X. But $H^p(G, I(X')) = \check{H}^p(X'/X, I)$, which (2.4) shows is zero because I is also injective as a presheaf.

Remark 2.21. (a) The same argument shows that if F is a sheaf for the flat topology on X, then there is a spectral sequence

$$H^p(G, H^q(X'_{\text{fl}}, F)) \Rightarrow H^{p+q}(X_{\text{fl}}, F).$$

(b) Let $X' \to X$ be an infinite Galois covering of X with Galois group G. For each finite quotient G_i of G corresponding to a Galois covering X_i of X, there is a spectral sequence

$$H^p(G_i, H^q((X_i)_{\text{et}}, F)) \Rightarrow H^{p+q}(X_{\text{et}}, F).$$

Since cohomology commutes with inverse limits of schemes (1.16) and with inverse limits of groups and direct limits are exact, by passing to the direct limit over these spectral sequences, we form a spectral sequence

$$H^p(G, H^q(X'_{\text{et}}, F)) \Rightarrow H^{p+q}(X_{\text{et}}, F).$$

Example 2.22. Let X be a regular integral quasi-compact scheme. We shall compute some of the cohomology groups of $\mathbb{G}_m$ relative to the étale topology on X (equivalently, the flat topology, see (3.9) below). Recall (II.3.9) that there is an exact sequence of sheaves,

$$0 \to \mathbb{G}_{m,X} \to g_*\mathbb{G}_{m,K} \to D_X \to 0$$

where g:spec $K \to X$ is the inclusion of the generic point of X into X and $D_X = \bigoplus_{v \in X_1} i_{v*}\mathbb{Z}$ where X_1 is the set of points of X of codimension one and i_v:spec $k(v) \hookrightarrow X$. Correspondingly, there is an exact sequence,

$$\cdots \to H^r(X_{\text{et}}, D_X) \to H^{r+1}(X_{\text{et}}, \mathbb{G}_{m,X}) \to H^{r+1}(X_{\text{et}}, g_*\mathbb{G}_{m,K}) \to \cdots,$$

and so we must compute the groups $H^r(X_{\text{et}}, D_X)$ and $H^r(X_{\text{et}}, g_*\mathbb{G}_{m,K})$.

The Leray spectral sequence for $i_v : v \hookrightarrow X$ is (1.18):

$$H^p(X, R^q i_{v*}\mathbb{Z}) \Rightarrow H^{p+q}(v, \mathbb{Z}).$$

Note that

$$H^0(v, \mathbb{Z}) = \mathbb{Z},\ H^1(v, \mathbb{Z}) = 0 \text{ (because } H^1(v, \mathbb{Z}) = \operatorname{Hom}_{\text{conts}}(G_v, \mathbb{Z}),$$

where $G_v = \operatorname{Gal}(k(v)_s/k(v))$, and $\mathbb{Z}$ has no finite subgroups) and $H^2(v, \mathbb{Z}) = \operatorname{Hom}(G_v, \mathbb{Q}/\mathbb{Z})$ (use the exact sequence

$$0 \to \mathbb{Z} \to \mathbb{Q} \to \mathbb{Q}/\mathbb{Z} \to 0; \qquad H^r(G_v, \mathbb{Q}) = 0$$

for all $r > 0$ because it is torsion and $\mathbb{Q}$ is uniquely divisible). The same reasoning that showed that $H^1(v, \mathbb{Z}) = 0$ shows that $R^1 i_{v*}\mathbb{Z} = 0$ (use (1.13)). Thus

$$\begin{cases} H^0(X, i_{v*}\mathbb{Z}) = \mathbb{Z} \\ H^1(X, i_{v*}\mathbb{Z}) = 0 \\ H^2(X, i_{v*}\mathbb{Z}) \hookrightarrow \operatorname{Hom}(G_v, \mathbb{Q}/\mathbb{Z}) \end{cases} \quad \text{and so} \quad \begin{cases} H^0(X, D_X) = \bigoplus_{v \in X_1} \mathbb{Z} \\ H^1(X, D_X) = 0 \\ H^2(X, D_X) \hookrightarrow \bigoplus_v \operatorname{Hom}(G_v, \mathbb{Q}/\mathbb{Z}). \end{cases}$$

(Since cohomology commutes with direct sums, see (3.6d) below.)

The Leray spectral sequence of $g:\operatorname{spec} K \to X$ is:

$$H^p(X, R^q g_* \mathbb{G}_{m,K}) \Rightarrow H^{p+q}(\operatorname{spec} K, \mathbb{G}_m).$$

The stalk of $R^q g_* \mathbb{G}_{m,K}$ at a geometric point $\bar{x}$ of X is $H^q(K_{\bar{x}}, \mathbb{G}_m)$ where $K_{\bar{x}}$ is the field of fractions of the strictly local ring of X at $\bar{x}$. Thus $R^1 g_* \mathbb{G}_{m,K} = 0$ by Hilbert's Theorem 90. It follows that

$$\begin{cases} H^0(X, g_* \mathbb{G}_{m,K}) = K^* \\ H^1(X, g_* \mathbb{G}_{m,K}) = H^1(K, \mathbb{G}_m) = 0 \\ H^2(X, g_* \mathbb{G}_{m,K}) \hookrightarrow H^2(K, \mathbb{G}_{m,K}) \end{cases}$$

Putting this together, we get exact sequences

$$0 \to \Gamma(X, \mathcal{O}_X^*) \quad \to K^* \to \bigoplus_{v \in X_1} \mathbb{Z} \to H^1(X, \mathbb{G}_m) \to 0,$$

$$0 \to H^2(X, \mathbb{G}_m) \to H^2(K, \mathbb{G}_{m,K}).$$

Thus

$$H^1(X, \mathbb{G}_m) = (\text{Divisors})/(\text{Principal Divisors}) = \operatorname{Pic} X = H^1(X_{\text{Zar}}, \mathcal{O}_X^*).$$

In fact $H^1(X_{\text{et}}, \mathbb{G}_m) = \operatorname{Pic}(X)$ for any scheme X (see (4.9) below).

To say more, we have to restrict X further.

Case (*a*) Let X have dimension 1, with perfect residue fields $k(v)$ for all $v \in X_1$.

In this case, $K_{\bar{v}}$ for $v \in X_1$ is the field of fractions of a Henselian discrete valuation ring with algebraically closed residue field. The classical argument that shows $H^2(L, \mathbb{G}_m) = 0$ when L is the field of fractions of a complete such ring carries over to show that $H^2(L, \mathbb{G}_m) = 0$ in this case also. For $H^2(L, \mathbb{G}_m)$ is the Brauer group of L; the valuation on L extends uniquely to a valuation on any central division algebra A over L (using the fact that the ring is Henselian); such an A contains a subfield L' such that $[A:L] = [L':L]^2$ and L' is unramified over L; hence $L = L'$ in this case, which shows that $A = L$ (compare Serre [7, XII.§2]). It follows that in this case, $R^2 g_* \mathbb{G}_m = 0$. As X has dimension 1, the points v in X_1 are closed, and so i_{v*} is exact (II.3.6). Thus $R^q i_{v*} = 0$, $q > 0$.

Hence there is an exact sequence

$$0 \to H^2(X, \mathbb{G}_m) \to H^2(K, \mathbb{G}_{m,K}) \to \bigoplus_v \mathbf{X}(G_v) \to H^3(X, \mathbb{G}_m) \to H^3(K, \mathbb{G}_{m,K})$$

where we have written $\mathbf{X}(G_v)$ for the group of continuous homomorphisms $G_v \to \mathbb{Q}/\mathbb{Z}$, that is, for the character group of G_v.

Case (*b*) Assume further that X is excellent [I.1.2]; for example, that K has characteristic zero or that X is an algebraic curve over a field.

In this case Lang has shown that $K_{\bar{v}}$ is C_1 (Shatz [2]), which implies that $H^r(K_{\bar{v}}, \mathbb{G}_m) = 0$, $r > 0$. Thus in this case $R^q g_* \mathbb{G}_m = 0$, $q > 0$, and the sequence extends to an infinite exact sequence,

$$\cdots \to H^r(X, \mathbb{G}_m) \to H^r(K, \mathbb{G}_{m,K}) \to \bigoplus_v H^{r-1}(k(v), \mathbb{Q}/\mathbb{Z}) \to H^{r+1}(X, \mathbb{G}_m) \to \cdots.$$

Case (*c*) Assume $X = \operatorname{spec} A$, where A is an excellent discrete valuation ring with perfect residue field. Then (b) gives an exact sequence,

$$\cdots \to H^r(X, \mathbb{G}_m) \to H^r(K, \mathbb{G}_m) \to H^{r-1}(k, \mathbb{Q}/\mathbb{Z}) \to \cdots.$$

If A is Henselian, this sequence breaks up into short exact sequences, which have noncanonical splittings,

$$0 \to H^r(X, \mathbb{G}_m) \to H^r(K, \mathbb{G}_m) \to H^{r-1}(k, \mathbb{Q}/\mathbb{Z}) \to 0.$$

To see this we have to define a section to the map

$$H^r(K, \mathbb{G}_m) \to H^r(k, \mathbb{Z}) \approx H^{r-1}(k, \mathbb{Q}/\mathbb{Z}), \; r \geq 2.$$

Choose a uniformizing parameter π for A and identify $\mathbb{Z}$ with the subgroup $\{\pi^n \mid n \in \mathbb{Z}\}$ of K_s^*. The composite of the inflation map $H^r(k, \mathbb{Z}) \to H^r(K, \mathbb{Z})$ with $H^r(K, \mathbb{Z}) \to H^r(K, \mathbb{G}_m)$ is the required map.

If A is Henselian, but not necessarily excellent, one still has the split exact sequence

$$0 \to H^2(X, \mathbb{G}_m) \to H^2(K, \mathbb{G}_m) \to \mathbf{X}(G(k_s/k)) \to 0.$$

In IV we shall show that, in this situation, $H^2(X, \mathbb{G}_m) \approx H^2(k, \mathbb{G}_m)$.

Case (*d*) Assume X is a smooth algebraic curve over an algebraically closed field k.

Tsen's theorem (Shatz [2, Thm. 24]) says that K is C_1, and so $H^r(K, \mathbb{G}_m) = 0$ for $r > 0$. As $H^r(k(v), \mathbb{Q}/\mathbb{Z}) = 0$ for $r \geq 1$, the exact sequence in (b) shows that $H^r(X, \mathbb{G}_m) = 0$ for $r > 1$.

Case (*e*) Assume X is regular, excellent, connected, of dimension 1, and $k(v)$ is finite for all $v \in X_1$.

Then $G(k(v)_s/k(v)) = G_v \approx \hat{\mathbb{Z}}$, and so $H^r(G_v, \mathbb{Q}/\mathbb{Z}) = \mathbb{Q}/\mathbb{Z}, \mathbb{Q}/\mathbb{Z}, 0, \ldots$ for $r = 0, 1, 2, \ldots$. Thus (b) gives:

$$\begin{cases} 0 \to H^2(X, \mathbb{G}_m) \to H^2(K, \mathbb{G}_m) \to \bigoplus_{v \in X_1} \mathbb{Q}/\mathbb{Z} \to H^3(X, \mathbb{G}_m) \to H^3(K, \mathbb{G}_m) \to 0 \\ H^r(X, \mathbb{G}_m) \approx H^r(K, \mathbb{G}_m), \; r \geq 4. \end{cases}$$

Case (*f*) Assume X is an open subscheme of spec R, where R is the ring of integers in an algebraic number field K.

Class field theory (Cassels-Fröhlich [1, VII.10]) shows that there exists an exact sequence

$$0 \to Br(K) \to \bigoplus_{v} Br(\hat{K}_v) \xrightarrow{\Sigma} \mathbb{Q}/\mathbb{Z} \to 0$$

where v runs through all the primes of K (including infinite primes) and $\hat{K}_v$ is the completion of K at v. By comparing this with the exact sequence in (e), and using the fact that there exist commutative diagrams

$$\begin{array}{ccc} Br(K) & \longrightarrow & Br(\hat{K}_v) \approx H^2(\hat{K}_v, \mathbb{G}_m) \\ \downarrow \approx & & \downarrow \approx \\ H^2(K, \mathbb{G}_m) & \longrightarrow & H^1(k(v), \mathbb{Q}/\mathbb{Z}) \end{array}$$

for all finite primes v of K (compare (c)), one obtains the following:

$$\begin{cases} 0 \to H^2(X, \mathbb{G}_m) \to \bigoplus_{v \in S} Br(K_v) \to \mathbb{Q}/\mathbb{Z} \to H^3(X, \mathbb{G}_m) \\ \quad \to H^3(K, \mathbb{G}_m) \to 0 \text{ is exact and} \\ H^r(X, \mathbb{G}_m) = H^r(K, \mathbb{G}_m), \ r \geq 4 \end{cases}$$

(where S = set of primes of K not corresponding to a point of X). For example, if $X = \operatorname{spec} \mathbb{Z}$, then S contains one element, and

$$H^2(X, \mathbb{G}_m) = 0,$$
$$H^3(X, \mathbb{G}_m) \approx \mathbb{Q}/\mathbb{Z}$$

(as $H^3(\mathbb{Q}, \mathbb{G}_m) = 0$ Cassels-Fröhlich [1, VII.11.4]).

Case (*g*) Let X be a complete smooth algebraic curve over a finite field k. It follows from the results proved in Serre [8, II] that $H^r(K, \mathbb{G}_m) = 0$ for $r > 2$. Thus class field theory for function fields (Artin-Tate [1]) implies, as above, that

$$\begin{cases} H^2(X, \mathbb{G}_m) = 0 \\ H^3(X, \mathbb{G}_m) \approx \mathbb{Q}/\mathbb{Z} \\ H^r(X, \mathbb{G}_m) = 0, \ r > 3. \end{cases}$$

Exercise 2.23. (a) Prove the results in (g) without using class field theory. (Hint: write $\bar{X} = X \otimes_k k_{sep}$ and use the Hochschild-Serre spectral sequence for $\bar{X}/X$; note that $H^1(k, \operatorname{Pic}^0(X)) = 0$ because $1 - F: \operatorname{Pic}^0 \to \operatorname{Pic}^0$, the identity morphism minus the Frobenius morphism, is a surjective map of group schemes (Lang's theorem, Mumford [4, p. 205]).)

(b) Let X be a complete smooth algebraic curve over a separably closed field K. Show that $H^r(X, \mathbb{G}_m) = 0$, $r > 1$. (Hint: use the Leray spectral sequence for $\pi: X_{fl} \to (\text{spec } k)_{fl}$; show that $R^q\pi_*\mathbb{G}_m = \mathbb{G}_m, \underline{\text{Pic}}_X, 0, \ldots$; now use (3.9) below.)

Exercise 2.24. (Mayer-Vietoris). Let $\mathscr{U} = \{U_0 \to X, U_1 \to X\}$ be an open Zariski covering of X, and let F be a sheaf on some site $(\mathbf{C}/X)_E$. Show that there is a long exact sequence,

$$\cdots \to H^{n-1}(U_0, F) \times H^{n-1}(U_1, F) \xrightarrow{\phi} H^{n-1}(U_{01}, F) \to H^n(X, F)$$
$$\to H^n(U_0, F) \times H^n(U_1, F) \cdots$$

where $\phi(s_0, s_1) = s_0 - s_1$. (Hint: use (2.7) and (2.2d)). Use (1.13) to extend this to a statement about higher direct images.

Exercise 2.25. Let A be a sheaf of rings on $(\mathbf{C}/X)_E$, and write $H^p(X, A, -)$ for the pth right derived functor of the functor $M \mapsto \Gamma(X, M)$: $\mathbf{S}(X, A) \to \mathbf{Ab}$. Show that the isomorphism $H^0(X, A, M) \to H^0(X, M)$ extends to a canonical isomorphism $H^p(X, A, M) \to H^p(X, M)$, $p \geq 0$. (Hint: show that the forgetful functor $\mathbf{S}(X, A) \to \mathbf{S}(X)$ maps injective sheaves to flabby sheaves; use (2.12).)

§3. Comparison of Topologies

The results of this section frequently allow one, when computing cohomology, to replace one site by a less complicated site.

Changing C/X

PROPOSITION 3.1. *Let* $\mathbf{C}/X$ *be a subcategory of* $\mathbf{C}'/X$, *and let*

$$f:(\mathbf{C}'/X)_E \to (\mathbf{C}/X)_E$$

be the morphism defined by the identity map on X.

(a) *The functor* f_* *is exact, and* $F \to f_*f^*F$ *is an isomorphism for any sheaf* F *on* $(\mathbf{C}/X)_E$.

(b) *The functor* $f^*:\mathbf{S}((\mathbf{C}/X)_E) \to \mathbf{S}((\mathbf{C}'/X)_E)$ *is faithfully full.*

(c) *The canonical maps*

$$H^i(X, f_*F') \to H^i(X, F') \quad \textit{and} \quad H^i(X, F) \to H^i(X, f^*F)$$

are isomorphisms for all $i \geq 0$ *and all sheaves* F' *on* $(\mathbf{C}'/X)_E$ *and* F *on* $(\mathbf{C}/X)_E$.

Proof. (a) The exactness of f_* is obvious. For any U in $\mathbf{C}/X$,

$$\Gamma(U, f^pF) = \Gamma(U, F)$$

because there is an initial object, namely, (id, U), in the category over which the limit $\Gamma(U, f^pF) = \varinjlim \Gamma(V, F)$ is formed. Since F is a sheaf on $(\mathbf{C}/X)_E$, forming the associated sheaf does not change the sections over any object U in $\mathbf{C}/X$. Thus

$$\Gamma(U, f_*f^*F) = \Gamma(U, f^*F) = \Gamma(U, f^pF) = \Gamma(U, F).$$

(b) This is immediate from the second part of (a).

(c) That the first map is an isomorphism follows from the exactness of f_*. The second follows from the fact that the composite of the maps

$$H^i(X, F) \to H^i(X, f^*F) \xrightarrow{\approx} H^i(X, f_*f^*F) \xrightarrow{\approx} H^i(X, F)$$

is the identity map.

Remark 3.2. (a) The proposition applies in particular to the small and big E-sites on a scheme X. Thus the small E-site gives the same cohomology groups as the big E-site.

(b) It is not always true that $F \approx f^*f_*F$ for a sheaf F on $(\mathbf{C}'/X)_E$. For example, if F is the sheaf on the big étale site on a reduced scheme X defined by α_p, then $f_*F = 0$ on the small étale site but $F \neq 0$.

Changing E

PROPOSITION 3.3. *Let $E_1 \supset E_2$ be classes of morphisms satisfying the conditions* (e) *of* (II.1), *let $\mathbf{C}_2/X$ be a subcategory of $\mathbf{C}_1/X$, and let $f:(\mathbf{C}_1/X)_{E_1} \to (\mathbf{C}_2/X)_{E_2}$ be the morphism induced by the identity map on X. Assume that for every U in $\mathbf{C}_2/X$ and every covering of U in the E_1-topology, there is a covering of U in the E_2-topology that refines it. Then $f_*:\mathbf{S}(X_{E_1}) \to \mathbf{S}(X_{E_2})$ is exact and hence $H^i(X_{E_2}, f_*F) \xrightarrow{\approx} H^i(X_{E_1}, F)$ for any sheaf F on X_{E_1}.*

Proof. The functor f_* simply restricts a sheaf on X_{E_1} to the coarser E_2-topology. Let $F \to F'' \to 0$ be exact as a sequence of sheaves on X_{E_1}, and let $s \in F''(U)$. There exists a covering (U_i) of U in the E_1-topology such that $s|U_i$ is in the image of $F(U_i) \to F''(U_i)$ for all i. Let (V_j) be a covering of U for the E_2-topology that refines the first covering. If $V_j \to U$ factors through $U_i \to U$, then

$$\begin{array}{ccc} F(V_j) & \longrightarrow & F''(V_j) \\ \uparrow & & \uparrow \\ F(U_i) & \longrightarrow & F''(U_i) \end{array}$$

commutes, which shows that $s|V_j$ is in the image of $F(V_j) \to F''(V_j)$. Thus f_* is right exact.

Examples 3.4. The above proposition applies in the following cases:

(a) $\begin{cases} E_1 = \text{class of all étale morphisms} \\ E_2 = (\text{ét}) \end{cases}$

(b) $\begin{cases} E_1 = (\text{ét}) \\ E_2 = \text{class of all separated étale morphisms or affine étale morphisms} \end{cases}$

(c) $\begin{cases} E_1 = \text{class of all smooth morphisms} \\ E_2 = (\text{ét}) \end{cases}$

This follows from (I.3.26).

(d) $\begin{cases} E_1 = (\text{fl}) \\ E_2 = \text{class of all flat morphisms of finite-type.} \end{cases}$

(e) $\begin{cases} E_1 = (\text{fl}) \\ E_2 = \text{class of all quasi-finite flat morphisms.} \end{cases}$

This follows from (I.2.25).

Noetherian Sites

A site is *Noetherian* if every covering $(U_i \to U)_{i \in I}$ has a finite subcovering, that is, there is a finite $J \subset I$ such that $(U_j \to U)_{j \in J}$ is still a covering. (It is not clear to the author why this is called a Noetherian site rather than a compact site.) If X is quasi-compact and $\mathbf{C}/X$ is a category of schemes of finite-type over X, then any site defined by a class E whose morphisms are open is Noetherian.

PROPOSITION 3.5. *Let* $(\mathbf{C}/X)_E$ *be a Noetherian site. Define categories* $\mathbf{P}(X_{E(f)})$ *and* $\mathbf{S}(X_{E(f)})$ *of presheaves and sheaves on* X *as before, except take the coverings to be finite surjective families of E-morphisms. Then the categories* $\mathbf{P}(X_{E(f)})$ *and* $\mathbf{P}(X_E)$, *and* $\mathbf{S}(X_{E(f)})$ *and* $\mathbf{S}(X_E)$, *are canonically equivalent.*

Proof. In fact, $\mathbf{P}(X_{E(f)}) = \mathbf{P}(X_E)$ because there is no mention of coverings in the definition of a presheaf. To prove the second equivalence, we shall show that if a presheaf F satisfies the conditions (S_1) and (S_2) for finite coverings, then it satisfies these conditions for all coverings. Let $s \in F(U)$ and let $(U_i \to U)_{i \in I}$ be a covering of U such that $s|U_i = 0$ for all i. Since there is a finite subset $J \subset I$ such that $(U_i \to U)_{i \in J}$ is a covering and $s|U_i = 0$ for all $i \in J$, it follows that $s = 0$. Thus F is a separated presheaf. Now suppose that for some covering $(U_i \to U)_{i \in I}$, there are given sections $s_i \in F(U_i)$ such that $s_i|U_i \times U_j = s_j|U_i \times U_j$ for all i and j. We know that for any finite subset $J \subset I$ such that $(U_i \to U)_{i \in J}$ is a covering, there is a unique $s \in F(U)$ such that $s|U_i = s_i$ for all $i \in J$. But if we choose an s for one such J, the uniqueness implies that $s|U_i = s_i$ for all $i \in I$.

Remark 3.6. (a) Let X be a quasi-compact scheme, and consider the Noetherian site $((\text{ét})/X)_{\text{et}(f)}$ where the (f) means we allow only finite coverings. On combining the above propositions, we see that $X_{\text{et}(f)}$ gives the same cohomology groups as the small étale site and the big étale site. A similar remark holds for the flat sites.

(b) On a Noetherian site, the presheaf direct limit of a pseudofiltered direct system of sheaves is a sheaf (II.2.17). Thus pseudofiltered direct limits and in particular direct sums of presheaves and sheaves agree on such a site.

(c) On a Noetherian site, a pseudofiltered direct limit $F = \varinjlim F_i$ of flabby sheaves is again flabby. Indeed, since direct limits commute with finite products, $\check{H}^q(\mathscr{U}/U, F) = 0$, $q > 0$, if $\mathscr{U}$ is a finite covering of U. But finite coverings are cofinal in the set of all coverings, and so $\check{H}^q(U, F) = 0$, $q > 0$, which implies that F is flabby (2.12).

(d) On a Noetherian site X_E, cohomology commutes with pseudofiltered direct limits of sheaves.

Indeed, let I be a small pseudofiltered category. The category of functors $\mathbf{Fun}(I, \mathbf{S}(X_E))$ inherits the properties of $\mathbf{S}(X_E)$ and, in particular, has enough injectives. Choose an $i_0 \in I$. The functor

$$(i_0)\colon \mathbf{Fun}(I, \mathbf{S}) \to \mathbf{S},\ F \mapsto F(i_0),$$

has a left adjoint by (II.2.2). Explicitly this left adjoint is the functor that sends a sheaf F to the functor

$$i \mapsto F(i) = \bigoplus_{\mathrm{Hom}_I(i_0, i)} F.$$

Since direct sums are exact in $\mathbf{S}$, this left adjoint is exact. Thus (i_0) preserves injectives.

Now let $\{F(i)\}$ be a direct system of sheaves indexed by I, that is, an element of $\mathbf{Fun}(I, \mathbf{S})$, and let $\{F(i)\} \to \{J^{\cdot}(i)\}$ be an injective resolution of $\{F(i)\}$. According to the above remark, $F(i_0) \to J^{\cdot}(i_0)$ is an injective resolution of $F(i_0)$ for each i_0. Write $\underrightarrow{F} = \varinjlim F(i)$ and $\underrightarrow{J}^{\cdot} = \varinjlim J^{\cdot}(i)$. According to (c) above, $\underrightarrow{J}^{\cdot}$ is a flabby resolution of $\underrightarrow{F}$. Since we are on a Noetherian site, for any U in $\mathbf{C}/X$ the functor $\mathbf{S} \to \mathbf{Ab}$, $F \mapsto \Gamma(U, F)$ commutes with direct limits (see (b) above). Thus $\Gamma(U, \underrightarrow{J}^{\cdot}) = \varinjlim \Gamma(U, J^{\cdot}(i))$ and, on taking homology, we get $H^i(U, \underrightarrow{F}) = \varinjlim H^i(U, F(i))$.

This remark applies in particular to X_{et} when X is quasi-compact. (In this case, Godement resolutions may be used to simplify the proof.)

Flat Cohomology versus Zariski Cohomology

The next proposition should be interpreted as saying that the Zariski topology is sufficiently fine to compute the true cohomology groups of quasi-coherent modules.

PROPOSITION 3.7. *Let $f: X_{\text{fl}} \to X_{\text{Zar}}$ be the continuous morphism induced by the identity map on X. Let F be a quasi-coherent $\mathcal{O}_X$-module and let $W(F)$ be the corresponding sheaf on X_{fl} (II.1.6). There are canonical isomorphisms*

$$H^i(X_{\text{Zar}}, F) \to H^i(X_{\text{fl}}, W(F))$$

Proof. Recall (1.18) that there is a spectral sequence

$$H^i(X_{\text{Zar}}, R^j f_* W(F)) \Rightarrow H^{i+j}(X_{\text{fl}}, W(F)).$$

Since obviously $f_* W(F) = F$, it suffices to show that $R^j f_* W(F) = 0$ for $j > 0$. As $R^j f_* W(F)$ is the sheaf associated with the presheaf $U \mapsto H^j(U_{\text{fl}}, W(F))$ on X_{Zar}, it suffices to show that for U affine, $H^j(U_{\text{fl}}, W(F)) = 0, j > 0$. Moreover, in proving this we may suppose that U is given the small E-site where E is the class of all affine flat morphisms of finite-type (see above). In this situation we show that $W(F)$ is flabby. For this it suffices (2.12b) to show that $\check{H}^j(\mathscr{V}/V, W(F)) = 0, j > 0$, for any V affine flat and of finite-type over U and any finite covering $\mathscr{V} = (V_i \to V)$. On replacing (V_i) by $(\coprod V_i)$, we reduce the problem to the following: let $V' \to V$ be a faithfully flat morphism of finite-type where V and V' are affine; then $H^j(V'/V, W(F)) = 0, j > 0$. If $V = \operatorname{spec} A$, $V' = \operatorname{spec} B$, and F is defined by the A-module M, then the Čech complex is

$$M \otimes_A B \to M \otimes_A B \otimes_A B \to \cdots \to M \otimes_A B^{\otimes r} \to \cdots$$

But, as was observed in (I.2.19), this sequence is exact, which completes the proof.

Remark 3.8. Exactly the same proof shows that $H^i(X_{\text{Zar}}, F) \xrightarrow{\approx} H^i(X_{\text{et}}, W(F))$.

Flat Cohomology versus Étale Cohomology

The étale topology is sufficiently fine to compute the true cohomology groups of a smooth group scheme.

THEOREM 3.9. *If G is a smooth, quasi-projective, commutative group scheme over a scheme X, then the canonical maps*

$$H^i(X_{\text{et}}, G) \to H^i(X_{\text{fl}}, G)$$

are isomorphisms.

Proof. Let $f: X_{\text{fl}} \to X_{\text{et}}$ be induced by the identity map. We have to show that $R^i f_* G = 0$ for $i > 0$, or equivalently (compare (1.13) and (1.17)) that $H^i(X_{\text{fl}}, G) = 0$, $i > 0$, for any strictly local scheme X.

LEMMA 3.10. *Let $X = \operatorname{spec} A$ where A is a local Henselian ring, and write X_E for the small E-site on X where E is the class of all finite flat* *morphisms. Then* $f_*:\mathbf{S}(X_{\mathrm{fl}}) \to \mathbf{S}(X_E)$, *where* f *is induced by the identity map on* X, *is exact. Thus* $H^i(X_E, F) \xrightarrow{\approx} H^i(X_{\mathrm{fl}}, F)$ *for all sheaves* F *on* X_{fl}.

Proof. This follows from (I.4.3), (I.4.2), and (3.3).

We shall show that G is flabby relative to this topology. According to (2.12), we must prove the following: if X is strictly local and $X' \to X$ is finite and faithfully flat, then $\check{H}^i(X'/X, G) = 0$ for $i > 0$. This will require several steps. Throughout, G is a smooth, quasi-projective group scheme over X.

Step 1. Let X be any affine scheme, X_0 a closed subscheme, and N the functor

$$\mathbf{Sch}/X \to \mathbf{Ab},\ Y \mapsto \ker(G(Y) \to G(Y_0))$$

where $Y_0 = X_0 \times Y$. If X_0 is defined by an ideal $\mathfrak{J}$ of square zero, then $N = W(M)$ for a coherent $\mathcal{O}_X$-module M.

Proof. Let $\omega_G = e^*\Omega^1_{G/X}$ where $e: X \to G$ is the identity section. We may take M to be the module $\underline{\mathrm{Hom}}_{\mathcal{O}_X}(\omega_G, \mathfrak{J})$. (See Demazure-Gabriel [1, I.4.2.4] or Giraud [2, VII.1].)

Step 2. Let X be a scheme and $X' \to X$ a faithfully flat finite morphism. Define $\underline{C}^{\cdot}(G)$ to be the complex of functors $\mathbf{Sch}/X \to \mathbf{Ab}$ such that, for any Y/X, $\underline{C}^{\cdot}(G)(Y)$ is the Čech complex $C^{\cdot}(Y'/Y, G)$ for the covering $Y' = Y \times_X X' \to Y$. Write $\underline{Z}^i$ for the functor

$$Y \mapsto \ker(d^i: \underline{C}^i(Y) \to \underline{C}^{i+1}(Y)),$$

Then $d^{i-1}: \underline{C}^{i-1} \to \underline{Z}^i$ is represented by a smooth morphism of schemes.

Proof. By definition $\underline{C}^i(G)$ is the functor

$$Y \mapsto G(Y' \times_Y Y' \times \cdots) = G(Y \times_X X' \times_X X' \times \cdots),$$

that is, it is $\pi_* G$ where $\pi: X' \times \cdots \times X' \to X$. As π is finite and faithfully flat, it is therefore represented by the Weil restriction of scalars of $G \times Y' \times \cdots \times Y'$ ([SGA. 1, VIII.7.7] or Demazure-Gabriel [1, I.1.6.6]). Also $\underline{Z}^i(G)$, being the kernel of a map $d^i: \underline{C}^i(G) \to \underline{C}^{i+1}(G)$ of group schemes, is representable. To show that $d^{i-1}: \underline{C}^{i-1}(G) \to \underline{Z}^i(G)$ is smooth, we have to show (I.3.22) that, for any affine X-scheme T and closed subscheme T_0 defined by an ideal of square zero, any $z \in \underline{Z}^i(G)(T)$ whose image z_0 in $\underline{Z}^i(G)(T_0)$ arises from an element c_0 of $\underline{C}^{i-1}(G)(T_0)$ itself arises from an element c of $\underline{C}^{i-1}(G)(T)$ that maps to c_0. Consider the diagram

$$\begin{array}{ccccccccc}
0 & \longrightarrow & C^{i-1}(T'/T, N) & \longrightarrow & \underline{C}^{i-1}(G)(T) & \longrightarrow & \underline{C}^{i-1}(G)(T_0) & \longrightarrow & 0 \\
 & & \downarrow{\scriptstyle d^{i-1}} & & \downarrow{\scriptstyle d^{i-1}} & & \downarrow{\scriptstyle d^{i-1}} & & \\
0 & \longrightarrow & Z^i(T'/T, N) & \longrightarrow & \underline{Z}^i(G)(T) & \longrightarrow & \underline{Z}^i(G)(T_0) & &
\end{array}$$

where N is the functor of T-schemes, $Y \mapsto \ker(G(Y) \to G(Y_0))$. The surjectivity of the upper right horizontal arrow follows from the smoothness of G. A diagram chase shows that we only have to prove that

$$d^{i-1}: C^{i-1}(T'/T, N) \to Z^i(T'/T, N)$$

is surjective, that is, that $\check{H}^i(T'/T, N) = 0$, $i \geq 1$. But this follows from Step 1 and (3.7).

Step 3. Let $X = \operatorname{spec} A$, with A Henselian, let $X' \to X$ be finite and faithfully flat, let X_0 be the closed point of X, and let $X'_0 = X' \times_X X_0$. The map

$$\check{H}^i(X'/X, G) \to \check{H}^i(X'_0/X_0, G)$$

is bijective for all $i \geq 1$.

Proof. It follows from the smoothness of G and the fact that $X' \times \cdots \times X'$ is a disjoint union of spectra of Henselian local rings that the maps

$$C^i(X'/X, G) \to C^i(X'_0/X_0, G)$$

are surjective (I.4.13). Thus we must show that the complex

$$\ker(C^{\cdot}(X'/X, G) \to C^{\cdot}(X'_0/X_0, G))$$

is exact. Let $z \in Z^i(X'/X, G)$ have image $z_0 = 0$ in $C^i(X'_0/X_0, G)$. We seek a $c \in C^{i-1}(X'/X, G)$ with $c_0 = 0$ such that $d^{i-1}(c) = z$. By Step 2, we know that $(d^{i-1})^{-1}(z)$ is a smooth subscheme of $\underline{C}^{i-1}$. But $(d^{i-1})^{-1}(z)$ has a section over X_0, namely the zero section, and as X is Henselian, this lifts to a section of $(d^{i-1})^{-1}(z)$ over X.

We may now prove the theorem. After Step 3, we only have to show that $\check{H}^i(X'/X, G) = 0$, $i \geq 1$, when X is the spectrum of a separably closed field. But as $\underline{C}^{i-1} \to \underline{Z}^i$ is smooth, we need only show that it is surjective as a map of schemes. For this it suffices to show that $\underline{C}^{i-1}(k) \to \underline{Z}^i(k)$ is surjective if k is an algebraically closed field, that is, that $\check{H}^i(X'/X, G) = 0$ when X is the spectrum of an algebraically closed field. But then $X' \to X$ has a section, which implies that these groups are zero, as one can construct a contracting homotopy.

Remark 3.11. (a) It follows from Step 3 of the proof that, for any smooth quasi-projective group scheme G over a Henselian ring A, $H^i(X, G) \xrightarrow{\approx} H^i(X_0, G_0)$ for $i \geq 1$, where $X = \operatorname{spec} A$, X_0 is the closed point of X, and $G_0 = G \times_X X_0$ is the closed fiber of G/X.

(b) The theorem (and the above remark) hold if G is represented by a smooth commutative algebraic space and hence, in particular, if G is a smooth commutative group scheme. The proof is identical to the above except that the reference to [SGA. 1] in Step 2 must be replaced by a reference to (V.1.4a). Thus the theorem holds for locally constructible sheaves on $(\mathbf{Sch}/X)_{et}$ (V.1.7d). This means that for any sheaf F on X_{et}, the canonical maps $F \to f_* f^* F$ and $H^i(X_{et}, F) \to H^i(X_{fl}, f^* F)$ are isomorphisms.

Étale Cohomology versus Complex Cohomology

The next theorem shows that étale and classical complex cohomology agree as well as it is reasonable to hope. Just as for the algebraic and classical fundamental groups, it is not reasonable to expect agreement for nontorsion coefficient groups. For example, if X is a smooth complete curve of genus g over $\mathbb{C}$, then $H^1(X(\mathbb{C}), \mathbb{Z}) \approx \mathbb{Z}^{2g}$, but

$$H^1(X_{et}, \mathbb{Z}) = \mathrm{Hom}_{\mathrm{Conts}}(\pi_1(X), \mathbb{Z}) = 0$$

as $\pi_1(X)$, being profinite, has only finite discrete quotients. However,

$$H^1(X(\mathbb{C}), \mathbb{Z}/n\mathbb{Z}) \approx (\mathbb{Z}/n\mathbb{Z})^{2g} \approx H^1(X_{et}, \mathbb{Z}/n\mathbb{Z}).$$

THEOREM 3.12. *Let X be a smooth scheme over $\mathbb{C}$. For any finite abelian group M, $H^i(X(\mathbb{C}), M) \approx H^i(X_{et}, M)$.*

Proof. We merely outline the proof. Details may be found in [SGA. 4, XI]. Two main results are needed, the first of which is due to Artin and the second to Grauert and Remmert.

An *elementary fibration* is a morphism of schemes $f: X \to S$ that may be embedded into a commutative diagram,

$$\begin{array}{ccccc} X & \xrightarrow{j} & \bar{X} & \xleftarrow{i} & Y \\ & {\scriptstyle f}\searrow & \downarrow{\scriptstyle \bar{f}} & \swarrow{\scriptstyle g} & \\ & & S & & \end{array}$$

in which:

(a) j is an open immersion, dense in each fiber, and $Y = \bar{X} - X$;

(b) $\bar{f}$ is smooth and projective, with geometrically irreducible fibers of dimension 1;

(c) g is finite and étale, and each fiber of g is nonempty. An *Artin neighborhood* relative to S is an S-scheme X for which there exists a sequence

$$X = X_n \xrightarrow{f_n} X_{n-1} \xrightarrow{f_{n-1}} \cdots \to X_0 = S$$

with each f_i an elementary fibration.

LEMMA 3.13. *Let X be a smooth scheme over an algebraically closed field k. Any closed point x of X is contained in an open subscheme U of X that is an Artin neighborhood relative to* spec k.

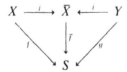

Proof. It is possible to embed an open neighborhood U of x into projective space $\mathbb{P}^r$ in such a way that the closure $\bar{U}$ of U is normal. Thus the singularities of $\bar{U}$ have codimension ≥ 2. After replacing the embedding $\bar{U} \hookrightarrow \mathbb{P}^r$ by a multiple $\bar{U} \hookrightarrow \mathbb{P}^R$, one finds there exists a particularly good projection map $\mathbb{P}^R \to \mathbb{P}^{n-1}$, $n = \dim X$, for $\bar{U}$. After blowing up $\bar{U}$ at the center of the projection map, one obtains an elementary fibration $U \to U_{n-1}$. This process may be continued, [SGA. 4, XI.3.3].

For $i = 0$, (3.12) simply asserts that $X(\mathbb{C})$ and X have the same number of connected components (relative to the complex and Zariski topologies respectively), which is well-known. (See, for example, Šafarevič [3, VII.2].) For $i = 1$, the theorem states that there is a one-to-one correspondence between the Galois coverings of $X(\mathbb{C})$ with automorphism group M and the similar coverings of X. This is a consequence of the following result.

LEMMA 3.14. (Riemann existence theorem). *Let X be a scheme locally of finite-type over $\mathbb{C}$, and let X^{an} be the associated complex analytic space. The functor $Y \mapsto Y^{\text{an}}$ defines an equivalence between the category of finite étale coverings Y/X and the category of similar coverings of X^{an}.*

Proof. If X is projective, this may be found in Serre [2]. The general case may be found in Grauert and Remmert [1]. See also [SGA. 4, XI.4.3]. A simpler proof, but one that assumes the resolution of singularities, may be found in [SGA. 1, XII]. See also the appendix to Katz [1].

For $i > 1$ we proceed as follows. Let X_{cx} denote the small *site* $(X^{\text{an}})_E$ where E is the class of all morphisms of complex analytic spaces that are local isomorphisms (of course, this is not quite a site in the sense of Chapter II as its objects are not schemes). Since the inclusion $U \hookrightarrow X(\mathbb{C})$ of an open subset (for the complex topology) is a local isomorphism, we have a morphism of sites $X_{cx} \to X(\mathbb{C})$. The arguments used earlier in this section (especially in the proof of (3.3)) show that this morphism induces isomorphisms on cohomology $H^i(X_{cx}, M) \approx H^i(X(\mathbb{C}), M)$. There is a morphism of sites $f: X_{cx} \to X_{\text{et}}$ under which the inverse image of a U étale over X is U^{an}. (As was observed in (I.3), the implicit function theorem implies that $U^{\text{an}} \to X^{\text{an}}$ is a local isomorphism.) There is a Leray spectral sequence

$$H^i(X_{\text{et}}, R^j f_* F) \Rightarrow H^{i+j}(X_{cx}, F).$$

Thus it remains to show that $R^j f_* F = 0$ for $j > 0$, which follows from the next lemma (and (1.13)).

LEMMA 3.15. *Let $\gamma \in H^i(X_{cx}, F)$, $i > 0$, where F is a locally constant torsion sheaf on X_{cx} with finite fibers (that is, that becomes finite and constant on the U_i of some étale covering $U_i \to X$). Then, for any $x \in X(\mathbb{C})$ there exists an étale morphism $U \to X$ whose image contains x and which is such that $\gamma | U_{cx} = 0$.*

Proof. We use induction on $n = \dim(X)$. Since the statement is local for the étale topology on X, we may assume that F is constant and that there is an elementary fibration $X \xrightarrow{f} S$. We use the same notations as in the definition of elementary fibration. All calculations will be relative to the topology of local analytic isomorphisms. An easy calculation shows that j_*F is the same constant sheaf on $\bar{X}$, R^1j_*F is the extension by zero of a constant sheaf on Y, and $R^ij_*F = 0$ for $i > 1$ (compare [VI.5.1]). As $\bar{f}$ is proper, the stalks of $R^i\bar{f}_*$ may be calculated on the fibers of $\bar{X}/S$, and as it is smooth, $R^i\bar{f}_*$ preserves locally constant torsion sheaves (compare [VI.4.2]). Now the spectral sequence $(R^i\bar{f}_*)(R^jj_*)F \Rightarrow R^{i+j}f_*F$ shows that f_*F is again a constant sheaf, R^1f_*F is a locally constant torsion sheaf with finite fibers, and $R^if_*F = 0$ for $i > 1$. The Leray spectral sequence

$$H^i(S_{cx}, R^jf_*F) \Rightarrow H^{i+j}(X_{cx}, F)$$

therefore reduces to a long exact sequence,

$$\cdots \to H^i(S_{cx}, f_*F) \to H^i(X_{cx}, F) \to H^{i-1}(S_{cx}, R^1f_*F) \to \cdots$$

By induction, there exists an étale map $U' \to S$ whose image contains $f(x)$ and which is such that $\gamma | U'_{cx} = 0 = \gamma' | U'_{cx}$ for any given $\gamma \in H^{i-1}(S_{cx}, R^1f_*F)$ and $\gamma' \in H^i(S_{cx}, f_*F)$ (provided $i > 1$). This completes the proof for $i > 1$ since we may take $U = U' \times X$ for a suitable U'. For $i = 1$ the lemma follows from (3.14).

Remark 3.16. A similar argument, which uses the theorems proved in Chapter VI, shows that for any $\gamma \in H^i(X_{et}, F)$ $(i > 0)$ and any $x \in X$, where X is a smooth scheme over an algebraically closed field k and F is a locally free sheaf of $\mathbb{Z}/n\mathbb{Z}$-modules with n prime to char (k), there exists a finite étale map $U \to U'$ with U' an open Zariski neighborhood of x such that $\gamma | U = 0$.

Exercise 3.17. The situation and notations are those of (1.16).

(a) Let $U_\alpha \in (\text{ét})/X$ be of the form $U_i \times_{X_i} X_\alpha$ for some $U_i \in (\text{ét})/X_i$; show that any covering of U_α has a refinement that arises by base change from a covering of $(U_i \times X_j)_{\text{et}}$ for some j such that $i \to j$. (The covering may be assumed to be finite; use (I.3.19).)

(b) Show that $\Gamma(U_\alpha, F_\alpha) = \varinjlim \Gamma(U_j, F_j)$ where $U_j = U_i \times_{X_i} X_j$; deduce that

$$\check{H}^p(\mathscr{U}_\alpha, F_\alpha) = \varinjlim \check{H}^p(\mathscr{U}_i \times X_j, F_j)$$

for any covering $\mathscr{U}_\alpha$ of the form $\mathscr{U}_i \times_{X_i} X_\alpha$, that $\check{H}^p(X_\infty, F_\infty) = \varinjlim \check{H}^p(X_j, F_j)$ and that, if the X_i and X_α satisfy the hypothesis of (2.17), then $H^p(X_\infty, F_\alpha) = \varinjlim H^p(X_j, F_j)$.

§4. Principal Homogeneous Spaces

In this section, we work almost entirely with the flat site. All group schemes will be flat and locally of finite-type, but not necessarily commutative. The existence of the identity section implies that they are faithfully flat. If G is such a group scheme over X, then an *action* of G on an X-scheme S is a morphism $S \times_X G \to S$ that induces an action (in the usual sense) of the group $G(T)$ on the set $S(T)$ for any X-scheme T. For example, the multiplication morphism $G \times G \to G$ defines an action of G on itself. There is an obvious notion of morphism, and isomorphism, of schemes with G-action.

PROPOSITION 4.1. *Let G act on a scheme S. The following are equivalent:*

(a) *the scheme S is faithfully flat and locally of finite-type over X, and $(s, g) \mapsto (s, sg): S \times_X G \to S \times_X S$ is an isomorphism;*

(b) *there is a covering $(U_i \to X)$ for the flat topology on X such that, for each i, $S_{(U_i)}$ is isomorphic with its $G_{(U_i)}$-action to $G_{(U_i)}$.*

Proof. (a) $\Rightarrow$ (b). Take the covering to be $(S \to X)$ and the isomorphism $G_{(S)} \to S_{(S)}$ to be the given one.

(b) $\Rightarrow$ (a). Let $U = \coprod U_i$. Then U is faithfully flat and locally of finite-type over X, and $S_{(U)} \approx G_{(U)}$, which implies that $S_{(U)}$ is faithfully flat and locally of finite-type over U. Now descent theory (I.2.24) implies that S has the same properties over X. Also, it is clear from (b), that the map $(S \times_X G)_{(U)} \to (S \times_X S)_{(U)}$ is an isomorphism, which by descent theory, shows that $S \times_X G \to S \times_X S$ is an isomorphism.

A scheme S, on which G acts, and that satisfies the statements in (4.1) is called a *principal homogeneous space* or *torsor* for G over X. A torsor S that is isomorphic to G (acting on itself) is called a *trivial* torsor. Equivalently S is trivial if and only if $S(X)$ is nonempty. (If $s \in S(X)$, $(g \mapsto sg)$: $G \to S$ is an isomorphism.) The trivial torsors form a distinguished element in the set $PHS(G/X)$ of all isomorphism classes of torsors for G over X. Form (b) of the definition of a torsor may be interpreted as saying that every torsor is locally trivial for the flat topology on X. If G is smooth, then every G-torsor is locally trivial for the étale topology on X, that is, there exists a covering $(U_i \to X)$ for the étale topology such that $S_{(U_i)} \approx G_{(U_i)}$. (This follows from (4.2) below, using (I.3.26).) For many groups G, the G-torsors that are locally trivial for the Zariski topology are uninteresting, and it was Serre who first suggested studying torsors that become trivial on some étale covering (Serre [5]).

PROPOSITION 4.2. *If G is smooth, respectively étale, respectively proper, . . . over X, then so also is any G-torsor.*

Proof. This follows from descent theory, using the fact that any torsor becomes isomorphic to G after a faithfully flat base change.

The rest of this section is devoted to computing $PHS(G/X)$, that is, to classifying isomorphism classes of G-torsors. For example, if G is a commutative finite group, then a G-torsor is a Galois covering of X with a given action of G, and so for such a G, $PHS(G/X) = \mathrm{Hom}(\pi_1(X), G)$. (See (I.5.4)). The computation of $PHS(G/X)$ requires three steps. Firstly, we define the concept of a sheaf being a G-torsor and investigate which sheaf torsors are representable by schemes. Secondly, we show that the sheaf torsors are classified by a certain cohomology group. Finally, we use the cohomological techniques to compute the cohomology group.

Let G be a group scheme over X, and let S be a sheaf of sets on $(\mathbf{LFT}/X)_{\mathrm{fl}}$ on which G acts. Then S is called a *principal homogeneous space* or *torsor* for G if it satisfies condition (b) of (4.1) (where now $G_{(U_i)}$ is identified with the sheaf defined on $(U_i)_{\mathrm{fl}}$ by G and $S_{(U_i)}$ with $S|U_i$). Clearly, the sheaf defined by a scheme S that is a G-torsor is a torsor in this sense, and two schemes S and S' are isomorphic as G-torsors if and only if they are isomorphic as sheaf torsors (by the Yoneda lemma [EGA. 0, 1.1.6]). The question of which sheaf torsors arise from schemes is, in general, quite delicate.

THEOREM 4.3. *A sheaf S on X_{fl} that is a G-torsor is representable in each of the following cases:*

(a) *G is affine over X;*

(b) *G is smooth and separated over X, and X has dimension ≤ 1;*

(c) *G is smooth and proper over X with geometrically connected fibers, and is regular;*

(d) *G is quasi-projective over X, and S becomes trivial on a scheme X' that is finite and faithfully flat over X;*

(e) *G is an abelian scheme that is projective over X, and S defines a torsion element of $\check{H}^1(X_{\mathrm{fl}}, G)$ (see (4.6) below for the cohomology class defined by a torsor).*

Proof. (a) Let $Y \to X$ be of finite-type and faithfully flat, and such that $S|Y \approx G|Y$ (as sheaves on Y_{fl}). It is easy to see that such a Y exists. The canonical isomorphism $p_1^*(S|Y) \to p_2^*(S|Y)$, where p_1 and p_2 are the projections $Y \times Y \rightrightarrows Y$, induces an isomorphism $\phi : p_1^*(G_{(Y)}) \to p_2^*(G_{(Y)})$ that satisfies the conditions of (I.2.23). Thus S is representable by an X-scheme.

(b), (c), (d), and (e) The proofs of these are similar except that the descent argument is more subtle. The cases (b), (c), and (e) (and much more) can be found in Raynaud [2, XI.3.1 and XII.2.3]. The case (d) follows from [SGA. 1, VIII.7.7].

Remark 4.4. (a) In the notes cited above, Raynaud also gives the following examples: an abelian scheme A over a normal local scheme X of dimension 2 and a nonrepresentable A-torsor that is locally trivial for the étale topology; a projective abelian scheme A over a local integral scheme X of dimension 1 and a representable A-torsor S that is not projective over X and does not define a torsion element of $\check{H}^1(X_{\text{fl}}, A)$. (Note that the property of being projective does not descend, unless there is an ample divisor compatible with the descent data.)

(b) It is perhaps more natural to ask that the sheaf-torsor be represented by an algebraic space (Artin [4, Cor. 7.2]).

Before relating sheaf torsors to cohomology, we extend some of our definitions to sheaves of noncommutative groups. Let G be a sheaf of groups on X_E, written multiplicatively, and let $\mathcal{U} = (U_i \to X)_{i \in I}$ be a covering. A 1-*cocycle* for $\mathcal{U}$ with values in G is a family $(g_{ij})_{I \times I}$, $g_{ij} \in G(U_{ij})$, such that

$$(g_{ij}|U_{ijk})(g_{jk}|U_{ijk}) = (g_{ik}|U_{ijk}).$$

Two cocycles g and g' are *cohomologous* if there is a family $(h_i)_I$, $h_i \in G(U_i)$, such that $g'_{ij} = (h_i|U_{ij})g_{ij}(h_j|U_{ij})^{-1}$. This is an equivalence relation, and the set of cohomology classes is written $\check{H}^1(\mathcal{U}/X, G)$. It is a set with a distinguished element (g_{ij}) where $g_{ij} = 1$ for all i and j. The set $\check{H}^1(X, G)$ is defined to be $\varinjlim H^1(\mathcal{U}/X, G)$ where the limit is taken over all coverings as in (2.2). When G is abelian, this definition of $\check{H}^1(\mathcal{U}/X, G)$ is the same as before, and since filtered direct limits of sets and abelian groups agree, the definition of $\check{H}^1(X, G)$ also agrees with the previous definition.

A sequence $1 \to G' \to G \to G'' \to 1$ of sheaves of groups is said to be *exact* if for every U in $\mathbf{C}/X$, $G'(U)$ is the kernel of the homomorphism $G(U) \to G''(U)$ and every $s \in G''(U)$ is locally liftable to a section of G.

PROPOSITION 4.5. *To any exact sequence of sheaves of groups*

$$1 \to G' \to G \to G'' \to 1,$$

there is associated an exact sequence of pointed sets

$$1 \to G'(X) \to G(X) \to G''(X) \xrightarrow{d} \check{H}^1(X, G') \to \check{H}^1(X, G) \to \check{H}^1(X, G'').$$

Proof. The map d is defined as follows: let $g \in G''(X)$, and let $(U_i \to X)$ be a covering of X such that there exist $g_i \in G(U_i)$ that map to $g|U_i$ under $G(U_i) \to G''(U_i)$; then

$$d(g)_{ij} = (g_i|U_{ij})^{-1}(g_j|U_{ij}).$$

The other maps are obvious. The proof that d is well-defined and the sequence is exact is routine, and hence omitted. (Compare Giraud [2, III.3.6].)

Let S be a sheaf torsor for G, and let $(U_i \to X)$ be a flat covering that trivializes S, so that $S(U_i)$ is nonempty, all i. Choose $s_i \in S(U_i)$; then there is a unique $g_{ij} \in G(U_{ij})$ such that $(s_i | U_{ij})g_{ij} = (s_j | U_{ij})$. Note that (omitting restriction signs), $s_i g_{ij} g_{jk} = s_k = s_i g_{ik}$, and so $g_{ij}g_{jk} = g_{ik}$, that is, (g_{ij}) is a 1-cocycle. If the choice of the s_i is changed, then $g = (g_{ij})$ is replaced by a cohomologous cocycle. Also the cohomology class is unaltered if S is replaced by an isomorphic torsor. Thus S defines an element $c(S) \in \check{H}^1(X, G)$.

PROPOSITION 4.6. *The map $S \mapsto c(S)$ defines a one-to-one correspondence between isomorphism classes of sheaf torsors for G and elements of $\check{H}^1(X, G)$ under which the trivial class corresponds to the distinguished element.*

Proof. We construct an inverse mapping. Let $\mathcal{U} = (U_i \to X)$ be a covering of X, and let $\underline{C}^0(\mathcal{U}, G)$ and $\underline{C}^1(\mathcal{U}, G)$ be the sheaves $V \mapsto \prod_i G(U_i \times V)$, $V \mapsto \prod_{i,j} G(U_{ij} \times V)$. That they are sheaves may be checked directly or by using the fact that $\underline{C}^0(\mathcal{U}, G) = \prod \pi_{i*}(G|U_i)$, where $\pi_i: U_i \to X$ etc. Let $d: \underline{C}^0 \to \underline{C}^1$ be the mapping $(h_i) \mapsto (h_i^{-1}h_j)$. Now fix a 1-cocycle g for G relative to $\mathcal{U}$. For any V, g restricts to an element $g|V$ of $\Gamma(V, \underline{C}^1(\mathcal{U}, G))$, and we define S to be the subsheaf of $\underline{C}^0(\mathcal{U}, G)$ such that $\Gamma(V, S)$ is the inverse image of $g|V$ for any V. There is an obvious action of G on S (on the right), namely, $((s_i), g) \mapsto (g^{-1}s_i)$. Suppose now that g is the trivial cocycle, that is, $g_{ij} = (g_i|U_{ij})^{-1}(g_j|U_{ij})$ for some $(g_i) \in \prod G(U_i)$. Then the map $G \to S$ that sends a section h of G over V to $(h^{-1}|V \times U_i)(g_i|V \times U_i)$ is an isomorphism that commutes with the action of G. Since g becomes trivial on each U_i, and the definition of S commutes with restriction, this shows that $S|U_i \approx G|U_i$, that is, that S is a G-torsor. Finally one checks that the 1-cocycle corresponding to S is the original cocycle g and that the torsor defined by the cocycle defined by a given torsor is isomorphic to the given torsor. This shows there is a one-to-one correspondence between isomorphism classes of torsors that become trivial on a given covering $\mathcal{U}$ and elements of $\check{H}^1(\mathcal{U}, G)$. Now pass to the limit over all coverings.

COROLLARY 4.7. *There is a canonical injection $PHS(G/X) \to \check{H}^1(X_{\mathrm{fl}}, G)$; if G is commutative, there is also an injection $PHS(G/X) \to H^1(X_{\mathrm{fl}}, G)$; if G/X satisfies* (a), (b), *or* (c) *of* (4.3), *then the map $PHS(G/X) \to \check{H}^1(X_{\mathrm{fl}}, G)$ is an isomorphism.*

Proof. These follow respectively from: (4.6) and the remarks preceeding (4.3); (2.10); (4.3).

Remark 4.8. (a) As we remarked earlier, if G is a smooth group scheme, then every G-torsor becomes trivial on some étale covering of X. Thus the proof of (4.6) shows that, for such a group scheme, the assertions of

(4.7) hold with the flat cohomology groups replaced by étale cohomology groups. This is not surprising in view of the comparison theorem (3.9).

(b) When G is commutative, it is possible to give a direct description of the composition law on sheaf torsors induced by the map $S \mapsto c(S)$ and the addition on $\check{H}^1(X_{\text{fl}}, G)$. If S is a sheaf of sets on which G acts, then S/G is defined to be the sheaf associated with the presheaf $U \mapsto S(U)/G(U)$. If S_1 and S_2 are both acted on by G, then $S_1 \vee S_2 = S_1 \vee^G S_2$ is defined to be $S_1 \times S_2/G$, where G acts on the product by $(s_1, s_2)g = (s_1 g^{-1}, s_2 g)$. If S_1 and S_2 are torsors, then $S_1 \vee S_2$ becomes a G-torsor with the action $(s_1, s_2)g = (s_1 g, s_2)(=(s_1, s_2 g))$ and $c(S_1 \vee S_2) = c(S_1) + c(S_2)$.

We now calculate some of the groups $\check{H}^1(X, G)$.

PROPOSITION 4.9. (Hilbert's Theorem 90). *The canonical maps*

$$H^1(X_{\text{Zar}}, \mathcal{O}_X^*) \to H^1(X_{\text{et}}, \mathbb{G}_m) \to H^1(X_{\text{fl}}, \mathbb{G}_m)$$

are isomorphisms; in particular $H^1(X_{\text{fl}}, \mathbb{G}_m) \approx \operatorname{Pic}(X)$.

Proof. We show only that $H^1(X_{\text{Zar}}, \mathcal{O}_X^*) \approx H^1(X_{\text{fl}}, \mathbb{G}_m)$ since the proof for the étale group is the same. From the Leray spectral sequence (1.18), we see that it suffices to show that $R^1 f_* \mathbb{G}_m = 0$, where f is the canonical morphism $X_{\text{fl}} \to X_{\text{Zar}}$. For this we must show that any element of $H^1(U_{\text{fl}}, \mathbb{G}_m)$, where U is a Zariski open subset of X, becomes zero in $H^1((U_i)_{\text{fl}}, \mathbb{G}_m)$ for the U_i in some Zariski covering of U. Clearly this reduces to showing that $H^1(U_{\text{fl}}, \mathbb{G}_m) = 0$ where $U = \operatorname{spec} A$, and A is local. We prove slightly more.

LEMMA 4.10. *Let* $U = \operatorname{spec} A$, *where* A *is a local ring; then*

$$\check{H}^1(U_{\text{fl}}, GL_n) = 0.$$

Proof. Here GL_n is the sheaf $U \mapsto GL_n(\Gamma(U, \mathcal{O}_U))$. Let $\alpha \in \check{H}^1(U_{\text{fl}}, GL_n)$. There is a U-scheme V, which we may assume to be faithfully flat, of finite-type, and affine, such that $\alpha \in \check{H}^1(V/U, GL_n)$. Let $B = \Gamma(V, \mathcal{O}_V)$, and let $\phi \in GL_n(V \times V) = GL_n(B \otimes_A B)$ be a cocycle corresponding to α. Then ϕ may be regarded as an isomorphism $(B \otimes_A B)^n \to (B \otimes_A B)^n$ or (compare I.2.21) $B^n \otimes_A B \to B \otimes_A B^n$. The fact that ϕ is a cocycle implies that, in its second manifestation, it satisfies the conditions of (I.2.21). Thus there is an A-module M giving rise to (B^n, ϕ). As B^n is flat and finitely generated as a B-module and B is faithfully flat over A, descent theory shows that M is flat and finitely generated over A. Thus M is free (I.2.9), $M \xrightarrow{\approx} A^n$. Now ϕ is a coboundary. (Compare the discussion of twisted-forms at the end of this chapter.)

Kummer Theory

For any scheme X and any integer n, the exact sequence (II.2.18b)

$$0 \to \mu_n \to \mathbb{G}_m \xrightarrow{n} \mathbb{G}_m \to 0$$

gives rise to an exact cohomology sequence

$$0 \to \mu_n(X) \to \Gamma(X, \mathcal{O}_X)^* \xrightarrow{n} \Gamma(X, \mathcal{O}_X)^* \to H^1(X, \mu_n) \to \operatorname{Pic}(X) \xrightarrow{n} \operatorname{Pic}(X).$$

This sequence has the following explicit interpretation: $H^1(X, \mu_n)$ is the set (modulo isomorphism) of pairs (L, ϕ) where L is an invertible sheaf on X and ϕ is an isomorphism $\mathcal{O}_X \to L^{\otimes n}$; the image of (L, ϕ) in $\operatorname{Pic}(X)_n$ is the class of L; if there exists an isomorphism $\psi: \mathcal{O}_X \to L$, that is, if (L, ϕ) maps to zero in $\operatorname{Pic}(X)$, then (L, ϕ) is the image of $a \in \Gamma(X, \mathcal{O}_X)^*$ where

$$\mathcal{O}_X \xrightarrow{\phi} L^{\otimes n} \xrightarrow{(\psi^{-1})^{\otimes n}} \mathcal{O}_X^{\otimes n} \xrightarrow{\text{can}} \mathcal{O}_X$$

is multiplication by a^{-1}. The μ_n-torsor corresponding to $(L, \phi) \in H^1(X, \mu_n)$ is $S = \mathbf{spec}\, B$ where B is the coherent $\mathcal{O}_X$-algebra $\bigoplus_{i=0}^{n-1} L^{\otimes i}$ with multiplication $B \otimes B \to B$ defined by:

$$L^{\otimes i} \otimes L^{\otimes j} \xrightarrow{\text{can}} L^{\otimes(i+j)}, \qquad i + j \leq n - 1,$$
$$L^{\otimes i} \otimes L^{\otimes j} \xrightarrow{\text{can}} L^{\otimes i+j} = L^{\otimes n-(n-i-j)} \xrightarrow{\phi^{-1} \otimes 1} L^{\otimes i+j-n}, \qquad i + j \geq n.$$

We examine this first in the case that $X = \operatorname{spec} A$ is affine and there is an isomorphism $\psi: A \to L$. Let $e = \psi(1)$; then $e^{\otimes i}$ generates $L^{\otimes i}$ and $e^{\otimes n}$ may be written $e^{\otimes n} = a\phi(1)$, $a \in A^*$. Thus B is a free A-module on $e_0, \ldots, e_{n-1}$, where $e_i = e^{\otimes i}$, and

$$e_i e_j = e_{i+j}, \qquad i + j \leq n - 1,$$
$$e_i e_j = a e_{i+j-n}, \qquad i + j \geq n,$$

that is, $B \simeq A[T]/(T^n - a)$. For any X-scheme Y,

$$S(Y) = \{c \in \Gamma(Y, \mathcal{O}_Y) \mid c^n = a\},$$

and we define an action of μ_n on S by requiring $S(Y) \times \mu_n(Y) \to S(Y)$ to be $(c, \zeta) \mapsto c\zeta$. Then S becomes isomorphic to μ_n as soon as a becomes an nth power. For a general X this implies that S is locally (for the Zariski topology on X) a μ_n-torsor. As the locally defined actions of μ_n on S patch, this shows that S is a torsor globally.

If n is invertible in $\Gamma(X, \mathcal{O}_X)$, that is, is not divisible by any residue field characteristic, and $\Gamma(X, \mathcal{O}_X)$ contains a primitive nth root of unity ζ, then μ_n is isomorphic (noncanonically) to the constant sheaf $\mathbb{Z}/n\mathbb{Z}$, and the above theory is referred to as Kummer theory. Roughly the isomorphism $\mu_n \to \mathbb{Z}/n\mathbb{Z}$ is defined by choosing a primitive nth root of 1 and mapping it to $1 \in \mathbb{Z}/n\mathbb{Z}$. However, it is better to observe that the conditions on X imply that it is a scheme over spec A, $A = \mathbb{Z}[1/n][\zeta]$, and that μ_n and $\mathbb{Z}/n\mathbb{Z}$ are isomorphic as group schemes over A, because

$$A[T]/(T^n - 1) \approx \prod_{i=0}^{n-1} A[T]/(T - \zeta^i) \approx A \times A \times \cdots \times A.$$

Thus, under these conditions, $H^1(X, \mu_n) \approx H^1(X, \mathbb{Z}/n\mathbb{Z}) \approx \pi^1(X, \mathbb{Z}/n\mathbb{Z}) =$ set of isomorphism classes of Galois covering of X with Galois group $\mathbb{Z}/n\mathbb{Z}$. Since μ_n is now étale, $H^1(X, \mu_n) = H^1(X_{\text{et}}, \mu_n)$.

PROPOSITION 4.11. *For any scheme X over $\mathbb{Z}[1/n][\zeta]$, there is an exact sequence*

$$0 \to \Gamma(X, \mathcal{O}_X)^*/\Gamma(X, \mathcal{O}_X)^{*n} \to H^1(X_{\text{et}}, \mathbb{Z}/n\mathbb{Z}) \to \operatorname{Pic}(X)_n \to 0.$$

Proof. This is a restatement of the above.

The first map, of course, sends $a \in \Gamma(X, \mathcal{O}_X)^*$ to **spec** B, where $B = \mathcal{O}_X[T]/(T^n - a)$, and the second map may be described in terms of invertible sheaves as above.

Consider now the case that X is a complete normal variety over a field k (containing a primitive nth root of 1). The above sequence may then be written,

$$0 \to k^*/k^{*n} \to \pi^1(X, \mathbb{Z}/n\mathbb{Z}) \to \operatorname{Pic}(X)_n \to 0.$$

An element $a \in k^*$ defines an extension k' of k, and the corresponding element of $\pi^1(X, \mathbb{Z}/n\mathbb{Z})$ is $X' = X \otimes_k k'$. On the other hand, consider an element X' of $\pi^1(X, \mathbb{Z}/n\mathbb{Z})$ such that k is algebraically closed in $R(X')$ or, equivalently, such that each component of X' is again a variety over k. Such a covering X' of X is called a *geometric covering* in Lang [1, pp. 289–290]. The ring $R(X')$ is a Galois extension of $k(X)$ and so, by the above proposition, corresponds to an element $a \in k(X)^*$, so that $R(X') = k(X)[T]/(T^n - a)$. Since X' is étale over X, the divisor of zeros and poles of a is of the form nD for some divisor D on X. The class of D in $\operatorname{Pic}(X)$ is the image of X' under the mapping $\pi^1(X, \mathbb{Z}/n\mathbb{Z}) \to \operatorname{Pic}(X)$. The covering X' of X is said to be of *Picard type* if D lies in the subgroup $\operatorname{Pic}^0(X)$ of $\operatorname{Pic}(X)$, that is, if D is algebraically equivalent to zero. An example of a geometric covering that is not of Picard type is given in Lang [1, p. 302].

Now let X be a complete smooth curve of genus g over an algebraically closed field k (of characteristic not dividing n). Recall that $\operatorname{Pic}(X) = \operatorname{Pic}^0(X) \oplus \mathbb{Z}$, and $\operatorname{Pic}^0(X) \approx J(k)$, where J is the Jacobian of X. As J

is an abelian variety of dimension g, $J(k)_n \approx (\mathbb{Z}/n\mathbb{Z})^{2g}$ (Mumford [4, p. 64]) and so

$$H^1(X, \mathbb{Z}/n\mathbb{Z}) \approx J(k)_n \approx (\mathbb{Z}/n\mathbb{Z})^{2g}.$$

Note that, when $k = \mathbb{C}$, this is in agreement with the complex cohomology groups.

Artin-Schreier Theory

Let X be a scheme of characteristic p, so that the canonical morphism $X \to \operatorname{spec} \mathbb{Z}$ factors through $\operatorname{spec} \mathbb{F}_p$. The exact sequence of sheaves (II.2.18)

$$0 \to \mathbb{Z}/p\mathbb{Z} \to \mathcal{O}_X \xrightarrow{F-1} \mathcal{O}_X \to 0$$

gives rise to an exact cohomology sequence

$$\Gamma(X, \mathcal{O}_X) \xrightarrow{F-1} \Gamma(X, \mathcal{O}_X) \to H^1(X, \mathbb{Z}/p\mathbb{Z}) \to H^1(X, \mathcal{O}_X) \xrightarrow{F-1} H^1(X, \mathcal{O}_X).$$

These groups may be regarded indifferently as étale or flat cohomology groups.

PROPOSITION 4.12. *For any scheme X over $\mathbb{F}_p$, there is an exact sequence.*

$$0 \to \Gamma(X, \mathcal{O}_X)/(F-1)\Gamma(X, \mathcal{O}_X) \to \pi^1(X, \mathbb{Z}/p\mathbb{Z}) \to H^1(X, \mathcal{O}_X)^F \to 0$$

where

$$H^1(X, \mathcal{O}_X)^F \overset{\text{df}}{=} \{a \in H^1(X, \mathcal{O}_X) \mid Fa = a\}.$$

Proof. Restatement of the above.

Thus, if X is affine, $X = \operatorname{spec} A$, then any $\mathbb{Z}/p\mathbb{Z}$-torsor X' is of the form $\operatorname{spec}(A[T]/(T^p - T + a))$, $a \in A$; the element $1 \in \mathbb{Z}/p\mathbb{Z}$ acts on $A' = A[t]$ by sending t to $t + 1$; X' is trivial if and only if $a = b^p - b$ for some $b \in A$.

Now let X' be a $\mathbb{Z}/p\mathbb{Z}$-torsor over X, where X is arbitrary. For any open affine covering $X = \bigcup U_i$, $U_i = \operatorname{spec} A_i$, there exist elements $a_i \in A_i$ such that

$$X' \times U_i = U_i' = \operatorname{spec}(A_i[T]/(T^p - T + a_i)).$$

The fact that $U_i' \times U_{ij} = U_j' \times U_{ij}$ etc. implies that $a_j - a_i = f_{ij}^p - f_{ij}$, where (f_{ij}) is a 1-cocycle for $\mathcal{O}_X$ relative to the covering (U_i). Thus an element X' in $\pi^1(X, \mathbb{Z}/p\mathbb{Z})$ is described by giving an affine covering $\mathcal{U} = (U_i)$ of X, an element $(a_i) \in C^0(\mathcal{U}, \mathcal{O}_X)$, and an element $(f_{ij}) \in Z^1(\mathcal{U}, \mathcal{O}_X)$ such that $f_{ij}^p - f_{ij} = a_j - a_i$ for all i and j. The image of the element in $H^1(X, \mathcal{O}_X)^F$ is the cohomology class of the 1-cocycle (f_{ij}). If this is zero, that is, if $f_{ij} = g_j - g_i$ for some $(g_i) \in C^0(\mathcal{U}, \mathcal{O}_X)$, then X' is the image of an element $a \in \Gamma(X, \mathcal{O}_X)$ such that $a = a_i - g_i^p + g_i$ in $\Gamma(U_i, \mathcal{O}_X)$, all i.

If X is a complete variety over an algebraically closed field k, then $\pi^1(X, \mathbb{Z}/p\mathbb{Z}) \approx H^1(X, \mathcal{O}_X)^F$ and $H^1(X, \mathcal{O}_X)$ is a finite-dimensional vector space over k, and so the following elementary lemma applies.

LEMMA 4.13. *Let V be a finite-dimensional vector space over an algebraically closed field k of characteristic p, and let $\phi: V \to V$ be an additive map such that $\phi(av) = a^p\phi(v)$ for all $a \in k$, $v \in V$. Then V decomposes into a direct sum, $V = V_s \oplus V_n$, where V_s and V_n are subspaces stable under ϕ and ϕ is bijective on V_s and nilpotent on V_n. Moreover, V_s has a basis $e = \{e_1, \ldots, e_\sigma\}$ such that $\phi(e_i) = e_i$ for all i. It follows that*

$$V^\phi = \{\textstyle\sum a_i e_i \mid a_i \in \mathbb{F}_p, e_i \in e\},$$

and $\phi - 1: V \to V$ is surjective.

Proof. See, for example [SGA. 7, XXII.1], or Mumford [4, p. 143].

Thus $H^1(X, \mathbb{Z}/p\mathbb{Z}) = \pi^1(X, \mathbb{Z}/p\mathbb{Z})$ has order p^σ where $\sigma \leq \dim_k(H^1(X, \mathcal{O}_X))$. If X is a nonsingular curve, then σ is determined by the Hasse-Witt matrix of X. If X is a complete intersection of dimension ≥ 2 or projective space of any dimension, then $\pi^1(X, \mathbb{Z}/p\mathbb{Z}) = 0$ because $H^1(X, \mathcal{O}_X) = 0$. Note that Kummer theory implies that $\pi^1(\mathbb{P}_k^m, \mathbb{Z}/n\mathbb{Z}) = 0$ for any n not divisible by p, and so $\pi_1(\mathbb{P}_k^m)$ has no abelian quotient. In fact, $\pi_1(\mathbb{P}_k^n) = 0$ [SGA. 1, XI.1]. If X is a connected subscheme of $\mathbb{P}_k^n$, then there is a relation between the vanishing of the cohomology groups $H^i(\mathbb{P}^n - X, M)$ for M a coherent sheaf, $i \geq n - t$ and the vanishing of $H^i(X, \mathbb{Z}/p\mathbb{Z})$, $i \leq t - 1$, which has been exploited to give examples of varieties which are not set-theoretic complete intersections (Boutot [1]).

The above results on $\pi^1(X, \mathbb{Z}/p\mathbb{Z})$ may be extended to $\pi^1(X, \mathbb{Z}/p^n\mathbb{Z})$ by replacing $\mathcal{O}_X$ with the Witt vectors of length n over $\mathcal{O}_X$ (Serre [4]). For a complete variety over a finite field, the groups $H^i(X, \mathbb{Z}/p\mathbb{Z})$ may be used to compute the zeta function of X "mod p" [SGA. 7, XXII].

Infinitesimal Coverings

Let X again be a scheme of characteristic p. The exact sequence (II.2.18) of sheaves on X_{fl},

$$0 \to \alpha_p \to \mathbb{G}_a \xrightarrow{F} \mathbb{G}_a \to 0$$

gives rise to an exact cohomology sequence

$$0 \to \Gamma(X, \mathcal{O}_X)/\Gamma(X, \mathcal{O}_X)^p \to H^1(X_{\text{fl}}, \alpha_p) \to H^1(X, \mathcal{O}_X) \xrightarrow{F} H^1(X, \mathcal{O}_X).$$

Thus the α_p-torsors have a similar description to the $\mathbb{Z}/p\mathbb{Z}$-torsors, which we leave to the reader to work out explicitly. The only other simple infinitesimal commutative flat group scheme over an algebraically closed field is μ_p, and we have already discussed μ_p-torsors.

In the case that X is a smooth variety over a perfect field, μ_p and α_p-torsors can be described in terms of the differentials. Recall Seshadri [1] that for such a variety there is a (p^{-1})-linear mapping C, called the *Cartier operator*, from the sheaf of closed differentials Ω^1_{cl} on X to the sheaf of all differentials Ω^1_X.

PROPOSITION 4.14. *Let X be a smooth variety over a perfect field k of characteristic p. Then*

$$H^1(X_{\mathrm{fl}}, \alpha_p) = \{\omega \in \Gamma(X, \Omega^1_X) \mid d\omega = 0,\ C\omega = 0\}, \text{ and}$$
$$H^1(X_{\mathrm{fl}}, \mu_p) = \{\omega \in \Gamma(X, \Omega^1_X) \mid d\omega = 0,\ C\omega = \omega\}.$$

Proof. There are exact sequences of sheaves on X_{et},

$$0 \to \mathcal{O}_X \xrightarrow{F} \mathcal{O}_X \xrightarrow{d} \Omega^1_{X,\mathrm{cl}} \xrightarrow{C} \Omega^1_X \to 0,$$

and

$$0 \to \mathcal{O}_X^* \xrightarrow{F} \mathcal{O}_X^* \xrightarrow{\mathrm{dlog}} \Omega^1_{X,\mathrm{cl}} \xrightarrow{C-1} \Omega^1_X \to 0$$

where dlog is the map $g \mapsto dg/g$. Except for the surjectivity of $C - 1$, the exactness of these sequences follows immediately from the standard properties of C (Seshadri [1]). To see that $C - 1$ is surjective, note that for any closed point $P \in X$ and any differential 1-form defined in a neighborhood of P, there is an open neighborhood U of P such that the form is a sum of terms $g(dx_i/x_i)$, where $1 + x_1, \ldots, 1 + x_m$ are local uniformizing parameters at P. But

$$(C - 1)(h^p(dx_i/x_i)) = (h - h^p)(dx_i/x_i),$$

and so $g(dx_i/x_i)$ is in the image of $C - 1$ when restricted to the étale covering $U' \to U$ of U defined by the equation $T^p - T + g = 0$.

Let $f: X_{\mathrm{fl}} \to X_{\mathrm{et}}$ be the obvious morphism. Since $R^if_*\mathbb{G}_a = 0$, $i > 0$ (3.7), the long exact sequence associated with $0 \to \alpha_p \to \mathbb{G}_a \xrightarrow{F} \mathbb{G}_a \to 0$ shows $R^if_*\alpha_p = 0$, $i \neq 1$, and $R^1f_*\alpha_p = \operatorname{coker}(\mathcal{O}_X \xrightarrow{F} \mathcal{O}_X)$. Thus the first exact sequence above gives $R^1f_*\alpha_p = \ker(\Omega^1_{X,\mathrm{cl}} \xrightarrow{C} \Omega^1_X)$, and hence

$$H^1(X_{\mathrm{fl}}, \alpha_p) = H^0(X_{\mathrm{et}}, R^1f_*\alpha_p) = \ker(H^0(\Omega^1_{X,\mathrm{cl}}) \xrightarrow{C} H^0(\Omega^1_X)).$$

The proof of the second equality is the same.

Remarks 4.15. (a) One can also say that $H^1(X, \alpha_p)$ and $H^1(X, \mu_p)$ are equal, respectively, to $\{\omega \mid \omega$ is locally exact$\}$ and $\{\omega \mid \omega$ is locally logarithmic$\}$.

(b) We leave it to the reader to explicitly describe the torsors corresponding to a given 1-form.

(c) Generalizations of the above result to nonconstant finite groups schemes over X may be deduced from Artin-Milne [1].

Relations to Pic

It has long been known that the Albanese of a variety plays the role of the abelian fundamental group of the variety. A more precise statement is given below (4.19). Recall (Mumford [4, §14]) that, for any finite flat commutative group scheme N over a scheme X, the sheaf $\underline{\mathrm{Hom}}_{X_{\mathrm{fl}}}(N, \mathbb{G}_m)$ is representable by a group scheme N' of the same type called the *Cartier dual* of N; $N \mapsto N'$ preserves exact sequences. Recall also that for any morphism $\pi: Y \to X$, $\underline{\mathrm{Pic}}_{Y/X}$ is defined to be the sheaf $R^1\pi_*\mathbb{G}_m$ on X_{fl}. Under the conditions of the following proposition $\underline{\mathrm{Pic}}_{Y/X}$ is representable by an algebraic space (Artin [6]) and by a scheme if π is also projective.

PROPOSITION 4.16. *Let $\pi: Y \to X$ be proper, flat, and such that $\pi_*\mathcal{O}_Y = \mathcal{O}_X$. For any finite flat commutative group scheme N over X, $R^1\pi_*(N_Y) \xrightarrow{\approx} \underline{\mathrm{Hom}}_X(N', \underline{\mathrm{Pic}}_{Y/X})$, where $N_Y = N \times Y$ and N' is the Cartier dual of N.* (All statements are relative to the flat sites.)

Proof. We first need a lemma.

LEMMA 4.17. *For any scheme X and any finite flat commutative group scheme N over X, $\underline{\mathrm{Ext}}^1_{X_{\mathrm{fl}}}(N, \mathbb{G}_m) = 0$.*

Proof. We have to show that for any U locally of finite-type over X, any element of $\mathrm{Ext}^1_{U_{\mathrm{fl}}}(N, \mathbb{G}_m)$ becomes zero on some flat covering of U. If N is killed by n, then any element of $\mathrm{Ext}^1_U(N, \mathbb{G}_m)$ arises from an element of $\mathrm{Ext}^1_U(N, \mu_n)$. Let E be an extension of N_U by μ_n, so $0 \to \mu_n \to E \to N_U \to 0$ is an exact sequence of sheaves. Then E may be regarded as a μ_n-torsor over the scheme N_U and so is representable by a finite flat scheme over U (or N_U) (4.3a). By increasing n if necessary, we may suppose that E is killed by n. Cartier duality gives an exact sequence $0 \to N'_U \to E' \to \mathbb{Z}/n\mathbb{Z} \to 0$. If E' has a section over U, then, as $nE' = 0$, this section may be used to define a section to $E' \to \mathbb{Z}/n\mathbb{Z}$. Thus the sequence splits as soon as E' has a section, but E' acquires a section on the faithfully flat covering $(E' \to U)$ of U.

Consider now the functor of sheaves on Y_{fl},

$$G \mapsto H(G) = \pi_* \underline{\mathrm{Hom}}(N'_Y, G) \approx \underline{\mathrm{Hom}}(N', \pi_*G).$$

This is left exact, and there are two spectral sequences arising from the two ways H has been expressed as a composite, namely,

$$R^p\pi_*(\underline{\mathrm{Ext}}^q(N'_Y, G)) \Rightarrow R^{p+q}H(G)$$

and

$$\underline{\mathrm{Ext}}^p(N', R^q\pi_*G) \Rightarrow R^{p+q}H(G).$$

Take $G = \mathbb{G}_{m,Y}$. Since $\pi_*\mathbb{G}_{m,Y} = \mathbb{G}_m$, the first sequence gives

$$0 \to R^1\pi_*N_Y \xrightarrow{\approx} R^1H(\mathbb{G}_m) \to \pi_* \underline{\mathrm{Ext}}^1(N'_Y, \mathbb{G}_m) = 0$$

and the second gives

$$\underline{\mathrm{Ext}}^1(N', \mathbb{G}_m) = 0 \to R^1H(\mathbb{G}_m) \to \underline{\mathrm{Hom}}(N', R^1\pi_*\mathbb{G}_m)$$
$$\to \underline{\mathrm{Ext}}^2(N', \mathbb{G}_m) \to R^2H(\mathbb{G}_m).$$

Thus we obtain a map $R^1\pi_*N_Y \to \underline{\mathrm{Hom}}(N', \underline{\mathrm{Pic}}_{Y/X})$ that is injective and is surjective if and only if $\underline{\mathrm{Ext}}^2(N', \pi_*\mathbb{G}_m) \to R^2H(\mathbb{G}_m)$ is injective. Since it suffices to check this last assertion locally for the flat topology, we may assume that there is a section $s: X \to Y$ to π. There is a morphism $\pi_*G \to s^*G$, functorial in G, that may be defined directly as follows: let $U \to X$ be étale; $\Gamma(U, \pi_*G) = \Gamma(Y_{(U)}, G)$, whereas $\Gamma(U, s^*G)$ is a limit, $\varinjlim \Gamma(V, G)$, over certain V étale over Y; as the section to $Y \to X$ induces a section to $Y_{(U)} \to U$, $Y_{(U)}$ is a V, and the map $\Gamma(U, \pi_*G) \to \Gamma(U, s^*G)$ is obvious. Note that if $G = \mathbb{G}_m$, both groups equal $\Gamma(U, \mathcal{O}_U^*)$. The morphism defines a map $R^2H(G) \to \underline{\mathrm{Ext}}^2(N', s^*G)$ whose composite with $\underline{\mathrm{Ext}}^2(N', \pi_*G) \to R^2H(G)$ is the map on $\underline{\mathrm{Ext}}^2(N', -)$ induced by $\pi_*G \to s^*G$. Since the map $\pi_*(\mathbb{G}_{m,Y}) \to s^*(\mathbb{G}_{m,Y})$ is an isomorphism, this completes the proof.

COROLLARY 4.18. *Let X be a complete variety over an algebraically closed field k. Then,*

$$H^1(X, \mu_n) \approx \{a \in \mathrm{Pic}(X) \mid na = 0\},$$

and

$$H^1(X, N) \approx \mathrm{Hom}(\mathrm{Lie}(N'), \mathrm{Lie}(\underline{\mathrm{Pic}}_X))$$

if N' is infinitesimal, where Lie (N') *and* Lie $(\underline{\mathrm{Pic}}_X)$ *are the p-*Lie *algebras defined by N and $\underline{\mathrm{Pic}}_X$.*

Proof. From the theorem, $H^1(X, N) \approx \mathrm{Hom}_k(N', \underline{\mathrm{Pic}}_X)$, and so it is only a question of interpreting the second term. For $N = \mu_n$ one only has to note that $N' = \mathbb{Z}/n\mathbb{Z}$, and for N' infinitesimal one may use the results in Mumford [4, §14, 15].

If A is any group (scheme), we write $A_n = \ker(n: A \to A)$, TA for the projective system of group (schemes) $(A_n)_{n \geq 1}$ with maps $A_n \xleftarrow{m} A_{mn}$ and T_lA for the projective system $(A_{l^n})_{n \geq 1}$ with maps $A_{l^n} \xleftarrow{l} A_{l^{n+1}}$.

COROLLARY 4.19. *Let X be a complete variety over an algebraically closed field k.*

(a) *If $\underline{\mathrm{Pic}}_X$ is smooth, and $\underline{\mathrm{NS}}_X \stackrel{df}{=} \underline{\mathrm{Pic}}_X/\underline{\mathrm{Pic}}_X^0$ is torsion free, then $H^1(X, N) \approx \mathrm{Hom}(T(\underline{\mathrm{Alb}}_X), N)$ for any finite commutative group scheme N over k.*

(b) *The following sequence is exact modulo p-groups, p = characteristic of k:*

$$0 \to (NS(X)_{\text{tors}})^* \to \pi_1(X)^{\text{ab}} \to T(\underline{\text{Alb}}_X(k)) \to 0$$

where $\underline{\text{Alb}}_X$ *is the Albanese variety of* X, $\pi_1(X)^{\text{ab}}$ *is the maximal abelian quotient of* $\pi_1(X)$, *and* $NS(X)^*_{\text{tors}}$ *is the Pontryagin dual of the (finite) torsion subgroup of* $NS(X)$.

Proof. (a) The conditions imply that

$$\text{Hom}_k(N', \underline{\text{Pic}}_X) = \text{Hom}_k(N', \underline{\text{Pic}}^0_X),$$

which clearly equals $\text{Hom}_k(N', (\underline{\text{Pic}}^0_X)_n)$ if N' is killed by n. Since $\underline{\text{Pic}}^0_X$ and $\underline{\text{Alb}}_X$ are dual abelian varieties (Mumford [4]), $(\underline{\text{Pic}}^0)_n$ is the Cartier dual of $(\underline{\text{Alb}}_X)_n$. Thus

$$\text{Hom}_k(N', (\underline{\text{Pic}}^0_X)_n) = \text{Hom}_k((\underline{\text{Alb}}_X)_n, N) = \text{Hom}_k(T(\underline{\text{Alb}}_X), N).$$

(b) Since $\text{Pic}^0(X)$ is divisible ($n: \underline{\text{Pic}}^0_X \to \underline{\text{Pic}}^0_X$ is finite and faithfully flat, (Mumford [4, §6])), the exact sequence

$$0 \to \text{Pic}^0(X) \to \text{Pic}(X) \to NS(X) \to 0$$

gives an exact sequence

$$0 \to \text{Pic}^0(X)_n \to \text{Pic}(X)_n \to NS(X)_n \to 0.$$

It is known that $NS(X)$ is finitely generated (compare (V.3.25) and (VI.11.7) below), and so we may take n large enough so that $NS(X)_n = NS(X)_{\text{tors}}$. According to (4.18), this last sequence may be rewritten (modulo p-groups)

$$0 \to \text{Pic}^0(X)_n \to \pi^1(X, \mathbb{Z}/n\mathbb{Z}) \to NS(X)_{\text{tors}} \to 0.$$

On applying the functor $\text{Hom}(-, \mathbb{Z}/n\mathbb{Z})$, we get an exact sequence,

$$0 \to (NS(X)_{\text{tors}})^* \to \pi_1(X)^{\text{ab}}/n\pi_1(X)^{\text{ab}} \to \text{Pic}^0(X)^*_n \to 0,$$

and $\text{Pic}^0(X)^*_n = \text{Alb}(X)_n$. On passing to the inverse limit, we obtain the required exact sequence.

COROLLARY 4.20. *For any abelian variety X over an algebraically closed field k, and any finite commutative group scheme N over k,* $H^1(X, N) = \text{Ext}^1_k(X, N)$.

Proof. Since $\underline{\text{Alb}}_X = X$, $\underline{\text{Pic}}_X$ is smooth, and $NS(X)$ is torsion free, we have $H^1(X, N) = \text{Hom}_k(T(X), N)$. On passing to the inverse limit in

$$(0 =)\ \text{Hom}_k(X, N) \to \text{Hom}_k(X_n, N) \to \text{Ext}^1(X, N) \xrightarrow{n} \text{Ext}^1(X, N),$$

we find that $\text{Hom}_k(T(X), N) = \text{Ext}^1_k(X, N)$.

Remark 4.21. The last corollary can be interpreted as saying that any N-torsor X' arises from an exact sequence $0 \to N \to X' \to X \to 0$ and hence, if it is a variety, it is an abelian variety.

Weak Mordell-Weil Theorem

The above theory has many arithmetic applications. As an example we prove the weak Mordell-Weil theorem.

THEOREM 4.22. *Let A be an abelian variety over a number field K. For any integer n, $A(K)/nA(K)$ is finite.*

Proof. There is an open subset U of the spectrum of the ring of integers in K such that A has good reduction at every point of U, that is, A is the generic fiber of an abelian scheme A' over U. We may assume that n is invertible on U. Then there is an exact sequence of sheaves on X_{et},

$$0 \to N \to A' \xrightarrow{n} A' \to 0$$

because multiplication by n is étale on A' (since it is on each fiber; compare (II.2.19)). The cohomology sequence of this reads:

$$\cdots \to A'(U) \xrightarrow{n} A'(U) \to H^1(U_{\text{et}}, N) \to H^1(U, A') \to \cdots.$$

Since $A = A' \times_U \operatorname{spec} K$, we know that $A'(K) = A(K)$, and it is easy to see that any morphism $\operatorname{spec} K \to A'$ factors through U and so $A'(K) = A'(U)$. Thus we are reduced to showing that $H^1(U, N)$ is finite. It is easily seen that we may replace U by an open subscheme and, using (2.20), that we may replace U by a Galois covering. Thus we need only consider the case that N is a constant sheaf, killed by n, and $\Gamma(U, \mathcal{O}_U)$ contains the nth roots of unity. Since cohomology commutes with direct sums, we may replace N by $\mathbb{Z}/n\mathbb{Z} \approx \mu_n$. The sequence in (4.11) becomes

$$0 \to \Gamma(U, \mathcal{O}_U)^*/\Gamma(U, \mathcal{O}_U)^{*n} \to H^1(U_{\text{et}}, \mathbb{Z}/n\mathbb{Z}) \to \operatorname{Pic}(U)_n \to 0.$$

The two fundamental theorems of algebraic number theory state that if $U = \operatorname{spec} R$, then R^* is finitely generated and R has finite ideal class group (that is, Picard group). These together imply that $H^1(U_{\text{et}}, \mathbb{Z}/n\mathbb{Z})$ is finite.

Remark 4.23. (a) The Mordell-Weil theorem states that, in the situation of (4.22), the group $A(K)$ is finitely generated. To deduce it from (4.22), one has to use the theory of heights; see, for example, Lang [3].

(b) A study of the groups $H^1(U, N)$ and $H^1(U, A)$ and an explanation of their relation to the Selmer and Tate-Šafarevič groups may be found in Mazur [1].

Twisted-forms

Let Y be a scheme (sheaf of modules, algebras, group scheme ...) over X. Another object Y' of the same type over X is a *twisted-form* of Y for the flat (respectively étale, Zariski) topology on X if there exists a covering $\mathcal{U} = (U_i \to X)$ for the flat (respectively étale, Zariski) topology such that $Y \times_X U_i \approx Y' \times_X U_i$ for all i. Any such twisted-form Y' defines an element $c(Y') \in \check{H}^1(\mathcal{U}/X, \underline{\mathrm{Aut}}\,(Y))$, where $\underline{\mathrm{Aut}}\,(Y)$ is the sheaf associated with the presheaf $U \mapsto \mathrm{Aut}_U\,(Y \times_X U)$, as follows: let ϕ_i be an isomorphism $Y \times U_i \to Y' \times U_i$; then (α_{ij}), where $\alpha_{ij} = \phi_i^{-1}\phi_j$, is a 1-cocycle representing $c(Y')$. The class $c(Y')$ is well-defined, and two twisted-forms Y' and Y'' are isomorphic over X if and only if $c(Y') = c(Y'')$. Any element of $\check{H}^1(\mathcal{U}/X, \underline{\mathrm{Aut}}\,(Y))$ defines a descent datum on $Y \times U$ where $U = \coprod U_i$, that is, an isomorphism ϕ satisfying the conditions of (I.2.23). The map $Y' \mapsto c(Y')$ thus defines an injection from the set of isomorphism classes of twisted-forms of Y that become trivial when restricted to $\mathcal{U}$ into $H^1(\mathcal{U}/X, \underline{\mathrm{Aut}}\,(Y))$, and this injection is surjective whenever every descent datum on $Y \times U$ arises from a twisted-form.

For example, $GL_n = \underline{\mathrm{Aut}}_{\mathcal{O}_X}(\mathcal{O}_X^n)$, and so $\check{H}^1(\mathcal{U}/X, GL_n)$ is equal to the set of isomorphism classes of $\mathcal{O}_X$-modules that on each U_i become isomorphic to $\mathcal{O}_{U_i}^n$. Descent theory shows that such a module is locally free (of rank n) as an $\mathcal{O}_X$-module on X_{Zar} in the usual sense. Thus, on passing to the limit over all coverings, we obtain isomorphisms $\check{H}^1(X_E, GL_n) \approx$ {isomorphism classes of locally free $\mathcal{O}_X$-modules of rank n}, where $E =$ (Zar), (ét), or (fl).

As another example, the twisted-forms of $\mathbb{P}_X^n$ for the étale topology are called *Brauer-Severi* schemes over X. It is known that

$$\underline{\mathrm{Aut}}\,(\mathbb{P}_X^n) = PGL_n = \underline{\mathrm{Aut}}\,(M_n(\mathcal{O}_X))$$

(Mumford [1, p. 20]). Thus there is an injection from the set of isomorphism classes of Brauer-Severi schemes over X into

$$\check{H}^1(X_{\mathrm{et}}, PGL_n)\,(= \check{H}^1(X_{\mathrm{fl}}, PGL_n)).$$

According to [SGA. 1, VIII.7.8] this is a bijection.

Exercise 4.24. (a). Show that for any Brauer-Severi scheme Y/X there is a dual Brauer-Severi scheme Y^0/X such that for any $U \to X$ with $Y \times U \approx \mathbb{P}_U^n$, $Y^0 \times U$ is canonically isomorphic to the dual projective space of hyperplanes in $Y \times U$. Moreover, $c(Y^0) = c(Y)^{-1}$.

(b) Suppose that $Y \to X$ admits a section, let D be the divisor on Y^0 that, on each geometric fiber $Y_{\bar{x}}^0$, is the divisor corresponding to the point of $Y_{\bar{x}}\,(= \mathbb{P}^n)$ defined by the section, and let L be the line-bundle on Y^0 corresponding to D. Show that $\pi_* L$, where π is the structure morphism

$Y^0 \to X$, is locally free and that Y^0 is isomorphic to the projective bundle $\mathbb{P}(\pi_* L)$ and $Y \approx \mathbb{P}(\pi_* L^{\check{}})$.

(c) Deduce that Y has a section over X if and only if $c(Y)$ is in the image of

$$\check{H}^1(X, GL_n) \to \check{H}^1(X, PGL_n).$$

(See also Serre [7, X.6].)

Exercise 4.25. Let X be a smooth proper variety over an algebraically closed field of characteristic $p \neq 0$. Show there exists a canonical map $H^2(X_{\text{fl}}, \mu_p) \to H^1(X, \Omega^1_{cl})$, which is injective if and only if every global differential 1-form on X is closed. (Use the methods of the proof of (4.14).) When is the composite $H^2(X_{\text{fl}}, \mu_p) \to H^1(X, \Omega^1_{cl}) \to H^2_{DR}(X)$ into algebraic deRham cohomology injective?

Comments on the Literature

Most of the material in the first three sections may again be found in Artin [1] and [SGA. 4]. A very thorough treatment of nonabelian and abelian first and second cohomology groups and of the explicit representations of them may be found in Giraud [2].

CHAPTER IV

The Brauer Group

The Brauer group of a field K may be defined to be the group of similarity classes of central simple algebras over K or, equivalently, the cohomology group

$$H^2(\mathrm{Gal}\,(K_{\mathrm{sep}}/K), K^*_{\mathrm{sep}}) = H^2((\mathrm{spec}\,K)_{\mathrm{et}}, \mathbb{G}_m).$$

Both of these definitions generalize to schemes, but they may give different groups. The first group, that of similarity classes of Azumaya algebras over X, is called the *Brauer group* $\mathrm{Br}\,(X)$ of X, and the second, $H^2(X_{\mathrm{et}}, \mathbb{G}_m)$, is called the *cohomological Brauer group* $\mathrm{Br}'\,(X)$ of X. There is always an injection $\mathrm{Br}\,(X) \hookrightarrow \mathrm{Br}'\,(X)$, which is known to be surjective in many cases. It would be useful to know exactly when this map is surjective, that is, when every cohomology class in $H^2(X_{\mathrm{et}}, \mathbb{G}_m)$ is represented by an Azumaya algebra, for much the same reasons as it is useful to know that every cohomology class in $H^1(X, \mathbb{G}_m)$ is represented by an invertible sheaf. From a geometric point of view, the Brauer group classifies the cohomology 2-classes that do not arise from an algebraic divisor, that is, it classifies the transcendental classes.

§1. The Brauer Group of a Local Ring

Throughout this section, R will be a commutative local ring with maximal ideal $\mathfrak{m}$ and A a not necessarily commutative algebra over R. We assume that A has an identity element and that the map $R \to A$, $r \mapsto r1$, identifies R with a subring of the center of A. Residue class maps modulo $\mathfrak{m}$ will be written $a \mapsto \bar{a}$. *Ideal* will mean two-sided ideal. We assume as known the theory of Brauer groups over fields (see, for example, Herstein [1, Chapter 4] or Blanchard [1]).

Let A° denote the opposite algebra to A, that is, the algebra with the multiplication reversed. A is called an *Azumaya algebra* over R if it is free of finite rank as an R-module and if the map $A \otimes_R A^\circ \to \mathrm{End}_{R\text{-mdl}}\,(A)$ that sends $a \otimes a'$ to the endomorphism $(x \mapsto axa')$ is an isomorphism. (Compare Bourbaki [2, II.5, Ex. 14].)

PROPOSITION 1.1. *Let A be an Azumaya algebra over R. Then A has center R; moreover, for any ideal $\mathfrak{J}$ of A, $\mathfrak{J} = (\mathfrak{J} \cap R)A$, and for any ideal $\mathfrak{J}$ of R, $\mathfrak{J} = (\mathfrak{J}A) \cap R$. Thus $\mathfrak{J} \mapsto \mathfrak{J} \cap R$ gives a bijection between the ideals of A and those of R.*

Proof. Let ϕ be an endomorphism of A as an R-module. As ϕ is multiplication by an element of $A \otimes A^{\circ}$, it follows that $\phi(ac) = \phi(a)c$ for any c in the center of A and that $\phi(\mathfrak{J}) \subset \mathfrak{J}$ for any ideal $\mathfrak{J}$ of A.

Let $a_1 = 1, a_2, \ldots, a_n$ be a basis for A as an R-module, and let $\chi_1, \ldots, \chi_n$ be R-linear endomorphisms of A such that $\chi_i(a_j) = \delta_{ij}$ (Kronecker delta). Let $c \in$ center (A), and write it as $c = \sum r_i a_i$, $r_i \in R$. Then

$$c = \chi_1(a_1)c = \chi_1(a_1 c) = \chi_1(\textstyle\sum r_i a_i) = r_1 \in R.$$

Now let $\mathfrak{J}$ be an ideal of A, and let $a \in \mathfrak{J}$, $a = \sum r_i a_i$, $r_i \in R$. Then $r_i = \chi_i(a) \in \mathfrak{J}$, and so $a \in (\mathfrak{J} \cap R)A$.

Finally, let $\mathfrak{J}$ be an ideal of R, and let $a \in \mathfrak{J}A$. Then $a = \sum r_i a_i$ for some $r_i \in \mathfrak{J}$, and $a \in R$ if and only if $r_i = 0$ for $i > 1$. Thus $\mathfrak{J}A \cap R = \mathfrak{J}$.

In particular, if $R = k$ is a field, then an Azumaya algebra over k is a central simple algebra over k (and conversely, by a well-known theorem).

PROPOSITION 1.2. (a) *If A is an Azumaya algebra over R and R' is a commutative local R-algebra, then $A \otimes_R R'$ is an Azumaya algebra over R' (we do not require the map $R \to R'$ to be local).*

(b) *If A is free of finite rank as an R-module and $\bar{A} = A \otimes (R/\mathfrak{m})$ is an Azumaya algebra over $R/\mathfrak{m}$, then A is an Azumaya algebra over R.*

Proof. For any R-algebra A that is free and of finite rank as an R-module and any (commutative) R-algebra R', there is a commutative diagram:

$$\begin{array}{ccc}
\phi \otimes R' : (A \otimes_R A^{\circ}) \otimes R' & \to & \mathrm{End}_{R\text{-mdl}}(A) \otimes R' \\
\downarrow \approx & & \downarrow \approx \\
\phi' : (A \otimes R') \otimes (A \otimes R')^{\circ} & \to & \mathrm{End}_{R'\text{-mdl}}(A \otimes R').
\end{array}$$

Thus, if ϕ is an isomorphism, then ϕ' is also. Conversely, it is easy to see that if $\phi \otimes (R/\mathfrak{m})$ is an isomorphism, then so also is ϕ. (It is surjective according to Nakayama's lemma and injective according to (1.11) below.)

COROLLARY 1.3. (a) *If A and A' are Azumaya algebras over R, then $A \otimes_R A'$ is an Azumaya algebra over R.*

(b) *The matrix ring $M_n(R)$ is an Azumaya algebra over R.*

Proof. Both parts follow from (1.2b) and the corresponding statements for fields.

We now define the *Brauer group* $Br(R)$ of R. Two Azumaya algebras A and A' over R are said to be *similar* if $A \otimes_R M_n(R) \approx A' \otimes_R M_{n'}(R)$

for some n and n'. Similarity is an equivalence relation, and if A_1 is similar to A_1' and A_2 to A_2', then $A_1 \otimes_R A_2$ is similar to $A_1' \otimes_R A_2'$ (because $M_n(R) \otimes M_m(R) \approx M_{nm}(R)$). Write $[A]$ for the similarity class of A. Then the similarity classes form a group under the law of composition $[A][A'] = [A \otimes_R A']$. The identity element is $[R]$, and $[A^\circ]$ is the inverse of $[A]$. This group is called the Brauer group of R. Clearly Br $(-)$ is a functor from local rings to **Ab**.

PROPOSITION 1.4. (Skolem-Noether). *Let A be an Azumaya algebra over R. Every automorphism of A as an R-algebra is inner, that is, of the form $a \mapsto uau^{-1}$ with u a unit in A.*

Proof. Let $\phi: A \to A$ be an automorphism of A. It is possible to make A into a left $A \otimes_R A^\circ$-module in two different ways, namely,

$$\begin{cases} (a_1 \otimes a_2^\circ)a = a_1 a a_2 \\ (a_1 \otimes a_2^\circ)a = \phi(a_1) a a_2. \end{cases}$$

We denote the resulting $A \otimes_R A^\circ$-modules by A and A' respectively. Both $\bar{A}' = A' \otimes_R R/\mathfrak{m}$ and $\bar{A}$ are simple left $\bar{A} \otimes_{\bar{R}} \bar{A}^\circ$-modules, and since $\bar{A} \otimes \bar{A}^\circ$ is a central simple algebra over $\bar{R}$, there is an isomorphism $\bar{\psi}: \bar{A} \to \bar{A}'$ of $\bar{A} \otimes_{\bar{R}} \bar{A}^\circ$-modules.

Next we show that A is projective as an $A \otimes A^\circ$-module. Since $A \otimes A^\circ \approx \mathrm{End}_{R\text{-mdl}}(A)$, it suffices to show that A is projective as an End (A)-module. As A is a free R-module, there exists a homomorphism $g: A \to R$ of R-modules such that $g(r) = r$ for all $r \in R$. The surjection $\mathrm{End}_{R\text{-mdl}}(A) \to A$, $f \mapsto f(1)$ has an $\mathrm{End}_{R\text{-mdl}}(A)$-module section $a \mapsto (a' \mapsto g(a')a)$, which shows that A is projective.

Now the map $A \to \bar{A} \xrightarrow{\bar{\psi}} \bar{A}'$ lifts to a homomorphism of $A \otimes A^\circ$-modules, $\psi: A \to A'$. The surjectivity of $\bar{\psi}$ implies that $\psi(A) + \mathfrak{m}A' = A'$, and Nakayama's lemma applied to A' as an R-module shows that ψ is surjective. Let $u = \psi(1)$; then for any $a \in A$, $\psi(a) = \psi(a1) = \phi(a)u$ and $\psi(a) = \psi(1a) = ua$. Hence $\phi(a)u = ua$ for all $a \in A$, and it remains for us to check that u is a unit. But if $a_0 \in A$ is such that $\psi(a_0) = 1$, then $1 = \psi(a_0) = \phi(a_0)u$, and $\phi(a_0) = u^{-1}$.

COROLLARY 1.5. *The automorphism group of $M_n(R)$ (as an R-algebra) is $PGL_n(R) = GL_n(R)/R^*$.*

Proof. $GL_n(R)$ is, by definition, the group of units of $M_n(R)$, and the inner automorphism defined by $U \in GL_n(R)$ is the identity map if and only if U is in the center of $M_n(R)$.

PROPOSITION 1.6. *If R is Henselian, then the canonical map* Br $(R) \to$ Br$(\bar{R})$, *where $\bar{R} = R/\mathfrak{m}$, is injective.*

Proof. Let A be an Azumaya algebra over R such that $\bar{A} \xrightarrow{\approx} M_n(R)$, and let $\varepsilon \in \bar{A}$ map to the matrix with 1 in the (1, 1)-position and 0 elsewhere. Note that ε is idempotent, $\varepsilon^2 = \varepsilon$. Choose an $a \in A$ such that $\bar{a} = \varepsilon$. Then $R[a]$ is a finite commutative R-subalgebra of A, and (I.4.2 and 4.3) imply that ε lifts to an idempotent e in $R[a]$. As $A = Ae \oplus A(1 - e)$, Ae is a finitely generated free R-module, and so it remains to show that $\phi: A \to \mathrm{End}_R(Ae)$, $\phi(a) =$ left multiplication by a, is an isomorphism. The kernel of ϕ is an ideal in A whose intersection with R is zero (as Ae is free over R); thus (1.1) ϕ is injective. The same argument shows that $\bar{\phi}$ is injective, and as $\bar{A}$ and $\mathrm{End}_{\bar{R}}(\bar{A}\varepsilon)$ have the same dimension, $\bar{\phi}$ is an isomorphism; now Nakayama's lemma shows that ϕ is surjective.

COROLLARY 1.7. *For any strictly local ring R, $Br(R) = 0$.*

Proof. The Brauer group of a separably closed field is zero.

COROLLARY 1.8. *If A is an Azumaya algebra over a Henselian local ring R, then there exists a finite étale faithfully flat homomorphism $R \to R'$ such that $A \otimes_R R' \approx M_n(R')$.*

Proof. This follows from (1.6), (I.4.4), and the fact that it is true for R a field.

Remark 1.9. We shall prove later (2.13) that the map $Br(R) \to Br(\bar{R})$ is an isomorphism for local Henselian rings R.

We say that an R-algebra R' *splits* an Azumaya R-algebra A if $A \otimes_R R' \approx M_n(R')$. According to (1.7), any A is split by a faithfully flat R-algebra.

THEOREM 1.10. *Let A be an Azumaya R-algebra of rank n^2.*

(a) *Let $a \in A$ and let R' be a faithfully flat R-algebra that splits A, for example, $\phi: A \otimes R' \xrightarrow{\approx} M_n(R')$; the characteristic polynomial $c_a(T)$ of the matrix $\phi(a \otimes 1)$ belongs to $R[T]$ and is independent of R' and ϕ; it is called the Cayley-Hamilton polynomial of a; $c_a(a) = 0$.*

(b) *There exists a commutative étale subalgebra R' of A of rank n, which is a direct summand of A as an R-module; such a subalgebra is called a maximal étale subalgebra.*

(c) *Any maximal étale subalgebra of A splits it.*

Proof. (a) We first remark that if ϕ_1, ϕ_2 are isomorphisms $A \otimes_R R' \rightrightarrows M_n(R')$, where R' is faithfully flat over R, then $\phi_1(a \otimes 1)$ and $\phi_2(a \otimes 1)$ have the same characteristic polynomials. Indeed, for any maximal ideal $\mathfrak{m}$ of R', (1.4) shows that there is a $u \in GL_n(R'_{\mathfrak{m}})$ such that $\phi_2(a \otimes 1) = u\phi_1(a \otimes 1)u^{-1}$ in $M_n(R_{\mathfrak{m}})$, and so the characteristic polynomials have the same image in $R'_{\mathfrak{m}}[T]$ for all $\mathfrak{m}$; this proves the remark.

Write $\phi = \phi_1$ and let $c_a(T)$ be the characteristic polynomial of $\phi(a \otimes 1)$. The remark applied to $R' \otimes R'$ shows that the images of $c_a(T)$ under the two maps $R'[T] \rightrightarrows R' \otimes_R R'[T]$ agree. As

$$R[T] \to R'[T] \rightrightarrows R'[T] \otimes_{R[T]} R'[T] = R' \otimes_R R'[T]$$

is exact (I.2.18), $c_a(T) \in R[T]$.

The independence assertion follows easily from the special case shown in the above remark. The final statement follows from the fact that $c_a(\phi(a \otimes 1)) = 0$ in $M_n(R')$.

(b) Choose an $a \in A$ such that $\bar{R}[\bar{a}]$ is a maximal étale subalgebra of $\bar{A}$. Then $\bar{R}[\bar{a}]$ has rank n over $\bar{R}$. Let $R' = R[T]/(c_a(T))$; it is an étale R-algebra of rank n, and there is a canonical map $R' \to A$, $T \mapsto a$. As $R' \otimes_R \bar{R} \xrightarrow{\approx} \bar{R}[\bar{a}] \hookrightarrow \bar{A}$ is injective, the following standard result shows that $R' \to A$ is injective and R' is a direct summand of A.

LEMMA 1.11. *Let M and N be finitely generated R-modules with N free. If $\phi: M \to N$ is an R-linear map such that $\bar{\phi}: \bar{M} \to \bar{N}$ is injective, where $\bar{M} = M \otimes_R \bar{R}$* etc, *then ϕ has a section. If $\bar{\phi}$ is an isomorphism, then so also is ϕ.*

Proof. Let $\phi': N \to M$ be such that $\bar{\phi}'\bar{\phi} = id_{\bar{M}}$, and let $\psi = \phi'\phi$. According to Nakayama's lemma, $\psi: M \to M$ is surjective. If M is regarded as an $R[T]$-module by means of ψ, then Atiyah-Macdonald [1, 2.5] shows that there exists a polynomial $f(T)$ such that $(1 - f(\psi)\psi)M = 0$, that is, such that $f(\psi)\psi = id_M$. Now $f(\psi)\phi': N \to M$ has the property that $(f(\psi)\phi')\phi = id_M$. The second statement is a consequence of the first and Nakayama's lemma.

Proof (*of* (1.10c)). Regard A as a right R'-module. The map

$$a_0 \otimes r' \mapsto (a \mapsto a_0 a r'): A \otimes R' \to \mathrm{End}_{R'\text{-mdl}}(A)$$

is well-known to be an isomorphism modulo $\mathfrak{m}$ and hence is an isomorphism.

§2. The Brauer Group of a Scheme

Let X be a scheme. An $\mathcal{O}_X$-algebra A is called an *Azumaya algebra* over X if it is coherent as an $\mathcal{O}_X$-module and if, for all closed points x of X, A_x is an Azumaya algebra over the local ring $\mathcal{O}_{X,x}$. The conditions imply that A is locally free and of finite rank as $\mathcal{O}_X$-module (I.2.9), and that for any point x of X, A_x is an Azumaya algebra over $\mathcal{O}_{X,x}$ (1.2a).

PROPOSITION 2.1. *Let A be an $\mathcal{O}_X$-algebra that is of finite-type as an $\mathcal{O}_X$-module. The following are equivalent:*

(a) *A is an Azumaya algebra over X;*

(b) *A is locally free as an $\mathcal{O}_X$-module and $A(x) \stackrel{df}{=} A_x \otimes k(x)$ is a central simple algebra over $k(x)$ for all x in X;*

(c) *A is locally free as an $\mathcal{O}_X$-module and the canonical homomorphism $A \otimes_{\mathcal{O}_X} A^\circ \to \underline{\mathrm{End}}_{\mathcal{O}_X\text{-mdl}}(A)$ is an isomorphism;*

(d) *there is a covering $(U_i \to X)$ for the étale topology on X such that for each i, there exists an r_i, for which $A \otimes_{\mathcal{O}_X} \mathcal{O}_{U_i} \approx M_{r_i}(\mathcal{O}_{U_i})$;*

(e) *there is a covering $(U_i \to X)$ for the flat topology on X such that for each i, there exists an r_i, for which $A \otimes_{\mathcal{O}_X} \mathcal{O}_{U_i} \approx M_{r_i}(\mathcal{O}_{U_i})$.*

Proof. (a) $\Leftrightarrow$ (b). This follows from (1.2b).

(a) $\Leftrightarrow$ (c). As A is locally free, $(A \otimes A^\circ)_x = A_x \otimes A_x^\circ$ and $(\mathrm{End}_{\mathcal{O}_X}(A))_x = \mathrm{End}_{\mathcal{O}_{X,x}}(A_x)$. Thus this equivalence follows from the definitions.

(a) $\Rightarrow$ (d). For any geometric point $\bar{x}$ of X, $A \otimes \mathcal{O}_{X,\bar{x}} \approx M_r(\mathcal{O}_{X,\bar{x}})$ (1.7). It follows that there exists an étale morphism $U \to X$ whose image contains x such that $A \otimes_{\mathcal{O}_X} \mathcal{O}_U \approx M_r(\mathcal{O}_U)$.

(d) $\Rightarrow$ (e). Trivial.

(e) $\Rightarrow$ (b). Let $U = \coprod U_i$. As U is faithfully flat over X, $A \otimes \mathcal{O}_U$ being flat as an $\mathcal{O}_U$-module implies that A is flat and hence locally free, as an $\mathcal{O}_X$-module. Also (e) implies that $A(x) \otimes_{k(x)} k' \approx M_r(k')$ for some extension field k' of $k(x)$, and it is well-known that this implies that $A(x)$ is a central simple algebra over $k(x)$.

Remark 2.2. (a) Let $X = \operatorname{spec} R$ be affine. An Azumaya algebra over X corresponds to an R-algebra A. The conditions in (2.1c) say exactly that A is projective and finitely generated as an R-module and that the canonical map $A \otimes_R A^\circ \to \mathrm{End}_{R\text{-mdl}}(A)$ is an isomorphism (note that $(A \otimes_R A^\circ)^\sim \approx \tilde{A} \otimes_{\mathcal{O}_X} \tilde{A}^\circ$ and $(\mathrm{End}_R(A))^\sim \approx \underline{\mathrm{End}}_{\mathcal{O}_X\text{-mdl}}(\tilde{A})$, because $\tilde{A}$ is coherent). Thus the notion of an Azumaya algebra over X corresponds exactly to that of a central separable algebra over R in the sense of Auslander-Goldman [1].

(b) Condition (2.1d) holds in a stronger form: there is a Zariski covering (U_i) of X and finite surjective étale maps $U_i' \to U_i$ such that, for all i, $A \otimes_{\mathcal{O}_X} \mathcal{O}_{U_i'} \approx M_{r_i}(\mathcal{O}_{U_i'})$. This follows from (1.10b,c).

We now define the Brauer group of X. Two Azumaya algebras A and A' over X are said to be *similar* if there exist locally free $\mathcal{O}_X$-modules E and E', of finite rank over $\mathcal{O}_X$, such that

$$A \otimes_{\mathcal{O}_X} \underline{\mathrm{End}}_{\mathcal{O}_X}(E) \approx A' \otimes_{\mathcal{O}_X} \underline{\mathrm{End}}_{\mathcal{O}_X}(E').$$

The similarity relation is an equivalence relation, because $\underline{\mathrm{End}}(E) \otimes \underline{\mathrm{End}}(E') \approx \underline{\mathrm{End}}(E \otimes E')$. Clearly the tensor product of two Azumaya algebras is an Azumaya algebra (use (1.3a)), and this operation is compatible with the similarity relation. The set of similarity classes of Azumaya algebras on X becomes a group under the operation $[A][A'] = [A \otimes A']$: the identity element is $[\mathcal{O}_X]$ and $[A]^{-1} = [A^\circ]$. This is the *Brauer group* $\mathrm{Br}(X)$ of X. $\mathrm{Br}(-)$ is a functor from schemes to abelian groups.

In relating $\mathrm{Br}(X)$ to the cohomology group $H^2(X_{\mathrm{et}}, \mathbb{G}_m)$, we shall need the following generalization of the Skolem-Noether theorem.

PROPOSITION 2.3. *Let A be an Azumaya algebra on X. Any automorphism ϕ of A is locally, for the Zariski topology on X, an inner automorphism, that is, there is a covering of X by open sets U_i such that $\phi | U_i$ is of the form $a \mapsto uau^{-1}$ for some $u \in \Gamma(U_i, A)^*$.*

Proof. Let $x \in X$, and let U be a neighborhood of x such that there exists $u \in \Gamma(U, A)^*$ with the property that $\phi_x(a) = u_x^{-1} a u_x$ for all $a \in A_x$ (these exist by (1.4)). Then the maps $\phi: A|U \to A|U$ and $(a \mapsto u^{-1}au)$: $A|U \to A|U$ are equal on some neighborhood $V \subset U$ of x.

Let GL_n be the functor **Sch** $\to$ **Gp** such that

$$GL_n(S) = GL_n(\Gamma(S, \mathcal{O}_S)) = M_n(\Gamma(S, \mathcal{O}_S))^*$$

for all schemes S. Then GL_n is representable by

$$\operatorname{spec}\left(\frac{\mathbb{Z}[T_{11}, \ldots, T_{nn}, T]}{(T \det(T_{ij}) - 1)}\right)$$

and so defines a sheaf on X for the flat, or any coarser, topology (II.1.7).

Let PGL_n be the functor **Sch** $\to$ **Gp** such that $PGL_n(S) = \operatorname{Aut}(M_n(\mathcal{O}_S))$ (automorphisms of $M_n(\mathcal{O}_S)$ as a sheaf of $\mathcal{O}_S$-algebras). Then PGL_n is also representable and so defines a sheaf on X for the flat (or coarser) topology. Indeed, any automorphism of $M_n(\mathcal{O}_S)$ as an $\mathcal{O}_S$-algebra may also be regarded as an endomorphism of $M_n(\mathcal{O}_S)$ as an $\mathcal{O}_S$-module. Thus PGL_n is a subfunctor of M_{n^2}. The condition that an endomorphism be an automorphism of algebras is described by polynomials, and hence PGL_n is represented by a closed subscheme of $M_{n^2} = \operatorname{spec} \mathbb{Z}[T_{11}, \ldots, T_{n^2n^2}]$. In fact, PGL_n is represented by spec S where S is the subring of elements of degree zero in $\mathbb{Z}[T_{11}, \ldots, T_{nn}, \det(T_{ij})^{-1}]$.

The next result is an immediate consequence of the Skolem-Noether theorem.

COROLLARY 2.4. *The following sequence*

$$1 \to \mathbb{G}_m \to GL_n \to PGL_n \to 1$$

is exact as a sequence of sheaves on X_{Zar}, X_{et}, or X_{fl}.

THEOREM 2.5. *There is a canonical injective homomorphism* $\operatorname{Br}(X) \to H^2(X_{\text{et}}, \mathbb{G}_m)$.

Proof. We give one proof based on Čech cohomology, which assumes X satisfies the hypotheses of (III.2.17), and sketch a second, based on Giraud's nonabelian cohomology, which is general.

Step 1. The set of isomorphism classes of Azumaya algebras of rank n^2 over X is equal to $\check{H}^1(X_{\text{et}}, PGL_n)$.

Proof. Since, by definition, an Azumaya algebra is a twisted-form of $M_n(\mathcal{O}_X)$ for the étale topology and $\underline{\operatorname{Aut}}(M_n(\mathcal{O}_X)) = PGL_n$, this is a special case of the theory discussed at the end of (III.4). The only problem is in seeing that any 1-cocycle does arise from an Azumaya algebra. A section of PGL_n may be regarded as an automorphism of M_n as an $\mathcal{O}_X$-module. Thus a 1-cocycle for PGL_n defines a 1-cocycle for GL_{n^2}, which we know (III.4) arises from a locally free $\mathcal{O}_X$-module of rank n^2. The fact that the 1-cocycle came from a 1-cocycle for PGL_n means that this locally free $\mathcal{O}_X$-module automatically has the structure of an Azumaya algebra.

Step 2. The set of isomorphism classes of locally free modules of rank n over X is equal to $\check{H}^1(X_{\text{et}}, GL_n)$; the map $\check{H}^1(X_{\text{et}}, GL_n) \to \check{H}^1(X_{\text{et}}, PGL_n)$ defined by the surjection $GL_n \to PGL_n$ sends an $\mathcal{O}_X$-module E to $\underline{\text{End}}_{\mathcal{O}_X}(E)$.

Proof. For the first statement see (III.4). For the second, let E be an $\mathcal{O}_X$-module and $\mathcal{U} = (U_i)$ a Zariski covering of X for which there exist isomorphisms $\phi_i : \mathcal{O}_{U_i}^n \to E|U_i$. Then E corresponds to the 1-cocycle $(\phi_i^{-1}\phi_j)$. Let $A = \underline{\text{End}}(E)$. There are isomorphisms $\psi_i : M_n(\mathcal{O}_{U_i}) \to A|U_i$, $\psi_i(a) = \phi_i a \phi_i^{-1}$. Thus A corresponds to the 1-cocycle $(\psi_i^{-1}\psi_j) = (\alpha_{ij})$, where $\alpha_{ij}(a) = \phi_i^{-1}\phi_j a \phi_j^{-1}\phi_i$, $a \in \Gamma(U_{ij}, M_n)$. This is the image of $(\phi_i^{-1}\phi_j)$ because $GL_n \to PGL_n$ maps u to the automorphism of M_n, $(a \mapsto uau^{-1})$.

Step 3. Assume that X satisfies the hypotheses of (III.2.17). There is an exact sequence of pointed sets,

$$\cdots \to \check{H}^1(X_{\text{et}}, \mathbb{G}_m) \to \check{H}^1(X_{\text{et}}, GL_n) \to \check{H}^1(X_{\text{et}}, PGL_n) \xrightarrow{d} \check{H}^2(\check{X}_{\text{et}}, \mathbb{G}_m);$$

moreover, the maps d are compatible for varying n, and $d(c(A \otimes A')) = dc(A)dc(A')$ where $c(A)$ denotes the class in $\check{H}^1(X, PGL_n)$ of an Azumaya algebra A.

Proof. The map d is defined as follows: let $\gamma \in \check{H}^1(X_{\text{et}}, PGL_n)$ be represented by a cocycle (c_{ij}) for the covering (U_i); after refining (U_i), we may assume (III.2.19) that each c_{ij} is the image of an element $c'_{ij} \in \Gamma(U_{ij}, GL_n)$; then $d(\gamma)$ is the class of the 2-cocycle (a_{ijk}) where

$$a_{ijk} = c'_{jk}(c'_{ik})^{-1}c'_{ij} \in \Gamma(U_{ijk}, \mathbb{G}_m).$$

The verification of the exactness and the other statements is routine and hence is omitted. (Compare Giraud [2, IV.3.5].)

We now define $\text{Br}(X) \hookrightarrow \check{H}^2(X, \mathbb{G}_m)$. Let A be an Azumaya algebra over X. If X is connected, then A has constant rank on X and so defines an element $c(A) \in \check{H}^1(X_{\text{et}}, PGL_n)$ for some n. The element $dc(A) \in \check{H}^2(X_{\text{et}}, \mathbb{G}_m)$ depends, according to Step 3, only on the similarity class of A. Thus we have an injection $\text{Br}(X) \to H^2(X_{\text{et}}, \mathbb{G}_m)$, which, according to Step 3 again, is a homomorphism. If X is not connected, then as it is quasi-compact, both $\text{Br}(X)$ and $\check{H}^2(X, \mathbb{G}_m)$ break up into products following the splitting of X into a disjoint sum of its connected components. Thus it suffices to define the map on each component.

We next sketch the general proof. Let $\phi:\mathbf{F} \to \mathbf{C}$ be a functor; for any object U of $\mathbf{C}$ write $\mathbf{F}(U)$ for the category whose objects are those u in $\mathbf{F}$ such that $\phi(u) = U$ and whose morphism are those f such that $\phi(f) = id_U$. Let $f:v \to u$ be a morphism in $\mathbf{F}$, and let $\phi(f:v \to u) = (g:V \to U)$; we say that f is *Cartesian* or that v is the *inverse image* $g^*(u)$ of u with respect to g, if for any v' in $\mathbf{F}(V)$,

$f' \mapsto ff':\mathrm{Hom}_{\mathbf{F}(V)}(v', v) \to \mathrm{Hom}_g(v', u) \overset{\mathrm{df}}{=} \{h \in \mathrm{Hom}(v', u) \mid \phi(h) = g\}$

is an isomorphism:

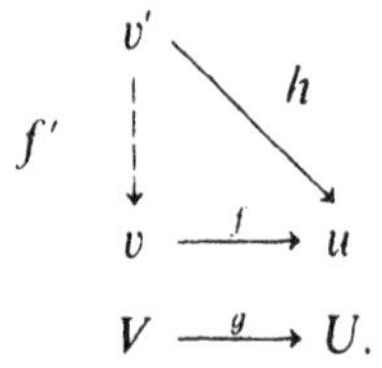

We say $\phi:\mathbf{F} \to \mathbf{C}$ is a *fibered category* if inverse images always exist and if the composite of two Cartesian morphisms is Cartesian. Then g^* can be made into a functor $\mathbf{F}(U) \to \mathbf{F}(V)$, and $(g_1g_2)^*$ is canonically isomorphic to $g_2^*g_1^*$.

Now consider a fibered category $\phi:\mathbf{F} \to \mathbf{C}/X$ where $(\mathbf{C}/X)_E$ is a site. Let $(U_i \xrightarrow{g_i} U)$ be a covering in $(\mathbf{C}/X)_E$. Any $u \in \mathbf{F}(U)$ gives rise to a family (u_i), $u_i = g_i^*u \in \mathbf{F}(U_i)$, and the inverse images of u_i and u_j on $U_{ij} = U_i \times_U U_j$ are isomorphic; moreover, the isomorphisms satisfy the cocycle condition on U_{ijk}, that is, there is a descent datum on the family (u_i). If conversely, every family (u_i), $u_i \in \mathbf{F}(U_i)$, with a descent datum arises from a $u \in \mathbf{F}(U)$ and if, moreover, for any $u_1, u_2 \in \mathbf{F}(U)$, the functor $(V \xrightarrow{g} U) \mapsto \mathrm{Hom}_{\mathbf{F}(V)}(g^*u_1, g^*u_2)$ is a sheaf on $(\mathbf{C}/U)_E$, then ϕ is a *stack* (*champ* in French), (Giraud [2, II.1.2.1]). A stack is a *gerbe* if:

(a) each $\mathbf{F}(U)$ is a groupoid; that is, all morphisms in $\mathbf{F}(U)$ are isomorphisms;

(b) there exists a covering (U_i) of X such that each $\mathbf{F}(U_i)$ is nonempty;

(c) for any U in $\mathbf{C}/X$, any two objects of $\mathbf{F}(U)$ are locally isomorphic, that is, their inverse images on some covering of U are isomorphic.

A gerbe is said to be *trivial* if $\mathbf{F}(X)$ is nonempty. A gerbe $\mathbf{F}$ is *bound* by a sheaf of groups G on $(\mathbf{C}/X)_E$ if for any U in $\mathbf{C}/X$ and any u in $\mathbf{F}(U)$ there are functorial isomorphisms $G(U) \xrightarrow{\approx} \mathrm{Aut}_{\mathbf{F}(U)}(u)$. Giraud defines $H^2(X_E, G)$ to be the set of gerbes bound by G, modulo G-equivalence (Giraud [2, IV.3.1.1]). To avoid confusion when G is abelian, we shall denote this set by $H_g^2(X_E, G)$. To prove the theorem we must show:

(i) there is a canonical isomorphism $H_g^2(X_E, G) \xrightarrow{\approx} H^2(X_E, G)$ when G is abelian;

(ii) there is a canonical injective homomorphism $\mathrm{Br}(X) \to H_g^2(X_{\mathrm{et}}, \mathbb{G}_m)$.

It is easy to describe the map in (ii). Associate with an Azumaya algebra A on X the category $\mathbf{F}_A$ over X_{et} such that an object of $\mathbf{F}_A(U)$ is a pair (E, α) where E is a locally free $\mathcal{O}_U$-module of finite rank and α is an isomorphism $\mathrm{End}(E) \xrightarrow{\approx} A \otimes \mathcal{O}_U$; a morphism $(E, \alpha) \to (E', \alpha')$ is an isomorphism $E \to E'$ such that the obvious diagram commutes. Descent theory shows that $\mathbf{F}_A$ is a stack, and (2.1d) shows that it is a gerbe. The map $\mathbb{G}_m(U) \to \mathrm{Aut}_U(E, \alpha)$ that sends an element of $\Gamma(U, \mathcal{O}_U^*)$ to multiplication on E by that element is an isomorphism. Thus $\mathbf{F}_A$ is bound by $\mathbb{G}_m$ and so defines an element of $H_g^2(X_{\mathrm{et}}, \mathbb{G}_m)$. Clearly the element is trivial if and only if $[A] = 0$ in $\mathrm{Br}(X)$.

For (i) it suffices to check:

(i′) for any exact sequence of abelian sheaves $0 \to G' \xrightarrow{u} G \xrightarrow{v} G'' \to 0$ there is an exact sequence of abelian groups

$$\ldots H^1(X, G'') \to H^1(X, G'') \to H_g^2(X, G') \to H_g^2(X, G) \to H_g^2(X, G'');$$

(i″) if G is an injective abelian sheaf, then $H_g^2(X, G) = 0$.

Let $u: G' \to G$ be a homomorphism of abelian sheaves, and let $\mathbf{F}'$ and $\mathbf{F}$ be gerbes bound by G' and G respectively. $\mathrm{HOM}_u(\mathbf{F}', \mathbf{F})$ is defined to be the fibered category over $\mathbf{C}/X$ whose fiber over U is all functors $\mathbf{F}'|U \to \mathbf{F}|U$ that preserve the fiberings, preserve Cartesian arrows, and that commute with the actions of G' on $\mathbf{F}'$ and G on $\mathbf{F}$ (Giraud [2, IV.2.3.2]). It is in fact a gerbe. The map $H_g^2(u): H_g^2(X, G') \to H_g^2(X, G)$ sends $\mathbf{F}'$ to $\mathrm{HOM}_u(\mathbf{F}', \mathbf{F}_0)$ where $\mathbf{F}_0$ is the trivial gerbe, that is, $\mathbf{F}_0(U) =$ set of torsors for $G|U$.

The group structure on $H_g^2(X, G)$ (G abelian) is defined by the maps

$$H_g^2(X, G) \times H_g^2(X, G) \to H_g^2(X, G \times G) \to H_g^2(X, G)$$

induced by $G \rightrightarrows G \times G \xrightarrow{+} G$.

The map $H^1(X, G'') \to H^2(X, G)$ is defined as follows: let P be a G''-torsor, and let $\mathbf{F}$ be the gerbe such that $\mathbf{F}(U)$ consists of all pairs (Q, α) with Q a G-torsor on U and α an isomorphism of G''-torsors $v_*(Q) \xrightarrow[\approx]{\alpha} P$. As $\mathbf{F}$ is bound by G' we may define its class in $H_g^2(X, G')$ to be the image of P. By definition, $\mathbf{F}$ is trivial if and only if P arises from a G-torsor.

The proof of (i′) now only involves straightforward, but tedious, checking (Giraud [2, IV.3.4]).

For (i″), let $\mathbf{F}$ be a gerbe for the injective sheaf G. There is a covering (U_i) of X such that $\mathbf{F}|U_i$ is trivial. Consider $G \to \pi_*\pi^*G$ where $\pi: \coprod U_i \to X$. As $\mathbf{F}$ maps to zero in $H_g^2(X, \pi_*\pi^*G)$ and as there is a section $G \leftrightarrows \pi_*\pi^*G$, $\mathbf{F}$ is zero in $H_g^2(X, G)$ (Giraud [2, IV.3.4.3]).

COROLLARY 2.6. *Let X be a regular integral scheme and let $K = R(X)$. The canonical map* $\mathrm{Br}(X) \to \mathrm{Br}(K)$ *is injective.*

Proof. This follows from the theorem, using the fact that $H^2(X, \mathbb{G}_m) \to H^2(K, \mathbb{G}_m)$ is injective (III.2.22).

PROPOSITION 2.7. *The image of* $\check{H}^1(X, PGL_n)$ *in* $H^2(X, \mathbb{G}_m)$ *is killed by* n. *Thus, if* X *has only finitely many connected components*, Br (X) *is torsion*.

Proof. We use the flat topology (and, by implication, Giraud [2, IV.3.4.5]). Consider the diagram,

$$\begin{array}{ccccccccc} & & 1 & & 1 & & & & \\ & & \downarrow & & \downarrow & & & & \\ 1 & \longrightarrow & \mu_n & \longrightarrow & \mathbb{G}_m & \xrightarrow{n} & \mathbb{G}_m & \longrightarrow & 1 \\ & & \downarrow & & \downarrow & & \| & & \\ 1 & \longrightarrow & SL_n & \longrightarrow & GL_n & \xrightarrow{\det} & \mathbb{G}_m & \longrightarrow & 1 \\ & & \downarrow & & \downarrow & & & & \\ & & PGL_n & = & PGL_n & & & & \\ & & \downarrow & & \downarrow & & & & \\ & & 1 & & 1 & & & & \end{array}$$

where SL_n is defined to be the kernel of the determinant mapping. The top row is exact (II.2.18), which implies that det is surjective, and the middle row is exact. A diagram chase now shows that the first column is exact.

On comparing the flat cohomology sequences of the first two columns, one finds that the map $H^1(X, PGL_n) \to H^2(X, \mathbb{G}_m)$ factors through $H^2(X, \mu_n)$, which is killed by n. (A much more complicated proof, but which avoids referring to Giraud [2] can be found in Knus and Ojanguren [1, IV.6.1].)

Remark 2.8. By making use of an example of Mumford, Grothendieck [4, II.1.11b] has shown that there exists a normal (but singular) surface X over the complex numbers such that $H^2(X_{\text{et}}, \mathbb{G}_m)$ is not torsion (and Br $(X) \to$ Br (K) is not injective). Thus, the map Br $(X) \to H^2(X, \mathbb{G}_m)$ cannot be surjective in this case. However, it is natural to ask the following question.

Question 2.9. Is Br $(X) \to H^2(X, \mathbb{G}_m)_{\text{tors}}$ surjective for quasi-compact X?

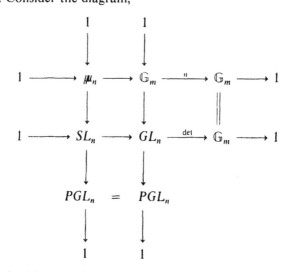

When X is a topological space and $\mathcal{O}_X$ is the sheaf of continuous functions into **C**, it is possible to mimic the above definitions and so define the Brauer group of X. In the case that X is a finite CW-complex it is known that the map $\mathrm{Br}(X) \to H^2(X, \mathcal{O}_X^*)_{\mathrm{tors}}$ is surjective (see Grothendieck [4, I.1]). We shall prove below that the same is true for certain classes of schemes. Apparently there is no example known where the map is not surjective. Positive answers to (2.9) are valuable because they enable cohomological techniques to be applied to the Brauer group and, from the point of view of the cohomology, they allow a cohomology class to be explicitly represented by an Azumaya algebra.

We write $\mathrm{Br}'(X)$ for $H^2(X, \mathbb{G}_m)$ and call it the *cohomological Brauer group*. In view of (2.6) and the fact that the Brauer group of a field is torsion, one may also ask.

Question 2.9′. If X is a regular scheme with only finitely many connected components, then is $\mathrm{Br}(X) \to \mathrm{Br}'(X)$ surjective?

Remark 2.10. For any scheme X, let $\underline{\mathrm{Br}}$ (respectively, $\underline{\mathrm{Br}}'$) be the sheaf on X_{Zar} associated with the presheaf $U \mapsto \mathrm{Br}(U)$ (respectively, $U \mapsto \mathrm{Br}'(U)$). If $f: X_{\mathrm{et}} \to X_{\mathrm{Zar}}$ is the obvious morphism, then $R^2 f_* \mathbb{G}_m = \underline{\mathrm{Br}}'$. Since $R^1 f_* \mathbb{G}_m = 0$ (III.4.9), the Leray spectral sequence gives the first row of the following diagram:

$$
\begin{array}{ccccccccc}
0 & \longrightarrow & H^2(X_{\mathrm{Zar}}, \mathcal{O}_X^*) & \longrightarrow & \mathrm{Br}'(X) & \longrightarrow & \Gamma(X, \underline{\mathrm{Br}}') & \longrightarrow & H^3(X_{\mathrm{Zar}}, \mathcal{O}_X^*) \\
 & & \uparrow & & \uparrow & & \uparrow & \nearrow_{\phi} & \\
0 & \longrightarrow & \mathrm{Br}(X_{\mathrm{Zar}}) & \longrightarrow & \mathrm{Br}(X) & \longrightarrow & \Gamma(X, \underline{\mathrm{Br}}). & &
\end{array}
$$

$\mathrm{Br}(X_{\mathrm{Zar}})$ is the subgroup of $\mathrm{Br}(X)$ generated by Azumaya algebras that are split by a Zariski covering of X. The first vertical map is induced by the other two. We leave it as an exercise to the reader to give an explicit description of ϕ in terms of Azumaya algebras. The question (2.9) can now be broken into three questions.

(a) (Local question) Is $\underline{\mathrm{Br}}(X) \to \underline{\mathrm{Br}}'(X)$ an isomorphism? Equivalently, is $\mathrm{Br}(\mathcal{O}_{X,x}) \to \mathrm{Br}'(\mathcal{O}_{X,x})$ an isomorphism for all $x \in X$?

(b) (Global question). Does every element of $\Gamma(X, \underline{\mathrm{Br}})$ that maps to zero in $H^3(X_{\mathrm{Zar}}, \mathcal{O}_X^*)$ arise from an element in $\mathrm{Br}(X)$?

(c) (Singular question). Is $\mathrm{Br}(X_{\mathrm{Zar}}) \to H^2(X_{\mathrm{Zar}}, \mathcal{O}_X^*)$ an isomorphism? (If X is regular, $H^2(X_{\mathrm{Zar}}, \mathcal{O}_X^*) = 0 = \mathrm{Br}(X_{\mathrm{Zar}})$ and $U \mapsto \mathrm{Br}'(U)$ is a sheaf.)

PROPOSITION 2.11. *Let* $X = \mathrm{spec}\, R$, *where* R *is a local ring, and let* $\gamma \in \mathrm{Br}'(X)$. *The following are equivalent:*

(a) $\gamma \in \mathrm{Br}(X)$;

(b) *there is a finite ètale surjective map* $Y \to X$ *such that* γ *maps to zero in* $\mathrm{Br}'(Y)$;

(c) *there is a finite flat surjective map* $Y \to X$ *such that* γ *maps to zero in* $\mathrm{Br}'(Y)$;

Proof. (a) $\Rightarrow$ (b). This follows from (1.10).

(b) $\Rightarrow$ (c). This is trivial.

(c) $\Rightarrow$ (a). Let $Y = \operatorname{spec} S$. The spectral sequence (III.2.7) for the covering $Y \to X$ gives an exact sequence,

$$\check{H}^0(Y/X, \underline{\text{Pic}}) \to \check{H}^2(Y/X, \mathbb{G}_m) \to \text{Br}'(Y/X) \to \check{H}^1(Y/X, \underline{\text{Pic}}) \to \check{H}^3(Y/X, \mathbb{G}_m)$$

where we have written $\text{Br}'(Y/X)$ for $\ker(\text{Br}'(X) \to \text{Br}'(Y))$. As S and $S \otimes S$ are semilocal rings, their Picard groups are zero, and the above sequence reduces to an isomorphism $\check{H}^2(Y/X, \mathbb{G}_m) \approx \text{Br}'(Y/X)$. Let $u \in (S \otimes S \otimes S)^*$ represent the class in $\check{H}^2(Y/X, \mathbb{G}_m)$ corresponding to γ in $\text{Br}'(Y/X)$, and let E be S regarded as a free R-module of rank n. Multiplication by u defines an $S \otimes S$-linear automorphism $\phi: S \otimes S \otimes E \to S \otimes S \otimes E$ that may be regarded as an element of $GL_n(S \otimes S)$. Writing ϕ_i for ϕ tensored with id in the ith place, we find that

$$\phi_1 \phi_2^{-1} \phi_3 : S \otimes S \otimes S \otimes E \to S \otimes S \otimes S \otimes E$$

is ϕ_4 = multiplication by $u \otimes 1$ (because $\phi_1 \phi_2^{-1} \phi_3 \phi_4^{-1}$ is the coboundary of the 2-cocycle u). As

$$u \otimes 1 \in (S \otimes S \otimes S)^* = \text{center}\,(\text{Aut}_{S \otimes S \otimes S}(S \otimes S \otimes S \otimes E)),$$

the image of ϕ in $PGL_n(S \otimes S)$ is a 1-cocycle and so defines an element $\gamma' \in \check{H}^1(X, PGL_n)$. The image of γ' under the boundary map $\check{H}^1(X, PGL_n) \to \check{H}^2(X, \mathbb{G}_m)$ is obviously γ. Thus any Azumaya algebra with class γ' will represent γ.

COROLLARY 2.12. *If X is the spectrum of a Henselian local ring, then* $\text{Br}(X) = \text{Br}'(X)$.

Proof. Any $\gamma \in \text{Br}'(X)$ satisfies (2.11b), according to (III.2.11) and (I.4.2).

COROLLARY 2.13. *If R is a Henselian local ring, then* $\text{Br}(R) \approx \text{Br}(R/\mathfrak{m})$.

Proof. This follows from (2.12) and (III.3.11a).

Remark 2.14. (a) A direct proof of (b) $\Leftrightarrow$ (a) of the proposition may be given as follows: the map $Y \to X$ may be assumed to be Galois, with Galois group G; the Hochschild-Serre spectral sequence for Y/X shows that γ corresponds to an element of $H^2(G, \Gamma(Y, \mathcal{O}_Y)^*)$; such an element defines a crossed product algebra over X in the same way as for Galois extensions of fields.

(b) The algebra representing γ in (2.11c) can be described by descent theory. Regard $E = S \otimes S$ as a free S-module by letting S act on the first factor and define an $S \otimes S$-linear isomorphism $\phi: S \otimes E \to E \otimes S$ by setting

$$\phi(x \otimes y \otimes z) = \sum a_i x \otimes c_i z \otimes b_i y$$

where

$$u = \sum a_i \otimes b_i \otimes c_i.$$

One checks easily that $\phi_2^{-1}\phi_3\phi_1$ is multiplication by $u_1u_2^{-1}u_3 = u_4$, and so

$$g \mapsto \phi^{-1}g\phi : \mathrm{End}_{S\otimes S}(E \otimes S) \to \mathrm{End}_{S\otimes S}(S \otimes E)$$

is a descent datum on $\mathrm{End}_S(E)$. The corresponding Azumaya algebra A over R, such that $A \otimes_R S \approx \mathrm{End}_S(E)$, represents γ.

(c) Let X be an arbitrary quasi-compact scheme, and let $Y \to X$ be a finite faithfully flat map. The proof of (c) $\Rightarrow$ (a) above can be extended to show that in this case also an element of $\mathrm{Br}'(X)$ that maps to zero in $\mathrm{Br}'(Y)$ arises from an element of $\mathrm{Br}(X)$. See Hoobler [1, Prop.3.1] or prove it in the affine case by comparing the sequence of Chase-Rosenberg [1, Thm. 7.6] with the exact sequence in the proof of (2.11).

PROPOSITION 2.15. *If X is a smooth variety over a field k, then* $\underline{\mathrm{Br}}(X) = \underline{\mathrm{Br}}'(X)$, *that is, every element γ of* $\mathrm{Br}'(X)$ *arises locally from an Azumaya algebra.*

Proof. Let $\gamma \in \mathrm{Br}'(U)$, where U is a Zariski open subset of X. For any point x of U we must show that there exists a Zariski open neighborhood V of x in U and a finite flat surjective map $V' \to V$ such that γ becomes zero in $\mathrm{Br}'(V')$. Clearly we may suppose that k is algebraically closed. If $n\gamma = 0$, then the Kummer sequence shows that γ arises from an element $\gamma' \in H^2(U, \mu_n)$. If the characteristic p of k does not divide n, then $\mu_n \approx \mathbb{Z}/n\mathbb{Z}$, and the existence of Artin neighborhoods implies the existence of a V' on which γ' becomes zero (III.3.16). Now suppose $p = n$. Then (compare III.4.14) there is an exact sequence

$$H^0(U, \Omega^1_{\mathrm{cl}}) \xrightarrow{C-1} H^0(U, \Omega^1) \to H^2(U, \mu_p) \to H^1(U, \Omega^1_{\mathrm{cl}}).$$

Since $H^1(U, \Omega^1_{\mathrm{cl}})$ may be computed using the Zariski topology (Ω^1_{cl} is a coherent sheaf over $\mathcal{O}_{X^{(p)}}$), the image of γ' in $H^1(U, \Omega^1_{\mathrm{cl}})$ becomes zero on some Zariski neighborhood of x. Thus we may assume that γ' arises from some element γ'' of $H^0(U, \Omega^1)$. But then (see the proof of (III.4.14)) there is a Zariski neighborhood V of x and a finite Artin-Schreier covering $V' \to V$ such that $\gamma''|V'$ is in the image of $H^0(V', \Omega^1_{\mathrm{cl}}) \xrightarrow{C-1} H^0(V', \Omega^1)$.

THEOREM 2.16. *Let X be a quasi-compact scheme, and let $\gamma \in$* $\mathrm{Br}'(X)$. *There exists an open subset U of X with* $\mathrm{codim}(X - U) > 1$ *such that $\gamma|U$ is represented by an Azumaya algebra on U. Moreover, if X is regular, then U may be chosen so that* $\mathrm{codim}(X - U) > 2$, *and if X is a smooth affine variety, then U may be taken to be X.*

Proof. We first (partially) answer the local question.

LEMMA 2.17. *Let R be a local ring of dimension ≤ 1. Then* $\mathrm{Br}(R) \xrightarrow{\approx} \mathrm{Br}'(R)$.

Proof. If R is of dimension 0, then it is a local Artin ring and hence Henselian. Thus the result follows from (2.12). Now suppose that R has dimension 1. There is an exact sequence of sheaves on spec R_{et} (compare II.3.9):

$$0 \to \mathbb{G}_{m,R} \to g_*\mathbb{G}_{m,K} \to \mathrm{Div} \to 0$$

where K is the total ring of fractions of R and Div is the sheaf of Cartier divisors. Since Div has support on the closed point spec k of R, there is an exact sequence,

$$H^1(k, \mathrm{Div}) \to \mathrm{Br}'(R) \to \mathrm{Br}(K).$$

Any element of $\mathrm{Br}(K)$ or $H^1(k, \mathrm{Div})$ is killed by a finite flat extension of K or k, and such extensions are induced by a finite flat extension of R. (For example, if L has basis $x_1 = 1, x_2, \ldots, x_n$ over K, and

$$x_i x_j = \sum c_{ijk} x_k$$

with $c_{ijk} \in K$, $dc_{ijk} \in R$, then $L = S \otimes_R K$ where $S \subset L$ has basis $1, dx_2, \ldots, dx_n$ over R.) Thus the lemma follows from (2.11).

Now let X be as in the theorem and let $x_1, \ldots, x_r$ be the generic points of the irreducible components of X. Since the local rings at the x_i have dimension 0, the lemma shows that there exist open neighborhoods $U_1, \ldots, U_r$ (which we may take to be disjoint) of $x_1, \ldots, x_r$ and an Azumaya algebra A_i on each U_i that represents $\gamma | U_i$. Now $U = U_1 \cup \cdots \cup U_r$ is an open dense subset of X, and we have an Azumaya algebra A on it representing $\gamma | U$.

Let $z \notin U$, $\dim(\mathcal{O}_{X,z}) = 1$. According to (2.17), there exists an Azumaya algebra A_z over $\mathcal{O}_{X,z}$ representing $\gamma | \mathrm{spec}(\mathcal{O}_{X,z})$. Let K be the total ring of fractions of $\mathcal{O}_{X,z}$. Then $A_z \otimes K$ is similar to $A \otimes_{\mathcal{O}_U} K$, and so after replacing them by matrix algebras over them if necessary, we may assume they are equal. Thus A extends over z. After a finite number of such extensions, we shall have $\mathrm{codim}(X - U) > 1$.

Before considering a regular scheme X, we recall some definitions and facts about orders (Reiner [1]).

Let R be an integral domain with field of fractions K, and let A_K be a finite K-algebra (not necessarily commutative). An *R-order* in A_K is a sub-R-algebra of A_K that contains a basis for A_K as a K-vector space and that is finitely generated as an R-module. If A is an R-order, then $A_\mathfrak{m}$ is an $R_\mathfrak{m}$-order for all maximal ideals of R and conversely. Thus, given an integral scheme X such that $R(X) = K$, we may define an $\mathcal{O}_X$-order in A_K to be a coherent $\mathcal{O}_X$-algebra A that, locally for the Zariski topology on X, is an order.

If R is integrally closed and A_K is an Azumaya algebra over K, then any subalgebra of A_K that contains a basis for A_K and is integral over R is an order. To show this, we only have to show that A is finitely generated as an R-module, but the reduced trace on A_K defines a nondegenerate bilinear form $A_K \times A_K \to K$, and so the usual proof in the commutative case (Atiyah-Macdonald [1, 5.17]) carries over. It follows that any sequence of orders $A_1 \subset A_2 \subset \cdots$ terminates because $A = \bigcup A_i$ is finitely generated. Thus there exist maximal orders. Clearly this result extends to orders over quasi-compact normal schemes.

Let X be a connected quasi-compact regular scheme, let $\gamma \in \mathrm{Br}'(X)$, and let A_K be the Azumaya algebra over $K = R(X)$ that represents the image of γ in $\mathrm{Br}'(K)$. Choose a maximal $\mathcal{O}_X$-order A in A_K. Clearly the set U of points $x \in X$ such that A_x is an Azumaya algebra over $\mathcal{O}_{X,x}$ is open and dense in X, and the injectivity of $\mathrm{Br}(U) \hookrightarrow \mathrm{Br}'(U) \hookrightarrow \mathrm{Br}'(K)$ implies that $A|U$ represents $\gamma|U$. It remains to show that $\operatorname{codim}(X - U) > 2$ (after possibly replacing A by a matrix algebra over it).

(It should be noted that A being a maximal order in A_K implies that $M_n(A)$ is a maximal order in $M_n(A_K)$. In proving this, we may assume that X is affine, $X = \operatorname{spec} R$. Let B be an order such that $M_n(A) \subset B \subset M_n(A_K)$, and let B_0 be the set of elements of A_K that occur as an entry in some matrix in B. From the matrix identity

$$e_{1k}(b_{ij})e_{l1} = \begin{pmatrix} b_{kl} & 0 \\ 0 & 0 \end{pmatrix},$$

it follows that B_0 is a ring. Since it is generated as an R-module by the entries in any finite set of generators for B and it contains A, it is an order. Thus $B_0 = A$ and $M_n(A) = B$.)

Suppose that there is a point x in $X - U$ such that $R = \mathcal{O}_{X,x}$ has dimension one. A_x is a maximal R-order in A_K, and, from the lemma, we know that there exists another R-order A' in A_K that is an Azumaya algebra representing $\gamma|\operatorname{spec} R$ (after possibly replacing A by a matrix algebra over it). The set $\mathfrak{J} = \{a \in A_K | aA_x \subset A'\}$ is a left ideal in A'. Since left ideals in Azumaya algebras over fields are principal (Herstein [1, 1.4.2]), Nakayama's lemma shows that $\mathfrak{J}$ is principal, $\mathfrak{J} = A'u$. One sees easily that $\mathfrak{J} \cap K \neq 0$, which shows that u is a unit in A_K. Since $\mathfrak{J}$ is a right A_x-module, $uA_x \subset \mathfrak{J} = A'u$, and so $A_x \subset u^{-1}A'u$. The maximality of A_x implies that $A_x = u^{-1}A'u$, and so A_x is an Azumaya algebra. Thus we may assume that $x \in U$.

Write $A^{\check{}} = \underline{\mathrm{Hom}}_{\mathcal{O}_X}(A, \mathcal{O}_X) \subset A_K^{\check{}} = \mathrm{Hom}_K(A_K, K)$. We may identify A_K with $A_K^{\check{}\check{}}$, and the maximality of A shows that the natural injection $A \hookrightarrow A^{\check{}\check{}}$ is an isomorphism.

Let x be a point of X such that $R = \mathcal{O}_{X,x}$ has dimension two, and consider a sequence

$$0 \to M \to F_0 \to A_x^{\check{}} \to 0$$

with F_0 a free and finitely generated R-module. On applying $\mathrm{Hom}_R(-, R)$ to this, we get an exact sequence,

$$0 \to A_x \to F_0^{\check{}} \to N \to 0$$

with N a submodule of $M^{\check{}}$. As R is an integral domain, $M^{\check{}}$ and N are torsion free, and so $N \hookrightarrow N \otimes_R K$. Thus N may be embedded in a free, finitely generated, R-submodule F_1 of $N \otimes_R K$. As R has homological dimension two, the existence of a sequence

$$0 \to A_x \to F_0^{\check{}} \to F_1 \to F_1/N \to 0$$

shows that A_x is free.

The canonical map $A_x \otimes A_x^0 \to \mathrm{End}_R(A_x)$ is injective because it is a submap of $A_K \otimes A_K^0 \to \mathrm{End}_K(A_K)$; let M be its cokernel. As $A_x \otimes A_x^0$ and $\mathrm{End}_R(A_x)$ are both free, M has homological dimension ≤ 1. Thus the maximal ideal of R is not an associated prime of M (see Serre [9, p. IV-36]). But $M \otimes R_{\mathfrak{p}} = 0$ for any prime $\mathfrak{p}$ of height ≤ 1 since we know that for such $\mathfrak{p}$, $A \otimes R_{\mathfrak{p}}$ is an Azumaya algebra. Thus no prime is associated with M, $M = 0$, A_x is an Azumaya algebra, and $x \in U$.

Finally we come to the case that X is a smooth affine variety, say $X = \operatorname{spec} R$. Let $\mathfrak{I}$ be the set of $f \in R$ such that $\gamma|\operatorname{spec} R_f$ is represented by an Azumaya algebra. If we can show that $\mathfrak{I}$ is an ideal, then (2.15) will show that it equals R, and the theorem will be proved. Obviously $fg \in \mathfrak{I}$ if $f \in \mathfrak{I}$, $g \in R$. Let f and $g \in \mathfrak{I}$; it remains to show that $f + g \in \mathfrak{I}$. By definition, there are Azumaya algebras A and B over R_f and R_g respectively representing $\gamma|\operatorname{spec} R_f$ and $\gamma|\operatorname{spec} R_g$. As A_g and B_f are similar on $\operatorname{spec} R_{fg}$, there exist finitely generated, locally free, R_{fg}-modules E and F such that $A_g \otimes \mathrm{End}(E) \approx B_f \otimes \mathrm{End}(F)$. The maps on the Grothendieck groups

$$K_0(R_f) \to K_0(R_{fg}), \qquad K_0(R_g) \to K_0(R_{fg})$$

are surjective; this is obvious for K_0 (finitely generated modules) but as R is regular, this group agrees with K_0 (finitely generated projective modules) according to Bass [1, IX.2.1]. Thus the classes of E and F in $K_0(R_{fg})$ extend to R_f and R_g. Since we may choose E and F to have arbitrarily high rank, the stability theorems (Bass [1, IX.4.1]) show that

E and F themselves extend to locally free, finitely generated, R_f and R_g-modules E' and F'. After replacing A and B by $A \otimes \mathrm{End}(E')$ and $B \otimes \mathrm{End}(F')$ respectively, we find that $A_g \approx B_f$, that is, A and B patch and γ is represented by an Azumaya algebra on spec $R_f \cup$ spec R_g. As this contains spec R_{f+g}, the proof is complete.

Remark 2.18. (a) If X is an affine regular scheme, then arguments similar to the above show that $\mathrm{Br}(X) \xrightarrow{\approx} H^2(X_{\mathrm{etf}}, \mathbb{G}_m)$ where (etf) is the class of all composites of finite étale morphisms and open immersions.

(b) The proof of the second case of (2.17) is essentially that used by Auslander and Goldman to prove that for any regular connected scheme X of dimension 2, $\mathrm{Br}(X)$ is the subgroup $\bigcap_{x \in X_1} \mathrm{Br}(\mathcal{O}_{X,x})$ of $\mathrm{Br}(R(X))$. Hoobler, to whom the proof of the third case of (2.17) is due, shows that (2.17) implies the same result for X a smooth affine variety. The result may be regarded as a noncommutative analogue of the purity of branch locus (I.3.7): if an Azumaya algebra ramifies then it does so on a set of pure codimension one.

(c) The proof of the third case of (2.17) shows that $\mathrm{Br}(X) = \mathrm{Br}'(X)$ if X is a smooth variety with a covering $X = X_1 \cup X_2$ where X_1, X_2 and $X_1 \cap X_2$ are open affine. This holds, for example, if X is a smooth projective curve over a smooth affine variety, a fact that has been exploited by Artin to generalize his result (Grothendieck [4, III.3.1]).

(d) G. Ofer and R. Hoobler have announced proofs that $\mathrm{Br}(X) = \mathrm{Br}'(X)$ for any affine scheme X.

Finally, we mention a question that, because of its relation to Tate's conjecture, is the most interesting concerning the Brauer group.

Question 2.19. (Artin). If X is proper over spec $\mathbb{Z}$, is $\mathrm{Br}(X)$ finite?

For X of dimension 1, class field theory shows that the answer is yes, but already the case of a surface over a finite field presents serious problems.

Exercise 2.20. (a) Show that $\mathrm{Br}(R) = 0$ for any ring R in which every element r satisfies an equation $r^n = r$, where n is an integer ≥ 2 depending on r. (Hint: show that R is a union of finite rings.)

(b) Show that for any perfect field k, $\mathrm{Br}(k) \approx \mathrm{Br}(k[T])$. (Hint: use (III.2.21) for $k_s[T]/k[T]$, and note that $\mathrm{Br}(k_s[T]) \subset \mathrm{Br}(k_s(T)) = 0$.)

(c) Show that if R is a Henselian discrete valuation ring, then $\mathrm{Br}(K) = \mathrm{Br}(\hat{K})$, where K and $\hat{K}$ are the fields of fractions of R and its completion, except possibly for the p-components when $\mathrm{char}(K) = p \neq 0$. (Hint: K and $\hat{K}$ have the same Galois groups (Serre [7, II]), so one only needs to examine $L^* \hookrightarrow \hat{L}^*$ for L a finite extension of K; $\hat{L}^*/L^*$ is uniquely divisible by any integer prime to p.)

(d) Show that for any field k, there is an exact sequence

$$0 \to \mathrm{Br}(k[T]) \to \mathrm{Br}(\tilde{K}) \to \mathbf{X}G(k_s/k) \to 0$$

where $\tilde{K}$ is the field of fractions of $(k[T]_{(T)})^h$ and the first map is induced by $T \mapsto T^{-1}$. (Hint: note that $\tilde{K}$ is the Henselization of $k(\mathbb{P}^1)$ at the point at infinity; use (III.2.23b), (III.1.25), and (III.1.28) to prove it in the case that $k = k_{\text{sep}}$; use (III.2.21) to prove it in general.) (Compare Yuan [1].)

(e) Let R be the ring of integers in a quadratic number field $\mathbb{Q}[\sqrt{d}]$; use elementary number theory (in particular, no class field theory) to show that $\text{Br}(R/\mathbb{Z}) = 0$. (Hint: note that

$$\text{Br}(R/\mathbb{Z}) = \ker(\text{Br}(\mathbb{Q}[\sqrt{d}]/\mathbb{Q}) \to \prod \text{Br}(\mathbb{Q}_p[\sqrt{d}]/\mathbb{Q}_p)$$

where $\mathbb{Q}_p = p$-adic numbers; by the periodicity of the cohomology of $G = \text{Gal}(\mathbb{Q}[\sqrt{d}]/\mathbb{Q})$ the exercise is equivalent to showing that $a \in \mathbb{Q}$ is a norm from $\mathbb{Q}[\sqrt{d}]$ if it is a norm from $\mathbb{Q}_p[\sqrt{d}]$, all p; now use the proof of case $n = 3$ of Serre [10, IV. Thm. 8]. This proof is due to Legendre.)

(f) Let X be a smooth variety over a field of characteristic p. Show that any p-torsion element in $\text{Br}'(X)$ becomes zero in $\text{Br}'(Y)$ for some finite flat morphism $Y \to X$. Deduce that such an element is in $\text{Br}(X)$ (2.14c). (Hint: use the Frobenius map.)

(g) Let A be an Azumaya algebra over a scheme X. Show that there exists a scheme Y_A over X such that, for any X-scheme U, $A|U$ is trivial if and only if Y_A has a point in U. (Hint: use (III.4.24).) (Compare Roquette [1].)

Comments on the Literature

Azumaya algebras were first studied over local rings by Azumaya [1], over arbitrary rings by Auslander and Goldman [1], and over schemes by Grothendieck [4]. Grothendieck was the first to give a satisfactory cohomological description of the Brauer group. We have largely followed these three sources. There is now a large literature on the Brauer group (see DeMeyer and Ingraham [1], Orzech and Small [1], Knus-Ojanguren, [1], and their bibliographies) much of which, unfortunately ignores the powerful methods introduced by Grothendieck. Yu. Manin has used the Brauer group to study the arithmetic and the geometry of cubic surfaces [1].

CHAPTER V

The Cohomology of Curves and Surfaces

The class of constant, or locally constant, sheaves is not closed under the formation of direct images with respect to proper maps (or even closed immersions). The smallest useful class of sheaves containing the finite constant sheaves and having this property is the class of constructible sheaves. The constructible sheaves on a scheme X form an abelian category, which is Noetherian if X is quasi-compact. From another viewpoint, the constructible sheaves are precisely those that can be represented by étale algebraic spaces of finite-type.

The Poincaré duality theorem holds for constructible sheaves on smooth curves (provided the torsion of the sheaves is prime to the characteristic). The Euler-Poincaré characteristic $\sum(-1)^i \dim_{\mathbb{F}_l} H^i(X, F)$ of a constructible sheaf F of $\mathbb{F}_l$-modules on a curve X can be expressed as a sum of local terms. The possibility of wild ramification makes this quite subtle in characteristic p.

The method of computing the cohomology of a surface is very similar to the classical one, and in particular one first constructs a Lefschetz pencil of curves on the surface. The results also are very similar provided the sheaves only have torsion that is prime to the characteristic; for example, the Picard-Lefschetz formula holds. We give a proof using étale cohomology of Castelnuovo's criterion for the rationality of a surface.

§1. Constructible Sheaves: Pairings

A sheaf F on X_{et} is *finite* if $F(U)$ is finite for all quasi-compact U, and it has *finite stalks* if $F_{\bar{x}}$ is finite for all geometric points $\bar{x}$ of X. A sheaf may be finite without having finite stalks and conversely. A sheaf F is *locally constant* or *twisted-constant* if there is a covering $(U_i \to X)$ such that $F|U_i$ is constant for all i.

PROPOSITION 1.1. *Let F be a locally constant sheaf on X_{et}. If F has finite stalks, then it is finite and is represented by a group scheme $\tilde{F}$ that is finite and étale over X.*

Proof. Let (U_i) be a covering of X_{et} such that the restrictions $F|U_i$ are constant. Clearly the abstract group corresponding to $F|U_i$ must be finite. For any quasi-compact scheme U, $(U_i \times_X U \to U)$ contains a finite subcovering. Therefore, $F(U)$ is a subgroup of a finite product $\prod F(U_i \times U)$ of finite groups.

Let $X' = \coprod U_i$; then $F|X'$ is represented by a finite étale group scheme $\tilde{F}'/X'$. The canonical isomorphism $p_1^*(F|X') \to p_2^*(F|X')$, where p_1 and p_2 are the projections $X' \times X' \to X'$, defines a descent datum on $\tilde{F}'$. According to descent theory (I.2.23), $\tilde{F}'$ with its descent datum arises from a group scheme $\tilde{F}$ over X which (I.2.24) is again finite and étale.

Remark 1.2. (a) The scheme $\tilde{F}$ should be considered to be the *éspace étalé* of F in the sense of Godement [1, II.1.2].

(b) Proposition (1.1) has an obvious converse: the sheaf defined by a finite étale group scheme is locally constant with finite stalks.

Let X be connected, and choose a geometric point $\bar{x}$ of it. Recall (I.5.3) that there is a category equivalence between finite $\pi_1(X, \bar{x})$-sets and finite étale X-schemes; according to (1.1), this extends to a category equivalence between finite $\pi_1(X, \bar{x})$-modules and locally constant sheaves F with finite stalks, under which a sheaf F corresponds to its stalk $F_{\bar{x}}$, and a module M corresponds to a sheaf F such that $F(U) = \mathrm{Hom}_{\pi_1}(\mathrm{Hom}_X(\bar{x}, U), M)$ for any scheme U finite and étale over X. (Note that $F(U) = M^{\pi_1(U,\bar{x})}$ if U is connected.) For such an F there is a finite étale morphism $X' \to X$ such that $\tilde{F} \times_X X'$ is a disjoint union of copies of X' and $F|X'$ is constant; for example, X' may be taken to be a Galois closure of $\tilde{F} \to X$ (compare I.5.4). If X is normal and connected with generic point $g: \eta \to X$, then a sheaf F on X with finite stalks is locally constant if and only if $F \to g_* g^* F$ is an isomorphism and the action of $\mathrm{Gal}(k(\bar{\eta})/k(\eta))$ on $F_{\bar{\eta}}$ factors through $\pi_1(X, \bar{\eta})$.

Exercise 1.3. Let X be a normal connected scheme of dimension one, and let F be a generically locally constant sheaf on X_{et} that corresponds to a family $(M, (\phi_v : M_v \to M^{I_v}))$ as in (II.3.16). Show:

(a) F has finite stalks if and only if M, and all M_v, are finite;

(b) F is finite if and only if M^H and $M_v^{H_v}$ are finite for all open subgroups $H \subset G$ and $H_v \subset D_v$;

(c) F is constant if and only if G acts trivially on M and ϕ_v is an isomorphism for all v;

(d) F is locally constant $\Leftrightarrow$ for all v, I_v acts trivially on M and ϕ_v is an isomorphism $\Leftrightarrow$ there exists a finite étale map $X' \to X$ such that $F|X'$ is constant;

(e) F is constructible (see below) if and only if it has finite stalks.

We next show that any sheaf F on X_{et} can be represented by an *éspace étalé*, $\tilde{F}$ provided $\tilde{F}$ is allowed to be an algebraic space rather than a scheme. Recall the following definitions of Artin [9] and Knutson [1].

(a) A pair of maps of X-schemes

$$R \overset{i_1}{\underset{i_2}{\rightrightarrows}} U$$

is an *equivalence relation* if, for all X-schemes T,

$$(i_1, i_2): R(T) \to U(T) \times_{X(T)} U(T)$$

is injective and $R(T) \subset U(T) \times_{X(T)} U(T)$ is an equivalence relation in the usual sense.

(b) A sheaf of sets A on $(\mathbf{Sch}/X)_{\text{et}}$ is an *algebraic space* over X if there is an equivalence relation $i_1, i_2: R \rightrightarrows U$ and a map $U \xrightarrow{u} A$ (regarding U as a sheaf of sets on $(\mathbf{Sch}/X)_{\text{et}}$) such that each map i_j is surjective and étale, $R \xrightarrow{(i_1, i_2)} U \times_X U$ is quasi-compact, and u identifies A with the quotient sheaf U/R, that is, A is the sheaf associated with the presheaf $A' = (T \mapsto U(T)/\sim)$. Since $R(T) \hookrightarrow (U \times_X U)(T)$ for all T and R is a sheaf, it follows that A' satisfies (S_1). Thus (compare the second proof of (II.2.11)), $A' \hookrightarrow A$ and an element s of $A(T)$ is described by giving an étale covering (T_i) of T plus sections $s_i' \in U(T_i)$ and $r_{ij} \in R(T_i \times T_j)$ such that $i_1 \circ r_{ij} = s_i' | T_{ij}$ and $i_2 \circ r_{ij} = s_j' | T_{ij}$. Equivalently, an element of $A(T)$ is described by a diagram

$$\begin{array}{ccccc} T' \times_T T' & \overset{p_1}{\underset{p_2}{\rightrightarrows}} & T' & \longrightarrow & T \\ \downarrow{\scriptstyle r} & & \downarrow{\scriptstyle s'} & & \downarrow{\scriptstyle s} \\ R & \overset{i_1}{\underset{i_2}{\rightrightarrows}} & U & \longrightarrow & A \end{array}$$

in which $T' \to T$ is étale and surjective of finite-type and $s'p_1 = i_1 r$, $s'p_2 = i_2 r$.

(c) A is *locally separated* if $R \to U \times_X U$ is an immersion and is *separated* if $R \to U \times_X U$ is a closed immersion.

(d) A is *reduced*, *normal*, etc. if there is a scheme U as in (b) that is reduced, normal, etc.

(e) A is *étale*, of *finite-type*, etc. over X if there is a scheme U as in (b) that is étale, of finite-type, etc. over X.

(f) Let A and A' be algebraic spaces; a map $s: A \to A'$ can be described by a commutative diagram

$$\begin{array}{ccccc} R & \rightrightarrows & U & \longrightarrow & A \\ \downarrow{\scriptstyle r} & & \downarrow{\scriptstyle s'} & & \downarrow{\scriptstyle s} \\ R' & \rightrightarrows & U' & \longrightarrow & A' \end{array}$$

where $R \rightrightarrows U$ and $R' \rightrightarrows U'$ are étale equivalence relations with quotients A and A'. The map s is *étale*, *surjective*, etc. if s' can be chosen to be étale, surjective, etc.

Exercise 1.4. (a) Let $\pi\colon Y \to X$ be finite, flat, and surjective. Show that if $V \to U$ is an étale map of Y-schemes such that $\pi_* V$ and $\pi_* U$ are representable (π_* computed on $(\mathbf{Sch}/X)_{\text{et}}$), then $\pi_* V \to \pi_* U$ is étale (use I.3.22). Deduce that if A is a quasi-compact algebraic space over Y, then $\pi_* A$ is representable by an algebraic space over X. (Choose R and U to be finite disjoint unions of affines and use Demazure-Gabriel [1, I.1.6.6].)

(b) Let $R \rightrightarrows U \to A$ be an algebraic space, as above. Show that $R \xrightarrow{\approx} U \times_A U$. Choose U to be a disjoint union of affines and, in particular, separated. Show that for any scheme V and any map $V \to A$, the map $U \times_A V \to V$ (of sheaves) is represented by an étale surjective map of schemes, that is, $U \times_A V$ is representable and the map is étale and surjective. (Let $V' \to V$ be an étale surjective map of finite-type such that $V \to A$ lifts to a map $V' \to U$. Show that each square in the following diagrams is Cartesian, and apply descent theory. For example, according to [EGA. IV.18.12.12], any quasi-finite, separated morphism is quasi-affine, and so $R \to U \times_X U$ is quasi-affine; [SGA. 1, VIII.7.9] now implies that $U \times_A V$ is representable. Also $R \to U$ being étale implies $U \times_A V' \to V'$ is étale and that $U \times_A V \to V$ is étale.

$$\begin{array}{ccccc}
 & & U \times_A V' & \longrightarrow & V' \\
 & & \downarrow & & \downarrow \\
U \times_A U = & & R & \xrightarrow[\text{étale}]{i_1} & U
\end{array}$$

$$\begin{array}{ccccccc}
 & & U \times_A V & \longrightarrow & U \times_X V & \longrightarrow & V \\
 & & \uparrow & & \uparrow & & \uparrow \ \text{covering} \\
 & & U \times_A V' & \longrightarrow & U \times_X V' & \longrightarrow & V' \\
 & & \downarrow & & \downarrow & & \\
R & = & U \times_A U & \xrightarrow[\text{quasi-affine}]{} & U \times_X U) & &
\end{array}$$

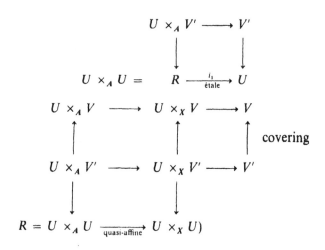

(c) Let $R \rightrightarrows U \to A$ be as in (b), regard A as a presheaf on $(\mathbf{Sch}/X)_{\text{fl}}$, and let A' be the associated sheaf. Show that $A \hookrightarrow A'$ and that all statements in (b) hold with A replaced by A'. Deduce that $A \xrightarrow{\approx} A'$, that is, that A is a sheaf on $(\mathbf{Sch}/X)_{\text{fl}}$. (Note that an element of $A'(V)$ is given by a diagram:

$$\begin{array}{ccccc} V' \times_V V' & \rightrightarrows & V' & \longrightarrow & V \\ \downarrow & & \downarrow & & \vdots \\ R & \rightrightarrows & U & \longrightarrow & A' \end{array}$$

with $V' \to V$ flat, surjective, and of finite-type. The proof of (b) is essentially unchanged. To show that $A \to A'$ is surjective, note that in the above diagram V' may be replaced by $U \times_{A'} V$.)

THEOREM 1.5. *Any sheaf F on X_{et} is represented (on (ét)/X) by a unique locally separated algebraic space $\tilde{F}$ that is étale over X.*

LEMMA 1.6. *Let U_1 and U_2 be affine schemes, étale over X, and let $s_1 \in F(U_1)$ and $s_2 \in F(U_2)$. The functor $U_1 \times_F U_2$:(ét)/$X \to$* **Sets**,

$$(U_1 \times_F U_2)(V) \stackrel{df}{=} \{(f_1, f_2) | f_i : V \to U_i,\ \text{res}_{f_1}(s_1) = \text{res}_{f_2}(s_2)\},$$

is represented by an open subscheme of $U_1 \times_X U_2$.

Proof. For any $(f_1, f_2) \in (U_1 \times_F U_2)(V)$, the morphism (f_1, f_2): $V \to U_1 \times_X U_2$ is étale (because V and $U_1 \times_X U_2$ are étale over X (I.3.6)) and so has open image V' (I.2.12). Moreover, $s_1 | V' = s_2 | V'$. It is now clear that $U_1 \times_F U_2$ is represented by the open subscheme

$$U = \bigcup_{(f_1, f_2)} V' \text{ of } U_1 \times_X U_2.$$

Proof (*of* 1.5). Let $U = \coprod V$, where the disjoint union is over all pairs (V, s) with V an affine scheme étale over X and $s \in F(V)$. The s define a canonical element c of $F(U) = \prod F(V)$, which we may regard as a map of sheaves of sets $U \to F$. From the lemma we know that $R \stackrel{df}{=} U \times_F U$ is an open subscheme of $U \times_X U$, and we let i_1 and i_2 be the composites

$$R \hookrightarrow U \times_X U \underset{p_2}{\overset{p_1}{\rightrightarrows}} U.$$

Then

$$R \underset{i_2}{\overset{i_1}{\rightrightarrows}} U$$

satisfies the conditions in (b) above, and we define $\tilde{F}$ to be the corresponding algebraic space. It is étale and locally separated over X as U is étale over X, and $R \hookrightarrow U \times_X U$ is an open immersion.

To show that F and $\tilde{F}$ are equal as sheaves on X_{et}, it suffices to show that they agree on affine schemes étale over X. Let V be such a scheme, and let $s \in F(V)$. The element s allows V to be regarded as a subscheme of U, and so we have a map (of sheaves) $V \to \tilde{F}$, that is, an element of

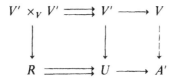

$\tilde{F}(V)$. To show that this defines an isomorphism $F(V) \to \tilde{F}(V)$, we construct an inverse. Let $s \in \tilde{F}(V)$ be described by the diagram

$$\begin{array}{ccccc} V' \times_V V' & \overset{p_1}{\underset{p_2}{\rightrightarrows}} & V' & \longrightarrow & V \\ \downarrow{\scriptstyle r} & & \downarrow{\scriptstyle s'} & & \downarrow{\scriptstyle s} \\ R & \overset{i_1}{\underset{i_2}{\rightrightarrows}} & U & \longrightarrow & \tilde{F} \end{array}$$

Let $s_0 = c|V'$, that is, $s_0 = \mathrm{res}_{s'}(c)$. Then

$$\mathrm{res}_{p_j}(s_0) = \mathrm{res}_{s'p_j}(c) = \mathrm{res}_{i_j r}(c) = \mathrm{res}_r(\mathrm{res}_{i_j}(c))).$$

But, from the definition of R, $\mathrm{res}_{i_1}(c) = \mathrm{res}_{i_2}(c)$, and so the two restrictions of s_0 to $V' \times_V V'$ are equal and $s_0 \in F(V)$. It is easy to see that $s \mapsto s_0 : \tilde{F}(V) \to F(V)$ is inverse to the earlier map.

This proves the existence of $\tilde{F}$. For the uniqueness, note that if $R \rightrightarrows U \to A$ is an étale algebraic space over X, then

$$\tilde{F}(A) \to \tilde{F}(U) \rightrightarrows \tilde{F}(R)$$

is exact (definition of A as quotient sheaf). Thus $\tilde{F}$ is uniquely determined as a functor on étale algebraic spaces over X and hence is uniquely determined (by the Yoneda lemma).

Remark 1.7 (a) In certain cases, $\tilde{F}$ will automatically be a scheme, for example, if it is of finite-type over a field (Artin [9, IV.4.8]).

(b) Let F be a sheaf on X_{et}. As $\tilde{F}$ is etale over X, the argument (II.3.1d) shows that $(\pi^*F)^{\sim} = \tilde{F} \times_X Y$ for any morphism $\pi: Y \to X$.

(c) Let f be the morphism $(\mathbf{Sch}/X)_{\mathrm{et}} \to X_{\mathrm{et}}$ defined by the identity map. Remark (b) shows that, for any sheaf F on X_{et}, the map $f^*F \to \tilde{F}$ that corresponds to the identity map under

$$\mathrm{Hom}_{(\mathbf{Sch}/X)_{\mathrm{et}}}(f^*F, \tilde{F}) \approx \mathrm{Hom}_{X_{\mathrm{et}}}(F, \tilde{F}|X_{\mathrm{et}}) = \mathrm{Hom}_{X_{\mathrm{et}}}(F, F)$$

is an isomorphism. It follows from (1.4c) that f^*F is also a sheaf on $(\mathbf{Sch}/X)_{\mathrm{fl}}$, and so the inverse images of F under $(\mathbf{Sch}/X)_{\mathrm{fl}} \to X_{\mathrm{et}}$ and $(\mathbf{Sch}/X)_{\mathrm{et}} \to X_{\mathrm{et}}$ are equal.

(d) If F is a sheaf on $(\mathbf{Sch}/X)_{\mathrm{et}}$ and $F_1 = F|X_{\mathrm{et}}$, then it is not true in general that $F = \tilde{F}_1$ on $\mathbf{Sch}/X$. From (c) this will be true if and only if $F \approx f^*f_*F$. Such a sheaf is said to be *locally constructible.* A sheaf F on X_{et} is *constructible* if $\tilde{F}$ is of finite-type over X; a sheaf F on $(\mathbf{Sch}/X)_{\mathrm{et}}$ is *constructible* if it is locally constructible and $\tilde{F}$ is of finite-type.

(e) Let G be an étale group scheme over X. The sheaf on X_{et} defined by G is constructible if and only if G is of finite-type over X.

(f) If $X = \operatorname{spec} A$, with A a local Artin ring, then F on X_{et} is constructible if and only if $\tilde{F}$ is a finite étale scheme over X. In proving this, we may assume that A has separably closed residue field. Let $\tilde{F} = U/R$ with U étale and of finite-type over X. Then U and R are disjoint unions of a finite number of copies of X, from which it is clear that U/R is a scheme.

PROPOSITION 1.8. *Let F be a sheaf on X_{et}. The following are equivalent:*

(a) *F is constructible;*

(b) *every irreducible closed subscheme Z of X contains a nonempty open subscheme U such that $F|U$ is locally constant and has finite stalks;*

(c) *every quasi-compact open $U \subset X$ is a finite union of locally closed subschemes, $U = \bigcup Z_i$, such that $F|Z_i$ is finite and $F|Z_i'$ is constant for some finite étale $Z_i' \to Z_i$;*

(d) *for any $x \in X$, $\tilde{F}_x \overset{df}{=} \tilde{F} \times_F \operatorname{spec} k(x)$ is a finite algebraic space over $k(x)$, and $x \mapsto h(x) \overset{df}{=}$ (order of $F_{\bar{x}}$) is a constructible function (that is, for any $t \in \mathbb{Z}$ and any quasi-compact open $U \subset X$, $h^{-1}(t) \cap U$ is a finite union of locally closed subsets).*

Proof. Since all the conditions are local on X, we may assume it to be quasi-compact (hence Noetherian).

(a) ⇒ (b). If z is the generic point of Z, then (1.7f) shows that there is a finite étale group scheme N on $\operatorname{spec} \mathcal{O}_{Z,z}$ and an isomorphism $\phi: N \overset{\approx}{\to} \tilde{F}|\operatorname{spec} \mathcal{O}_{Z,z}$. Then N extends to a finite group scheme over an open neighborhood of z and, as $\tilde{F}$ is of finite-type, ϕ extends to an isomorphism over an open subneighborhood. Now we may apply (1.2b).

(b) ⇒ (c). Suppose that (c) is false, and let Z be minimal among the closed subschemes of X for which it is false. Clearly Z will be irreducible, and we may take it to be reduced. According to (b) and (1.1), there exists an étale map $U \to Z$ with U affine such that $F|U$ is constant and finite. Since the generic fiber of this map is finite, (I.1.12) shows that there is an open subscheme V of Z such that $U \times_Z V \to V$ is finite. As $F|U \times V$ is constant and (by the minimality of Z) $F|Z - V$ satisfies (c), we have a contradiction.

(c) ⇒ (d). h is constant on Z if Z is connected and there exists a finite étale map $Z' \to Z$ such that $F|Z'$ is constant.

(d) ⇒ (a). By the argument used in (b) ⇒ (c) (that is, by Noetherian induction) we need only show that every closed irreducible subscheme Z of X contains a nonempty open subscheme U such that $F|U$ is constructible. If z is the generic point of Z, then $\tilde{F}|\operatorname{spec} \mathcal{O}_{Z,z}$ is a finite étale group scheme. Thus there is a finite étale group scheme N over an open subscheme U of Z and a morphism $\phi: N \to \tilde{F}|U$ that is an isomorphism over $\operatorname{spec} \mathcal{O}_{Z,z}$. After contracting U, we may assume that ϕ is a local isomorphism (compare [EGA. I.6.6.4]) and that the stalks of N and F have the same (constant) number of points at any geometric point of U; but then ϕ is an isomorphism.

Remark 1.9. (a) For an example of a nonconstructible sheaf with finite stalks, see Artin [9, VII.2.10].

(b) Let $\phi: F \to F'$ be a map of sheaves on X_{et}. Since the following diagram is Cartesian in either the category of algebraic spaces over X or the category of sheaves of sets over X,

$$\begin{array}{ccc} \ker(\phi) & \longrightarrow & X \\ \big\downarrow & & \big\downarrow \scriptstyle\text{zero-section} \\ \tilde{F} & \longrightarrow & \tilde{F}', \end{array}$$

$\ker(\phi)$ is constructible (that is, of finite-type) if F is.

Note that

$$\psi = \phi \times 1: \tilde{F} \times_X \tilde{F}' \to \tilde{F}' \times_X \tilde{F}'$$

is an étale map of algebraic spaces (analogue of (I.3.6)), and so its image R is an open algebraic subspace of $\tilde{F}' \times_X \tilde{F}'$. Thus $R \rightrightarrows \tilde{F}'$, where the maps are induced by the projections $\tilde{F}' \times \tilde{F}' \to \tilde{F}'$, is an étale equivalence relation on the algebraic space $\tilde{F}'$ and $\tilde{F}'/R = \operatorname{coker}(\phi)$. One sees easily that $\tilde{F}' \to \tilde{F}'/R$ is a surjective, étale map of algebraic spaces and that therefore $\operatorname{coker}(\phi)$ is constructible if F' is.

Finally, $\operatorname{im}(\phi) \approx \operatorname{coker}(\ker(\phi) \to F) \approx \ker(F' \to \operatorname{coker}(\phi))$ is constructible if either F or F' is. From these remarks it follows that the constructible sheaves form an abelian category.

(c) Since $\mathbf{S}((\mathbf{Sch}/X)_{\text{et}}) \underset{f^*}{\overset{f_*}{\rightleftarrows}} \mathbf{S}(X_{\text{et}})$ are exact, it follows from (b) that if $\phi: F \to F'$ is a map of locally constructible sheaves on $(\mathbf{Sch}/X)_{\text{et}}$, then $\ker(\phi)$ is constructible if F is, $\operatorname{coker}(\phi)$ is constructible if F' is, and $\operatorname{im}(\phi)$ is constructible if either F or F' is.

Exercise 1.10. (a) Let F be a constructible sheaf on X_{et}. Show that F is locally constant if and only if for any pair $\bar{x}_0$, $\bar{x}_1$ of geometric points X with x_0 in the closure of x_1, there exists a common étale neighborhood U of $\bar{x}_0$ and $\bar{x}_1$ such that $F(U) \to F_{\bar{x}_0}$ and $F(U) \to F_{\bar{x}_1}$ are isomorphisms, or (equivalently) if and only if for some (or every) choice of a map $\mathcal{O}_{X,\bar{x}_0} \to \mathcal{O}_{X,\bar{x}_1}$, the cospecialization map $F_{\bar{x}_0} \to F_{\bar{x}_1}$ is an isomorphism.

(b) Show that any locally constructible torsion sheaf on $(\mathbf{Sch}/X)_{\text{et}}$ is a direct limit of constructible sheaves. (It suffices to show that any torsion sheaf F on X_{et} is such a limit; show that

$$F = \varinjlim_{(s,U)} \operatorname{im}(g_*(\mathbb{Z}/(n)) \to F)$$

where U is affine, $g: U \to X$ is étale, $s \in F(U)$, n is the order of s, and the map is induced by $(1 \mapsto s):(\mathbb{Z}/(n))_U \to F|U$; to show that $g_*(\mathbb{Z}/(n))$ is constructible, show that $X = \bigcup X_i$ with each X_i locally closed and such that $U \times X_i \to X_i$ is finite (compare (1.8).)

We next consider the category $\mathbf{S}(X_{\mathrm{et}}, \mathbb{Z}/(n))$ of sheaves of $\mathbb{Z}/(n)$-modules. Recall (III.2.25) that the cohomology of a sheaf of $\mathbb{Z}/(n)$-modules agrees with its cohomology as a sheaf of $\mathbb{Z}$-modules. The same is not true of Exts. For example, if X is the spectrum of a separably closed field, so that $\mathbf{S}(X_{\mathrm{et}}, \mathbb{Z}/(n))$ may be identified with the category of $\mathbb{Z}/(n)$-modules, then $\operatorname{Ext}^1_{\mathbb{Z}/(n)}(\mathbb{Z}/(n), \mathbb{Z}/(n)) = 0$, but $\operatorname{Ext}^1_{\mathbb{Z}}(\mathbb{Z}/(n), \mathbb{Z}/(n)) \neq 0$, since

$$0 \to \mathbb{Z}/(n) \to \mathbb{Z}/(n^2) \to \mathbb{Z}/(n) \to 0$$

does not split. A sheaf F of $\mathbb{Z}/(n)$-modules is *locally constant* or *constructible* if it is so as a sheaf of abelian groups. It is *locally free* (of *finite rank*) if there is a covering $(U_i \to X)$ such that $F|U_i$ is the constant sheaf defined by a free module (of finite rank) for each i. If F is locally constant and constructible, we define $\check{F} = \underline{\operatorname{Hom}}(F, \mathbb{Z}/(n))$. Then $F \mapsto \check{F}$ is exact (according to (II.3.20b)) and takes locally free sheaves to locally free sheaves. These last definitions and remarks extend in an obvious way to $\mathbf{S}(X, \Lambda)$ where Λ is a finite $\mathbb{Z}/(n)$-algebra.

Let X be a scheme whose residue characteristics are prime to n. Then $\boldsymbol{\mu}_n$ is a locally free sheaf of $\mathbb{Z}/(n)$-modules of rank 1 on X_{et}, and we define:

$$(\mathbb{Z}/(n))(r) = \begin{cases} \boldsymbol{\mu}_n \otimes \cdots \otimes \boldsymbol{\mu}_n \ (r \text{ copies}) & \text{if } r > 0; \\ \mathbb{Z}/(n) & \text{if } r = 0; \\ (\mathbb{Z}/(n))(-r)^{\vee} & \text{if } r < 0. \end{cases}$$

For any sheaf F of $\mathbb{Z}/(n)$-modules, the rth (*Tate*) *twist* of F is $F(r) = F \otimes \mathbb{Z}/(n)(r)$; thus $F(r)$ is locally isomorphic (noncanonically) to F. If X is a variety over a separably closed field, then $F(r) \approx F$ (noncanonically) and

$$H^s(X, F) \otimes (\mathbb{Z}/(n))(r) \xrightarrow{\approx} H^s(X, F(r))$$

(canonically) as may be seen most easily by looking at the Čech groups.

Let l be a prime number. An *l-adic sheaf* (or sheaf of $\mathbb{Z}_l$-*modules*) on X_{et} is a projective system $F = (F_n)_{n \in \mathbb{N}}$ of sheaves such that, for any n, the given map $F_{n+1} \to F_n$ is isomorphic to the canonical map $F_{n+1} \to F_{n+1} \otimes_{\mathbb{Z}} \mathbb{Z}/(l^n)$, that is, $F_{n+1} \to F_n$ induces an isomorphism $F_{n+1}/l^n F_{n+1} \xrightarrow{\approx}$

F_n. By induction, $F_m/l^nF_m \xrightarrow{\sim} F_n$ for all $m > n$. It follows that $F_0 = 0$ and F_n is killed by l^n, that is, that it is a $\mathbb{Z}/(l^n)$-module. The cohomology groups of an l-adic sheaf F are defined to be $H^r(X, F) = \varprojlim H^r(X, F_n)$. Note that $H^r(X, F) \neq H^r(X, \varprojlim F_n)$ in general, even for $r = 1$ and X the spectrum of a field. The ring $\mathbb{Z}_l \stackrel{df}{=} \varprojlim \mathbb{Z}/(l^n)$ acts on $H^r(X, F)$ and on

$$\operatorname{Hom}(F, F') \stackrel{df}{=} \varprojlim_m \varinjlim_n \operatorname{Hom}(F_n, F'_m) = \varprojlim_m \operatorname{Hom}(F_m, F'_m)$$

for any l-adic sheaves F and F'. An l-adic sheaf F is *twisted-constant* or *constructible* if each F_n has the corresponding property. Note that F being twisted-constant does not imply that there exists a finite surjective étale $U \to X$ such that $F|U$ is constant. In particular, if X is connected with geometric point $\bar{x}$, then the action of $\pi_1(X, \bar{x})$ on $F_{\bar{x}} \stackrel{df}{=} \varprojlim (F_n)_{\bar{x}}$ need not factor through a finite quotient. However, the action is continuous when $F_{\bar{x}}$ is given the l-adic topology, and $F \mapsto F_{\bar{x}}$ induces an equivalence between the category of twisted-constant constructible sheaves on X and the category of $\pi_1(X, \bar{x})$-modules that are finitely generated over $\mathbb{Z}_l$. If F is a constructible l-adic sheaf on X, then every quasi-compact open $U \subset X$ is a finite union of locally closed schemes, $U = \bigcup Z_i$, such that each $F_n|Z_i$ is finite and twisted-constant (exercise!). An l-adic sheaf F is *locally free of finite rank* r if each F_n is locally free of rank r over $\mathbb{Z}/(l^n)$. (In [SGA. 4 1/2, Rapport 2.1] "$\mathbb{Z}_l$-sheaf" is redefined to mean constructible $\mathbb{Z}_l$-sheaf, and a twisted-constant constructible $\mathbb{Z}_l$-sheaf is said to be *lisse*, that is, smooth.)

Let X be a scheme whose residue characteristics are prime to l. The sheaves $\mathbb{Z}_l(r) \stackrel{df}{=} ((\mathbb{Z}/(l^n))(r))_n$ are locally free of rank 1, and for any l-adic sheaf we define $F(r) = (F_n(r))_n$. As before, if X is a variety over a separably closed field k, this twisting commutes with the formation of cohomology. If $k = \mathbb{C}$ and X is smooth, then (III.3.12) shows that

$$H^r(X_{\text{et}}, F) \xrightarrow{\sim} \varprojlim H^r(X(\mathbb{C}), F_n) = H^r(X(\mathbb{C}), F)$$

when the F_n are constant finite sheaves. In fact there is such an isomorphism when X is any variety over $\mathbb{C}$ and F is constructible [SGA. 4, XVI.4].

The category of *constructible sheaves of* $\mathbb{Q}_l$-*vector-spaces* is defined to be the quotient of the category of constructible sheaves of $\mathbb{Z}_l$-modules by its subcategory of torsion objects, that is, those killed by some power of l. More concretely, every constructible sheaf F of $\mathbb{Z}_l$-modules gives rise to a sheaf $F \otimes \mathbb{Q}_l$ of $\mathbb{Q}_l$-vector-spaces; all constructible sheaves of $\mathbb{Q}_l$-vector-spaces arise in this way, and

$$\operatorname{Hom}(F \otimes \mathbb{Q}_l, F' \otimes \mathbb{Q}_l) = \operatorname{Hom}(F, F') \otimes_{\mathbb{Z}_l} \mathbb{Q}_l.$$

We say that a constructible sheaf of $\mathbb{Q}_l$-vector-spaces is *twisted-constant* if it is isomorphic to a sheaf $F \otimes \mathbb{Q}_l$ with F twisted-constant and constructible. If X is connected with geometric point $\bar{x}$, then

$$F \otimes \mathbb{Q}_l \mapsto (F \otimes \mathbb{Q}_l)_{\bar{x}} \stackrel{df}{=} F_{\bar{x}} \otimes \mathbb{Q}_l$$

is an equivalence between the category of twisted-constant, constructible sheaves of $\mathbb{Q}_l$-vector-spaces and the category of finite-dimensional $\mathbb{Q}_l$-vector-spaces on which $\pi_1(X, \bar{x})$ acts continuously. Since every such vector space contains a stable lattice (because $\pi_1(X, \bar{x})$ is compact) every twisted-constant constructible sheaf of $\mathbb{Q}_l$-vector-spaces is of the form $F \otimes \mathbb{Q}_l$ with F locally free of finite rank. We write $(F \otimes \mathbb{Q}_l)(r)$ for $F(r) \otimes \mathbb{Q}_l$ and $(F \otimes \mathbb{Q}_l)^{\vee}$ for $\check{F} \otimes \mathbb{Q}_l \stackrel{df}{=} (\check{F}_n) \otimes \mathbb{Q}_l$.

If Ω is a finite field extension of $\mathbb{Q}_l$, and A is the integral closure of $\mathbb{Z}_l$ in $\mathbb{Q}_l$, then the above discussion may be extended to sheaves of A-modules and Ω-vector-spaces. For example, a sheaf of A-modules is a projective system $F = (F_n)$ where F_n is an $A/\mathfrak{m}^n$-module and $F_{n+1} \otimes (A/\mathfrak{m}^n) = F_n$.

Usually, in what follows, we prove results only for sheaves of $\mathbb{Z}/(n)$-modules (or $A/\mathfrak{m}^n$-modules) and leave it to the reader to check that they remain true for sheaves of $\mathbb{Z}_l$-modules and $\mathbb{Q}_l$-vector-spaces. For this, the following lemma is useful.

LEMMA 1.11. *Let $F = (F_n)$ be an l-adic sheaf on X_{et} such that each F_n is flat as a sheaf of $\mathbb{Z}/(l^n)$-modules and $H^r(X, F_n)$ is finite for all r and n. Then $H^r(X, F)$ is finitely generated as a $\mathbb{Z}_l$-module and there are exact sequences,*

$$0 \to H^r(X_{\text{et}}, F)^{(l^n)} \to H^r(X_{\text{et}}, F_n) \to H^{r+1}(X_{\text{et}}, F)_{l^n} \to 0$$

for all r and n.

Proof. We shall need to use the fact that inverse limits of exact sequences of finite abelian groups are exact. One way of seeing this is to note that the category of profinite (that is, compact and totally disconnected) abelian groups is dual to the category of discrete torsion abelian groups (by trivial Pontryagin duality), and so the statement claimed is dual to the statement that direct limits preserve exact sequences of (torsion) abelian groups. On tensoring

$$0 \to \mathbb{Z}/(l^n) \xrightarrow{l^s} \mathbb{Z}/(l^{n+s}) \to \mathbb{Z}/(l^s) \to 0$$

with F_{n+s}, we obtain an exact sequence

$$0 \to F_n \to F_{n+s} \to F_s \to 0.$$

For varying n these sequences are compatible in the sense that

$$\begin{array}{ccccccccc} 0 & \longrightarrow & F_{n+1} & \xrightarrow{l^s} & F_{n+1+s} & \longrightarrow & F_s & \longrightarrow & 0 \\ & & \downarrow & & \downarrow & & \| & & \\ 0 & \longrightarrow & F_n & \xrightarrow{l^s} & F_{n+s} & \longrightarrow & F_s & \longrightarrow & 0 \end{array}$$

commutes. On forming the cohomology sequence for each n and passing to the inverse limit over all n, we obtain an exact sequence,

$$\cdots \to H^r(F) \xrightarrow{l^s} H^r(F) \to H^r(F_s) \to H^{r+1}(F) \xrightarrow{l^s} H^{r+1}(F) \to \cdots,$$

which gives the required sequences.

As $H^r(F)$ is an inverse limit of l-power-torsion finite groups, no nonzero element of it is divisible by all powers of l. Thus $\varprojlim H^{r+1}(F)_{l^n} = 0$ (the transition maps are $H^{r+1}(F)_{l^n} \xleftarrow{l} H^{r+1}(F)_{l^{n+1}}$) and $\varprojlim H^r(F)^{(l^n)} \xrightarrow{\approx} H^r(F)$; it follows that $H^r(F)$ is generated by any subset that generates it mod l.

If A is a ring and M is an A-module that is finitely generated (or of finite length), then $[M]$ denotes the class of M in the Grothendieck group of the corresponding category of A-modules. If A is a sheaf of rings on some site X_E and F is a sheaf of A-modules, then $H^r(X_E, F)$ is a $\Gamma(X, A)$-module, and we write $\chi(X_E, F)$ for the *Euler-Poincaré characteristic*

$$\sum (-1)^r [H^r(X_E, F)]$$

when this is defined. For example, when $\Gamma(X, A)$ is a field K and the $H^r(X_E, F)$ are finite-dimensional vector spaces we may identify $[H^r(X_E, F)]$ with $\dim_K (H^r(X_E, F))$ and $\chi(X_E, F)$ with

$$\sum (-1)^r \dim_K (H^r(X_E, F));$$

when $\Gamma(X, A)$ is $\mathbb{Z}$ and the $H^r(X_E, F)$ are finite, then we may identify $[H^r(X_E, F)]$ with the order of $H^r(X_E, F)$ and $\chi(X_E, F)$ with

$$\prod [H^r(X_E, F)]^{(-1)^r}.$$

The *l-adic Betti numbers* of a variety over a separably closed field are defined to be

$$\beta_r(X, l) = \dim_{\mathbb{Q}_l} (H^r(X_{\text{et}}, \mathbb{Q}_l))$$

and the *l-adic Euler-Poincaré characteristic* is

$$\chi(X_{\text{et}}, l) = \sum (-1)^r \beta_r(X, l).$$

If the groups $H^r(X_{\text{et}}, \mathbb{F}_l)$ are finite, then

$$\chi(X_{\text{et}}, l) = \chi(X_{\text{et}}, \mathbb{F}_l).$$

Finally in this section we prepare for the Poincaré duality theorems to be proved in the next section and in (VI.11) by describing a canonical pairing between Ext groups. Recall that in an abelian category **A** with enough injectives, $\mathrm{Ext}^r_{\mathbf{A}}(B, C)$ is usually defined by taking an injective resolution $C \to I^{\cdot}$ of C and forming the rth cohomology group of the complex

$$\mathrm{Hom}(B, I^0) \to \mathrm{Hom}(B, I^1) \to \cdots$$

that is, $\mathrm{Ext}^r(B, C) = H^r(\mathrm{Hom}(B, I^{\cdot}))$. We need another description. Recall that a *map of degree r* of complexes, $\phi: B^{\cdot} \to C^{\cdot}$, is a family of maps $\phi^s: B^s \to C^{s+r}$ such that $\phi^{s+1} d^s_B = (-1)^r d^{r+s}_C \phi^s$; alternatively it is a map (of degree zero) $B^{\cdot} \to C^{\cdot}[r]$ where $C^{\cdot}[r]$ is $C^{\cdot}$ "shifted r places to the left", that is, $(C^{\cdot}[r])^s = C^{r+s}$ and $d_{C^{\cdot}[r]} = (-1)^r d_{C^{\cdot}}$. Two such maps ϕ and ψ are *homotopic* if there exists a family of maps $k^s: B^s \to C^{s+r-1}$ such that $\phi^s - \psi^s = dk^s + k^{s+1}d$. For example, if $B^{\cdot}$ is a complex with $B^0 = B$ and $B^s = 0$ for $s \neq 0$, then a map $B^{\cdot} \to I^{\cdot}$ of degree r is simply a map $\phi: B \to I^r$ such that $d \circ \phi = 0$. Moreover, ϕ is homotopic to zero if and only if there exists a map $\psi: B \to I^{r-1}$ such that $d \circ \psi = \phi$. Thus, if $I^{\cdot}$ is an injective resolution of C, the group of homotopy classes of maps $B \to I^{\cdot}[r]$ is precisely $\mathrm{Ext}^r(B, C)$. Equally easy arguments show that if $B \to B^{\cdot}$ is any resolution of B, that is, $0 \to B \to B^0 \to B^1 \to \cdots$ is exact but the B^s need not be injective, then the group of homotopy classes of maps $B^{\cdot} \to I^{\cdot}[r]$ is isomorphic to the group of homotopy classes of maps $B \to I^{\cdot}[r]$ and therefore to $\mathrm{Ext}^r(B, C)$.

This enables us to define a canonical pairing

$$\mathrm{Ext}^r_{\mathbf{A}}(A, B) \times \mathrm{Ext}^s_{\mathbf{A}}(B, C) \to \mathrm{Ext}^{r+s}_{\mathbf{A}}(A, C),$$

namely, we choose injective resolutions $B \to B^{\cdot}$ and $C \to C^{\cdot}$ of B and C, interpret the Ext groups as homotopy classes of maps $A \to B^{\cdot}[r]$, $B^{\cdot}[r] \to C^{\cdot}[r+s]$, and $A \to C^{\cdot}[r+s]$, and define the pairing to be composition. Of course, there are many things to be checked, for example, that composites of homotopic maps are homotopic, but they are easy (Cartier [1]).

One also checks that the pairings are functorial in A, B, and C and are compatible, in an obvious sense, with short exact sequences in any position. The pairings may also be described as follows: interpret $\gamma \in \mathrm{Ext}^r(A, B)$ as an r-fold extension (Cartier [1, III.3]),

$$0 \to B \to E^1 \to E^2 \to \cdots \to E^r \to A \to 0$$

if $r > 0$; the map $\mathrm{Ext}^s(B, C) \to \mathrm{Ext}^{r+s}(A, C)$ induced by pairing with γ is, up to sign, the r-fold boundary map defined by the exact sequence ($r > 0$) or the map induced by $\gamma: A \to B$ ($r = 0$).

We shall be mainly interested in this pairing when $\mathbf{A} = \mathbf{S}(X_{et})$ (or $\mathbf{S}(X_{et}, \mathbb{Z}/(n))$) and $A = \mathbb{Z}$ (or $\mathbb{Z}/(n)$). Then the pairing reads:

$$H^r(X, F) \times \mathrm{Ext}^s_X(F, G) \to H^{r+s}(X, G).$$

Also consider the situation where $j: X \hookrightarrow \bar{X}$ is an open immersion. For sheaves F and G on X we have a canonical pairing,

$$H^r(\bar{X}, j_!F) \times \mathrm{Ext}^s_{\bar{X}}(j_!F, j_!G) \to H^{r+s}(\bar{X}, j_!G).$$

As $\mathrm{Hom}_{\bar{X}}(j_!F, -) = \mathrm{Hom}_X(F, j^*(-))$ and j^* is exact and preserves injectives, $\mathrm{Ext}^s_{\bar{X}}(j_!F, j_!G)$ may be replaced by $\mathrm{Ext}^s_X(F, j^*j_!G)$. Moreover, $G \xrightarrow{\approx} j^*j_!G$, and so we have a canonical pairing

$$H^r(\bar{X}, j_!F) \times \mathrm{Ext}^s_X(F, G) \to H^{r+s}(\bar{X}, j_!G).$$

If $\bar{X}$ is a complete variety, this may be written (III.1.29) as

$$H^r_c(X, F) \times \mathrm{Ext}^s_X(F, G) \to H^{r+s}_c(X, G).$$

In the course of the proofs of the Poincaré duality theorem, we shall need to know how these pairings behave relative to étale maps, and to describe this, we need a trace map.

LEMMA 1.12. *Let $\pi: X' \to X$ be a finite étale map of constant degree d. For any sheaf F on X there is a trace map $tr: \pi_*\pi^*F \to F$, functorial in F, such that $\phi \mapsto tr \circ \pi_*(\phi)$ is an isomorphism $\mathrm{Hom}_{X'}(F', \pi^*F) \to \mathrm{Hom}_X(\pi_*F', F)$ for any F' on X'. Thus $\pi_* = \pi_!$, that is, π_* is left adjoint to π^*, and tr is the adjunction map. The composites*

$$F \to \pi_*\pi^*F \xrightarrow{tr} F \text{ and } H^r(X, F) \to H^r(X', F|X') \xrightarrow{tr} H^r(X, F)$$

are multiplication by d.

Proof. Let $X'' \to X$ be a Galois morphism, with Galois group G, which factors into $X'' \twoheadrightarrow X' \to X$. Then $X'' \to X'$ is Galois with Galois group $H \subset G$. For any U étale over X we have

$$\Gamma(U, F) \hookrightarrow \Gamma(U', F) \hookrightarrow \Gamma(U'', F)$$

and

$$\Gamma(U, F) \xrightarrow{\approx} \Gamma(U'', F)^G,$$

where $U' = U \times X'$, $U'' = U \times X''$. For $s \in \Gamma(U, \pi_*\pi^*F) \stackrel{df}{=} \Gamma(U', F)$ we define

$$tr(s) = \sum_{\sigma \in G/H} \sigma(s|U'');$$

as this is fixed by G it may be regarded as an element of $\Gamma(U, F)$. Clearly tr defines a map $\pi_*\pi^*F \to F$ whose composite with $F \to \pi_*\pi^*F$ is multiplication by d.

If X' is a disjoint union of d copies of X, then it is obvious that $\mathrm{Hom}_{X'}(F', \pi^*F) \xrightarrow{\approx} \mathrm{Hom}_X(\pi_*F', F)$, and one may reduce the question to that case by passing to a finite étale covering of X, for example to $X'' \to X$, and using the fact that $\underline{\mathrm{Hom}}$ is a sheaf.

In

$$H^r(X, F) \xrightarrow{\text{res}} H^r(X', \pi^*F) \xrightarrow{\text{can}} H^r(X, \pi_*\pi^*F) \xrightarrow{tr} H^r(X, F)$$

the composite of the first two maps is induced by $F \to \pi_*\pi^*F$, and the composite of all three is induced by $(F \to \pi_*\pi^*F \xrightarrow{tr} F) =$ multiplication by d.

PROPOSITION 1.13. *Let $\pi: X' \to X$ be a separated étale map and let $X' \overset{j}{\hookrightarrow} \bar{X}' \overset{\bar{\pi}}{\hookrightarrow} X$ be a factorization of π with j an open immersion and $\bar{\pi}$ finite. Then $\pi_! = \bar{\pi}_* j_!$ and we define $tr: \pi_!\pi^*F \to F$ to be the adjunction map. For any sheaves F on X and F' on X',*

$$\mathrm{Ext}^r_{X'}(F', \pi^*F) \xrightarrow{\text{can}} \mathrm{Ext}^r_X(\pi_!F', \pi_!\pi^*F) \xrightarrow{tr} \mathrm{Ext}^r_X(\pi_!F', F)$$

is an isomorphism and the diagram

$$\begin{array}{ccccc} \text{can}: H^r(\bar{X}', j_!F') & \times & \mathrm{Ext}^s_{X'}(F', \pi^*F) & \longrightarrow & H^{r+s}(\bar{X}', j_!\pi^*F) \\ \uparrow \approx & & \downarrow \approx & & \downarrow tr \\ \text{can}: H^r(X, \pi_!F') & \times & \mathrm{Ext}^s_X(\pi_!F', F) & \longrightarrow & H^{r+s}(X, F) \end{array}$$

commutes.

Proof. According to (II.3.18), $\pi_!F$ and $\bar{\pi}_*j_!F$ are the sheaves associated with presheaves F_1 and F_2 respectively, where F_1 and F_2 have the following descriptions. Let U be étale over X and assume it to be connected and quasi-compact. Write $U' \hookrightarrow \bar{U}'$ for $U \times_X (X' \hookrightarrow \bar{X}')$, and let $U' = U'_1 \cup \cdots \cup U'_r$ be the decomposition of U' as a disjoint union of connected components. Then $\Gamma(U, F_s) = \bigoplus F(U'_i)$ where the sum is over those U'_i that map isomorphically onto U in case $s = 1$ (see I.3.12) and over those U'_i that are finite over U (and hence open and closed in $\bar{U}'$) in case $s = 2$. Clearly there is a map $F_1 \hookrightarrow F_2$ that is an isomorphism if U is strictly local. On passing to the associated sheaves, we obtain an isomorphism $\pi_!F \xrightarrow{\approx} \pi_*j_!F$.

By definition

$$\mathrm{Hom}_{X'}(F', \pi^*F) \xrightarrow{\approx} \mathrm{Hom}_X(\pi_!F', F).$$

Since π^* is exact and preserves injectives, $\mathrm{Ext}^r_{X'}(F', \pi^*F)$ may be regarded as the rth right derived functor of $F \mapsto \mathrm{Hom}_{X'}(F', \pi^*F)$, and the uniqueness of derived functors shows that the Exts are isomorphic.

Let F_1 be a sheaf on X, let F_2 and F_3 be sheaves on $\bar{X}'$, and let $\phi: F_1 \to \bar{\pi}_* F_2$ and $\psi: F_2 \to F_3$ be morphisms. Then

$$(F_1 \xrightarrow{\text{can}} \bar{\pi}_* \bar{\pi}^* F_1 \xrightarrow{\bar{\pi}_*(\psi\phi')} \bar{\pi}_* F_3) = (F_1 \xrightarrow{\bar{\pi}_*(\psi)\phi} \bar{\pi}_* F_3)$$

where $\phi': \bar{\pi}^* F_1 \to F_2$ corresponds by adjointness to ϕ. Indeed, ϕ' is the unique map $\bar{\pi}^* F_1 \to F_2$ such that $\bar{\pi}_*(\phi') \circ \text{can} = \phi$. Thus

$$\begin{array}{ccccc}
\mathrm{Hom}_{\bar{X}'}(\bar{\pi}^* F_1, F_2) & \times & \mathrm{Hom}_{\bar{X}'}(F_2, F_3) & \longrightarrow & \mathrm{Hom}_{\bar{X}'}(\bar{\pi}^* F_1, F_3) \\
\uparrow \approx & & \downarrow \bar{\pi}_* & & \downarrow \approx \\
\mathrm{Hom}_X(F_1, \bar{\pi}_* F_2) & \times & \mathrm{Hom}_X(\bar{\pi}_* F_2, \bar{\pi}_* F_3) & \longrightarrow & \mathrm{Hom}_X(F_1, \bar{\pi}_* F_3)
\end{array}$$

commutes, where the pairings are composition.

Let $F \to I^{\cdot}$ and $j_! F' \to J^{\cdot}$ be injective resolutions of F and $j_! F'$. The above remark shows that the upper part of the following diagram commutes:

$$\begin{array}{ccccc}
\mathrm{Hom}_{\bar{X}'}(\mathbb{Z}, J^{\cdot}[r]) & \times & \mathrm{Hom}_{\bar{X}'}(J^{\cdot}[r], j_! \pi^* I^{\cdot}[r+s]) & \longrightarrow & \mathrm{Hom}_{\bar{X}'}(\mathbb{Z}, j_! \pi^* I^{\cdot}[r+s]) \\
\uparrow \approx & & \downarrow & & \downarrow \approx \\
\mathrm{Hom}_X(\mathbb{Z}, \bar{\pi}_* J^{\cdot}[r]) & \times & \mathrm{Hom}_X(\bar{\pi}_* J^{\cdot}[r], \pi_! \pi^* I^{\cdot}[r+s]) & \longrightarrow & \mathrm{Hom}_X(\mathbb{Z}, \pi_! \pi^* I^{\cdot}[r+s]) \\
\| & & \downarrow tr & & \downarrow tr \\
\mathrm{Hom}_X(\mathbb{Z}, \bar{\pi}_* J^{\cdot}[r]) & \times & \mathrm{Hom}_X(\bar{\pi}_* J^{\cdot}[r], I^{\cdot}[r+s]) & \longrightarrow & \mathrm{Hom}_X(\mathbb{Z}, I^{\cdot}[r+s])
\end{array}$$

The lower part of the diagram commutes because of the functoriality of tr. On passing to homotopy classes of maps and using the fact that $\pi_! F' \to \bar{\pi}_* J^{\cdot}$ is an injective resolution of $\pi_! F'$, we get the required diagram.

Variant 1.14. Let $\pi: X' \to X$ be a finite étale map of constant degree d that fits into a Cartesian square

$$\begin{array}{ccc}
X' & \xrightarrow{j'} & \bar{X}' \\
\downarrow \pi & & \downarrow \bar{\pi} \\
X & \xrightarrow{j} & \bar{X}
\end{array}$$

with $\bar{\pi}$ finite and j and j' open immersions. For any sheaf F on X we have

$$(F \to \pi_* \pi^* F \xrightarrow{tr} F) = \text{multiplication by } d.$$

Since $\bar{\pi}_* j'_! = j_! \pi_*$ (look on stalks), this shows that

$$(j_! F \to \bar{\pi}_* j'_! \pi^* F \xrightarrow{tr} j_! F) = \text{multiplication by } d.$$

If $\bar{X}$ is a complete variety, this implies that there are maps such that

$$(H_c^r(X, F) \xrightarrow{\text{res}} H_c^r(X', F|X') \xrightarrow{tr} H_c^r(X, F)) = \text{multiplication by } d.$$

Moreover, the same argument as above shows that

$$\begin{array}{ccccc} H_c^r(X', F') & \times & \operatorname{Ext}_{X'}^s(F', \pi^* F) & \longrightarrow & H_c^{r+s}(X', \pi^* F) \\ \uparrow \wr & & \downarrow \wr & & \downarrow tr \\ H_c^r(X, \pi_* F') & \times & \operatorname{Ext}_X^s(\pi_* F', F) & \longrightarrow & H_c^{r+s}(X, F) \end{array}$$

commutes.

It will also be useful to have pairings between cohomology groups. To construct them we use Godement resolutions (III.1.20c)

LEMMA 1.15. *Let X be a scheme, let F be a sheaf on X_{et}, let $C^0(F)$ be the zeroth term of the Godement resolution of F, and let $Z = \operatorname{coker}(F \to C^0(F))$. Then*

$$0 \to F \to C^0(F) \to Z \to 0$$

is split-exact on stalks, that is, for all geometric points $\bar{x} \to X$,

$$0 \to F_{\bar{x}} \to C^0(F)_{\bar{x}} \to Z_{\bar{x}} \to 0$$

is a split-exact sequence of abelian groups.

Proof. Recall $C^0(F) = u_* u^*(F) = \prod_x u_{x*} F_{\bar{x}}$ where $u: \coprod_{x \in X} \bar{x} \to X$. For any $x \in X$ and any étale neighborhood U of $\bar{x}$ there is a map $\Gamma(U, C^0(F)) \to F_{\bar{x}}$ that sends an element of $\Gamma(U, C^0(F))$ to its $\bar{x}^{th}$ component (more precisely, to its $(\bar{x} \to U)^{th}$ component). On passing to the direct limit over all U, we obtain a map $C^0(F)_{\bar{x}} \to F_{\bar{x}}$ that is the required section to $F_{\bar{x}} \to C^0(F)_{\bar{x}}$.

PROPOSITION 1.16. *Let $\mathbf{A}$ be an abelian category with tensor products, and let f_1, f_2, and f_3 be left exact functors $\mathbf{S}(X_{\text{et}}) \to \mathbf{A}$ such that any induced sheaf is f_i-acyclic, $i = 1, 2, 3$. Suppose that there is given a morphism of bi-functors $f_1(F_1) \otimes f_2(F_2) \to f_3(F_1 \otimes F_2)$. Then there is a unique family of morphisms of bi-functors*

$$(R^r f_1 F_1) \otimes (R^s f_2 F_2) \to R^{r+s} f_3(F_1 \otimes F_2),$$

written $(\gamma_1, \gamma_2) \mapsto \gamma_1 \cup \gamma_2$, such that

(a) *for $r = 0 = s$ it is the given morphism;*

(b) *if $0 \to F'_1 \to F_1 \to F''_1 \to 0$ is split-exact on stalks, then for $\gamma_1 \in R^r f_1 F''_1$ and $\gamma_2 \in R^s f_2 F_2$ we have $(d\gamma_1) \cup \gamma_2 = d(\gamma_1 \cup \gamma_2)$;*

(c) *if $0 \to F'_2 \to F_2 \to F''_2 \to 0$ is split-exact on stalks, then for $\gamma_1 \in R^r f_1 F_1$ and $\gamma_2 \in R^s f_2 F''_2$ we have $\gamma_1 \cup (d\gamma_2) = (-1)^r d(\gamma_1 \cup \gamma_2)$.*

Proof. Note that a sequence that is split-exact on stalks remains exact when tensored with any sheaf (because the sequences of stalks remains exact), which shows that (b) and (c) make sense. The proof is by induction and makes repeated use of the short exact sequence in (1.15). We omit it (see Bredon [1, II.6.2]).

Example 1.17. (a) The natural pairing

$$\Gamma(X, F_1) \times \Gamma(X, F_2) \to \Gamma(X, F_1 \otimes F_2)$$

induces a unique family of pairings

$$H^r(X, F_1) \times H^s(X, F_2) \to H^{r+s}(X, F_1 \otimes F_2).$$

(b) The natural pairing

$$\pi_* F_1 \times \pi_* F_2 \to \pi_*(F_1 \otimes F_2),$$

defined for any morphism $\pi: Y \to X$, induces a unique family of pairings

$$R^r\pi_* F_1 \times R^s\pi_* F_2 \to R^{r+s}\pi_*(F_1 \otimes F_2).$$

(c) The natural pairing

$$\Gamma_{Z_1}(X, F_1) \times \Gamma_{Z_2}(X, F_2) \to \Gamma_{Z_1 \cap Z_2}(X, F_1 \otimes F_2)$$

defined when Z_1 and Z_2 are closed subschemes of X, induces a unique family of pairings

$$H^r_{Z_1}(X, F_1) \times H^s_{Z_2}(X, F_2) \to H^{r+s}_{Z_1 \cap Z_2}(X, F_1 \otimes F_2).$$

Any of these pairings will be referred to as the *cup-product* pairing. More generally, if we are given a pairing of sheaves $F_1 \times F_2 \to F_3$, that is, a map $F_1 \otimes F_2 \to F_3$, then any one of the above pairings when composed with the map induced by $F_1 \otimes F_2 \to F_3$ will also be referred to as cup-product.

Remark 1.18. If the tensor product $F_1 \otimes F_2$ is identified, as usual, with $F_2 \otimes F_1$, then the uniqueness part of (1.16) shows that $\gamma_1 \cup \gamma_2 = (-1)^{rs}\gamma_2 \cup \gamma_1$ for $\gamma_1 \in R^r f_1 F_1$ and $\gamma_2 \in R^s f_2 F_2$.

Remark 1.19. (a) For any sheaves F and G on X_{et} and any covering $\mathcal{U}$ of X, there is a pairing of Čech complexes

$$C^{\cdot}(\mathcal{U}, F) \times C^{\cdot}(\mathcal{U}, G) \to C^{\cdot}(\mathcal{U}, F \otimes G)$$

that sends $f = (f_{i_0 \cdots i_r}) \in C^r(\mathcal{U}, F)$ and $g = (g_{i_0 \cdots i_s}) \in C^s(\mathcal{U}, G)$ to $f \cup g$ with

$$(f \cup g)_{i_0 \cdots i_{r+s}} = f_{i_0 \cdots i_r} \otimes g_{i_r \cdots i_{r+s}}.$$

After passing to the direct limit over all coverings $\mathcal{U}$ and taking cohomology, we obtain a pairing

$$\check{H}^r(X, F) \times \check{H}^s(X, G) \to \check{H}^{r+s}(X, F \otimes G).$$

When Čech cohomology agrees with derived functor cohomology, it is easy to check that these pairings satisfy the same conditions as the pairings in (1.17a) and hence equal them.

(b) Write $C^{\cdot}(F)$ for the Godement resolution of a sheaf F. It is shown in Godement [1, II.6] that there are canonical pairings of complexes

$$C^{\cdot}(F_1) \times C^{\cdot}(F_2) \to C^{\cdot}(F_1 \otimes F_2)$$

when F_1 and F_2 are sheaves in the usual sense on a topological space. The same argument shows that there are such pairings in our situation also. On applying the functors $\Gamma(X, -)$ or π_* and taking cohomology, we get pairings

$$H^r(X, F_1) \times H^s(X, F_2) \to H^{r+s}(X, F_1 \otimes F_2)$$
$$R^r\pi_*F_1 \times R^s\pi_*F_2 \to R^{r+s}\pi_*(F_1 \otimes F_2).$$

These may easily be shown to satisfy the conditions of (1.16) and hence to agree with the pairings of (1.17a) and (1.17b). (An explicit description of the pairings of complexes is not needed for this—it suffices to know they exist.)

Finally we show how to derive the cup-product pairing of (1.17a) from the canonical pairing on Exts. For any sheaves F and G on X, there are canonical maps

$$G \to \underline{\mathrm{Hom}}\,(F, F \otimes G) \text{ and } H^s(X, \underline{\mathrm{Hom}}\,(F, F \otimes G)) \to \mathrm{Ext}^s\,(F, F \otimes G),$$

the second being the edge morphism in the spectral sequence

$$H^s(X, \underline{\mathrm{Ext}}^t\,(F, F \otimes G)) \Rightarrow \mathrm{Ext}^{s+t}\,(F, F \otimes G).$$

On composing, we get a canonical map $H^s(X, G) \to \mathrm{Ext}^s_X\,(F, F \otimes G)$ and hence a pairing that makes

$$\begin{array}{ccc}
H^r(X, F) \times H^s(X, G) & \longrightarrow & H^{r+s}(X, F \otimes G) \\
\| \qquad\qquad \downarrow & & \| \\
H^r(X, F) \times \mathrm{Ext}^s_X\,(F, F \otimes G) & \longrightarrow & H^{r+s}(X, F \otimes G)
\end{array}$$

commute.

PROPOSITION 1.20. *The pairing defined above equals the cup-product pairing of* (1.17a).

Proof. According to (1.16), it suffices to show that the pairings agree for $r = 0 = s$ and that the new pairing behaves well with respect to short exact sequences in G and in F. Of these, the first two are obvious, and the third becomes so once the maps have been elucidated. We begin with the boundary map on the Exts.

Let $\mathbf{A}$ be any abelian category with enough injectives, and let $0 \to A \to B \to C \to 0$ be an exact sequence in $\mathbf{A}$. It is possible to choose injective resolutions $A^{\cdot}$, $B^{\cdot}$, and $C^{\cdot}$ of A, B, and C such that $0 \to A^{\cdot} \xrightarrow{\alpha} B^{\cdot} \to C^{\cdot} \to 0$ is exact. Let $C_{\alpha}^{\cdot}$ be the *mapping cone* of α, that is, $C_{\alpha}^{\cdot} = B^{\cdot} \oplus A^{\cdot}[1]$ with

$$d_{C_\alpha} = \begin{pmatrix} d_B & \alpha[1] \\ 0 & d_{A[1]} \end{pmatrix}.$$

There is an exact sequence $0 \to B^{\cdot} \to C_{\alpha}^{\cdot} \to A^{\cdot}[1] \to 0$ and an obvious map $C_{\alpha}^{\cdot} \to C^{\cdot}$. On writing out the cohomology sequences of these two short exact sequences of complexes and applying the five-lemma, one finds that $C_{\alpha}^{\cdot} \to C^{\cdot}$ is a quasi-isomorphism. As $C_{\alpha}^{\cdot}$ is bounded below and consists of injectives, $C_{\alpha}^{\cdot} \to C^{\cdot}$ has a homotopy inverse (Hartshorne [1, I.4.5]), and therefore there is a map $C^{\cdot} \xrightarrow{\beta} A^{\cdot}[1]$ (well-defined up to homotopy). The boundary map $\operatorname{Ext}^r(D, C) \to \operatorname{Ext}^{r+1}(D, A)$ merely composes a homotopy class of maps $D \to C^{\cdot}[r]$ with $\beta[r]$.

We next describe $H^s(X, G) \to \operatorname{Ext}^s(F, F \otimes G)$. Let $F^{\cdot}$ and $(F \otimes G)^{\cdot}$ be injective resolutions of F and $F \otimes G$, and let $C^{\cdot}(G)$ be the Godement resolution of G. From (1.15) one sees easily that $F \otimes C^{\cdot}(G)$ is a resolution of $F \otimes G$ and hence there is a map $F \otimes C^{\cdot}(G) \to (F \otimes G)^{\cdot}$. An element γ of $H^s(X, G)$ may be represented by a map $\mathbb{Z} \to C^{\cdot}(G)[s]$, which induces a map $F \to F \otimes C^{\cdot}(G)[s]$. On composing, we obtain a map $\gamma^F : F \to (F \otimes G)^{\cdot}[s]$ that represents the image of γ in $\operatorname{Ext}^s(F, F \otimes G)$. We may also regard γ^F as a map $F^{\cdot} \to (F \otimes G)^{\cdot}[s]$.

Finally let $0 \to F_1 \to F_2 \to F_3 \to 0$ be a sequence of sheaves that is split-exact on stalks, and let $\gamma_2 \in H^s(X, G)$. The diagram

$$\begin{array}{ccc} F_3^{\cdot} & \xrightarrow{\beta_1} & F_1^{\cdot}[1] \\ \downarrow{\scriptstyle \gamma_2^{F_3}} & & \downarrow{\scriptstyle \gamma_2^{F_1}[1]} \\ (F_3 \otimes G)^{\cdot} & \xrightarrow{\beta_2} & (F_1 \otimes G)^{\cdot}[1] \end{array}$$

commutes up to homotopy. Consider an element of $H^r(F_3)$ represented by $\gamma_1 : \mathbb{Z} \to F_3^{\cdot}[r]$. Then

$$(d\gamma_1) \cup \gamma_2 = \gamma_2^{F_1}[1] \circ (\beta_1 \circ \gamma_1) = \beta_2 \circ (\gamma_2^{F_3} \circ \gamma_1) = d(\gamma_1 \cup \gamma_2)$$

as required.

Remark 1.21. When X is an open subvariety of a complete variety, the canonical pairing on Exts defines a cup-product pairing

$$H_c^r(X, F) \times H^s(X, G) \to H_c^{r+s}(X, F \otimes G)$$

with properties similar to the above.

§2. The Cohomology of Curves

Throughout this section, X will be a smooth projective curve over a field k. In the next theorem we write $\mathrm{Ext}^r_U(F, F')$ for $\mathrm{Ext}^r_{\mathbf{S}(U_{\mathrm{et}}, \mathbb{Z}/(n))}(F, F')$.

THEOREM 2.1. (Poincaré duality). *Assume that k is algebraically closed and that n is prime to* char (k).

(a) *For any nonempty open subscheme U of X there is a canonical isomorphism $\eta(U): H^2_c(U, \mu_n) \xrightarrow{\approx} \mathbb{Z}/(n)$. ($H^*_c$ is as in* (III.1.29)).

(b) *For any constructible sheaf F of $\mathbb{Z}/(n)$-modules on such a U, the groups $H^r_c(U, F)$ and $\mathrm{Ext}^r_U(F, \mu_n)$ are finite for all r and zero for $r > 2$; the canonical pairings*

$$H^r_c(U, F) \times \mathrm{Ext}^{2-r}_U(F, \mu_n) \to H^2_c(U, \mu_n) \underset{\approx}{\xrightarrow{\eta}} \mathbb{Z}/(n)$$

are nondegenerate.

Proof. (a) We first define $\eta(X)$. Recall that the degree map

$$\deg: \mathrm{Pic}(X) \to \mathbb{Z}$$

is surjective with kernel $J_X(k)$, where J_X is the Jacobian of X and that $J_X(k)$ is divisible by n (Mumford [4, p. 42]). As $H^2(X, \mathbb{G}_m) = 0$ (III.2.22d), the Kummer sequence gives a diagram,

$$\begin{array}{ccccccccc}
 & & H^1(X, \mathbb{G}_m) & \xrightarrow{n} & H^1(X, \mathbb{G}_m) & \longrightarrow & H^2(X, \mu_n) & \longrightarrow & 0 \\
 & & \downarrow{\scriptstyle \deg} & & \downarrow{\scriptstyle \deg} & & \downarrow{\scriptstyle \eta(X)} & & \\
0 & \longrightarrow & \mathbb{Z} & \xrightarrow{n} & \mathbb{Z} & \longrightarrow & \mathbb{Z}/(n) & \longrightarrow & 0,
\end{array}$$

and the snake lemma shows that $\eta(X)$ is an isomorphism.

If $U \subset X$ is open, then the exact sequence (III.1.30)

$$\cdots \to H^r_c(U, \mu_n) \to H^r(X, \mu_n) \to H^r(X, i_* i^* \mu_n) \to \cdots,$$

where i is the closed immersion $X - U \hookrightarrow X$, shows that $H^2_c(U, \mu_n) \to H^2(X, \mu_n)$ is an isomorphism, and we define $\eta(U)$ to be its composite with $\eta(X)$. Thus $\eta(U)$ is the unique isomorphism making

$$\begin{array}{ccccc}
H^1_c(U, \mathbb{G}_m) & \xrightarrow{\text{canon}} & H^1(X, \mathbb{G}_m) & \xrightarrow{\deg} & \mathbb{Z} \\
\downarrow & & & & \downarrow{\scriptstyle \text{canon.}} \\
H^2_c(U, \mu_n) & & \xrightarrow{\eta(U)} & & \mathbb{Z}/(n)
\end{array}$$

commute.

If $V \subset U \subset X$, then clearly

$$\begin{array}{ccc} H_c^2(V, \mu_n) & \xrightarrow{\eta(V)} & \mathbb{Z}/(n) \\ \downarrow & & \| \\ H_c^2(U, \mu_n) & \xrightarrow{\eta(U)} & \mathbb{Z}/(n) \end{array}$$

commutes.

It will be necessary to have a second description of $\eta(U)$. Let $i: z \hookrightarrow U$ be the inclusion of some closed point into U, and let $g: \operatorname{spec} K \hookrightarrow U$ be the generic point. Then $R^s g_* \mathbb{G}_{m,K} = 0$ for $s > 0$ (Tsen's theorem (III.2.22)) and $i^! g_* = 0$, and so it follows from the spectral sequence

$$R^r i^! R^s g_* \mathbb{G}_{m,K} \Rightarrow R^{r+s}(i^! g_*) \mathbb{G}_{m,K} = 0$$

that $(R^r i^!)(g_* \mathbb{G}_{m,K}) = 0$ for all r. Similar arguments show that $(R^s i^!)(i_{x*}\mathbb{Z}) = 0$ for all s and all closed points $i_x: x \hookrightarrow X$ not equal to z and $(R^s i^!) i_* \mathbb{Z} = \mathbb{Z}, 0$ respectively for $s = 0, \geq 1$. Thus, from

$$0 \to \mathbb{G}_m \to g_* \mathbb{G}_{m,K} \to \bigoplus i_{x*} \mathbb{Z} \to 0$$

we find that $R^s i^! \mathbb{G}_m = 0, \mathbb{Z}, 0$ respectively for $s = 0, 1, \geq 1$; the spectral sequence

$$H^r(z, R^s i^! \mathbb{G}_m) \Rightarrow H_z^{r+s}(U, \mathbb{G}_m)$$

now shows $H_z^r(U, \mathbb{G}_m) = 0$ for $r \neq 1$ and that there is an isomorphism $\zeta: H_z^1(U, \mathbb{G}_m) \to \mathbb{Z}$. If we identify $H_z^1(U, \mathbb{G}_m)$ with $H_z^1(X, \mathbb{G}_m)$ (excision, III.1.27), then ζ is the unique isomorphism making

$$\begin{array}{ccccccc} \Gamma(X - z, \mathbb{G}_m) & \longrightarrow & H_z^1(X, \mathbb{G}_m) & \longrightarrow & H^1(X, \mathbb{G}_m) & \longrightarrow & H^1(X - z, \mathbb{G}_m) \\ \downarrow = & & \downarrow \zeta & & \downarrow \approx & & \downarrow \approx \\ \Gamma(X - z, \mathcal{O}_X^*) & \xrightarrow{\text{ord}_z} & \mathbb{Z} & \longrightarrow & \operatorname{Pic}(X) & \longrightarrow & \operatorname{Pic}(X - z) \end{array}$$

commute, where $\mathbb{Z} \to \operatorname{Pic}(X)$ maps 1 to the divisor class of z. As

$$H_z^1(U, \mathbb{G}_m)\,(= H_z^1(X, \mathbb{G}_m)) \to H^1(X, \mathbb{G}_m)$$

is the composite

$$H_z^1(U, \mathbb{G}_m) \xrightarrow{\text{(III.1.29)}} H_c^1(U, \mathbb{G}_m) \to H^1(X, \mathbb{G}_m)$$

and as ζ is the composite

$$H_z^1(U, \mathbb{G}_m) \to H^1(X, \mathbb{G}_m) \xrightarrow{\deg} \mathbb{Z},$$

we see that $\eta(U)$ is the unique map making

$$\begin{array}{ccccc} H^1_z(U, \mathbb{G}_m) & & \xrightarrow{\approx} & & \mathbb{Z} \\ \downarrow & & & & \downarrow \\ H^2_z(U, \mu_n) & \xrightarrow{\approx} & H^2_c(U, \mu_n) & \xrightarrow{\eta(U)} & \mathbb{Z}/(n) \end{array}$$

commute.

(b) For a sheaf F on U we write $\phi^r(U, F)$ or simply $\phi^r(F)$ for the map $\mathrm{Ext}^{2-r}_U(F, \mu_n) \to H^r_c(U, F)^{\check{}}$ ($\check{}$ = dual) defined by the pairing in (b). The proof will be in several steps.

Step 1. Let $\pi: U' \to U$ be a Galois morphism of curves; for a sheaf F on U', $\phi^r(U', F)$ is an isomorphism if and only if $\phi^r(U, \pi_* F)$ is an isomorphism.

Proof. The map extends to a finite flat map $\pi: X' \to X$ of the smooth projective closures of U' and U. Thus the step will follow from (1.14) once we have shown that

$$\begin{array}{ccc} H^2_c(U', \mu_n) & \xrightarrow{\eta(U')} & \mathbb{Z}/(n) \\ \downarrow{\scriptstyle tr} & & \| \\ H^2_c(U, \mu_n) & \xrightarrow{\eta(U)} & \mathbb{Z}/(n) \end{array}$$

commutes. There is a commutative diagram

$$\begin{array}{ccc} H^1_c(U', \mathbb{G}_m) & \longrightarrow & \mathrm{Pic}\,(X') \\ \downarrow{\scriptstyle tr} & & \downarrow{\scriptstyle tr} \\ H^1_c(U, \mathbb{G}_m) & \longrightarrow & \mathrm{Pic}\,(X) \end{array}$$

where $tr: \mathrm{Pic}\, X' \to \mathrm{Pic}\,(X)$ is induced by the map on divisors

$$\sum n_z z \mapsto \sum n_z \bar{\pi}(z).$$

(Essentially the commutativity follows from the equality $tr(\mathrm{div}\,(f)) = \mathrm{div}\,(\mathrm{norm}\,(f))$.) Since $tr: \mathrm{Pic}\,(X') \to \mathrm{Pic}\,(X)$ preserves degrees, the required commutativity is obvious from the first description of $\eta(U)$.

Step 2. The theorem is true if F has support on a proper closed subscheme of U.

Proof. We may take the sheaf to be $i_* F$, where $i: z \hookrightarrow U$ is the inclusion of a single point.

$$H^r_c(U, i_* F) = H^r(X, j_! i_* F) = H^r(X, i_* F) \approx H^r(z, F) = 0$$

for $r \neq 0$. Thus we must show that $\mathrm{Ext}^s_U(i_*F, \mu_n) = 0$ for $s \neq 2$. As $R^s i^! \mathbb{G}_m = \mathbb{Z}, 0$ respectively for $s = 1, \neq 1$ (see the proof of (a)), $R^s i^! \mu_n = \mathbb{Z}/(n), 0$ respectively for $s = 2, \neq 2$, and so the spectral sequence

$$\mathrm{Ext}^r_z(F, R^s i^! \mu_n) \Rightarrow \mathrm{Ext}^{r+s}_U(i_*F, \mu_n)$$

shows that

$$\mathrm{Ext}^2_U(i_*F, \mu_n) \xrightarrow{\approx} \mathrm{Hom}_z(F, \mathbb{Z}/(n))$$

and

$$\mathrm{Ext}^s_U(i_*F, \mu_n) = \mathrm{Ext}^{s-2}_z(F, \mathbb{Z}/(n)) = 0$$

for $s \neq 2$.

Consider the diagram

$$\begin{array}{ccccccc}
H^0_c(U, i_*F) & \times & \mathrm{Ext}^2_U(i_*F, \mu_n) & \longrightarrow & H^2_c(U, \mu_n) & \xrightarrow{\eta(U)} & \mathbb{Z}/(n) \\
\uparrow \approx & & \| & & \uparrow \approx & & \| \\
H^0_z(U, i_*F) & \times & \mathrm{Ext}^2_U(i_*F, \mu_n) & \longrightarrow & H^2_z(U, \mu_n) & & \| \\
\downarrow \approx & & \downarrow \approx & & \downarrow \approx & & \| \\
H^0(z, F) & \times & \mathrm{Hom}_z(F, \mathbb{Z}/(n)) & \longrightarrow & H^0(z, \mathbb{Z}/(n)) & = & \mathbb{Z}/(n).
\end{array}$$

The pairings are the canonical Ext pairings defined in the last section (note that $H^r_z(U, G) = \mathrm{Ext}^r_U(i_*\mathbb{Z}/(n), G)$), and the two isomorphisms at lower center are as defined above. Since the bottom pairing is obviously nondegenerate, it remains to show that the diagram commutes. That the top two pairings agree can be seen by interpreting everything as an Ext on X (for example, $H^0_z(U, i_*F) \rightarrow H^0_c(U, i_*F)$ is the map

$$\mathrm{Ext}^0_X(i_*\mathbb{Z}/(n), j_! i_*F) \rightarrow \mathrm{Ext}^0_X(\mathbb{Z}/(n), j_! i_*F)$$

induced by $\mathbb{Z}/(n) \rightarrow i_*(\mathbb{Z}/(n)_z)$); that the bottom two pairings agree follows from the map $\mathrm{Ext}^2_U(i_*F, \mu_n) \rightarrow \mathrm{Hom}_z(F, \mathbb{Z}/(n))$ being functorial in F; that the rectangle at right commutes follows from the second description of $\eta(U)$.

Step 3. Let V be an open subvariety of U; $\phi^r(U, F)$ is an isomorphism if $\phi^r(V, F|V)$ is an isomorphism and $\phi^{r+1}(V, F|V)$ is surjective; $\phi^r(V, F|V)$ is an isomorphism if $\phi^r(U, F)$ is an isomorphism and $\phi^{r-1}(U, F)$ is injective.

Proof. Let $j: V \hookrightarrow U$ be the given open immersion and $i: Z \hookrightarrow U$ the complementary closed immersion. The exact sequence of sheaves on U,

$$0 \rightarrow F_V \rightarrow F \rightarrow F_Z \rightarrow 0,$$

where $F_V = j_!(F|V)$ and $F_Z = i_* i^* F$, gives rise to an exact commutative diagram,

$$\begin{array}{ccccccccc}
\cdots \longrightarrow & \operatorname{Ext}_U^{2-r}(F_Z, \mu_n) & \longrightarrow & \operatorname{Ext}_U^{2-r}(F, \mu_n) & \longrightarrow & \operatorname{Ext}_U^{2-r}(F_V, \mu_n) & \longrightarrow \cdots \\
& \downarrow{\scriptstyle \phi^r(F_Z)} & & \downarrow{\scriptstyle \phi^r(F)} & & \downarrow{\scriptstyle \phi^r(F_V)} & \\
\cdots \longrightarrow & H_c^r(U, F_Z)^\check{} & \longrightarrow & H_c^r(U, F)^\check{} & \longrightarrow & H_c^r(U, F_V)^\check{} & \longrightarrow \cdots
\end{array}$$

Almost by definition $H_c^r(U, F_V) = H_c^r(V, F|V)$, and $\operatorname{Ext}_U^r(F_V, \mu_n) = \operatorname{Ext}_V^r(F|V, j^*\mu_n)$ because j^* is exact, preserves injectives, and is adjoint to $j_!$. Moreover, $\phi^r(U, F_V)$ can be identified with $\phi^r(V, F|V)$ because of the compatibility of $\eta(V)$ with $\eta(U)$. As $\phi^r(F_Z)$ is an isomorphism for all r (Step 2), the five-lemma implies the claim.

Step 4. The group $H^r(U, F) = 0$ for $r > 2$; thus $H_c^r(U, F) = 0$ for $r > 2$.

Proof. Let g:spec $K \hookrightarrow U$ be the inclusion of the generic point, and consider a constructible sheaf F_0 on spec K. According to (III.1.15), $(R^s g_* F_0)_{\bar{x}} = H^s(K_{\bar{x}}, F_0)$ for any $x \in X$, where $K_{\bar{x}}$ is the field of fractions of the strictly local ring at x. As $K_{\bar{x}} = K_{\text{sep}}$, if x is the generic point, any section of $R^s g_* F_0$ has support on a proper closed subset of U for $s > 0$, and as $K_{\bar{x}}$ is of transcendence degree one over (the algebraically closed field) k, Tate's theorem (Shatz [2, Thm. 28, p. 119]) shows that $R^s g_* F_0 = 0$ for $s > 1$. Thus, in the Leray spectral sequence $H^r(U, R^s g_* F_0) \Rightarrow H^{r+s}(K, F_0)$, the left hand terms are zero except possibly for $s = 0$ or $(r, s) = (0, 1)$. As $H^r(K, F_0) = 0$ for $r > 1$ (Shatz [2, Thm. 24]), this shows that $H^r(U, g_* F_0) = 0$ for $r > 2$. We apply this with $F_0 = g^* F$. As the kernel and cokernel of $F \to g_* g^* F$ have support on a proper closed subset of U, $H^r(U, F) = H^r(U, g_* g^* F)$ for $r > 1$, which completes the proof of the first statement. The second is the special case of the first with $(U, F) = (X, j_! F)$.

Step 5. If F is locally constant, then $\operatorname{Ext}_U^r(F, \mu_n) = 0$ for $r > 2$.

Proof. Obviously F is pseudocoherent, and so $\underline{\operatorname{Ext}}^s(F, \mu_n) = 0$ for $s > 0$ (III.1.31). Thus

$$\operatorname{Ext}^r(F, \mu_n) = H^r(U, \underline{\operatorname{Hom}}(F, \mu_n)) = 0$$

for $r > 2$.

Step 6. If $\phi^{r_0}(X, \mathbb{Z}/(n))$ is an isomorphism and if $\phi^r(U, F)$ is an isomorphism for all U, all locally constant F, and all $r < r_0$, then $\phi^{r_0}(U, F)$ is an isomorphism for all U and all locally constant F.

Proof. Step 3 shows that $\phi^{r_0}(U, \mathbb{Z}/(n))$ is an isomorphism for all U. First assume that F is constant. Then there is an exact sequence

$$0 \to F \to F_0 \to F_1 \to 0$$

with F_0 free, that is, of the form $\mathbb{Z}/(n) \oplus \cdots \oplus \mathbb{Z}/(n)$, and F_1 constant. Consider

$$\begin{array}{ccccccccc}
\cdot & \longrightarrow & \cdot & \longrightarrow & \cdot & \longrightarrow & \cdot & \longrightarrow & \cdot \\
\downarrow{\scriptstyle \phi^{r_0}(F_1)} & & \approx\downarrow{\scriptstyle \phi^{r_0}(F_0)} & & \downarrow{\scriptstyle \phi^{r_0}(F)} & & \approx\downarrow{\scriptstyle \phi^{r_0-1}(F_1)} & & \approx\downarrow{\scriptstyle \phi^{r_0-1}(F_0)} \\
\cdot & \longrightarrow & \cdot & \longrightarrow & \cdot & \longrightarrow & \cdot & \longrightarrow & \cdot
\end{array}$$

Since $\phi^{r_0}(\mathbb{Z}/(n))$ is an isomorphism, so also is $\phi^{r_0}(F_0)$, and the five-lemma shows $\phi^{r_0}(F)$ is surjective. The same argument applied to F_1 shows that $\phi^{r_0}(F_1)$ is surjective, and now the five-lemma argument shows that $\phi^{r_0}(F)$ is injective. If F is merely locally constant, that is (1.3d) there is a finite étale map (that we may choose to be Galois) $\pi: U' \to U$ such that π^*F is constant, then we embed F in the sequence

$$0 \to F \to \pi_*\pi^*F \to F_1 \to 0.$$

As F_1 is locally constant, and Step 1 shows that $\phi^{r_0}(\pi_*\pi^*F)$ is an isomorphism, the same five-lemma argument may be applied.

Step 7. If F is locally constant, then $\phi^r(U, F)$ is an isomorphism.

Proof. By Step 5, $\phi^r(U, F)$ is an isomorphism for $r < 0$. For $(U, F, r) = (X, \mathbb{Z}/(n), 0)$ the pairing is

$$\mathbb{Z}/(n) \times H^2(X, \mu_n) \to H^2(X, \mu_n) \approx \mathbb{Z}/(n),$$

which is obviously nondegenerate. Thus $\phi^0(X, \mathbb{Z}/(n))$ is an isomorphism and the assertion, for $r = 0$, follows from Step 6.

Let $\gamma \in H^1(X, \mathbb{Z}/(n))$ and let $\pi: X' \to X$ be the Galois covering corresponding to γ (III.4). Then $\gamma \mapsto 0$ under

$$H^1(X, \mathbb{Z}/(n)) \to H^1(X, \pi_*\pi^*\mathbb{Z}/(n))\,(= H^1(X', \mathbb{Z}/(n)))$$

and an easy diagram chase, starting from the exact sequence

$$0 \to \mathbb{Z}/(n) \to \pi_*\pi^*\mathbb{Z}/(n) \to F_1 \to 0,$$

shows that the image of γ under

$$\phi^1(X, \mathbb{Z}/(n))^{\check{}}: H^1(X, \mathbb{Z}/(n)) \to \mathrm{Ext}^1\,(\mathbb{Z}/(n), \mu_n)^{\check{}}$$

is nonzero. Thus $\phi^1(X, \mathbb{Z}/(n))^{\check{}}$ is injective, and $\phi^1(X, \mathbb{Z}/(n))$ is surjective. It is an isomorphism because $H^1(X, \mathbb{Z}/(n))$ and $H^1(X, \mu_n)$ have the same finite order n^{2g} (recall $\mathbb{Z}/(n) \approx \mu_n$). Thus the assertion, for $r = 1$, follows from Step 6.

For $(U, F, r) = (X, \mathbb{Z}/(n), 2)$ the pairing is

$$H^2(X, \mathbb{Z}/(n)) \times \mu_n(k) \to H^2(X, \mu_n) \approx \mathbb{Z}/(n),$$

which is obviously nondegenerate. Thus the case $r = 2$ is proven. For $r > 2$, Step 4 shows that ϕ^r is the map $0 \xrightarrow{\approx} 0$.

We now prove the theorem. According to (1.8), there is an open subset V of U such that $F|V$ is locally constant. Step 7 shows that $\phi^r(V, F|V)$ is an isomorphism for all r, and it follows, by Step 3, that $\phi^r(U, F)$ is an isomorphism for all r. Finally it is easy to trace through the steps and see that the finiteness of the groups $H^r(X, \mathbb{Z}/(n))$ (compare the proof of Step 7) implies that of the groups $H_c^r(U, F)$.

PROPOSITION 2.2. (a) *Let $\tilde{X} = \operatorname{spec} k\{t\}$ where $k\{t\}$ is the strictly local ring at the origin of $\mathbb{A}_k^1$ with k algebraically closed; let z be the closed point of $\tilde{X}$, and let F be a constructible sheaf of $\mathbb{Z}/(n)$-modules on $\tilde{X}$ with n prime to the characteristic of k. Then there is a canonical isomorphism $\eta: H_z^2(\tilde{X}, \mu_n) \xrightarrow{\approx} \mathbb{Z}/(n)$ and the canonical pairing*

$$H_z^r(\tilde{X}, F) \times \operatorname{Ext}_{\tilde{X}}^{2-r}(F, \mu_n) \to H_z^2(\tilde{X}, \mu_n) \approx \mathbb{Z}/(n)$$

is nondegenerate. (The Ext *is in* $\mathbf{S}(\tilde{X}, \mathbb{Z}/(n))$*).*

(b) *Let U, k, F and n be as in (2.1) with F locally constant on U. For any open immersion $U \overset{j}{\hookrightarrow} V \subset X$ the pairings*

$$H_c^r(V, j_*F) \times H^{2-r}(V, j_*\check{F}(1)) \to H_c^2(V, \mu_n) \approx \mathbb{Z}/(n)$$

are nondegenerate.

(c) *For any twisted-constant, constructible sheaf F of $\mathbb{Q}_l$-vector spaces, $l \neq \operatorname{char}(k)$, on $U \subset X$ and any open immersion $U \overset{j}{\hookrightarrow} V \subset X$, the pairings*

$$H_c^r(V, j_*F) \times H^{2-r}(V, j_*\check{F}(1)) \to H_c^2(V, \mathbb{Q}_l(1)) \approx \mathbb{Q}_l$$

are nondegenerate pairings of finite-dimensional vector spaces.

Proof. (a) This may be deduced from (2.1) by interpreting $\tilde{X}$ as $\operatorname{spec} \mathcal{O}_{X,z}$ for some curve X and $F = F_0|\tilde{X}$ for some constructible sheaf F_0 on X and showing

$$H_z^r(\tilde{X}, F) = \varprojlim H_c^r(U, F)$$
$$\operatorname{Ext}_{\tilde{X}}^s(F, \mu_n) = \varinjlim \operatorname{Ext}_U^s(F, \mu_n)$$

where the limits are over the étale neighborhoods of z.

Alternatively we may proceed directly. The pairing is of the form defined in Section 1 because $H_z^r(\tilde{X}, F) = \operatorname{Ext}_{\tilde{X}}^r(i_*\mathbb{Z}, F)$ where $i: z \hookrightarrow \tilde{X}$ is the inclusion of the closed point. As in the proof of (2.1a), we may show that $R^r i^! \mu_n = \mathbb{Z}/(n)$ for $r = 2$ and is zero otherwise. Thus the spectral sequence

$$H^r(z, R^s i^! \mu_n) \Rightarrow H_z^{r+s}(\tilde{X}, \mu_n)$$

gives an isomorphism

$$H_z^2(\tilde{X}, \mu_n) \xrightarrow{\approx} H^0(z, \mathbb{Z}/(n)) = \mathbb{Z}/(n)$$

that we take to be η.

The proposition is trivial if F has support on z (compare Step 2 of the proof of (2.1)), and so we may assume the sheaf is of the form $j_!F$ where $j: u \hookrightarrow \tilde{X}$ is the inclusion of the generic point and F is the sheaf on u corresponding to the $I = \mathrm{Gal}(k(u)_s/k(u))$-module M. From the exact sequence (III.1.25)

$$0 \to H_z^0(\tilde{X}, j_!F) \to H^0(\tilde{X}, j_!F) \to H^0(u, F) \to \cdots$$

and the fact that $\tilde{X}$ is strictly local, we obtain isomorphisms

$$H^{r-1}(u, F) \xrightarrow{\approx} H_z^r(\tilde{X}, j_!F)$$

for all $r \geq 0$. Also

$$\mathrm{Ext}_{\tilde{X}}^s(j_!F, \mu_n) \approx \mathrm{Ext}_u^s(F, \mu_n) \approx H^s(u, \check{F}(1)).$$

Thus we must show that the cup-product pairings

$$H^r(I, M) \times H^{1-r}(I, \check{M}) \to H^1(I, \mathbb{Z}/(n)) \approx \mathbb{Z}/(n)$$

are nondegenerate. Let $I_1 \subset I$ be the first ramification group, and let $I' = I/I_1$. Since I_1 is a pro-p-group where p is the characteristic exponent of k and M has order prime to p, the Hochschild-Serre spectral sequence shows that $H^r(I, M) = H^r(I', M)$. But (compare I.5.2e) I' is generated topologically by a single element, and so the assertion is trivial (Serre [8, I-31]).

(b) We have to show

$$\mathrm{Ext}_V^s(j_*F, \mu_n) \approx H^s(V, j_*\check{F}(1)).$$

From the spectral sequence

$$H^r(V, \underline{\mathrm{Ext}}^s(j_*F, \mu_n)) \Rightarrow \mathrm{Ext}_V^{r+s}(j_*F, \mu_n)$$

we see that it suffices to show that $\underline{\mathrm{Hom}}(j_*F, \mu_n) \approx j_*\check{F}(1)$ and that $\underline{\mathrm{Ext}}^s(j_*F, \mu_n) = 0$ for $s > 0$. Since $j_*\mu_n = \mu_n$ and $\underline{\mathrm{Hom}}(j_*F, j_*\mu_n) = j_*\underline{\mathrm{Hom}}(F, \mu_n)$, the first is obvious. The second may be checked locally on V_{et}, and so, with the notations of (a) and its proof, it suffices to show that $\mathrm{Ext}_{\tilde{X}}^s(j_*F, \mu_n) = 0$ for $s > 0$. But $\mathrm{Ext}_{\tilde{X}}^s(j_*F, \mu_n)$ is dual to

$$H_z^{2-s}(\tilde{X}, j_*F),$$

and there is an exact sequence

$$0 \to H_z^0(\tilde{X}, j_*F) \to H^0(\tilde{X}, j_*F) \xrightarrow{\approx} H^0(\tilde{X} - z, F) \to H_z^1(\tilde{X}, j_*F) \to 0$$

(c) On passing to an inverse limit in (b) and tensoring with $\mathbb{Q}_l$, one obtains (c).

COROLLARY 2.3. *Let k be a finite field, and let n be prime to* char (k). *For any nonempty open subscheme U of X there is a canonical isomorphism $\eta(U): H_c^3(U, \mu_n) \xrightarrow{\approx} \mathbb{Z}/(n)$, and for any locally constant, constructible sheaf F of $\mathbb{Z}/(n)$-modules the canonical pairings*

$$H_c^r(U, F) \times H^{3-r}(U, \check{F}(1)) \to H_c^3(U, \mu_n) \xrightarrow{\approx} \mathbb{Z}/(n)$$

are nondegenerate pairings of finite groups.

Proof. Let $\Gamma = \mathrm{Gal}(k_s/k)$, and let σ be the canonical topological generator of Γ. For any finite n-torsion Γ-module M, $H^0(\Gamma, M)$ and $H^1(\Gamma, M)$ are respectively the kernel M^Γ and cokernel M_Γ of $\sigma - 1$: $M \to M$, and $H^r(\Gamma, M) = 0$ for $r \geq 2$. Moreover, if $M^\vee = \mathrm{Hom}(M, \mathbb{Z}/(n))$, then the canonical pairings

$$H^r(\Gamma, M) \times H^{1-r}(\Gamma, M^\vee) \to H^1(\Gamma, \mathbb{Z}/(n)) \approx \mathbb{Z}/(n)$$

are nondegenerate.

Let $\bar{X} = X \otimes_k k_s$ and $\bar{U} = U \otimes_k k_s$. The Hochschild-Serre spectral sequences for $\bar{X}/X$ and $\bar{U}/U$ (III.2.21b) give short exact sequences,

$$0 \to H_c^{r-1}(\bar{U}, F)_\Gamma \to H_c^r(U, F) \to H_c^r(\bar{U}, F)^\Gamma \to 0$$
$$0 \to H^{r-1}(\bar{U}, F)_\Gamma \to H^r(U, F) \to H^r(\bar{U}, F)^\Gamma \to 0.$$

In particular,

$$H_c^3(U, \mu_n) \approx H_c^2(\bar{U}, \mu_n)_\Gamma \xrightarrow[\approx]{\eta(U)} \mathbb{Z}/(n),$$

which defines $\eta(U)$. To prove the corollary one only has to check that

$$\begin{array}{ccccccccc} 0 & \longrightarrow & H_c^{r-1}(\bar{U}, F)_\Gamma & \longrightarrow & H_c^r(U, F) & \longrightarrow & H_c^r(\bar{U}, F)^\Gamma & \longrightarrow & 0 \\ & & \downarrow\approx & & \downarrow & & \downarrow\approx & & \\ 0 & \longrightarrow & (H^{3-r}(\bar{U}, \check{F}(1))^\Gamma)^\vee & \longrightarrow & H^{3-r}(U, \check{F}(1))^\vee & \longrightarrow & (H^{2-r}(\bar{U}, \check{F}(1))_\Gamma)^\vee & \longrightarrow & 0 \end{array}$$

commutes and apply the five-lemma.

Remark 2.4. (a) Let U/k, F and n be as in (2.1) and assume that U is affine. For some $j: V \hookrightarrow U$, $F|V$ is locally constant, and then (2.2b) shows that $H^2(U, j_*(F|V))$ is dual to $H_c^0(U, j_*(\check{F}|V)(1))$, which is zero since no section of $j_*(\check{F}|V)$ has support on a proper closed subset of U (compare III.1.29a). Since the kernel and cokernel of $F \to j_*j^*F$ have support in dimension zero, it follows that $H^2(U, F) = 0$. It follows (1.10b) that $H^2(U, F) = 0$ for any torsion sheaf F whose torsion is prime to the characteristic of k. Now let U be any (not necessarily smooth) affine curve over k, and let F be a torsion sheaf on U as before. The normalization map $\pi: U' \to U$ is finite and so

$$H^r(U, \pi_*\pi^* F) = H^r(U', \pi^* F) = 0$$

for $r \geq 2$. As the kernel and cokernel of $F \to \pi_*\pi_* F$ have support on the singular set of U, this shows that $H^r(U, F) = 0$ for $r \geq 2$.

(b) Recall that a twisted-constant constructible sheaf F of $\mathbb{Q}_l$-vector spaces on U corresponds to a $\pi_1(U, \bar{u})$-module $F_{\bar{u}}$ and that $H^0(U, F) = (F_{\bar{u}})^{\pi_1(U,\bar{u})}$. Proposition (2.2c) can be used to give a similarly explicit description of $H_c^2(U, F)$, namely,

$$H_c^2(U, F) = H^0(U, \check{F}(1))^\vee = (\check{F}_{\bar{u}}(1)^{\pi_1})^\vee = (F_{\bar{u}})_{\pi_1}(-1),$$

where $^\vee$ means $\mathbb{Q}_l$-linear dual and $(F_{\bar{u}})_{\pi_1}$ is the largest quotient of $F_{\bar{u}}$ on which $\pi_1(U, \bar{u})$ acts trivially. (To see that $(V^{\pi_1})^\vee = (V^\vee)_{\pi_1}$, note that the space of invariants V^{π_1} and the space of co-invariants V_{π_1} satisfy obvious universal mapping properties and that $V \mapsto V^\vee$ defines an autoequivalence of the category of finite-dimensional $\mathbb{Q}_l$-vector-spaces on which π_1 acts continuously.)

(c) Recall (II.3.17) that if U is a scheme over a field k, k' is a purely inseparable field extension of k, and $U' = U \otimes_k k'$, then $V \mapsto V' = V \times_U U'$ defines an equivalence between the categories of schemes étale over U and over U'. (This may also be deduced from (I.3.23) and descent theory, for $U \otimes_k k'$ is the closed subscheme of $U' \otimes_k k'$ defined by a nilpotent ideal.) Thus the functor $F \mapsto F' = F|U'$ is an equivalence of categories $\mathbf{S}(U_{\text{et}}) \to \mathbf{S}(U'_{\text{et}})$.

From this we may conclude that the above theorem and corollaries hold if the field k is only separably closed. For let $U \subset X$ be defined over such a k, let $k' = k_{al}$, and let $U' = U \otimes_k k'$ etc. Then $F \mapsto F'$ maps constructible sheaves on U to constructible sheaves on U', $\mu_{n,U}$ to $\mu_{n,U'}$, and $(\mathbb{Z}/(n))_U$ to $(\mathbb{Z}/(n))_{U'}$. Thus $H_c^r(U', F') = H_c^r(U, F)$, $\text{Ext}_{U'}^r(F'_1, F'_2) = \text{Ext}_U^r(F_1, F_2)$ etc. from which the conclusion follows. (Note: one needs to use caution in applying the above type of argument, for example, $\mathbb{G}'_m \neq \mathbb{G}_{m'}$.)

(d) Artin and Verdier have proved a theorem similar to (2.3) in the case that X is the spectrum of the ring of integers in a number field. It states: there is a canonical isomorphism $\eta: H^3(X, \mathbb{G}_m) \to \mathbb{Q}/\mathbb{Z}$ (modulo 2-torsion) and for any constructible sheaf F the pairings

$$H^r(X, F) \times \text{Ext}_X^{3-r}(F, \mathbb{G}_m) \to H^3(X, \mathbb{G}_m) \xrightarrow{\approx} \mathbb{Q}/\mathbb{Z}$$

are nondegenerate pairings of finite groups (modulo 2-torsion). The proof is essentially a refinement, using class-field theory, of the proof of (2.1). A discussion of the theorem may be found in Mazur [2] and the statement of a generalization, due to Artin and Mazur, on p. 240 of Mazur [1].

(e) Artin and the author have extended (2.1) to a theorem in which the sheaf may have p-torsion where $p = \text{char}(k) \neq 0$ (Artin-Milne [1]).

(f) The cup-product pairing

$$H^1(X, \mu_n) \times H^1(X, \mu_n) \to H^2(X, \mu_n \otimes \mu_n) \approx \mu_n(k)$$

agrees with the e_n-pairing $J_n(k) \times J_n(k) \to \mu_n(k)$. (See Mumford [4, §20] for the definition of the e_n-pairing and [SGA. $4\frac{1}{2}$, Dualité §3] for the agreement.)

The next theorem restates a result of A. Weil. For any morphism $\phi: X \to X$ we write $Tr_l^r(\phi)$, for the trace of the induced map $H^r(X, \mathbb{Q}_l) \to H^r(X, \mathbb{Q}_l)$. Recall that if k is algebraically closed, then $H^0(X, \mathbb{Q}_l) = \mathbb{Q}_l$,

$$H^1(X, \mathbb{Q}_l(1)) = V_l(J_X(k)) \overset{df}{=} \text{Hom}(\mathbb{Q}_l/\mathbb{Z}_l, J(k)) \otimes_{\mathbb{Z}_l} \mathbb{Q}_l,$$

and

$$H^2(X, \mathbb{Q}_l(1)) = \varprojlim (\text{Pic}(X)/l^n \text{Pic}(X)) \otimes \mathbb{Q}_l = \mathbb{Q}_l.$$

THEOREM 2.5. (Lefschetz trace formula). *Assume that k is algebraically closed. Let $\phi: X \to X$ be a nonconstant morphism and let $(\Gamma_\phi . \Delta)$ be the algebraic number of fixed points of ϕ (where Γ_ϕ is the graph of ϕ and Δ the diagonal of $X \times X$). Then, for any $l \neq$ char (k),*

$$(\Gamma_\phi . \Delta) = \sum (-1)^r Tr_l^r(\phi)$$

Proof. First note that $Tr^0(\phi) = 1$. As

$$H^r(X, \mathbb{Q}_l(1)) = H^r(X, \mathbb{Q}_l) \otimes \mathbb{Q}_l(1),$$

$Tr_l^r(\phi)$ is also the trace of the endomorphism of $H^r(X, \mathbb{Q}_l(1))$ defined by ϕ. Thus, as $H^1(X, \mathbb{Q}_l(1)) = V_l(J(k))$, $Tr_l^1(\phi)$ may be identified with the trace of the endomorphism $J(\phi)$ of J defined by ϕ (Mumford [4, p. 182]). Also $Tr_l^2(\phi) = \deg(\phi)$ as ϕ multiplies the degree of a divisor by $\deg(\phi)$. Thus we have to show that

$$(\Gamma_\phi . \Delta) = 1 - Tr(J(\phi)) + \deg(\phi),$$

but this is a special case of the equality $\sigma(X) = Tr(\tau)$ of Lang [2, VI. Thm. 6]. (For a different(?) proof, see (VI.12) below.)

COROLLARY 2.6. *Let $k = \mathbb{F}_q$ and let $F: \bar{X} \to \bar{X}$ be the Frobenius endomorphism of $\bar{X} = X \otimes_k k_s$. Then the zeta function of X, $Z(X, t)$ equals*

$$\frac{P_1(X, t)}{P_0(X, t) P_2(X, t)}$$

where $P_r(X, t) = \det(1 - Ft \mid H^r(\bar{X}, \mathbb{Q}_l))$ (any $l \neq$ char (k)).

Proof. Recall that $Z(X, t)$ is the formal power series

$$\exp\left(\sum_{n>0} \nu_n(X)t^n/n\right)$$

where $\nu_n(X)$ is the number of points of X with coordinates in $k_n = \mathbb{F}_{q^n}$. By F we mean the $\bar{k}$-morphism $F_{X/k} \otimes 1$ where $F_{X/k}$ is the k-morphism $X \to X$ that is the identity map on the underlying topological space of X and is the qth power map on $\mathcal{O}_X$. It acts on $\bar{X}(k_s)$ by raising the coordinates of any point to the qth power. Clearly $\nu_n(X)$ is the number of fixed points of $F^n: \bar{X}(k_s) \to \bar{X}(k_s)$. But this is also the algebraic number of fixed points of F^n because every point of $\Gamma_{F^n}.\Delta$ occurs with multiplicity one. (If π is a local uniformizing parameter at a point x of $\bar{X}^0$ that is fixed by F^n, then the multiplicity of (x, x) in $\Gamma_{F^n}.\Delta$ is the order of the zero of $\pi^{q^n} - \pi$ at x.) Thus (2.5) shows that

$$\nu_n(X) = Tr_l^0(F^n) - Tr_l^1(F^n) + Tr_l^2(F^n)$$

and the corollary follows from the next lemma.

LEMMA 2.7. *Let α be an endomorphism of a finite-dimensional vector space V; then* $\log(\det(1 - \alpha t|V)^{-1}) = \sum_{n>0} Tr(\alpha^n|V)t^n/n$.

Proof. If V has dimension one and α acts as multiplication by a, then the formula is simply the identity

$$\log(1 - at) = -\sum \frac{a^n t^n}{n}.$$

The general formula is a sum of $\dim(V)$ such identities, as may be seen by choosing a basis for V relative to which the matrix is triangular.

Now assume that k is algebraically closed and that X is the quotient $X = Y/G$ of a smooth projective curve Y by a finite group of automorphisms G; this simply means that $k(Y)$ is Galois over $k(X)$ with Galois group G and that Y is the normalization of X in $k(Y)$. The group G acts on $H^1(Y, \mathbb{Q}_l(1)) = V_l(J_Y(k))$, and (2.5) will enable us to determine the character of this representation. Let G_y be the decomposition group at a closed point y of Y, that is, $G_y = \{\sigma \in G \mid y\sigma = y\}$. For any $\sigma \neq 1$, $\sigma \in G_y$, we define $i_y(\sigma)$ to be the order of the zero of $\sigma(\pi) - \pi$ at y, where π is a local uniforming parameter at y. Equivalently, $i_y(\sigma)$ is the multiplicity of (y, y) in $\Gamma_\sigma.\Delta$. The *Artin character* $a_y: G_y \to \mathbb{Z}$ is defined by the formulas:

$$a_y(\sigma) = -i_y(\sigma) \qquad \sigma \neq 1,$$
$$a_y(1) = \sum_{\sigma \neq 1} i_y(\sigma).$$

It is known (Serre [11, 19]) that a_y may be realized as the character of a representation of G_y on a vector space over $\mathbb{Q}_l$ for any $l \neq \operatorname{char}(k)$. Clearly $a_y \neq 1$ if and only if $Y \to X$ is ramified at y. The character a_y of G_y induces a character of G, called the *Artin character* a_x of G at x, where x is the image of y in X. An easy calculation shows that:

$$a_x(\sigma) = -\sum_{y \mapsto x} i_y(\sigma) = \sum_{y \mapsto x} a_y(\sigma), \qquad \sigma \neq 1,$$

$$a_x(1) = -\sum_{\sigma \neq 1} a_x(\sigma),$$

where we put $i_y(\sigma) = 0 = a_y(\sigma)$ for $\sigma \notin G_y$. Thus a_x depends only on x. We write r for the character of the regular representation of G, so that $r(\sigma) = 0$ for $\sigma \neq 1$ and $r(1) = [G]$.

COROLLARY 2.8. *The character ψ of the representation of G on $H^1(Y, \mathbb{Q}_l)$ is given by the formula,*

$$\psi(\sigma) = 2 + (2g_X - 2)r(\sigma) + \sum_{x \in X^0} a_x(\sigma)$$

where g_X is the genus of X.

Proof. If $\sigma \neq 1$, then (2.5) states that $2 - \psi(\sigma) = (\Gamma_\sigma . \Delta)$ where

$$(\Gamma_\sigma . \Delta) = \sum_{y \in Y^0} i_y(\sigma) = -\sum_{x \in X^0} a_x(\sigma).$$

If $\sigma = 1$, then the formula is simply the Hurwitz genus formula $2g_Y - 2 = n(2g_X - 2) + \deg(D_{Y/X})$ since $\psi(1) = 2g_Y$, $r(1) = n =$ the degree of $Y \to X$ and the discriminant

$$D_{Y/X} = \sum a_x(1)x$$

(Serre [7, IV.1. Prop. 4] and Hartshorne [2, IV.2.4]).

Remark 2.9. Let A_y be a $\mathbb{Q}_l[G_y]$-module, $l \neq \operatorname{char}(k)$, whose character is a_y, and let A_x be the induced $\mathbb{Q}_l[G]$-module, so that the representation on A_x has character a_x. The above corollary shows that

$$\sum_{r=0}^{2} (-1)^r [H^r(Y, \mathbb{Q}_l)] = (2 - 2g_X)[\mathbb{Q}_l[G]] - \sum_{x \in X^0} [A_x]$$

where $[*]$ denotes the class of a finitely generated $\mathbb{Q}_l[G]$-module in the Grothendieck group $R_{\mathbb{Q}_l}(G)$ of the category of such modules.

We next prove a formula (due to Grothendieck in the general case and Ogg and Šafarevič in the tame case) that expresses the Euler-Poincaré characteristic of a constructible sheaf of $\mathbb{F}_l$-modules, $l \neq \operatorname{char}(k)$, as a sum of local terms. As before, let k be algebraically closed of characteristic exponent p, and let $X = Y/G$ where G is a finite group of automorphisms of Y. For any point y of Y^0, the *Swan character* $p_y: G_y \to \mathbb{Z}$ is defined to be

$$p_y = a_y - u_y = a_y - r_y + 1$$

where a_y and u_y are the Artin and augmentation characters of G_y. It is known (Serre [11, 19.2]) that there is a unique projective $\mathbb{Z}_l[G_y]$-module P_y whose character is p_y. The Swan character at y is zero if and only if $Y \to X$ is tamely ramified at y.

Let F be a constructible sheaf of $\mathbb{F}_l$-modules on X, $l \neq \operatorname{char}(k)$, and assume that $F|Y$ is generically constant, that is, that if $g: \eta \to X$ is the generic point of X, then the representation of $\operatorname{Gal}(k(X)_s/k(X))$ on the stalk $F_{\bar{\eta}}$ factors through $G = \operatorname{Gal}(k(Y)/k(X))$. Given an F, it is easy to find a suitable Y, for we may take Y to be the normalization of X in any sufficiently large finite Galois extension of $k(X)$. For any point $x \in X^0$, we define (the exponent of) the *wild conductor* of F at x to be

$$\alpha_x(F) = \dim(\operatorname{Hom}_{G_y}(P_y, F_{\bar{\eta}}))$$

for any $y \in Y$ mapping to x. (Here, $\dim = \dim_{\mathbb{F}_l}$.) It has the following properties (Serre [11, 19.3] and [7, VI]):

(a) it is independent of the choice of y (easy);

(b) it is independent of the choice of Y;

(c) it is additive, that is, $0 \to F' \to F \to F'' \to 0$ exact implies $\alpha_x(F') + \alpha_x(F'') = \alpha_x(F)$ (because P_y is projective);

(d)

$$\alpha_x(F) = \sum_{i=1}^{\infty} \frac{g_i}{g_0} \dim(F_{\bar{\eta}}/F_{\bar{\eta}}^{G_i})$$

where $G_0 = G_y$, the G_i are the higher ramification groups and $g_i = [G_i]$; in particular $\alpha_x(F) = 0$ if and only if $F_{\bar{\eta}}$ is tamely ramified at x, that is, G_1 acts trivially on $F_{\bar{\eta}}$;

(e) if $F = \pi_* F'$, where $\pi: X' \to X$ is a finite map such that $k(X')$ is separable over $k(X)$, then

$$\alpha_x(F) = \sum_{x' \mapsto x} (\alpha_{x'}(F') + \dim(F'_{\bar{\eta}})) - \dim F_{\bar{\eta}} + \operatorname{ord}_x(D_{X'/X}) \dim F'_{\bar{\eta}}$$

where $D_{X'/X}$ is the discriminant of X'/X. (This follows from Serre [7, VI.2, Prop. 4].)

The (exponent of the) *conductor* of F at x is defined to be

$$c_x(F) = \dim F_{\bar{\eta}} - \dim F_x + \alpha_x(F)$$

where F_x is the stalk of F at x.

LEMMA 2.10. (a) $\dim F_x = \sum(-1)^r \dim H_x^r(X, F)$.

If $F \approx g_* g^* F$, *then:*

(b) $\dim F_x = \dim (F_{\bar\eta}^{G_y})$ *for any* y *mapping to* x;

(c) $c_x(F) = \sum_{i=0}^{\infty} g_i/g_0 \dim (F_{\bar\eta}/F_{\bar\eta}^{G_i})$ *with the notation of* (d) *above;*

(d) $c_x(F) = 0$ *if and only if* $F_{\bar\eta}$ *is unramified at* x, *that is,* G_y *acts trivially on* $F_{\bar\eta}$ *for one (or any)* y *mapping to* x.

Proof. (b) is obvious, (c) follows from (b) and property (d) of α_x, and (d) follows from (c), and so it remains to prove (a). Let $\tilde X = \operatorname{spec} \mathcal{O}_{X,x}^h$ and let $\tilde K$ be the field of fractions of $\tilde X$. By excision $H_x^r(X, F) \approx H_x^r(\tilde X, F)$, and there is an exact sequence (III.1.25),

$$0 \to H_x^0(\tilde X, F) \to H^0(\tilde X, F) \to H^0(\tilde K, F) \to H_x^1(\tilde X, F) \to \cdots$$

Since $\tilde X$ is strictly local, $H^r(\tilde X, F) = 0$ for $r > 0$ while $H^0(\tilde X, F) = F_x$. Thus

$$\sum(-1)^r \dim H_x^r(X, F) = \dim F_x - \sum(-1)^r \dim H^r(\tilde K, F).$$

Let $I = \operatorname{Gal}(\tilde K_{\text{sep}}/\tilde K)$, let $I_1 \subset I$ be the first ramification group, and let $I' = I/I_1$. Since I_1 is a pro-p-group, the Hochschild-Serre spectral sequence shows that $H^r(\tilde K, F) \approx H^r(I', F_{\bar\eta}^{I_1})$ for all r. But (compare I.5.2e), $I' \approx \prod_{l \neq p} \mathbb{Z}_l$ is generated topologically by a single element σ. Thus (Serre [7, XIII.1]) $H^0(\tilde K, F)$ and $H^1(\tilde K, F)$ are the kernel and cokernel respectively of $F_{\bar\eta} \xrightarrow{\sigma - 1} F_{\bar\eta}$ whereas $H^r(\tilde K, F) = 0$ for $r > 1$. Since $F_{\bar\eta}$ is finite, this shows that

$$\sum(-1)^r \dim H^r(\tilde K, F) = 0.$$

Remark 2.11. If $F = g_* g^* F$ and G_y has order prime to l, then $c_x(F) = \dim \operatorname{Hom}_G (A_x', F_{\bar\eta})$ where A_x' is a lattice in A_x. Indeed $p_y = a_y - r_y + 1$ that implies

$$[P_y \otimes \mathbb{Q}_l] = [A_y] - [\mathbb{Q}_l[G_y]] + [\mathbb{Q}_l]$$

in $R_{\mathbb{Q}_l}(G_y)$ and that (Serre [11, 15])

$$[P_y/lP_y] = [A_y'/lA_y'] - [\mathbb{F}_l[G_y]] + [\mathbb{F}_l]$$

in $R_{y_l}(G_y)$. Thus

$$\begin{aligned} \dim \operatorname{Hom}_G (A_x', F_{\bar\eta}) &= \dim \operatorname{Hom}_{G_y} (A_y', F_{\bar\eta}) \\ &= \dim \operatorname{Hom}_{G_y} (P_y, F_{\bar\eta}) + \dim F_{\bar\eta} \\ &\quad - \dim F_{\bar\eta}^{G_y} \text{ (Serre [11, 15.5])}, \\ &= c_x(F) \end{aligned}$$

according to (2.10b).

THEOREM 2.12. *For any constructible sheaf F of $\mathbb{F}_l$-modules on X,*

$$\chi(X,F) \overset{df}{=} \sum (-1)^r \dim H^r(X,F) = (2-2g)\dim F_{\bar\eta} - \sum_{x\in X^0} c_x(F)$$

provided $l \neq \mathrm{char}\,(k)$ *and k is algebraically closed* (g = *genus of* X).

Proof. We first need a lemma.

LEMMA 2.13. *Let $\pi: X' \to X$ be a finite map with $k(X')$ separable over $k(X)$; the theorem is true for F' on X' if and only if it is true for $F = \pi_* F'$ on X.*

Proof. As π_* is exact, $\chi(X',F') = \chi(X,F)$ and for any $x \in X_0$

$$\sum_{x'\mapsto x}\sum_r (-1)^r \dim H^r_{x'}(X',F') = \sum_r (-1)^r \dim H^r_x(X,F).$$

Thus we have to show that if $S \subset Y$ is the branch locus of Y/X

$$(2-2g)\dim F_{\bar\eta} - \sum_{x\in\pi(s)} (\alpha_x(F) + \dim(F_{\bar\eta})) = (2-2g')\dim F'_{\bar\eta} - \sum_{x'\in S} (\alpha_{x'}(F') + \dim F'_{\bar\eta}).$$

As $\dim F_{\bar\eta} = \deg\pi \dim F'_{\bar\eta}$, and $2-2g' = (2-2g)\deg\pi \quad \deg(D_{X'/X})$ according to the Hurwitz genus formula, this follows from property (e) of α.

If F has support on a proper closed subscheme $i: Z \hookrightarrow X$ of X, then $\chi(X,F) = \dim H^0(Z, i^*F)$, $F_{\bar\eta} = 0$, $c_x(F) = 0$ if $x \notin Z$ and $c_x(F) = \dim F_x = \dim H^0(x, i_x^* F)$ if $x \in Z$. Thus the theorem is obvious in this case. Since both sides of the equation to be checked are additive in F, we may assume that $F = g_* g^* F$ in the remainder of the proof.

We now fix a Y such that $X = Y/G$ and consider only those sheaves F such that $F = g_* g^* F$ and $F|Y$ is generically constant. The correspondence $F \leftrightarrow F_{\bar\eta}$ induces an equivalence between the category of such sheaves and the category of finite $\mathbb{F}_l[G]$-modules. Each side of the equation to be proved therefore defines a homomorphism $R_{\mathbb{F}_l}(G) \to \mathbb{Z}$. To show that these homomorphisms are equal it suffices to show that they agree on a set of generators for $R_{\mathbb{F}_l}(G)$ or even, as $\mathbb{Z}$ is torsion free, on a set of generators for $R_{\mathbb{F}_l}(G) \otimes \mathbb{Q}$. But it is known that $R_{\mathbb{F}_l}(G) \otimes \mathbb{Q}$ is generated by classes of the form $[\mathrm{Ind}_G^H(M)]$ where H runs through the cyclic subgroups of G of order prime to l and $\mathrm{Ind}_G^H(M)$ is the G-module induced by an H-module M. (Artin's theorem (Serre [11, 12.5]) shows that $R_K(G) \otimes \mathbb{Q}_l$ is generated by $\{\mathrm{Ind}_G^H(M)\}$, H cyclic, when K has characteristic zero; Serre [11, 16.1, Thm. 33] extends this to characteristic l; finally Serre [11, 8.3, Prop. 26] shows that the cyclic groups of order divisible by l contribute nothing.) Thus we may assume that $F_{\bar\eta}$ is of this form, $F_{\bar{\eta}} = \mathrm{Ind}_G^H(M)$. If $X' = Y/H$ and π' is the morphism $X' \to X$ then $F = \pi'_* F'$ where F' is the unique sheaf on X' with $F'_{\bar{\eta}} = M$ and $F' = g'_* g'^* F'$. By the lemma it suffices to prove the theorem for F', that is, we may assume that $X = Y/G$ where G is cyclic of order prime to l and Y is such that $F|Y$ is generically constant.

Let A'_x be a lattice in A_x. On applying Serre [11,15.2] to (2.9), and noting that A_x and its dual, $\mathrm{Hom}_{\mathbb{Q}_l}(A_x, \mathbb{Q}_l)$, have the same trace, we find that

$$\sum_{r=0}^{2} (-1)^r [H^r(Y, \mathbb{F}_l)] = (2 - 2g_X)[\mathbb{F}_l[G]] - \sum_{x \in X^0} [\mathrm{Hom}_{\mathbb{Z}_l}(A'_x, \mathbb{F}_l)]$$

in $R_{\mathbb{F}_l}(G)$. We may tensor this with $F_{\bar{\eta}}$, and obtain an equality

$$\sum_{r=0}^{2} (-1)^r [H^r(Y, \mathbb{F}_l) \otimes_{\mathbb{F}_l} F_{\bar{\eta}}] = (2 - 2g_X)[\mathbb{F}_l[G] \otimes_{\mathbb{F}_l} F_{\bar{\eta}}]$$
$$- \sum_{x \in X^0} [\mathrm{Hom}_{\mathbb{Z}_l}(A'_x, F_{\bar{\eta}})]$$

in $R_{\mathbb{F}_l}(G)$. As G has order prime to l, $H^r(G, N) = 0$ for $r > 0$ and any $\mathbb{F}_l[G]$-module N. Thus $H^0(G, -)$ is exact and defines a homomorphism $R_{\mathbb{F}_l}(G) \to R_{\mathbb{F}_l}(1) = \mathbb{Z}$. On applying it to the last equality, we find that

$$\sum_{r=0}^{2} (-1)^r [(H^r(Y, \mathbb{F}_l) \otimes F_{\bar{\eta}})^G] = (2 - 2g_X)[F_{\bar{\eta}}]$$
$$- \sum_{x \in X^0} [\mathrm{Hom}_{\mathbb{Z}_l[G]}(A'_x, F_{\bar{\eta}})].$$

According to (2.11), in order to complete the proof, we must show that the left hand side equals $\sum_{r=0}^{2} (-1)^r [H^r(X, F)]$.

Let $\bar{F}$ be the constant sheaf on Y defined by the $\mathbb{F}_l$-vector space $F_{\bar{\eta}}$. Note that $H^r(Y, \mathbb{F}_l) \otimes F_{\bar{\eta}} \xrightarrow{\approx} H^r(Y, \bar{F})$, as may be most easily seen on the Čech groups, and that $H^r(Y, \bar{F}) \xrightarrow{\approx} H^r(X, \pi_* \bar{F})$ as Y is finite over X. There is a natural action of G on $\pi_* \bar{F}$: for any U étale over X, $\Gamma(U, \pi_* \bar{F}) = \Gamma(U \times Y, \bar{F}) \xrightarrow{\approx} \Gamma(U \times Y, \mathbb{F}_l) \otimes F_{\bar{\eta}}$, and we let G act on $\Gamma(U \times Y, \mathbb{F}_l) \otimes F_{\bar{\eta}}$ through its action on Y and on $F_{\bar{\eta}}$. This action induces an action of G on $H^r(X, \pi_* \bar{F})$ and it is clear, from interpreting the cohomology groups as Čech groups, that $H^r(Y, \mathbb{F}_l) \otimes F_{\bar{\eta}} \xrightarrow{\approx} H^r(X, \pi_* \bar{F})$ is an isomorphism of G-modules. Moreover, $F \xrightarrow{\approx} (\pi_* \bar{F})^G$. Since $\mathbb{F}_l[G] = \mathbb{F}_l \oplus (\sigma - 1)\mathbb{F}_l[G]$, where σ generates G, taking G-invariants commutes with any additive functor. Thus $H^r(X, \pi_* \bar{F})^G \approx H^r(X, (\pi_* \bar{F})^G) \approx H^r(X, F)$, and the proof is complete.

COROLLARY 2.14. *If F and F' are constructible sheaves of $\mathbb{F}_l$-modules on X_{et} that are locally isomorphic, that is, $F|U_i \approx F'|U_i$ for all U_i in some covering of X_{et}, then $\chi(X, F) = \chi(X, F')$.*

Proof. One only has to check that $c_x(F) = c_x(F')$ for all closed $x \in X$, but this is easy.

Remark 2.15. (a) Deligne has partially extended (2.12) to higher dimensions. In particular (2.14) is true for any proper variety.

(b) Let A_K be an abelian variety over $K = k(X)$, and let A be the Néron minimal model of A over X. Then (2.12) can be used to show (Raynaud [1]) that

$$H^1(X, A)(l) \approx (\mathbb{Q}_l/\mathbb{Z}_l)^{r_0} \oplus M$$

with M a finite group and

$$r_0 + r = 4d_0 + 2d(2g - 2) + \sum_{x \in X^0} c_x(A_l^0)$$

where $r = \operatorname{rank}_{\mathbb{Z}}(A(K)/B(k))$, $B = K/k$-trace of A_K, $d_0 =$ dimension of B, $d =$ dimension of A_K, $g =$ genus of X, and $A_l = \ker(l: A^0 \to A^0)$ where A^0 is the maximal open subscheme of A with connected fibers. Grothendieck [SGA. 7, IX.4.6] has shown that $\alpha_x(A_l)$ is independent of l (l prime to char (k)). It is easy to see that the tame conductor, $\beta_x(A) \stackrel{df}{=} c_x(A_l) - \alpha_x(A_l)$, is equal to $2\lambda + \mu$ where λ is the number of copies of $\mathbb{G}_a$ in a composition series for A_x and μ is the number of copies of $\mathbb{G}_m$; hence it also is independent of l. Note that if A is an abelian scheme over $\mathbb{P}^1$ then, in order to have $r + r_0 \geq 0$, one must have $d = d_0$, and so A_K is isogenous to $B \otimes_k K$.

(c) If one assumes the theorems of Chapter VI, it is possible to derive formulas for the Euler-Poincaré characteristic of a family of varieties. Let $\pi: Y \to X$ be a proper map with connected fibers. The sheaves $F^r = R^r\pi_*\mathbb{F}_l$ are constructible (VI.2.1), and the Leray spectral sequence for π shows that

$$\chi(Y, l) = \chi(Y, \mathbb{F}_l) = \sum_r (-1)^r \chi(X, F^r).$$

Thus

$$\chi(Y, l) = \chi(X, l)\chi(Y_{\bar{\eta}}, l) - \sum_{r,x} (-1)^r c_x(F^r) \qquad \text{according to (2.12)}$$

$$= \chi(X, l)\chi(Y_{\bar{\eta}}, l) + \sum_x \Big(\chi(Y_x, l) - \chi(Y_{\bar{\eta}}, l) - \sum_r (-1)^r \alpha_x(F^r)\Big).$$

In particular, if π is smooth, this says that

$$\chi(Y, l) = \chi(X, l)\chi(Y_{\bar{\eta}}, l) \qquad \text{according to (VI.4.2),}$$

that is, the Euler-Poincaré characteristics multiply. If char $(k) = 0$,

$$\chi(Y, l) = \chi(X, l)\chi(Y_{\bar{\eta}}, l) + \sum_x \big(\chi(Y_x, l) - \chi(Y_{\bar{\eta}}, l)\big).$$

(For a topological proof of this last formula, see Beauville [1, VI.4].)

These formulas are useful only when it is possible to compute the terms

$$\sum_r (-1)^r c_x(F^r) = c_x(Y/X).$$

When $Y = A$, the complete Néron minimal model of an abelian curve A_K over $K = k(X)$, the Ogg-Tate formula shows that $c_x(Y/X)$ can be expressed in terms of $\operatorname{ord}_x(\Delta)$ and n where Δ is the discriminant of the Weierstrass minimal equation of $A_K/\mathcal{O}_{X,x}$ and n is the number of components of A_x (Ogg [2]), and when Y is smooth and the fibers of π have only isolated singularities, Deligne has shown that

$$c_x(Y/X) = \sum_{y \mapsto x} \mu(Y/X, y)$$

where $\mu(Y/X, y)$ is the Milnor number of Y/X at a point y singular in Y_x [SGA. 7, XVI]. See also Iverson [1].

(d) Let X_0 be a complete smooth curve over a finite field k_0, and let F be a sheaf of $\mathbb{F}_l$-modules on X_0 such that the groups $H^r(X, F)$ are finite, where $X = X_0 \otimes k$, $k = k_{0,\mathrm{sep}}$ (for example, F constructible). Then it follows immediately from the Hochschild-Serre spectral sequence for X/X_0 that $\chi(X_0, F) = 0$ (compare the proof of (2.3)). Tate has proved an analogous result when X_0 is the spectrum of the ring of integers in a number field. See Tate [1, Thm. 2.2] for the statement and Bašmakov [1] for the proof.

We now reinterpret (2.12) in terms of the tame fundamental group. Let $U \neq X$ be an open dense subset of X and write π_1^t for the tame fundamental group $\pi_1^t(U, \bar{\eta})$ (I.5.2e). Thus $\pi_1^t = \operatorname{Gal}(K_t/K)$, where K_t is the union of the fields L, $K \subset L \subset k(\bar{\eta}) = K_{\mathrm{sep}}$, which are finite over K and such that the normalization X_L of X in L is unramified over U and tamely ramified over $X - U$. There is an exact sequence

$$1 \to \pi_1^w \to \pi_1(U) \to \pi_1^t \to 1$$

where $\pi_1^w = 1$ if $\operatorname{char}(k) = 0$ and π_1^w has no finite quotients of order prime to p if $\operatorname{char}(k) = p \neq 0$. Write $X_t = \varprojlim X_L$ and $U_t = \varprojlim U_L$.

LEMMA 2.16. *If F is a constructible sheaf of $\mathbb{Z}/(n)$-modules on U (n prime to* char (k)) *such that $F|U_t$ is constant, then $H^r(U_t, F|U_t) = 0$ for $r > 0$.*

Proof. Since $H^r(U_t, F) = \varinjlim H^r(U_L, F)$, (2.4a) shows that $H^r(U_t, F) = 0$ for $r > 1$. As $F|U_t$ is constant, $H^1(U_t, F) = \operatorname{Hom}(\pi_1^w, F_{\bar{\eta}}) = 0$.

LEMMA 2.17. *Let F be a sheaf on U satisfying the conditions of* (2.16), *and let M be $F_{\bar{\eta}}$ regarded as a π_1^t-module. Then $H^r(\pi_1^t, M) = H^r(U, F)$ for all r, and $H^r(\pi_1^t, M) = 0$ for $r > 1$.*

Proof. The last lemma shows that the Hochschild-Serre spectral sequence $H^r(\pi_1^t, H^s(U_t, F)) \Rightarrow H^{r+s}(U, F)$ gives isomorphisms $H^r(\pi_1^t, M) \approx H^r(U, F)$. The second statement follows from the first and from (2.4a).

THEOREM 2.18. *For any finite π_1^t-module M of order prime to* char (k),

$$\chi(\pi_1^t, M) \overset{df}{=} [H^0(\pi_1^t, M)]/[H^1(\pi_1^t, M)] = [M]^{(2-2g-s)}$$

where s is the number of points in $X - U$ (and $[*]$ denotes the order of $*$).

Proof. Since both sides of the formula are multiplicative in M, we may assume that M is killed by some prime l, and so is an $\mathbb{F}_l$-module. Let F_0 be the sheaf on $K = k(\eta)$ corresponding to the module M, and let $F = g_* F_0$. Then $H^r(\pi_1^t, M) = H^r(U, F|U)$ for all r. From the exact sequence,

$$0 \to \bigoplus_{x \in X-U} H_x^0(X, F) \to H^0(X, F) \to H^0(U, F) \to \bigoplus_{x \in X-U} H_x^1(X, F) \to \cdots$$

we find that

$$\chi(U, F|U) = \chi(X, F) - \sum_{x \in X-U} \sum_r (-1)^r \dim H_x^r(X, F).$$

As $M = F_{\bar\eta}$ is tamely ramified for all $x \in X$, $\alpha_x(F) = 0$ for all x, and as M is unramified for all $x \in U$, $c_x(F) = 0$ for $x \in U$. Thus (2.12) shows that

$$\chi(U, F|U) = (2 - 2g)\dim M - \sum_{x \in X-U} \dim M = (2 - 2g - s)\dim M,$$

which completes the proof.

Remark 2.19. (a) Write $X - U = \{x_1, \ldots, x_s\}$. Grothendieck has shown (Grothendieck [3, no. 182, p.27] or [SGA. I, XIII.2.12]) that $\pi_1^t = \pi_1^t(X - U)$ has $2g + s$ generators $u_1, v_1, u_2, v_2, \ldots, u_g, v_g, \sigma_1, \ldots, \sigma_s$ such that:

(i) $u_1 v_1 u_1^{-1} v_1^{-1} \cdots u_g v_g u_g^{-1} v_g^{-1} \sigma_1 \cdots \sigma_s = 1$, and

(ii) σ_i generates the tame inertia group of X_t/X at $x_i' \in X_t$, for suitable choices of x_i' mapping to x_i. Moreover, for any finite group G of order prime to p and elements $\bar u_i, \bar v_j, \bar\sigma_m$ of G satisfying the above relation, there is a unique homomorphism $\pi_1^t \to G$ sending u_i to $\bar u_i$, v_j to $\bar v_j$, and σ_m to $\bar\sigma_m$. Note that if $s = 0$, this is essentially (I.5.2j), and if $g = 0$, it is obvious from (I.5.2f) that the σ_i generate π_1 (and from the parenthetical remark, that any $s - 1$ of the σ_i still generate).

Theorem (2.18) implies that for any finite π_1^t-module M of order prime to char k and any $a_1, \ldots, a_{2g+s-1} \in M$, there is a unique 1-cocycle f for π_1^t in M such that

$$f(u_1) = a_1, \ldots, f(v_g) = a_{2g}, \ldots, f(\sigma_{s-1}) = a_{2g+s-1}.$$

Indeed the map

$$f \mapsto (f(u_1), \ldots, f(\sigma_{s-1})): Z^1(\pi_1^t, M) \xrightarrow{\alpha} M^{2g+s-1}$$

is obviously injective and so

$$\chi(\pi_1',M) \overset{df}{=} \frac{[H^0]}{[H^1]} = \frac{[M]}{[Z^1]} \geq [M]^{2-2g-s}$$

with equality if, and only if, α is surjective. Thus (2.18) shows that α is an isomorphism.

(b) It is not difficult to prove directly that the above map α is an isomorphism, and then reverse the above argument to deduce (2.18), that is, recover a weak form of (2.12). This approach, using the structure of π_1', is the original approach, due independently to Šafarevič [1] and Ogg [1].

Exercise 2.20. (Lefschetz formula with constant coefficients). Let U be an open subscheme of X, and assume that k is algebraically closed and n is prime to char (k).

(a) Show that the groups $H_c^r(U, \mathbb{Z}/(n))$ are free $\mathbb{Z}/(n)$-modules for all r.

(b) Let $\phi \neq id$ be an endomorphism of X that preserves U and $X - U$. Assume that the fixed points of ϕ have multiplicity one, that is, that Γ_ϕ intersects Δ transversally. Show that $\sum_{r=0}^{2}(-1)^r Tr(\phi \mid H_c^r(U, \mathbb{Z}/(n))) =$ number of fixed points of ϕ in U (mod n).

Exercise 2.21. (Artin L-series). Let Y be a complete smooth curve over the finite field $k = \mathbb{F}_q$; let G be a finite group of k-automorphisms of Y, and let $X = Y/G$. For any $y \in Y^0$, G_y and I_y denote the decomposition and inertia groups at y, so there is an exact sequence

$$1 \to I_y \to G_y \to \mathrm{Gal}\,(k(y)/k(x)) \to 1.$$

Write $\bar{Y} = Y \otimes_k \bar{k}$ where $\bar{k} = k_{al}$; then G acts on $\bar{Y}$ and $\bar{Y}/G = \bar{X} = X \otimes \bar{k}$.

(a) Show that $G_y = \{\sigma \mid \text{for all } \bar{y} \in \bar{Y}^0 \text{ mapping to } y,\ \bar{y}\sigma = F^n(\bar{y}) \text{ some } n\}$ and $I_y = \{\sigma \mid \text{for all } \bar{y} \in \bar{Y}^0 \text{ mapping to } y,\ \bar{y}\sigma = \bar{y}\}$ where $F: \bar{Y} \to \bar{Y}$ is the Frobenius (as in (2.6)).

Write f_y for the canonical generator of Gal $(k(y)/k(x))$, that is, $f_y(a) = a^{q^{\deg(x)}}$; we may identify f_y with an element of G_y/I_y. Let Ω be a field containing $\mathbb{Q}_l$ $(l \nmid q)$ and let $\rho: G \to \mathrm{Aut}_\Omega(V)$ be a finite-dimensional representation of G. The *Artin L-series* of ρ is the formal power series

$$L(Y, \rho, t) = \prod_{x \in X^0} \frac{1}{\det(1 - t^{\deg(x)}\rho(f_y) \mid V^{I_y})}$$

where, for each $x \in X^0$, a choice is made of a $y \in Y^0$ mapping to x.

(b) Check that the term corresponding to x in $L(Y, \rho, t)$ is independent of the choice of y.

(c) Write

$$\rho^{av}(f_y) = \frac{1}{[I_y]} \sum_{\substack{\sigma \mapsto f_y \\ \sigma \in G_y}} \rho(\sigma).$$

Show that

$$\det(1 - t\rho(f_y)|V^{I_y}) = \det(1 - t\rho^{av}(f_y)|V);$$

thus

$$L(Y, \rho, t) = \prod_{x \in X^0} \frac{1}{\det(1 - t^{\deg(x)}\rho^{av}(f_y)|V)}.$$

(Hint:

$$P = \frac{1}{[I_y]} \sum_{\sigma \in I_y} \rho(\sigma)$$

is a projection $V \to V^{I_y}$; $\rho(\sigma)P = P\rho(\sigma) = \rho^{av}(f_y)$ if $\sigma \mapsto f_y$.)

(d) Let χ be the character of ρ, and for any $x \in X(\mathbb{F}_{q^n})$ (that is, point of $X \otimes \mathbb{F}_{q^n}$ of degree 1), write

$$\chi(x) = \frac{1}{[I_y]} \sum_{\substack{\sigma \mapsto f_y \\ \sigma \in G_y}} \chi(\sigma)$$

where $y \in Y \otimes \mathbb{F}_{q^n}$ maps to x. Show that $\chi(x) = Tr(\rho^{av}(f_y)|V)$ and deduce that

$$L(Y, \rho, t) = \exp\left(\sum_{n > 0} \frac{\nu_n(Y, \chi)t^n}{n}\right)$$

where $\nu_n(Y, \chi) = \sum_{x \in X(\mathbb{F}_{q^n})} \chi(x)$. (Hint: use (2.7)).

(e) Show that

$$\nu_n(Y, \chi) = \frac{1}{[G]} \sum_{\sigma \in G} \chi(\sigma^{-1})L(\sigma F^n)$$

where $L(\sigma F^n)$ is the number of fixed points of σF^n on $\bar{Y}$. (Hint: note that the actions of $\sigma \in G$ and F on $\bar{Y}$ commute; show that for $x \in X(\mathbb{F}_{q^n})$,

$$\chi(x) = \frac{1}{[G]} \sum_{\sigma \in G} \chi(\sigma^{-1})L_x(\sigma F^n)$$

where $L_x(\sigma F^n)$ is the number of y, fixed by σF^n, which map to x.)

(f) Show that:

$$L(Y, \rho_1 \oplus \rho_2, t) = L(Y, \rho_1, t)L(Y, \rho_2, t);$$
$$L(Y, \rho, t) = Z(X, t) \text{ if } \rho = \text{trivial representation};$$
$$L(Y, r, t) = Z(Y, t) \text{ if } r = \text{regular representation}.$$

Deduce that

$$Z(Y, t) = Z(X, t)(\prod L(Y, \rho, t)^{\dim(\rho)})$$

where the product is over the nontrivial, irreducible representations ρ of G.

(g) Show that if ρ is irreducible and nontrivial, then

$$L(Y, \rho, t) = \det(1 - Ft|H^1(\bar{Y}, V^\vee)^G) = \det(1 - Ft|(H^1(\bar{Y}, \mathbb{Q}_l) \otimes V^\vee)^G)$$

where $(V^\vee, \rho^\vee)$ is the contragredient representation, that is, $\rho^\vee(\sigma) = \rho(\sigma^{-1})^{\text{transpose}}$ as matrices; in particular $L(Y, \rho, t)$ is a polynomial in t that divides the numerator of $Z(Y, t)$. (Hint:

$$\begin{aligned} \nu_n(Y, \chi) &= \frac{1}{[G]} \sum_{\sigma \in G} Tr(\rho(\sigma^{-1}) | V) \sum_r (-1)^r \, Tr(\sigma F^n | H^r(\mathbb{Q}_l)) \\ &= \sum_r (-1)^r \, Tr\left(\left(\frac{1}{[G]} \sum \rho^\vee(\sigma) \otimes \sigma\right) \circ (1 \otimes F^n) | V^\vee \otimes H^r(\mathbb{Q}_l)\right), \end{aligned}$$

and $[G]^{-1} \sum \rho^\vee(\sigma) \otimes \sigma$ is a projection $V^\vee \otimes H^r \to (V^\vee \otimes H^r)^G$. For the second statement note that $H^1(\bar{Y}, V^\vee)^G = \mathrm{Hom}_G(V, H^1(\bar{Y}, \Omega))$.)

(h) Grothendieck [3, no. 279] defines L-series by the formula:

$$L^{\text{Groth}}(Y, \rho, t) = \prod_{x \in X^0} \frac{1}{\det(1 - t^{\deg(x)} \rho(f_y^{-1}) | V^{I_y})}.$$

Show that $V^{I_y} \xrightarrow{\approx} V_{I_y} \approx (\check{V}^{I_y})^\vee$ where $V_{I_y} = V / \sum_{\sigma \in I_y} (\sigma - 1)V$ and deduce that

$$L^{\text{Groth}}(Y, \rho, t) = L^{\text{Artin}}(Y, \rho^\vee, t);$$

thus

$$L^{\text{Groth}}(Y, \rho, t) = \det(1 - Ft | H^1(\bar{Y}, V)^G)$$

when ρ is irreducible and nontrivial.

(Part (g) shows, in particular, that $L(Y, \rho, q^{-s})$ is an analytic function on the whole complex plane; the analogous functions for number fields are known to be meromorphic on the whole plane and E. Artin conjectured that they too are analytic; the conjecture is known only in special cases.)

§3. The Cohomology of Surfaces

Throughout this section, X will be a smooth complete (hence projective) surface over an algebraically closed field k. In order to study the cohomology of X we fiber it over $\mathbb{P}^1$.

THEOREM 3.1. (Existence of Lefschetz pencils). *There exists a surface X^* that is obtained from X by blowing up a finite number of points and a map $\pi: X^* \to \mathbb{P}^1$ satisfying the following condition:* (L) *π is proper, flat, and has a section; the generic fiber of π is a smooth curve and the closed fibers are connected with at most a single node as singularity.*

Proof. We merely sketch the proof; details may be found in Igusa [1] or [SGA. 7, XVII]. Suppose that X is embedded in $\mathbb{P}^m$, and let $\check{\mathbb{P}}^m$ be the dual projective space so that $\check{\mathbb{P}}^m(k)$ is the set of hyperplanes in $\mathbb{P}^m$. Let V be the closed subvariety of $X \times \check{\mathbb{P}}^m$ of pairs (x, H) such that H contains the tangent plane at x. From the Jacobian criterion of smoothness one sees that this condition is equivalent to x being a singularity of $X.H$ or $H \supset X$. The image of V under the projection $\phi: X \times \check{\mathbb{P}}^m \to \check{\mathbb{P}}^m$ is the *dual variety* $\check{X}$ of X; thus, unless X is contained in a hyperplane, $\check{X}(k)$ consists of those H for which $X.H$ is singular. Since the fibers of $V \to X$ are all irreducible of dimension $m - 3$, V is irreducible, complete, and of dimension $m - 1$ (unless $X = \mathbb{P}^m$, when it is empty) from which it follows that $\check{X}$ is irreducible, complete, and has codimension ≥ 1 in $\check{\mathbb{P}}^m$. The map $\phi: V \to \check{\mathbb{P}}^m$ is unramified at (x, H) if and only if x is a node on the curve $X.H$ [SGA. 7, XVII.3.3]. Either ϕ is generically unramified or becomes so if the embedding $X \hookrightarrow \mathbb{P}^m$ is replaced by its square $X \hookrightarrow \mathbb{P}^m \hookrightarrow \mathbb{P}^{\binom{m+2}{2}-1}$ [SGA. 7, XVII.3.7]. Thus we may assume this, and deduce that the set F of H in $\check{X}$ such that $X.H$ has more than a single node as singularity is closed of codimension ≥ 2 in $\check{\mathbb{P}}^m$.

Let D be a line in $\check{\mathbb{P}}^m$. If H_0 and H_∞ are any two distinct hyperplanes in D, then the elements of D are of the form $\alpha H_0 + \beta H_\infty$, $\alpha, \beta \in k$, and so $\bigcap\{H \mid H \in D\} = H_0 \cap H_\infty$. This linear space is called the *axis* of D.

According to Bertini's Theorem (Hartshorne [2, III.7.9.1]), $H.X$ is irreducible for a general hyperplane H. Thus we may fix an H_0 in $\check{\mathbb{P}}^m$ such that $H_0.X$ is smooth (that is, $H_0 \notin \check{X}$) and irreducible. Since $\widetilde{H_0.X}$ has codimension at least one in $\check{\mathbb{P}}^m$, there is an open set of H_∞ in $\check{\mathbb{P}}^m$ such that H_∞ cuts $H_0.X$ transversally. Also, since there is an $(m - 1)$-dimensional family of lines in $\check{\mathbb{P}}^m$ passing through H_0, and only an $(m - 2)$-dimensional family meets F, we may choose an H_∞ so that the line D joining H_0 to H_∞ does not meet F. Then:

(a) the axis of D cuts X transversally;

(b) all H in some open dense subset U of D cut X transversally;

(c) for all H in $D - U$, $H.X$ is smooth except for a single node. Condition (a) implies that there are only finitely many points $Q_1, \ldots, Q_\nu$ where two curves $X.H$ may meet (namely, the points of X on the axis of D) and that the curves cross transversally there. On blowing up the Q_i, we get a surface X^* and a morphism $\pi: X^* \to D = \mathbb{P}^1$ such that the fiber over H is $X.H$. (The graph of π is the graph of the obvious rational map $X \to \mathbb{P}^1$.) Moreover, π is flat by (I.2.6b) and is proper because X is complete.

Conditions (b) and (c) will imply that the fibers have the correct form once we show that they are geometrically connected. As π is faithfully flat, $\pi_* \mathcal{O}_{X^*}$ is a torsion free $\mathcal{O}_{\mathbb{P}^1}$-module, and as π is proper, it is finitely generated. Thus it is locally free, and, as at least one closed fiber is reduced and connected, its rank must be one. Hence $\mathcal{O}_{\mathbb{P}^1} \xrightarrow{\approx} \pi_* \mathcal{O}_{X^*}$ and the Zariski connectedness theorem [EGA. III.4.3.4] shows that all fibers are geometrically connected. (Compare Hartshorne [2, III.11.3].)

Let E be the exceptional curve on X^* obtained from the blowing up of one of the Q_i. Note that such an E exists because $X \cap H_0 \cap H_\infty$ cannot be empty. Each closed fiber of $\pi: X^* \to \mathbb{P}^1$ cuts E (transversally) at exactly one point. Thus π induces an isomorphism $E \xrightarrow{\approx} \mathbb{P}^1$, and the inverse isomorphism is a section for π.

Remark 3.2. (a) The morphism π may have reducible fibers. For example, if X is a quadratic surface in $\mathbb{P}^3$, then the fibers of $X^* \to \mathbb{P}^1$ are closed connected subschemes of dimension 1 and degree 2 in $\mathbb{P}^2$; such a subscheme is singular if and only if it is reducible. Igusa [1] shows, however, that if the embedding $X \hookrightarrow \mathbb{P}^m$ is replaced by its cube, then this does not occur, that is, it is possible to choose D such that $\pi: X^* \to \mathbb{P}^1$ has irreducible fibers.

(b) For any line D in $\check{\mathbb{P}}^m$ satisfying the conditions (a), (b), and (c) of the proof of the theorem, the family of curves $(X.H)_{H \in D}$ is called a *Lefschetz pencil* on X. We also refer to the map $\pi: X^* \to \mathbb{P}^1$ as a Lefschetz pencil.

PROPOSITION 3.3. *Let S be a smooth curve over k and let $\pi: X \to S$ be a proper flat map. Assume that $\mathcal{O}_S \xrightarrow{\approx} \pi_*\mathcal{O}_X$ (as sheaves on S_{Zar}) and that for the closed point s of S, X_s is reduced. Then:*

(a) *$k \xrightarrow{\approx} H^0(X_s, \mathcal{O}_{X_s})$;*

(b) *$p_a(X_s) = p_a(X_\eta)$ where X_η is the generic fiber of π and $p_a \overset{df}{=} \dim_k H^1(\mathcal{O})$ is the arithmetic genus.*

Proof. The condition $\mathcal{O}_S \approx \pi_*\mathcal{O}_X$ implies that all fibers are geometrically connected (Zariski connectedness theorem, compare Hartshorne [2, III.11.3]). Thus, as X_s is reduced, $H^0(X_s, \mathcal{O}_{X_s}) = k$, which proves (a). Since the function $s \mapsto \chi(X_s, \mathcal{O}_{X_s})$ is constant on S (Mumford [4, p. 50]), (b) follows from (a).

Remark 3.4. Conclusion (a) of (3.3) is equivalent to π being cohomologically flat in dimension zero at s [EGA. III.7], that is, if we replace S by spec $\mathcal{O}_{S,s}$, then $\mathcal{O}_S \xrightarrow{\approx} \pi_*\mathcal{O}_X$ universally (equivalently, $\mathcal{O}_S \xrightarrow{\approx} \pi_*\mathcal{O}_X$ on S_{fl}).

THEOREM 3.5. *Let S be a smooth curve over k, let $\pi: X \to S$ satisfy (L) and have irreducible fibers, and let T be the finite set of points $s \in S$ such that $\pi^{-1}(s) = X_s$ is singular; let n be prime to char (k).*

(a) *For any geometric point $\bar{s}$ of S, the canonical maps*

$$(R^r\pi_*\mu_n)_{\bar{s}} \to H^r(X_{\bar{s}}, \mu_n)$$

are isomorphisms.

(b) *$R^0\pi_*\mu_n = \mu_n$; $R^1\pi_*\mu_n$ is constructible and $R^1\pi_*\mu_n | S - T$ is locally constant; $R^2\pi_*\mu_n = \mathbb{Z}/(n)$; $R^r\pi_*\mu_n = 0$ for $r > 2$.*

(c) *$R^r\pi_*\mu_n \xrightarrow{\approx} g_*g^*R^r\pi_*\mu_n$, where $g: \eta \hookrightarrow S$ is the generic point of S.*

Proof. Recall (III.1.15) that $(R^r\pi_*\mu_n)_{\bar{s}} = H^r(\tilde{X}, \mu_n)$ where $\tilde{X} = X \times_S \operatorname{spec} \mathcal{O}_{S,\bar{s}}$. If $s = \eta$, then $\tilde{X} = X \otimes_K K_{\text{sep}} = X_{\bar{\eta}}$, and so the canonical maps are the identity maps. Otherwise they are the maps $H^r(\tilde{X}, \mu_n) \to H^r(X_{\bar{s}}, \mu_n)$ that restrict a class on $\tilde{X}$ to the closed fiber of $\tilde{X}/\operatorname{spec} \mathcal{O}_{S,\bar{s}}$. That these maps are isomorphisms is a special case of the proper base change theorem (VI.2). A proof of this case of the theorem may be found in Artin [1, IV.5].

It follows from the hypotheses that $\pi_*\mathcal{O}_X = \mathcal{O}_S$ (as sheaves on S_{et}) and hence that $\pi_*\mathbb{G}_m = \mathbb{G}_m$ and $\pi_*\mu_n = \mu_n$.

Write $F = R^1\pi_*\mu_n$; the stalk $F_{\bar{s}}$ is $\operatorname{Pic}(X_{\bar{s}})_n$, which is finite. If we choose, for s a closed point of S, an embedding $\mathcal{O}^{\text{sh}}_{S,s} \hookrightarrow k(\bar{\eta})$, then the induced map $\phi_s: F_s \to F_{\bar{\eta}}^{I_s}$ (where $I_s \subset \operatorname{Gal}(k(\bar{\eta})/k(\eta))$ is the inertia group corresponding to the embedding) may be identified with

$$(\operatorname{Pic} X_s)_n \xleftarrow[\approx]{\text{restrict}} \operatorname{Pic}(\tilde{X})_n \xrightarrow[\approx]{\text{restrict}} \operatorname{Pic}(\tilde{X}_{\bar{\eta}})_n = \operatorname{Pic}(X_{\bar{\eta}})_n^{I_s}$$

(where $\bar{\eta}$ = generic point of $\tilde{S}$). Thus ϕ_s is an isomorphism whose inverse p_s sends a divisor class on $\tilde{X}_{\bar{\eta}}$ to the intersection of $\tilde{X}_s = X_s$ with the Zariski closure of the divisor class. If $s \notin T$, then X_s and $X_{\bar{\eta}}$ are smooth curves of the same genus g according to (3.3). Thus $[F_s] = n^{2g} = [F_{\bar{\eta}}]$ and we find that $F_s \xrightarrow{\approx} F_{\bar{\eta}}^{I_s} = F_{\bar{\eta}}$. It follows that F is constructible and $F|S - T$ is locally constant (1.10).

For any complete integral curve Y over an algebraically closed field, $H^2(Y, \mu_n) \approx \mathbb{Z}/(n)$; if Y is smooth this is proved in Section 2 and the general case follows by comparing the cohomology of Y with its normalization (compare (2.4a)). The map $R^1\pi_*\mathbb{G}_m \to R^2\pi_*\mu_n$ factors through the degree map $R^1\pi_*\mathbb{G}_m = \underline{\operatorname{Pic}}_{X/S} \to \mathbb{Z}$ (since it does on each stalk) and the resulting map $\mathbb{Z}/(n) \to R^2\pi_*\mu_n$ is an isomorphism (since it is on each stalk). Finally $R^r\pi_*\mu_n = 0$ for $r > 2$ because $(R^r\pi_*\mu_n)_{\bar{s}} = H^r(X_{\bar{s}}, \mu_n) = 0$ for $r > 2$. Part (c) is obvious for $r \neq 1$, and $F \xrightarrow{\approx} g_*g^*F$ because ϕ_s: $F_s \xrightarrow{\approx} F_{\bar{\eta}}^{I_s}$ for all s (compare (II.3.16d)).

We next consider, in the situation of (3.5), the structure of $F = R^1\pi_*\mu_n$ at the $s \in T$. The group of *vanishing cycles* at a closed point $s \in S$ is $V_s \stackrel{\text{df}}{=} \ker(F_s \to \operatorname{Pic}(X'_s))$ where X'_s is the normalization of X_s. Since $F_s = \operatorname{Pic}(X_s)_n$, $F_s = 0$ for $s \notin T$. We usually identify V_s and F_s with the subgroups $\phi_s(V_s)$ and $\phi_s(F_s) = F_{\bar{\eta}}^{I_s}$ of $F_{\bar{\eta}}$.

Note that as $F_{\bar{\eta}} = (J(k(\bar{\eta})))_n$, where J is the Jacobian of X_η, there is a canonical, nondegenerate, skew-symmetric pairing $e_n: F_{\bar{\eta}} \times F_{\bar{\eta}} \to \mu_n(k)$ (Mumford [4, §20]). This pairing may be identified with the cup-product pairing on $H^1(X_{\bar{\eta}}, \mu_n)$ (2.4f). For any $\sigma \in \operatorname{Gal}(k(\bar{\eta})/k(\eta))$,

$$e_n(\sigma\gamma_1, \sigma\gamma_2) = \sigma e_n(\gamma_1, \gamma_2) = e_n(\gamma_1, \gamma_2).$$

THEOREM 3.6. *Let $\pi: X \to S$ be as in (3.5) and let $s \in T$. Then $F_s \approx (\mathbb{Z}/(n))^{2g-1}$ where $g = p_a(X_\eta)$, $V_s \approx \mathbb{Z}/(n)$, and V_s is the exact annihilator of F_s under the e_n-pairing $e_n: F_{\bar\eta} \times F_{\bar\eta} \to \mu_n(k)$.*

Proof. We need a lemma.

LEMMA 3.7. *Let Y be a complete curve over k and assume that Y has a node at Q and no other singularity. Let $\psi: Y' \to Y$ be the normalization of Y, and let Q_1 and Q_2 be the points of Y' mapping to Q. Then there is an exact sequence,*

$$0 \to \mathbb{G}_m \to J_Y \to J_{Y'} \to 0$$

(J_Y = generalized Jacobian of Y, Serre [6]) in which the second map is induced by the functoriality of Pic, and the first map may be described as follows: let $a \in \mathbb{G}_m(k) = k^*$; there exists a function $f \in k(Y') = k(Y)$ such that $f(Q_1) = a$ and $f(Q_2) = 1$; the image of a in $J_Y(k)$ is the class of the divisor (f).

Proof. This may be found in Serre [6, V.13]. Alternatively one may deduce it from the cohomology sequence of

$$0 \to \mathbb{G}_{m,Y} \to \psi_* \mathbb{G}_{m,Y'} \to \mathbb{G}_{m,Q} \to 0$$

(Serre [6, IV.4. Ex. 6]) using the fact that $J_Y(k) = \ker(H^1(Y, \mathbb{G}_m) \xrightarrow{\deg} \mathbb{Z})$.

Proof of (3.6). From the exact sequence in the lemma (with Y replaced by X_s) we get an exact sequence,

$$0 \to \mu_n(k) \to F_s \to (J_{X'_s}(k))_n \to 0.$$

Since X'_s has genus $g - 1$ (Serre [6, IV]), this shows that F_s is isomorphic to $(\mathbb{Z}/(n))^{2g-1}$ and V_s is the subgroup $\mu_n(k)$ of F_s.

Let $\gamma_1, \gamma_2 \in F_{\bar\eta} = J_{X_\eta}(k(\bar\eta))_n$. Then $e_n(\gamma_1, \gamma_2) = f_2(D_1)/f_1(D_2)$ where D_i is a divisor representing γ_i and $(f_i) = nD_i$ for $i = 1, 2$. (Compare Igusa [2, p. 755]; also Serre [6, Prop. 7, p. 46].) From this description of e_n it is clear that if $\gamma_1, \gamma_2 \in F_{\bar\eta}^{I_s}$ correspond to $\gamma'_1, \gamma'_2 \in F_s = J_{X_s}(k)_n$, then $e_n(\gamma_1, \gamma_2) = e'_n(\gamma'_1, \gamma'_2)$ where e'_n is the pairing on $J_{X'_s}(k)_n$. Thus if $\gamma_1 \in V_s$, $e_n(\gamma_1, \gamma_2) = e'_n(0, \gamma_2) = 1$, and so V_s and F_s annihilate each other. As the e_n-pairing is nondegenerate and V_s, F_s, and $F_{\bar\eta}$ have orders n, n^{2g-1}, and n^{2g} respectively, it follows that V_s and F_s are exact annihilators.

COROLLARY 3.8. *F is tamely ramified at any $s \in T$.*

Proof. We must show that there is a subgroup H of I_s, of finite index prime to char(k), which acts trivially on $F_{\bar\eta}$. For $\gamma_0, \gamma \in F_{\bar\eta}$ and $\tau \in I_s$,

$$e_n(\tau(\gamma_0), \tau(\gamma)) = e_n(\gamma_0, \gamma).$$

Thus, for any $\gamma \in F_s = F_{\bar{\eta}}^{I_s}$,

$$e_n(\tau(\gamma_0) - \gamma_0, \gamma) = e_n(\tau(\gamma_0), \gamma) - e_n(\gamma_0, \gamma) = 0,$$

that is, $\tau(\gamma_0) - \gamma_0$ is orthogonal to F_s. Now (3.6) shows that $\tau(\gamma_0) - \gamma_0 \in V_s$, and it follows that if we choose γ_0 to generate $F_{\bar{\eta}}/F_s$, then $\tau \mapsto \tau(\gamma_0) - \gamma_0$ is a homomorphism $I_s \to V_s \approx \mathbb{Z}/(n)$ whose kernel is the required H.

COROLLARY 3.9. *$\chi(X, l) = \chi(X_\eta, l)\chi(S, l) + t$, where t is the order of T.*

Proof. Note that

$$\chi(X, l) = \chi(X, \mathbb{F}_l) = \chi(X, \mu_l) = \chi(S, \mu_l) - \chi(S, F) + \chi(S, \mathbb{Z}/(l)),$$

according to the Leray spectral sequence. We use (2.12) to compute these last three numbers. Clearly (2.10) $c_s(\mu_l) = 0 = c_s(\mathbb{Z}/(l))$ for all s, and so

$$\chi(S, \mu_l) = \chi(S, l) = \chi(S, \mathbb{Z}/(l)).$$

Also $c_s(F) = 0$ for $s \notin T$ and

$$c_s(F) = \dim F_{\bar{\eta}} - \dim F_{\bar{\eta}}^{I_s} + 0 = 1$$

for $s \in T$, from which it follows that $\chi(S, F) = 2g\chi(S, l) - t$.

Remark 3.10. We sketch a geometric proof of (3.8). The problem is to show that after making a finite extension $k(\eta')$ of $k(\eta)$ which is tamely ramified over s, the inertia group at a point s' lying over s acts trivially on $F_{\bar{\eta}}$.

Let $t \in k(\eta)$ be a uniformizing parameter at s; let $k(\eta')$ be the field obtained by adjoining an nth root of t to $k(\eta)$, and let S' be the normalization of S in $k(\eta')$. Then $X \times_S S'$ is a normal surface whose singularities may be resolved to give a smooth surface X' that fits into a commutative diagram:

$$\begin{array}{ccc} X & \longleftarrow & X' \\ {\scriptstyle \pi}\big\downarrow & & \big\downarrow{\scriptstyle \pi'} \\ S & \longleftarrow & S'. \end{array}$$

If s' is the unique point of S' mapping to s, then the fiber $X'_{s'}$ is of the form

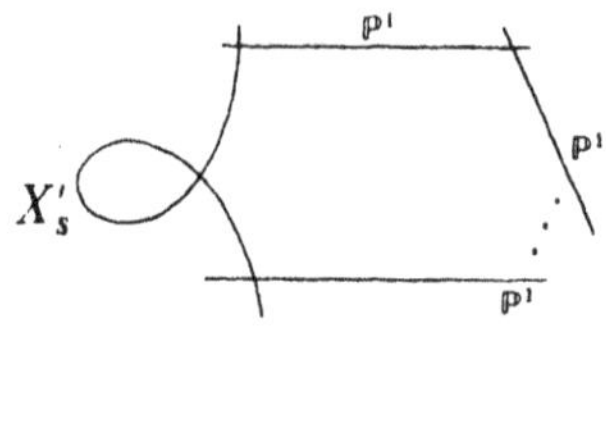

where X'_s is the normalization of X_s and there are $n-1$ copies of $\mathbb{P}^1$. Write $X'_{s'} = X_1 \cup \cdots \cup X_n$ where $X_1 = X'_s$ and $X_i = \mathbb{P}^1$ for $i > 1$. Then $(X_i.X_i) = -2$ for all i and $(X_i.X_j) = 1$ if $j \equiv i \pm 1 \bmod n$ and is zero otherwise (compare Artin [1, p. 116–117]). It is easy to show (using induction and the Mayer-Vietoris sequence, for example) that there is an exact sequence,

$$0 \to \mathbb{G}_m \to J_{X'_{s'}} \to J_{X'_s} \to 0$$

from which it follows that $F'_{s'}$ has order n^{2g-1} as before. However, there is an exact sequence

$$0 \to J_{X'_{s'}}(k) \to J_{X'_{\eta'}}(k(\eta')) \to \varepsilon_{\text{tors}} \to 0$$

where

$$\varepsilon = \operatorname{coker}\left(D \mapsto ((D.X_1), \ldots, (D.X_n)) : \bigoplus \mathbb{Z}X_i \to \mathbb{Z}^n\right)$$

(Artin [1, IV.4.5]). A computation with determinants shows that $[\varepsilon_{\text{tors}}] = n$, and so $F'_{\eta'}$ has order $n^{2g-1}.n = n^{2g}$. As $F'_{\eta'} = (F'_{\bar{\eta}'})^{I_{s'}}$, this proves what we want.

In order to apply these results to an arbitrary surface, we must know how the cohomology behaves when a point is blown up, that is, when we pass from X to X^* as in (3.1).

PROPOSITION 3.11. *Let $q: X' \to X$ be the map obtained by blowing up X at a closed point P, and let n be prime to* char (k). *Then $R^r q_* \mu_n = \mu_n$, 0, $i_{P*}\mathbb{Z}/(n)$, 0 respectively for $r = 0, 1, 2, \geq 3$, where i_P is the inclusion map $P \hookrightarrow X$.*

Proof. Obviously $q_*\mu_n \xrightarrow{\approx} \mu_n$ and $R^r q_* \mu_n$ is the sheaf associated with the presheaf $U \mapsto H^r(U', \mu_n)$ where U' is obtained from U by blowing up at all points of U lying over P. As $U' = U$ if the image of U does not contain P, it follows that $(R^r q_* \mu_n)_{\bar{x}} = 0$ if $r > 0$ and $x \neq P$. One may now use the proper base change theorem (VI.2) to show that $(R^r q_* \mu_n)_{\bar{x}} = H^r(\mathbb{P}^1_k, \mu_n)$ (as $X'_{\bar{x}} = \mathbb{P}^1_k$) from which the result follows. However, it is possible to gain more insight by considering $R^r q_* \mathbb{G}_m$ directly. In passing from U to U', one obtains one extra prime divisor for each new exceptional curve, and these divisors are independent in Pic (U') (Hartshorne [2, V.3.2]); thus Pic (U') = Pic $(U) \oplus (\bigoplus_{P'} \mathbb{Z})$, where the sum is over the points P' of U' mapping to P. It follows that $\Gamma(U, R^1 q_* \mathbb{G}_m) = \bigoplus_{P'} \mathbb{Z} = \Gamma(U, i_{P*}\mathbb{Z})$, and so $R^1 q_* \mathbb{G}_m = i_{P*}\mathbb{Z}$. Assume that U is connected. The maps spec $k(U) \to U - \{P' \mid P' \mapsto P\} \to U' \to U$ define maps Br $(U) \to$ Br $(U') \to$ Br $(U - \{P'\}) \to$ Br $(k(U))$, all of which are injective (III.2.22). As Br $(U) \to$ Br $(U - \{P'\})$ is an isomorphism (IV.2), we see that Br (U) = Br (U'), and it follows that $R^2 q_* \mathbb{G}_m = 0$. Now the Kummer sequence gives an exact sequence

$$0 \to R^1\pi_*\mu_n \to i_{P*}\mathbb{Z} \xrightarrow{n} i_{P*}\mathbb{Z} \to R^2\pi_*\mu_n \to 0$$

from which it follows that $R^1\pi_*\mu_n = 0$ and $R^2\pi_*\mu_n = i_{P*}(\mathbb{Z}/(n))$.

THEOREM 3.12. *$\chi(X_{\text{et}}, l) = c_2(X) = 12\chi(X, \mathcal{O}_X) - c_1(X)^2$, where $c_i(X)$ denotes the ith Chern class of the tangent bundle on X, and we have identified the class of a zero cycle with its degree ($l \neq$ char (k)).*

Proof. The second equality, $12\chi(X, \mathcal{O}_X) = c_1^2 + c_2$, is part of the Riemann-Roch theorem for X (Borel-Serre [1]). If $q: X' \to X$ is the map obtained by blowing up X at a closed point P, then $K_{X'} = q^{-1}(K_X) + E$, where $K_{X'}$ and K_X are the canonical classes on X' and X respectively and $E = q^{-1}(P)$ (Hartshorne [2, V.3.3]). Thus

$$c_1(X')^2 = c_1(X)^2 + E.E = c_1(X)^2 - 1.$$

As $\chi(X, \mathcal{O}_X)$ is invariant under blowing up (Hartshorne [2, V.3.5]), this shows that $12\chi(X, \mathcal{O}_X) - c_1(X)^2$ increases by one in passing from X to X'. Since the same is true of $\chi(X, l)$, according to (3.11), this shows that in proving the theorem we may assume that there is a Lefschetz pencil $\pi: X \to \mathbb{P}^1$ with irreducible fibers. Following (3.9), we have to show that $c_2(X) = -2(2g - 2) + t$ where g is the genus of X_η and t is the number of singular fibers.

I claim that (with $S = \mathbb{P}^1$)

$$c_2(X) = c_1(\Omega_X^2 \otimes (\pi^*\Omega_S^1)^{-1}) \,.\, \pi^*c_1(\Omega_S^1) + c_2(\Omega_X^1 \otimes (\pi^*\Omega_S)^{-1}).$$

Indeed, if we write

$$1 + c_1(\Omega_X^1)t + c_2(\Omega_X^1)t^2 = (1 + \gamma_1 t)(1 + \gamma_2 t)$$

and

$$1 + c_1(\Omega_S^1)t = 1 + \gamma t,$$

then the equality becomes,

$$\gamma_1\gamma_2 = (\gamma_1 + \gamma_2 - \gamma)\gamma + (\gamma_1 - \gamma)(\gamma_2 - \gamma).$$

(Compare Hartshorne [2, Appendix A3].) Let $c_1(\Omega_S^1) = -2s$ with $s \in S$ such that X_s is nonsingular; then

$$\begin{aligned} c_1(\Omega_X^2 \otimes (\pi^*\Omega_S^1)^{-1}).\pi^*c_1(\Omega_S^1) &= (K_X + 2X_s).(-2X_s) \\ &= -2(K_X + X_s).X_s \\ &= -2(2g - 2) \end{aligned}$$

by the adjunction formula (Hartshorne [2, V.1.5]). The canonical map $\pi^*\Omega_S^1 \to \Omega_X^1$ defines a section σ of $\Omega_X^1 \otimes (\pi^*\Omega_S)^{-1}$, and according to Grothendieck [2, Cor. to Thm. 2] $c_2(\Omega_X^1 \otimes (\pi^*\Omega_S^1)^{-1})$ is the cycle of zeros of σ. In terms of local parameters t_1 and t_2 at a point P of X, one sees easily that σ is the map

$$\left(\frac{\partial\pi}{\partial t_1}, \frac{\partial\pi}{\partial t_2}\right)$$

(where π is regarded as a function on X). Thus the component at P of the cycle of zeros of σ is

$$\dim_k\left(\mathcal{O}_{X,P}\middle/\left(\frac{\partial\pi}{\partial t_1}, \frac{\partial\pi}{\partial t_2}\right)\right),$$

which is zero unless P is a node and then is 1. This completes the proof.

Remark 3.13. (a) (3.12) was proved by Artin about 1960 by essentially the above methods. It also follows from the Lefschetz fixed point formula (VI.12) applied to the identity map, for this shows that $\chi(X, l) = (\Delta.\Delta)$, the self-intersection number of the diagonal on $X \times X$, and it is known that $(\Delta.\Delta) = c_2(X)$.

(b) (3.12) can be used to show that the l-adic Betti numbers $\beta_i(X, l)$ are independent of l ($\neq$char (k)). For $\beta_0(X, l) = 1 = \beta_4(X, l)$ (see (3.23) below) and $\beta_1(X, l) = \dim(\underline{\mathrm{Pic}}_X^0)$. By the Poincaré duality theorem (VI.11) or directly from (3.23), $\beta_3(X, l) = \beta_1(X, l)$. Finally, (3.12) shows that

$$\beta_2(X, l) = c_2(X) + 2\beta_1(X, l) - 2,$$

which is independent of l.

Let $\pi: X \to S = \mathbb{P}^1$ be a Lefschetz pencil with irreducible fibers, and let

$$F(-1) = F \otimes \mu_n^{\vee} = R^1\pi_*\mathbb{Z}/(n),$$

so that $F(-1)_{\bar{\eta}} = H^1(X_{\bar{\eta}}, \mathbb{Z}/(n))$ and $F(-1)_s = H^1(X_s, \mathbb{Z}/(n))$. For any $s \in T$, $V_s \approx \mu_n(k)$, and so

$$V_s(-1) \approx \mu_n(k) \otimes \mu_n(k)^{\vee} = \mathbb{Z}/(n)$$

has a canonical generator. The image of this generator in $F(-1)_{\bar{\eta}}$ will be written δ_s and called the *canonical vanishing cycle at* s. Note that δ_s depends on the choice of the embedding $\mathcal{O}_{S,s}^{sh} \hookrightarrow k(\bar{\eta})$ (as the cospecialization map $\phi_s: F_s \to F_{\bar{\eta}}$ does) and is only defined up to sign (because the map $\mathbb{G}_m \to J_Y$ of (3.7) is replaced by its reciprocal if Q_1 and Q_2 are interchanged). The e_n-pairing $F_{\bar{\eta}} \times F_{\bar{\eta}} \to \mu_n(k)$ induces a pairing

$$F(-1)_{\bar{\eta}} \times F(-1)_{\bar{\eta}} \to (\mathbb{Z}/(n))(-1),$$

which we write $(\gamma, \gamma') \mapsto (\gamma . \gamma')$; it may be identified with the canonical pairing

$$H^1(X_{\bar\eta}, \mathbb{Z}/(n)) \times H^1(X_{\bar\eta}, \mathbb{Z}/(n)) \to H^2(X_{\bar\eta}, \mathbb{Z}/(n))$$

of (1.20) (see (2.4f)). It is nondegenerate and skew-symmetric, and the exact annihilator of $F_s(-1)$ is $V_s(-1) = \langle \delta_s \rangle$. We write ε_s for the character $I_s \to \mu_n(k)$ of the inertia group at s such that $\sigma(t^{1/n}) = \varepsilon_s(\sigma)t^{1/n}$ for t a uniformizing parameter at s. (Compare (I.5.1e); the ε_s for increasing n define an isomorphism of the tame quotient I_s^t of I_s onto $\varprojlim_{p \nmid n} \mu_n(k)$, where $p = \operatorname{char} \exp(k)$).

THEOREM 3.14. (Picard-Lefschetz formula). *There exists a unit* $\lambda_X = (\lambda_X(n)) \in \varprojlim_{p \nmid n} \mathbb{Z}/(n)$ *such that*

$$\sigma(\gamma) = \gamma - \lambda_X(n)\varepsilon_s(\sigma)(\gamma . \delta_s)\delta_s,$$

for all

$$\sigma \in I_s \quad \textit{and} \quad \gamma \in F(-1)_{\bar\eta} = H^1(X_{\bar\eta}, \mathbb{Z}/(n)).$$

Proof. Choose $\gamma_s \in F(-1)_{\bar\eta}$ such that $(\gamma_s . \delta_s)$ has order n, and write $\sigma(\gamma_s) - \gamma_s = a(\sigma)\delta_s$ for $\sigma \in I_s$ (compare the proof of (3.8)). Then $\sigma \mapsto a(\sigma) \in \mathbb{Z}/(n)$ is a homomorphism that factors through I_s^t, and hence may be written $a(\sigma) = \lambda_X(n)(\gamma_s . \delta_s)\varepsilon(\sigma)$ for some $\lambda_X(n) \in \mu_n(k)$. If γ_s is replaced by γ_s', with $\gamma_s' = u\gamma_s + \gamma$, $u \in (\mathbb{Z}/(n))^*$, $\gamma \in F_s'$, then

$$a'(\sigma)\delta_s \overset{df}{=} \sigma(\gamma_s') - \gamma_s' = u\sigma(\gamma_s) - u\gamma_s = ua(\sigma).$$

As $(\gamma_s' . \delta_s) = u(\gamma_s . \delta_s)$, this shows that $\lambda_X(n)$ is independent of the choice of γ_s. It is a unit in $\mathbb{Z}/(n)$, for otherwise $F_{\bar\eta}^{I_s}$ would have index less than n in $F_{\bar\eta}$. Clearly the $\lambda_X(n)$ are compatible for varying n in the sense that $\lambda_X(n) = \lambda_X(rn) \bmod n$.

Remark 3.15. (a) It can be shown that $\lambda_X(n)$ depends only on the map $\tilde X \to \tilde S$ where $\tilde S = \operatorname{spec} \mathcal{O}_{S,s}^{sh}$ and $\tilde X = \operatorname{spec} \mathcal{O}_{X,x_0}^{sh}$, where x_0 is the node on X_s. Moreover Artin's approximation theorem can be used to show that $\tilde X \to \tilde S$ is isomorphic to the map obtained from

$$\operatorname{spec} k[T][T_1, T_2]/(T_1T_2 - T) \to \operatorname{spec} k[T]$$

by Henselizing at the origin in both schemes [SGA. 7, XV.1.3.2]. Thus $\lambda(X, n) = \lambda(n)$ is independent of X. Finally, by lifting this last map to characteristic zero (assuming one is not already in characteristic zero), one may use the classical results to show that $\lambda(n) = 1$. Henceforth we assume this (as it simplifies the formulas). See [SGA. 7, XV]. Thus the formula reads: $\sigma(\gamma) = \gamma - \varepsilon_s(\sigma)(\gamma . \delta_s)\delta_s$.

(b) Over $\mathbb{C}$, the above formula is due to Picard (without the precise form of the coefficient of δ_s) and Lefschetz (in essentially this form).

(c) There is an exact sequence,

$$0 \to F'_s \to F'_{\bar\eta} \to (\mathbb{Z}/(n))(-1) \to 0,$$

that is,

$$0 \to H^1(X_s, \mathbb{Z}/(n)) \to H^1(X_{\bar\eta}, \mathbb{Z}/(n)) \to (\mathbb{Z}/(n))(-1) \to 0$$

in which the second map is $\gamma \mapsto (\gamma.\delta_s)$.

(d) On passing to the limit over powers of l ($l \neq$ char (k)) and tensoring with $\mathbb{Q}_l$, one obtains statements similar to the above, except that the $\mathbb{Z}/(n)$-sheaves are replaced by $\mathbb{Q}_l$-sheaves. For example, $H^1(X_s, \mathbb{Q}_l) = H^1(X_{\bar\eta}, \mathbb{Q}_l)^{I_s}$, there exists a canonical vanishing cycle $\delta_s \in H^1(X_{\bar\eta}, \mathbb{Q}_l)$, and $\sigma(\gamma) = \gamma - \varepsilon_s(\sigma)(\gamma.\delta_s)\delta_s$ for $\sigma \in I_s$, $\gamma \in H^1(X_{\bar\eta}, \mathbb{Q}_l)$.

The above is usually called local Lefschetz theory; we now give a brief sketch of global Lefschetz theory. As before $\pi: X \to S = \mathbb{P}^1$ is a Lefschetz pencil with irreducible fibers, and we shall work only with sheaves of $\mathbb{Q}_l$-vector spaces. The representation of $\pi_1^t(\mathbb{P}^1 - T, \bar\eta)$ on $F'_{\bar\eta} = H^1(X_{\bar\eta}, \mathbb{Q}_l)$ is called the *monodromy* of the pencil. The *space of vanishing cycles* E on X is the subspace of $H^1(X_{\bar\eta}, \mathbb{Q}_l)$ generated by the vanishing cycles δ_s, $s \in T$. Since π_1^t is generated by the images of the I_s, $s \in T$ (we make a suitable choice of I_s for $s \in T$), the Picard-Lefschetz formula shows that E is stable under the action of π_1^t. It also shows that the space $E^\perp$ orthogonal to E under the canonical pairing on $H^1(X_{\bar\eta}, \mathbb{Q}_l)$ is $H^1(X_{\bar\eta}, \mathbb{Q}_l)^{\pi_1}$. As

$$\Gamma(S, R^1\pi_*\mathbb{Q}_l) = \Gamma(\eta, g^*R^1\pi_*\mathbb{Q}_l) = (R^1\pi_*\mathbb{Q}_l)_{\bar\eta}^{\pi_1} = H^1(X_{\bar\eta}, \mathbb{Q}_l)^{\pi_1}$$

(3.5) and the Leray spectral sequence gives an isomorphism $H^1(X, \mathbb{Q}_l) \xrightarrow{\approx} \Gamma(S, R^1\pi_*\mathbb{Q}_l)$ (see below (3.22)), we see that $E^\perp$ may be identified with the subspace $H^1(X, \mathbb{Q}_l)$ of $H^1(X_{\bar\eta}, \mathbb{Q}_l)$.

Remark 3.16. It is known that $E \cap E^\perp = 0$, that is, that no nonzero vanishing cycle is fixed by π_1. This fact is called *Hard Lefschetz* (or, more descriptively by the French, *Lefschetz vache*). It was proved over $\mathbb{C}$ by Hodge. For surfaces it is not particularly difficult to prove in nonzero characteristic (see Kleiman [1, 2.A.10]), but for varieties in general it has only fairly recently been proved by Deligne [3]; see also Messing [1]. The statement of Hard Lefschetz in Kleiman [1, 1.4] is as follows: (*HL*) the map

$$L_X: H^1(X, \mathbb{Q}_l) \to H^3(X, \mathbb{Q}_l(1)),$$

which takes an element to its cup-product with the cohomology class in $H^2(X, \mathbb{Q}_l(1))$ of a smooth hyperplane section, is an isomorphism. According to the Poincaré duality theorem (VI.11), (HL) is equivalent to

$$H^1(X, \mathbb{Q}_l) \times H^1(X, \mathbb{Q}_l) \to H^4(X, \mathbb{Q}_l(1)) \approx \mathbb{Q}_l(-1),$$

$(a, b) \mapsto L_X(a) \cup b$, being nondegenerate. But this pairing agrees with the pairing induced on $H^1(X, \mathbb{Q}_l)$ as a subspace of $H^1(X_{\bar{\eta}}, \mathbb{Q}_l)$ by our canonical pairing

$$H^1(X_{\bar{\eta}}, \mathbb{Q}_l) \times H^1(X_{\bar{\eta}}, \mathbb{Q}_l) \to \mathbb{Q}_l(-1).$$

Since $H^1(X, \mathbb{Q}_l) = E^{\perp}$, (HL) is obviously equivalent to $E \cap E^{\perp} = 0$.

It follows from (HL) that $H^1(X_{\bar{\eta}}, \mathbb{Q}_l)$ breaks up into an orthogonal direct sum, $H^1(X_{\bar{\eta}}, \mathbb{Q}_l) = H^1(X, \mathbb{Q}_l) \oplus E$.

THEOREM 3.17. (Conjugacy of the vanishing cycles). *For any pair δ_s, $\delta_{s'}$ of vanishing cycles there is a $\sigma \in \pi_1'(\mathbb{P}^1 - T, \bar{\eta})$ such that $\sigma\delta_s = \pm\delta_{s'}$.*

Proof. (sketch). We revert to the notations of the proof of (3.1); thus $X \subset \mathbb{P}^m$, $\check{X} \subset \check{\mathbb{P}}$ is the dual variety, D $(=\mathbb{P}^1 = S)$ is a line in $\check{\mathbb{P}}^m$, $\pi\colon X^* \to D$ is the Lefschetz pencil, and $D \cap \check{X}$ $(=T)$ is the set above which the fibers are singular.

LEMMA 3.18. *The inclusion $D - D \cap \check{X} \hookrightarrow \check{\mathbb{P}}^m - \check{X}$ induces a surjection $\pi_1'(D - D \cap \check{X}, \bar{d}) \twoheadrightarrow \pi_1'(\check{\mathbb{P}}^m - \check{X}, \bar{d})$ for $\bar{d}$ a closed point of $D - D \cap \check{X}$.*

Proof. Let G be the subvariety of the Grassmanian of lines in $\check{\mathbb{P}}^m$ comprising those lines that pass through $\bar{d}$, intersect $\check{X}$ only in its smooth subvariety, and intersect it there transversally. Consider the diagram

$$\begin{array}{ccccc}
\check{\mathbb{P}}^m & & \longleftarrow & & \check{\mathbb{P}}^m_{k(\bar{\xi})} \\
\uparrow & & & & \uparrow \\
D & \longrightarrow & \tilde{D} & \longleftarrow & D_{\bar{\xi}} \\
\downarrow & & \downarrow & & \downarrow \\
s & \longrightarrow & \tilde{G} & \longleftarrow & \bar{\xi}
\end{array}$$

where s is the point of G corresponding to D, $\tilde{G} = \operatorname{spec} \mathcal{O}^{sh}_{G,s}$, $\bar{\xi}$ is a geometric point lying over the generic point ξ of G, $\tilde{D}$ is the *universal line* over $\tilde{G}$, and $D_{\bar{\xi}} = \tilde{D} \times \bar{\xi}$. From this we get a diagram of maps of tame fundamental groups,

$$\begin{array}{ccccc}
\pi_1'(\check{\mathbb{P}}^m - \check{X}, \bar{d}) & & \longleftarrow & & \\
\uparrow & & & \searrow & \\
\pi_1'(D - D \cap \check{X}, \bar{d}) & \xrightarrow{\approx} & \pi_1'(\tilde{D} - \tilde{D} \cap \check{X}, \bar{d}) & \longleftarrow & \pi_1'(D_{\bar{\xi}} - D_{\bar{\xi}} \cap \check{X}, \bar{d})
\end{array}$$

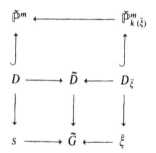

defined, and commutative, up to conjugacy. The lower-left map is an isomorphism because $\tilde{D}$ is proper over $\tilde{G}$ and the scheme being removed from $\tilde{D}$, $\tilde{D} \cap \check{X}$, is a divisor with normal crossings [SGA. 1, XIII.2, esp. 2.10]. We have, therefore, a specialization map

$$\pi'_1(D_{\bar{\xi}} - D_{\bar{\xi}} \cap \check{X}, \bar{d}) \to \pi'_1(D - D \cap \check{X}, \bar{d})$$

whose composite with

$$\pi'_1(D - D \cap \check{X}, \bar{d}) \to \pi'_1(\check{\mathbb{P}}^m - \check{X}, \bar{d})$$

is the canonical map

$$\pi'_1(D_{\bar{\xi}} - D_{\bar{\xi}} \cap \check{X}, \bar{d}) \to \pi'_1(\check{\mathbb{P}}^m - \check{X}, \bar{d}).$$

It remains to show that this last map is surjective or, equivalently, that the dual map on π^1 is injective. Let $U \to \check{\mathbb{P}}^m - \check{X}$ be a finite étale map with U connected; since $\check{\mathbb{P}}^m - \check{X}$ is irreducible, U will also be irreducible. A variant of Bertini's theorem now shows that $U \times (D_{\bar{\xi}} - \check{X})$ is connected, which completes the proof. (The variant states: let $f: U \to \mathbb{P}^m$ be a nonconstant morphism with U an irreducible scheme of finite-type over an algebraically closed field; if L is a generic linear space in $\mathbb{P}^m$ of codimension $<\dim(\overline{f(U)})$, then $U_L = f^{-1}(L)$ is geometrically irreducible. (Compare Zariski [4, I.6]. Apply this with $f = U \to \check{\mathbb{P}}^m - X \hookrightarrow \check{\mathbb{P}}^m$ and L the generic linear space of dimension 1.)

LEMMA 3.19. *The action of* $\pi'_1(D - D \cap \check{X}, \bar{d})$ *on* $H^1(X_{\bar{\eta}}, \mathbb{Q}_l)$ *factors through* $\pi'_1(\check{\mathbb{P}}^m - \check{X}, \bar{d})$.

Proof. The square

$$\begin{array}{ccc} X^* & \hookrightarrow & Y \subset X \times \check{\mathbb{P}}^m \\ \downarrow{\scriptstyle \pi} & & \downarrow{\scriptstyle \pi'} \\ D & \hookrightarrow & \check{\mathbb{P}}^m, \end{array}$$

where $Y = \{(x, H) \mid x \text{ lies on } H\}$ is Cartesian, that is $X^* = Y \cap D$. As π' is smooth over $\pi'^{-1}(\check{\mathbb{P}}^m - \check{X})$, the base change theorems (VI.2, 4) show that $R^1\pi'_*\mathbb{Q}_l$ is a twisted-constant sheaf on $\check{\mathbb{P}}^m - \check{X}$. Thus all of its stalks are equal and are acted on by $\pi_1(\check{\mathbb{P}}^m - \check{X}, \bar{d})$ (in fact, by $\pi'_1(\check{\mathbb{P}}^m - \check{X}, \bar{d})$); for example, $(R^1\pi'_*\mathbb{Q}_l)_{\bar{\eta}} = H^1(X_{\bar{\eta}}, \mathbb{Q}_l)$ is so acted on, and the action agrees with that of $\pi'_1(D - D \cap \check{X}, \bar{d})$.

We now prove the theorem. Changing the embedding $\mathcal{O}^{sh}_{D,s} \to k(\bar{\eta})$ corresponds to choosing a different point on the normalization of D in

$k(\bar{\eta})$ lying over s. If $\rho \in \mathrm{Gal}\,(k(\bar{\eta})/k(\eta))$ maps one point to the other, then the map

$$I_s \xrightarrow{j_s} \pi_1^t(D - D \cap \check{X}, \bar{\eta}) \subset \mathrm{Gal}\,(k(\bar{\eta})/k(\eta))$$

is replaced by its conjugate, $\sigma \mapsto \rho^{-1} j_s(\sigma)\rho$, and δ_s is replaced by $\rho^{-1}\delta_s$; the latter statement may either be checked directly from the definition of δ_s or by noting that

$$\rho^{-1}\sigma\rho(\gamma) = \rho^{-1}(\rho(\gamma) - \varepsilon_s(\sigma)(\rho\gamma . \delta_s)\delta_s) = \gamma - \varepsilon_s(\sigma)(\gamma . \rho^{-1}\delta_s)\rho^{-1}\delta_s$$

and that the Picard-Lefschetz formula determines δ_s uniquely (up to sign). Thus we have to show that the maps

$$\prod_{l \neq p} \mathbb{Z}_l(1) \underset{\approx}{\xrightarrow{\varepsilon_s^{-1}}} I_s^t \xrightarrow{j_s} \pi_1^t(D - D \cap \check{X}, \bar{\eta}) \to \mathrm{Aut}\,(H^1(X_{\bar{\eta}}, \mathbb{Q}_l)),$$

defined for $s \in T = D \cap \check{X}$, are all conjugate under conjugation by elements from $\pi_1^t(D - D \cap \check{X}, \bar{\eta})$. From the lemmas we have a diagram (commutative up to conjugacy):

$$\begin{array}{ccccccc}
 & & & & \pi_1^t(D_\xi - D_\xi \cap \check{X}) & & \\
 & & & \nearrow & & \searrow & \\
\prod_{l \neq p} \mathbb{Z}_l(1) & \longrightarrow & \pi_1^t(D_{\bar{\xi}} - D_{\bar{\xi}} \cap \check{X}) & & \longrightarrow & & \pi_1^t(\check{\mathbb{P}}^m - \check{X}) \\
 & \searrow_{j_s\varepsilon_s^{-1}} & & \searrow & & \nearrow & \downarrow \\
 & & & & \pi_1^t(D - D \cap \check{X}) & & \mathrm{Aut}\,(H^1(X_{\bar{\eta}}, \mathbb{Q}_l))
\end{array}$$

Thus it suffices to show that the maps $\prod_{l \neq p}\mathbb{Z}_l(1) \to \pi_1^t(D_{\bar{\xi}} - D_{\bar{\xi}} \cap \check{X})$, defined (analogously to $j_s\varepsilon_s^{-1}$) for $s \in D_{\bar{\xi}} \cap \check{X}$, become conjugate when composed with

$$\pi_1^t(D_{\bar{\xi}} - D_{\bar{\xi}} \cap \check{X}) \to \pi_1^t(D_\xi - D_\xi \cap \check{X}).$$

But Bertini's thoerem shows that, because $\check{X}$ is irreducible and D_ξ is generic, $D_\xi \cap \check{X}$ is irreducible, that is, consists of one point. Thus all points of $D_{\bar{\xi}} \cap \check{X}$ lie over a single point of D_ξ, and hence are permuted by $\mathrm{Gal}\,(k(\bar{\xi})/k(\xi))$. This shows that the maps are conjugate as $\mathrm{Gal}\,(k(\bar{\xi})/k(\xi))$ is a quotient of $\pi_1^t(D_\xi - D_\xi \cap \check{X})$. (See [SGA. 7, XVIII.6] for more details.)

COROLLARY 3.20. *The action of* $\pi_1^t(\mathbb{P}^1 - T, \bar{\eta})$ *on* $E/E \cap E^\perp$ *is absolutely irreducible.*

Proof. Let $F \subset E \otimes \mathbb{Q}_{l,al}$ be a subspace that is stable under the monodromy. If $F \not\subset (E \cap E^\perp) \otimes \mathbb{Q}_{l,al}$, there exists a $\gamma \in F$ and an $s \in T$

such that $(\gamma.\delta_s) \neq 0$. Then $\sigma\gamma - \gamma = -\varepsilon_s(\sigma)(\gamma.\delta_s)\delta_s$ for $\sigma \in I_s$, which shows that $\delta_s \in F$. After (3.17), all the δ_s are in F, and $F = E$.

Remark 3.21. If we choose an isomorphism $\mathbb{Q}_l(-1) \approx \mathbb{Q}_l$, the pairing on $H^1(X_{\bar{\eta}}, \mathbb{Q}_l)$ induces a nondegenerate skew-symmetric pairing

$$\psi: E/E \cap E^{\perp} \times E/E \cap E^{\perp} \to \mathbb{Q}_l.$$

As the monodromy respects this pairing, we obtain a representation

$$\rho: \pi_1(\mathbb{P}^1 - T, \bar{\eta}) \to Sp(E/E \cap E^{\perp}, \psi),$$

where Sp denotes the symplectic group. The theorem of Každan-Margulis says that (for any Lefschetz pencil with odd fiber dimension) the image of ρ is open. This theorem, which follows almost directly from the irreducibility of the monodromy (3.20), plays an important role in Deligne's proof of the Riemann hypothesis (Deligne [2, 5.9]). (This paper also contains a summary of Lefschetz theory for pencils of any dimension.)

We next give an explicit description of the cohomology groups $H^r(X, \mathbb{Z}/(n))$ when $\pi: X \to S = \mathbb{P}^1$ is a Lefschetz pencil with irreducible fibers. To simplify the notation we write $\Lambda = \mathbb{Z}/(n)$, so that $\Lambda(1) = \mu_n$ and $R^1\pi_*\Lambda(1) = F$. The Leray spectral sequence for π is

$$H^r(S, R^s\pi_*\Lambda(1)) \Rightarrow H^{r+s}(X, \Lambda(1)).$$

As $R^0\pi_*\Lambda(1) = \Lambda(1)$, $R^1\pi_*\Lambda(1) = F$, $R^2\pi_*\Lambda(1) = \Lambda$, and $R^s\pi_*\Lambda(1) = 0$ for $s > 2$ (3.5), and as any constant sheaf on S has zero first cohomology group, this gives

$$\Lambda(1) = H^0(S, \Lambda(1)) \xrightarrow{\approx} H^0(X, \Lambda(1)),$$

$$0 \to H^1(X, \Lambda(1)) \to H^0(S, F) \xrightarrow{\partial} H^2(S, \Lambda(1)) \underset{\alpha^*}{\overset{\pi^*}{\leftrightarrows}} H^2(X, \Lambda(1))' \to H^1(S, F) \to 0,$$

$$0 \to H^2(X, \Lambda(1))' \to H^2(X, \Lambda(1)) \leftrightarrows H^0(S, \Lambda) \to H^2(S, F) \to H^3(X, \Lambda(1)) \to 0,$$

$$H^4(X, \Lambda(1)) \xrightarrow{\approx} H^2(S, \Lambda) \approx \Lambda(-1),$$

and

$$H^r(X, \Lambda(1)) = 0, r > 4.$$

The map

$$\alpha^*: H^2(X, \Lambda(1))' \to H^2(S, \Lambda(1))$$

is that induced by the section α to π; thus $\alpha^*\pi^* = (\pi\alpha)^* = 1$, and so

$$H^2(S, \Lambda(1))' = H^1(S, F) \oplus H^2(S, \Lambda(1)).$$

Recall (2.1) that $H^2(S, \Lambda(1)) = \Lambda$, and so we may write

$$H^2(X, \Lambda(1))' = H^1(S, F) \oplus \langle \gamma_E \rangle$$

where γ_E is the element of $H^2(X, \Lambda(1))$ corresponding to the canonical generator of $H^2(S, \Lambda(1))$ and $\langle \gamma_E \rangle$ denotes the free Λ-module that γ_E generates. γ_E may also be described as the cohomology class, defined via the map $\mathrm{Pic}(X) = H^1(\mathbb{G}_m) \to H^2(\Lambda(1))$, of the divisor $E = \alpha(S)$. The edge homomorphism

$$H^2(X, \Lambda(1)) \to H^0(S, R^2\pi_*\Lambda(1)) = H^0(S, \Lambda)$$

can be interpreted as the map $\mathrm{Pic}(X)^{(n)} \to \Lambda$ that sends a divisor class to the degree of its intersection with E. If we write γ_F for the cohomology class of any smooth fiber of π, then the map

$$(1 \mapsto \gamma_F): H^0(S, \Lambda) \to H^2(X, \Lambda(1))$$

is a section to the edge homomorphism. Thus

$$H^2(X, \Lambda(1)) = H^1(S, F) \oplus \langle \gamma_E \rangle \oplus \langle \gamma_F \rangle.$$

We have shown: (with $F = R^1\pi_*\Lambda(1)$)

$$\begin{aligned}
H^0(X, \Lambda(1)) &= \Lambda(1), \\
H^1(X, \Lambda(1)) &= H^0(S, F), \\
H^2(X, \Lambda(1)) &= H^1(S, F) \oplus \langle \gamma_E \rangle \oplus \langle \gamma_F \rangle, \\
H^3(X, \Lambda(1)) &= H^2(S, F), \\
H^4(X, \Lambda(1)) &= \Lambda(-1), \\
H^r(X, \Lambda(1)) &= 0, \qquad r > 4.
\end{aligned}$$

It remains to compute the cohomology of F on S. By excision, $H^r_T(S, F) \approx \bigoplus_{s \in T} H^r_s(\tilde{S}_{(s)}, F)$ where $\tilde{S}_{(s)} = \mathrm{spec}\, \mathcal{O}^{sh}_{S,s}$. Moreover (see the proof of (2.10)), $H^r_s(\tilde{S}_{(s)}, F) = 0$ for $r \neq 2$ and $H^2_s(\tilde{S}_{(s)}, F) = H^1(\tilde{K}_{(s)}, F)$ where $\tilde{K}_{(s)}$ is the field of fractions of $\tilde{S}_{(s)}$; thus $H^2_s(\tilde{S}_{(s)}, F) = H^1(I_s, F)$, which equals $H^1(I^t_s, F)$ as F has order prime to char (k). Also (2.17) shows that

$$H^r(S - T, F) = H^r(\pi^t_1(S - T, \bar{\eta}), F_{\bar{\eta}}),$$

which is zero for $r > 1$. The cohomology sequence for the pair $(S, S - T)$ now shows that

$$\begin{aligned}
H^0(S, F) &= H^0(S - T, F) = H^0(\pi^t_1, F_{\bar{\eta}}) \\
H^1(S, F) &= \ker\left(H^1(\pi^t_1, F_{\bar{\eta}}) \to \bigoplus_{s \in T} H^1(I^t_s, F_{\bar{\eta}})\right) \\
H^2(S, F) &= \mathrm{coker}\left(H^1(\pi^t_1, F_{\bar{\eta}}) \to \bigoplus_{s \in T} H^1(I^t_s, F_{\bar{\eta}})\right).
\end{aligned}$$

THEOREM 3.22. *If $\pi: X \to \mathbb{P}^1$ is a Lefschetz pencil with irreducible fibers and n is prime to* char (k), *then:*

$$H^0(X, \Lambda(1)) = \Lambda(1),$$
$$H^1(X, \Lambda(1)) = F_{\bar\eta}^{\pi_1},$$
$$H^2(X, \Lambda(1)) = \langle\gamma_E\rangle \oplus \langle\gamma_F\rangle \oplus \ker\,(H^1(\pi_1^t, F_{\bar\eta}) \to \bigoplus_{s\in T} H^1(I_s^t, F_{\bar\eta})),$$
$$H^3(X, \Lambda(1)) = \operatorname{coker}\,(H^1(\pi_1^t, F_{\bar\eta}) \to \bigoplus_{s\in T} H^1(I_s^t, F_{\bar\eta})),$$
$$H^4(X, \Lambda(1)) = \Lambda(-1),$$

and

$$H^r(X, \Lambda(1)) = 0, \qquad r > 4,$$

where

$$F_{\bar\eta} = H^1(X_{\bar\eta}, \Lambda(1)) = \ker\,(n: J_{X_{\bar\eta}}(k) \to J_{X_{\bar\eta}}(k)),$$

$T \subset \mathbb{P}^1$ is the set over which the fibers are singular, $\pi_1^t = \pi_1^t(\mathbb{P}^1 - T, \bar\eta)$, and γ_E and γ_F are the cohomology classes of a section and of a smooth fiber respectively.

Proof. This is a restatement of the above.

To simplify the statement and proof of the next theorem, we choose an isomorphism

$$\prod_{l\neq p} \mathbb{Z}_l(1) \approx \prod_{l\neq p} \mathbb{Z}_l,$$

that is, we choose generators ζ_n of $\mu_n(k)$ for each n which satisfy $\zeta_n = \zeta_{rn}^r$. Thus we may identify Λ with $\Lambda(1)$, I_s^t with $\prod_{l\neq p} \mathbb{Z}_l$, and the e_n-pairing with a pairing

$$(\gamma, \gamma') \mapsto (\gamma . \gamma'): H^1(X_{\bar\eta}, \Lambda) \times H^1(X_{\bar\eta}, \Lambda) \to \Lambda,$$

etc. We write $T = \{s_1, \ldots, s_t\}$ and $I_{s_i} = I_i$, and we choose the embeddings $I_i \hookrightarrow \operatorname{Gal}(k(\bar\eta)/k(\eta))$ in such a way that if σ_i denotes the canonical generator of I_i^t, then $\sigma_1, \ldots, \sigma_t$ generate $\pi_1^t(\mathbb{P}^1 - T, \bar\eta)$ and $\sigma_1 \cdots \sigma_s = 1$ (see (2.19)).

THEOREM 3.23. *With the above notations,*

$$H^0(X, \Lambda) = \Lambda = H^4(X, \Lambda)$$

and $H^1(X, \Lambda)$, $H^2(X, \Lambda)/(\langle\gamma_E\rangle \oplus \langle\gamma_F\rangle)$, $H^3(X, \Lambda)$ are the cohomology groups of the complex,

$$F_{\bar\eta} \xrightarrow{\alpha} \Lambda^t \xrightarrow{\beta} F_{\bar\eta}$$

where

$$\Lambda^t = \Lambda \oplus \cdots \oplus \Lambda \ (t \text{ copies}),$$
$$\alpha(\gamma) = ((\gamma.\delta_1), \ldots, (\gamma.\delta_t)),$$
$$\beta(a_1, \ldots, a_t) = a_1\delta_1 + a_2\sigma_1(\delta_2) + \cdots + a_t\sigma_1\sigma_2 \cdots \sigma_{t-1}(\delta_t).$$

Proof. With the above notations, the Picard-Lefschetz formula becomes

$$\sigma_i(\gamma) = \gamma - (\gamma.\delta_i)\delta_i, \qquad \gamma \in F_{\bar{\eta}}.$$

Thus γ is fixed by σ_i if and only if $(\gamma.\delta_i) = 0$. As the σ_i generate π'_1, this shows that $\ker(\alpha) = F_{\bar{\eta}}^{\pi'_1} = H^1(X, \Lambda)$.

Recall (2.19a) that there is a unique cocycle $(f(\sigma))$ for π'_1 in $F_{\bar{\eta}}$ with arbitrarily assigned values $f(\sigma_1), \ldots, f(\sigma_{t-1}) \in F_{\bar{\eta}}$. Alternatively, there is a unique cocycle f for any set $f(\sigma_1), \ldots, f(\sigma_t)$ of elements of $F_{\bar{\eta}}$ such that

$$f(\sigma_1) + \sigma_1 f(\sigma_2) + \cdots + \sigma_1 \cdots \sigma_{t-1} f(\sigma_t) = 0.$$

Since

$$H^1(I_i, F_{\bar{\eta}}) = F_{\bar{\eta}}/(\sigma_i - 1)F_{\bar{\eta}},$$

such a cocycle f maps to zero in all $H^1(I_i, F_{\bar{\eta}})$ if and only if there exist $\gamma_1, \ldots, \gamma_t \in F_{\bar{\eta}}$ such that $f(\sigma_i) = \sigma_i(\gamma_i) - \gamma_i$ for all i. Equivalently f maps to zero if and only if there exist $a_1, \ldots, a_t \in \Lambda$ such that $f(\sigma_i) = a_i\delta_i$. Thus every such cocycle f defines an element $a(f) = (a_1, \ldots, a_t) \in \Lambda^t$. The cocycle condition on f translates into the following condition on $a(f)$:

$$a_1\delta_1 + a_2\sigma_1(\delta_2) + \cdots + a_t\sigma_1 \cdots \sigma_{t-1}(\delta_t) = 0.$$

Thus the image of $(f \mapsto a(f))$ is precisely the kernel of β. Moreover, $a(f)$ is in the image of α if and only if there exists a $\gamma \in F_{\bar{\eta}}$ such that $\sigma_i(\gamma) - \gamma = f(\sigma_i)$ for all i, that is, if and only if f is a coboundary. This shows that

$$\begin{aligned} \ker(\beta)/\operatorname{im}(\alpha) &= \ker\Big(H^1(\pi'_1, F_{\bar{\eta}}) \to \bigoplus_{s \in T} H^1(I'_s, F_{\bar{\eta}})\Big) \\ &= H^2(X, \Lambda(1))/(\langle\gamma_E\rangle \oplus \langle\gamma_F\rangle). \end{aligned}$$

Finally, one checks that the map

$$\gamma \mapsto (0, \ldots, 0, \gamma(\operatorname{mod}(\sigma_t - 1)F_{\bar{\eta}})) : F_{\bar{\eta}} \to \oplus H^1(I_i, F_{\bar{\eta}})$$

induces an isomorphism

$$\operatorname{coker}(\beta) \xrightarrow{\approx} \operatorname{coker}\left(H^1(\pi'_1, F_{\bar{\eta}}) \to \bigoplus_{s \in T} H^1(I'_s, F_{\bar{\eta}})\right).$$

Remark 3.24. Theorems (3.22) and (3.23) both remain valid if Λ is interpreted as being $\mathbb{Z}_l$ or $\mathbb{Q}_l$, $l \neq \operatorname{char}(k)$.

We next apply étale cohomology to the study of the *Néron-Severi* group, $NS(X) \stackrel{\mathrm{df}}{=} \operatorname{Pic}(X)/\operatorname{Pic}^0(X)$, of X.

THEOREM 3.25. (Severi, Néron). *$NS(X)$ is finitely generated.*

Proof. Recall that a divisor (or divisor class) D is *numerically equivalent* to zero if the degree of intersection $(D.D')$ of D with any divisor (or class) D' is zero. Thus the group $\operatorname{Pic}^n(X)$ of such classes contains $\operatorname{Pic}^0(X)$ and is the kernel of the intersection pairing

$$(D, D') \mapsto (D.D'): \operatorname{Pic}(X) \times \operatorname{Pic}(X) \to \mathbb{Z}.$$

We write $N(X) = \operatorname{Pic}(X)/\operatorname{Pic}^n(X)$ and note that there is a nondegenerate pairing $(\alpha, \alpha') \mapsto (\alpha.\alpha'): N(X) \times N(X) \to \mathbb{Z}$. Also the Kummer sequence (with $l \neq \operatorname{char}(k)$),

$$\cdots \to \operatorname{Pic}(X) \xrightarrow{l} \operatorname{Pic}(X) \to H^2(X, \mu_l) \to \cdots,$$

together with the finiteness of $H^2(X, \mu_l)$ (3.23), shows that $\operatorname{Pic}(X)/l\operatorname{Pic}(X)$ and, *a fortiori*, $N(X)/lN(X)$ are finite.

Our first lemma will show that $N(X)$ is finitely generated and free and the second that $\ker(NS(X) \to N(X))$ is finite.

LEMMA 3.26. *Let N be an abelian group on which there is a nondegenerate pairing $(\alpha, \beta) \mapsto \alpha.\beta: N \times N \to \mathbb{Z}$. Assume that, for some prime l, N/lN is finite, and let $\hat{N} = \varprojlim N/l^nN$. Then:*

(a) *$N \to \hat{N}$ is injective;*

(b) *N is finitely generated and free.*

Proof. (a) The kernel of $N \to \hat{N}$ is the set of elements that are divisible by all powers of l. But if α is divisible by all powers of l, so also is $\alpha.\beta$ for any β, which implies that $\alpha.\beta = 0$, and $\alpha = 0$.

(b) As N/l^nN is finite for all n, $\hat{N}$ is a finitely generated $\mathbb{Z}_l$-module. Since $N \twoheadrightarrow \hat{N}/l\hat{N}$, we may choose $\alpha_1, \ldots \alpha_n \in N$ that generate $\hat{N}/l\hat{N}$ and hence also $\hat{N}$. Consider the map

$$N \to \mathbb{Z}^n, \beta \mapsto (\beta.\alpha_1, \ldots, \beta.\alpha_n).$$

If β is in the kernel of this map, then $\beta.\alpha$ is divisible by all powers of l for any α, because α can be approximated arbitrarily closely in the l-adic topology by a $\mathbb{Z}$-linear-combination of α_i's. Thus $N \hookrightarrow \mathbb{Z}^n$, which proves (b).

LEMMA 3.27. *$\operatorname{Pic}^n(X)/\operatorname{Pic}^0(X)$ is finite.*

Proof. Let H be a hypersurface section of X of sufficiently high degree that,

(a) $\deg(H) > \deg(K)$, where K is the canonical class, and

(b) $\frac{1}{2}(H.H - K) + p_a(X) + 1 > 0$.

If H' is a divisor on X that is numerically equivalent to H, then (a) shows that $\deg(K - H') < 0$ and hence $H^0(\mathcal{O}_X(K - H')) = 0$. From the Riemann-Roch theorem, namely,

$$\dim H^0(\mathcal{O}_X(H')) + \dim H^0(\mathcal{O}_X(K - H')) \geq \left(\frac{H'.H' - K}{2}\right) + p_a(X) + 1,$$

and (b), it follows that $\dim H^0(\mathcal{O}_X(H')) > 0$. We have shown that every divisor numerically equivalent to H is linearly equivalent to a positive divisor. But the family of all positive divisors of given degree is parametrized by a finite set of components of the Chow variety of cycles on X (Main theorem of Chow and van der Waerden [1]). Thus if we choose one divisor D_i from each of these components, every divisor H' numerically equivalent to H is algebraically equivalent to one of the D_i, and every divisor numerically equivalent to zero is algebraically equivalent to a divisor $D_i - H$. This completes the proof of the lemma (and the theorem).

COROLLARY 3.28. *Let $\rho(X)$ denote the rank of $NS(X)$; then*

$$\rho(X) \leq \beta_2(X, l) = c_2(X) + 2q - 2 \qquad (l \neq \operatorname{char}(k))$$

where $q = \dim(\underline{\mathrm{Pic}}^0_{X/k})$ and $\beta_2(X, l) = \dim_{\mathbb{Q}_l} H^2(X, \mathbb{Q}_l)$.

Proof. The equality follows from the first equality of (3.12) (see (3.13b)). For the inequality, note that as $\mathrm{Pic}^0(X) = \underline{\mathrm{Pic}}^0_X(k)$ is divisible, the Kummer sequence gives an injection $NS(X)/l^n NS(X) \hookrightarrow H^2(X, \mu_n)$. In the limit this becomes $NS(X) \otimes_{\mathbb{Z}} \mathbb{Z}_l \hookrightarrow H^2(X, \mathbb{Z}_l(1))$ as $NS(X)$ is finitely generated. Thus elements of $NS(X)$ that are linearly independent over $\mathbb{Z}$ have images in $H^2(X, \mathbb{Z}_l(1))$ that are linearly independent over $\mathbb{Z}_l$, which completes the proof.

Remark 3.29. (a) The proof of (3.25) shows more than we have stated, namely, that the torsion elements of $NS(X)$ are precisely the divisor classes that are numerically equivalent to zero.

(b) For an exposition of (3.27) in terms of schemes, see [SGA. 6, XIII.5].

(c) The statement $\rho(X) \leq c_2(X) - 2q + 2$ of (3.28) does not involve étale cohomology and was proved first without it in Igusa [4]. The proof we have given is very close to his; he replaces $R^1\pi_*\mu_n$ by the points of order n on the family of Jacobians of the pencil and uses the cohomology of $\pi_1^t(\mathbb{P}^1 - T, \bar{\eta})$ (together with Grothendieck's description of π_1^t) instead of the étale cohomology of $\mathbb{P}^1 - T$.

(d) As $\mathrm{Br}(X)_l$ is finite, the l-primary component of $\mathrm{Br}(X)$ is a group of the form $(\mathbb{Q}_l/\mathbb{Z}_l)^{\rho_0(X,l)} \oplus F$ where F is finite. Thus $T_l\mathrm{Br}(X) \stackrel{df}{=} \varprojlim \mathrm{Br}(X)_{l^m}$ is a finitely generated $\mathbb{Z}_l$-module of rank $\rho_0(X, l)$. The Kummer sequence gives an exact sequence,

$$0 \to NS(X) \otimes \mathbb{Z}_l \to H^2(X, \mathbb{Z}_l(1)) \to T_l\mathrm{Br}(X) \to 0.$$

Thus $\rho + \rho_0 = \beta_2(X)$, and ρ_0 is independent of l ($l \neq \text{char}(k)$) (3.13b). On being tensored with $\mathbb{Q}_l$, the sequence becomes

$$0 \to NS(X) \otimes \mathbb{Q}_l \to H^2(X, \mathbb{Q}_l(1)) \to V_l\text{Br}(X) \to 0.$$

$V_l\text{Br}(X)$ can be regarded as the vector space of transcendental (that is, nonalgebraic) cycles on X and the map $H^2(X, \mathbb{Q}_l(1)) \to \mathbb{Q}_l^{\rho_0}$, deduced from the above by choosing a basis for $V_l\text{Br}(X)$, as an algebraically defined period map.

As a final application, we give Artin's proof in characteristic p of Castelnuovo's criterion of rationality of a surface. Recall that X is said to be rational if it is birationally equivalent to $\mathbb{P}^2$. The nth *plurigenus* of X is $P_n(X) = \dim_k H^0(X, (\Omega^2_{X/k})^{\otimes n})$ and the *arithmetic genus* is

$$p_a(X) = \chi(X, \mathcal{O}_X) - 1 = -\dim_k H^1(\mathcal{O}_X) + \dim_k H^2(\mathcal{O}_X).$$

THEOREM 3.30. (Castelnuovo's Criterion). *X is rational if and only if $P_2(X) = 0 = p_a(X)$.*

Proof. As $p_a(\mathbb{P}^2) = 0 = P_n(\mathbb{P}^2)$ $(n \geq 0)$ and p_a and P_n $(n \geq 0)$ are birational invariants, the necessity is clear (see Serre [3]). In this report, it is shown conversely that a minimal surface X for which $P_2(X) = 0 = p_a(X)$ is either rational or else: Pic $(X) = \mathbb{Z}$ and is generated by the class of the canonical divisor K; $|-K|$ contains an irreducible curve; $0 < K^2 \leq 5$. (See also Beauville [1, Chapter V].)

If $k = \mathbb{C}$, the exponential sequence $0 \to \mathbb{Z} \to \mathcal{O}_X \xrightarrow{\exp} \mathcal{O}_X^* \to 0$ gives an exact sequence of complex cohomology groups,

$$H^1(X, \mathcal{O}_X) \to H^1(X, \mathcal{O}_X^*) \to H^2(X, \mathbb{Z}) \to H^2(X, \mathcal{O}_X).$$

As $-K$ is positive, $P_2 = 0$ implies $P_1 = 0$, which, by Serre duality, implies $H^2(\mathcal{O}_X) = 0$. Also $p_a = 0$ implies that $H^1(\mathcal{O}_X) = 0$, and so $H^2(X, \mathbb{Z}) \approx$ Pic (X). Thus, if X is not rational, then $\beta_2(X) = 1$, $\chi(X, \mathbb{Z}) = 3$, and $K^2 = 9$ by Noether's formula: $K^2 = 12(1 + p_a) - \chi(X, \mathbb{Z})$. We have a contradiction.

Zariski completed the proof for k of characteristic $p \neq 0$ in Zariski [5] and [6]. The following argument is due to M. Artin (unpublished). (The proof in Kurke [1] is incorrect.)

We fix a prime $l \neq \text{char}(k)$. It follows from (3.29d) that Br $(X)(l)$, the l-primary component of Br (X), is finite if and only if $\rho(X) = \beta_2(X)$. As the second condition is preserved under blowing up (3.11), so also is the first; in particular Br $(X)(l)$ is finite for any rational surface as $\rho(\mathbb{P}^2) = 1 = \beta_2(\mathbb{P}^1)$.

As in the proof when $k = \mathbb{C}$, we assume X to be not rational and derive a contradiction. The Riemann-Roch theorem shows that $l(-K) \geq (K^2) + p_a + 1 \geq 2$, and so $|-K|$ contains a pencil of curves each of arithmetic genus

$$p_a = p_a(-K) = 1 + \tfrac{1}{2}(-K).(K - K) = 1.$$

There are three possibilities: the generic member of the pencil (as a curve over $k(\mathbb{P}^1)$) is not regular and hence its normalization has genus 0; the generic member is regular but nonsmooth, and almost all curves in the pencil are rational curves with a cusp; the generic member is smooth and almost all curves are smooth of genus one (Zariski [3]). In the first case the surface may be shown to be rational by the argument of Serre [3, p. 5]. In the remaining two cases we blow up at base points of the pencil to obtain a surface X^* and a mapping $\pi: X^* \to \mathbb{P}^1$ whose fibers are the curves of the pencil.

In the spectral sequence

$$H^r(\mathbb{P}^1, R^s\pi_*\mathbb{G}_m) \Rightarrow H^{r+s}(X^*, \mathbb{G}_m),$$

$H^2(\mathbb{P}^1, \mathbb{G}_m) = 0$ by (III.2.22) and $R^2\pi_*\mathbb{G}_m$ is l-torsion free because $R^2\pi_*\mu_l = \mathbb{Z}/(l)$ and $\underline{\mathrm{Pic}}_{X^*/\mathbb{P}^1} \to R^2\pi_*\mu_l$ is surjective; compare (3.5). Thus

$$\mathrm{Br}\,(X^*)(l) \approx H^1(\mathbb{P}^1, \mathrm{Pic}_{X^*/\mathbb{P}^1})(l).$$

If we define $A = \underline{\mathrm{Pic}}^0_{X^*/\mathbb{P}^1}$ by the exact sequence $0 \to A \to \mathrm{Pic}_{X^*/\mathbb{P}^1} \xrightarrow{\deg} \mathbb{Z} \to 0$, then, as $H^1(\mathbb{P}^1, \mathbb{Z}) = 0$, $H^1(\mathbb{P}^1, A)$ differs from $H^1(\mathbb{P}^1, \underline{\mathrm{Pic}}_{X^*/\mathbb{P}^1})$ by a finite group. Thus, if we can show that $H^1(\mathbb{P}^1, A)(l)$ is finite, we may deduce that $\mathrm{Br}\,(X^*)(l)$ is finite, that $\mathrm{Br}\,(X)(l)$ is finite, and that therefore $\beta_2(X) = \rho(X) = 1$. As before, this gives a contradiction (using (3.12)).

As usual, we let η be the generic point of $\mathbb{P}^1$. The Jacobian fibration $A' \to \mathbb{P}^1$ for $X^*/\mathbb{P}^1$ is the complete minimal model over $\mathbb{P}^1$ of the curve over $k(\eta)$, which is the Jacobian of X^*_η (completed with a point if X^*_η is singular). (See Šafarevič [2].) A is the open subscheme of A' obtained from A' by removing all points that are singular in their fibers. Any element of $H^1(\mathbb{P}^1, A)$ is represented by an A-torsor Y that, being locally isomorphic to A over $\mathbb{P}^1_{\mathrm{et}}$, may be completed as A by adding singular points to certain fibers to a complete smooth surface Y'. Thus Y' is an A'-torsor over $\mathbb{P}^1$ that is trivial if and only if it has a section or, equivalently, Y is trivial. We shall show that any nontrivial Y' is a rational surface. This will complete the proof, for if every Y' is trivial then $H^1(\mathbb{P}^1, A)(l) = 0$; otherwise there is a rational Y'. Then $\mathrm{Br}\,(Y')(l)$ is finite and, as $A = \underline{\mathrm{Pic}}^0_{Y'/\mathbb{P}^1}$, it follows (compare the above paragraph) that $H^1(\mathbb{P}^1, A)(l)$ is finite.

Thus consider a nontrivial torsor for A', $\pi': Y' \to \mathbb{P}^1$. As $R^0\pi'_*\mathcal{O}_Y = \mathcal{O}_{\mathbb{P}^1} = R^0\pi_*\mathcal{O}_{X^*}$ and $R^1\pi'_*\mathcal{O}_{Y'} = (\text{tangent space to } A) = R^1\pi_*\mathcal{O}_{X^*}$, the Leray spectral sequences for π and π' show that $p_a(Y') = p_a(X^*) = 0$. Also, as Y' and X^* are locally isomorphic over $\mathbb{P}^1_{\mathrm{et}}$, they have the same fibers, and the description of the canonical class of an elliptic or quasi-elliptic surface in terms of its fibers and $R^1\pi_*\mathcal{O}$ (Bombieri-Mumford [1]) shows that the fibers of π' are anticanonical, that is, $K_{Y'} = -(\text{fiber of } \pi')$. Let n be the least positive integer such that there exists a divisor D on Y' with $(D.-K) = (D.\text{fiber}) = n$. From the Riemann-Roch theorem we see that

$$\chi(\mathcal{O}(D)) = \tfrac{1}{2}(D.D - K) + 1 = \tfrac{1}{2}((D^2) + n) + 1$$

is positive if $(D^2) \geq -n$. Adding or subtracting a fiber $-K$ from D leaves $(D.-K)$ unchanged but replaces (D^2) by $(D \pm K)^2 = D^2 \mp 2n$. Thus there is a D with $\chi(\mathcal{O}(D)) > 0$ and $(D.-K) = n$. In fact there is a positive such D, for $H^0(\mathcal{O}(D)) \neq 0$ or $H^0(\mathcal{O}(K - D)) \neq 0$, and the latter is absurd for $(K - D.(\text{fiber})) = -n < 0$. We now subtract off a fiber from D as often as possible while keeping $H^0(\mathcal{O}(D)) \neq 0$; thus we may choose $D \geq 0$ and such that $H^0(\mathcal{O}(D + K)) = 0$. The Riemann-Roch theorem applied to $D + K$ shows that $(D^2) < n$. If we replace D by some irreducible component that has positive intersection with a fiber, then we obtain a curve D that still has $(D.-K) = n$ and $(D^2) < n$. Now

$$0 \leq p_a(D) = \tfrac{1}{2}((D^2) + (D.K)) + 1 \leq 0,$$

and so $p_a(D) = 0$ and $(D^2) = n - 2$. If $n = 1$, we would have, contrary to our assumption, a section. Thus $n \geq 2$ and $(D^2) \geq 0$. The existence of such a curve on Y' implies that Y' is rational (Serre [3, p. 06]).

Comments on the Literature

Constructible sheaves are treated in [SGA. 4, IX.] and Artin [9]; in the latter they are described in terms of algebraic spaces. Our exposition of the Ogg-Šafarevič-Grothendieck formula (2.12) follows that in Raynaud [1]; a generalization for complexes of sheaves may be found in [SGA. 5, X]. The cohomology of a surface over $\mathbb{C}$ was analyzed, by means of Lefschetz pencils, in Lefschetz [1]. Some of his work was extended to characteristic p in Igusa [1], [2], [3], and [4]. Artin interpreted Igusa's work in terms of étale cohomology and worked out the cohomology of surfaces in characteristic p (unpublished, except for Artin [1, IV.]). A useful comparison of the classical results with those in characteristic p may be found in Mumford's appendixes in the second edition of Zariski [1]. The theory and cohomology of Lefschetz pencils for higher dimensional varieties and all characteristics is developed in detail in [SGA. 7].

CHAPTER VI

The Fundamental Theorems

A smooth complete variety of dimension d over a separably closed field behaves cohomologically as a smooth manifold of d complex or $2d$ real dimensions; for example, its étale cohomology groups vanish in dimensions greater than $2d$ and satisfy a Poincaré duality theorem. For appropriately torsion sheaves, higher direct images relative to a proper morphism commute with arbitrary base changes, and higher direct images relative to an arbitrary quasi-compact morphism commute with smooth base changes. It follows that the cohomology groups of the restriction of a finite constant sheaf to the generic geometric fiber of a smooth proper map agree with the cohomology groups of its restriction to a special geometric fiber. The higher direct images of a constructible sheaf relative to a proper map are again constructible; in particular, a constructible sheaf on a complete variety has finite cohomology groups. The cohomology groups of an affine variety vanish in dimensions greater than the dimension of the variety; as there is a Gysin sequence, this implies that the weak Lefschetz theorem holds, that is, the cohomology groups of a smooth projective variety agree with those of a smooth hyperplane section except in the middle dimensions. The Künneth formula holds for products of complete varieties, and there is a cycle map that associates cohomology classes to algebraic cycles in such a way that algebraically equivalent cycles are assigned the same class. With these results, it is possible in a purely formal way to prove a Lefschetz fixed point formula and derive the rationality and functional equation of the zeta function of a smooth projective variety over a finite field in the manner suggested by A. Weil. From a more general fixed point formula it is possible to prove the rationality of very general L-series arising from representations of Galois groups.

§1. Cohomological Dimension

Let l be a prime number. A sheaf F is *torsion* (respectively *l-torsion*) if, for all quasi-compact U, $F(U)$ is torsion (respectively l-torsion, that is, each element is killed by a power of l). The *l-cohomological dimension*

$\mathrm{cd}_l\,(\mathbf{C}/X)_E$ of a site $(\mathbf{C}/X)_E$ is the smallest integer n (or ∞) such that $H^i(X_E, F) = 0$ for all $i > n$ and all l-torsion sheaves F. We write $\mathrm{cd}_l\,(A)$ for $\mathrm{cd}_l\,(\mathrm{spec}\,A)_{\mathrm{et}}$ etc. We shall need the following theorem of Tate: if K is a field of transcendence degree n over a field k, then $\mathrm{cd}_l\,(K) \leq \mathrm{cd}_l\,(k) + n$. See Shatz [2, Thm. 28, p. 119].

THEOREM 1.1. *If X is a scheme of finite-type over a separably closed field k, then* $\mathrm{cd}_l\,(X_{\mathrm{et}}) \leq 2\dim(X)$.

Proof. The proof will be by induction on $n = \dim(X)$. The theorem is obvious for $n = 0$, and we assume it to hold for $n - 1$. Let F be an l-torsion sheaf on X. If F has support in dimension $\leq n - 1$, that is, $F = \bigcup i_* F_0$ where the $i: Z \hookrightarrow X$ are closed subschemes of dimension $\leq n - 1$, then (III3.6d) $H^i(X, F) = \bigcup H^i(Z, F_0) = 0$ for $i > 2n - 2$. Let $\{x_1, \ldots, x_r\}$ be the generic points of the irreducible components of X and $g_s: x_s \hookrightarrow X$ the inclusion maps. The kernel and cokernel of the canonical map $F \to \bigoplus g_{s*} g_s^* F$ have support in dimension $\leq n - 1$, and so the above remark reduces us to considering a sheaf of the form $g_{s*}F$. We may assume that X is irreducible and (II.3.17) reduced.

LEMMA 1.2. *The sheaf $R^j g_* F$ has support in dimension $\leq n - j$.*

Proof. Let $x \in X$, let $A = \mathcal{O}_{X,x}^{sh}$, let $A_1, \ldots, A_s$ be the quotients of A by its minimal ideals, and let K_r be the field of fractions of A_r. Then (III.1.15) the stalk of $R^j g_* F$ at x is $\bigoplus H^j(K_r, F \mid K_r)$.

SUBLEMMA 1.3. *For any $x \in X$ there exists a separably closed field k', a scheme X' of finite-type over k', and a closed point x' of X' such that* $\mathcal{O}_{X,x}^{sh} \approx \mathcal{O}_{X',x'}^{sh}$.

Proof. We may assume that X is affine. Let Y be the closure of $\{x\}$ in X. By the Noether normalization theorem there is a map $X \to \mathbb{A}_k^d$ that induces a finite map $Y \to \mathbb{A}_k^d$, where $d = \dim(Y)$. Then $k' = k(\mathbb{A}_k^d)_{\mathrm{sep}}$, $X' = X \times_{\mathbb{A}_k^d} \mathrm{spec}\,k'$, and $x' =$ an inverse image of x in X', satisfy the conditions.

Now (1.3) implies (1.2) because, if $d = \dim \overline{\{x\}}$, K_r is of transcendence degree $\leq n - d$ over k' and so $H^j(K_r, F \mid K_r) = 0$ for $j > n - d$, that is, for $d > n - j$, according to Tate's theorem.

From (1.2) and the induction hypothesis we find that $H^i(X, R^j g_* F) = 0$ for $i > 2(n - j)$, $j \neq 0$. Thus the Leray spectral sequence $H^i(X, R^j g_* F) \Rightarrow H^{i+j}(k(X), F)$ gives an isomorphism $H^i(X, g_* F) \approx H^i(k(X), F)$ for $i > 2n$, and the last group is zero.

COROLLARY 1.4. *Let X be a scheme of finite-type over a field k. For all $l \neq \mathrm{char}\,(k)$,* $\mathrm{cd}_l\,(X_{\mathrm{et}}) \leq \mathrm{cd}_l\,(k) + 2\dim(X)$.

Proof. This is immediate from (1.1) and the Hochschild-Serre spectral sequence

$$H^i(k, H^j(X', F')) \Rightarrow H^{i+j}(X, F),\ X' = X \otimes_k k_s.$$

Remark 1.5. (a) We shall show in (7.2) below that $\mathrm{cd}_l(X_{\mathrm{et}}) \le \dim(X)$ for all primes l if X is an affine scheme of finite-type over a separably closed field k.

(b) For $p = \mathrm{char}(k)$, $\mathrm{cd}_p(X_{\mathrm{et}}) \le \dim(X) + 1$ if X is a scheme of finite-type over a field k and $\mathrm{cd}_p(X_{\mathrm{et}}) \le \dim(X)$ if X is separated and k is separably closed.

In proving the first assertion, one shows that it is only necessary to consider the sheaf $F = j_!\mathbb{Z}/(p)$ where $j: U \hookrightarrow X$ is an open immersion; then one uses the Artin-Schreier sequence and the vanishing of the Zariski cohomology of coherent sheaves for $i > \dim(X)$. For the second statement one uses Chow's lemma to show that X may be assumed to be quasi-projective; then one uses the fact that a map of vector spaces $1 - \phi: V \to V$ with ϕ p-linear is étale when V is regarded as a vector group and hence is surjective when V is finite-dimensional and k is separably closed [SGA. 4, X.5].

(c) If X is a scheme of characteristic p, then $\mathrm{cd}_p(X_{\mathrm{fl}}) = \infty$ except in rare cases. For example Shatz [1] shows that $\mathrm{cd}_p(X_{\mathrm{fl}}) = \infty$ when X is the spectrum of a complete local (Noetherian) ring A, except when A is a perfect field.

However, one can show that $H^i(X_{\mathrm{fl}}, N) = 0$ for $i > \mathrm{cd}_p(X_{\mathrm{et}}) + 1$ if N is a flat p-torsion commutative group scheme over X; for over any affine subscheme of X such a group scheme can be embedded in an exact sequence $0 \to N \to G^0 \to G^1 \to 0$ with G^0 and G^1 smooth group schemes. According to (III.3.9), $R^if_*G^j = 0$ for $i > 0$, where $f: X_{\mathrm{fl}} \to X_{\mathrm{et}}$ is the obvious morphism, which implies that $R^if_*N = 0$ for $i > 1$. The assertion now follows from the Leray spectral sequence for f.

(d) The results in (b) and (c) cannot be improved; for example, if X is a curve over a finite field, one may have $H^2(X_{\mathrm{et}}, \mathbb{Z}/(p)) \neq 0$, $H^3(X_{\mathrm{fl}}, \mu_p) \neq 0$.

§2. The Proper Base Change and Finiteness Theorems

The proper base change theorem states that the higher direct images $R^i\pi_*F$ of a torsion sheaf F relative to a proper morphism $\pi: Y \to X$ commute with base change; in particular, the stalk of $R^i\pi_*F$ at $\bar{x}$ is $H^i(Y_{\bar{x}}, F|Y_{\bar{x}})$, where $Y_{\bar{x}}$ is the geometric fiber of π over $\bar{x}$. The finiteness theorem states that the higher direct images of a constructible sheaf on Y_{et} are again constructible; in particular, the cohomology groups of such a sheaf on a complete variety are finite. The following theorem contains both results.

THEOREM 2.1. *If* $\pi : Y \to X$ *is proper and* F *is a constructible sheaf on* $(\mathbf{Sch}/Y)_{\mathrm{et}}$, *then* $R^i\pi_* F$ *is a constructible sheaf on* $(\mathbf{Sch}/X)_{\mathrm{et}}$, *for* $i \geq 0$.

Before considering the proof of (2.1), we list some of its consequences.

COROLLARY 2.2. *If* $\pi : Y \to X$ *is proper, and* F *is a locally constructible torsion sheaf on* $(\mathbf{Sch}/Y)_{\mathrm{et}}$, *then* $R^i\pi_* F$ *is a locally constructible sheaf on* $(\mathbf{Sch}/X)_{\mathrm{et}}$, *for* $i \geq 0$.

Proof. (2.1) states that this is so if F is constructible. Since higher direct images commute with direct limits of sheaves, (compare (III.3.6d)) and every locally constructible torsion sheaf on $(\mathbf{Sch}/Y)_{\mathrm{et}}$ is a direct limit of constructible sheaves (V.1.10b), the corollary follows.

We next express (2.2) in terms of the small étale site. For any square

$$\begin{array}{ccc} Y & \xleftarrow{f'} & Y' \\ \downarrow{\scriptstyle \pi} & & \downarrow{\scriptstyle \pi'} \\ X & \xleftarrow{f} & X' \end{array}$$

and sheaf F on Y_{et}, there is a map $f^* R^i\pi_* F \to R^i\pi'_*(f'^* F)$ that, when the square is Cartesian, is called the *base change morphism*. Since f^* and f_* are adjoint, to define it we have to give a morphism of functors $R^i\pi_* \to f_*(R^i\pi'_*)f'^*$. This we take to be the composite of:

$$R^i\pi_* \to (R^i\pi_*)f'_*f'^* \quad (\text{induced by } id \to f'_*f'^*)$$
$$(R^i\pi_*)f'_*f'^* \to R^i(\pi f')_* f'^* = R^i(f\pi')_* f'^*$$
$$R^i(f\pi')_* f'^* \to f_*(R^i\pi'_*)f'^*.$$

A similar argument applied to

$$\begin{array}{ccc} Y_{\mathrm{et}} & \xleftarrow{g'} & (\mathbf{Sch}/Y)_{\mathrm{et}} \\ \downarrow{\scriptstyle \pi} & & \downarrow{\scriptstyle \pi^b} \\ X_{\mathrm{et}} & \xleftarrow{g} & (\mathbf{Sch}/X)_{\mathrm{et}} \end{array}$$

shows that there is a "universal" base change morphism $g^*(R^i\pi_* F) \to R^i\pi^b_*(g'^* F)$ that gives the previous morphism when restricted from $\mathbf{S}((\mathbf{Sch}/X)_{\mathrm{et}})$ to $\mathbf{S}(X'_{\mathrm{et}})$. The base change morphism can also be described in terms of Godement resolutions; see Altman-Hoobler-Kleiman [1].

COROLLARY 2.3. (Proper base change theorem). *Let*

$$\begin{array}{ccc} Y & \xleftarrow{f'} & Y' \\ \downarrow{\scriptstyle \pi} & & \downarrow{\scriptstyle \pi'} \\ X & \xleftarrow{f} & X' \end{array}$$

be Cartesian. If π is proper and F is a torsion sheaf on Y_{et}, then the base change morphism $f^*(R^i\pi_* F) \to R^i\pi'_*(f'^* F)$ is an isomorphism for all i.

Proof. One checks directly from (III.3.1) that $R^i\pi_* F \xrightarrow{\approx} g_*(R^i\pi_*^b)g'^* F$. Thus $(R^i\pi_*^b)g'^* F$ being locally constructible means precisely that the universal base change morphism $g^*(R^i\pi_* F) \to R^i\pi_*^b(g'^* F)$ is an isomorphism.

Remark 2.4. Both (2.2) and (2.3) fail for nontorsion sheaves, as is shown by the example in [SGA. 4, XII.2].

COROLLARY 2.5. *Let $\pi: Y \to X$ be proper; let $\bar{x} \to X$ be a geometric point of X, and let $Y_{\bar{x}} \to \bar{x}$ be the geometric fiber of π over $\bar{x}$. If F is a torsion sheaf on Y_{et}, then there is a canonical isomorphism $(R^i\pi_* F)_{\bar{x}} \to H^i(Y_{\bar{x}}, F \mid Y_{\bar{x}})$ for $i \geq 0$. If, moreover, all fibers of Y/X have dimension $\leq n$, then $R^i\pi_* F = 0$ for $i > 2n$; if X has characteristic p and F is p-torsion, then $R^i\pi_* F = 0$ for $i > n$.*

Proof. The first assertion is a special case of (2.3); the second follows from the first and Section 1.

COROLLARY 2.6. (Invariance of cohomology with respect to change of base field). *Let $k \subset K$ be separably closed fields and let X be a scheme proper over k. If F is a torsion sheaf on X_{et}, then $H^i(X, F) \xrightarrow{\approx} H^i(X_K, F \mid X_K)$, where $X_K = X \otimes_k K$, for $i \geq 0$.*

Proof. This is a special case of (2.2) since $\Gamma(\text{spec } k, F) \approx \Gamma(\text{spec } K, F)$ for any locally constructible sheaf F on (**Sch**/spec k).

COROLLARY 2.7. *Let S be the spectrum of a Henselian ring A, let s_0 be the closed point of S, let $\pi: X \to S$ be a proper map, and let $X_0 \to s_0$ be the closed fiber of X/S. If F is a torsion sheaf on X_{et}, then there is a canonical isomorphism $H^i(X, F) \xrightarrow{\approx} H^i(X_0, F_0)$, where $F_0 = F \mid X_0$, for $i \geq 0$.*

Proof. Let $\bar{S} = \text{spec } A^{\text{sh}}$, and let $\bar{s}_0$ be the closed point of $\bar{S}$. Then (2.5) states that $H^i(\bar{X}, F \mid \bar{X}) \xrightarrow{\approx} H^i(\bar{X}_0, F_0 \mid \bar{X}_0)$, where $\bar{X} = X \times \bar{S}$. Let $G = \text{Gal}\,(k(\bar{s}_0)/k(s_0)) = \text{Gal}\,(A^{\text{sh}}/A)$. The corollary follows from considering the Hochschild-Serre spectral sequences,

$$H^i(G, H^j(\bar{X}, F \mid \bar{X})) \Rightarrow H^{i+j}(X, F)$$
$$H^i(G, H^j(\bar{X}_0, F_0 \mid \bar{X}_0)) \Rightarrow H^{i+j}(X_0, F)$$

since we have shown that the left-hand terms are isomorphic.

The above corollaries are all consequences of the local constructibility of $R^i\pi_* F$ in the statement of (2.1).

COROLLARY 2.8. *If X is proper over a field k and if F is a constructible sheaf on X_{et}, then $H^i(X, F)$ is finite for $i \geq 0$.*

Proof. If $G = \mathrm{Gal}(k_s/k)$, then (2.1) shows that $H^i(X \otimes k_s, F)$ is a finite G-module, all i. The corollary now follows from the Hochschild-Serre spectral sequence for $X \otimes k_s/X$.

Proof of (2.1). The proof given in [SGA. 4, XII, XIII, XIV] is long and difficult. However, it may be simplified by the use of algebraic spaces and Artin's approximation theorem. Since Artin has given a reasonably elementary exposition of the simplified proof in Artin [9, VII.] we only sketch it. As Artin, we shall always assume that X is of finite-type over an algebraically closed field k. (See also [SGA. $4\frac{1}{2}$, Arcata IV].)

LEMMA 2.9. *A sheaf F on $(\mathbf{Sch}/X)_{\mathrm{et}}$ is locally constructible if and only if the following hold:*

(a) *F is locally of finite-type (that is, $F(\varinjlim A_i) = \varinjlim F(A_i)$ for any filtered direct system of $\mathcal{O}_X$-algebras (A_i));*

(b) *for any complete local $\mathcal{O}_X$-algebra $\hat{A}$ with residue field k, the canonical map $F(\hat{A}) \to F(k)$ is an isomorphism.*

Proof. See Artin [9, VII.7.2].

Remark 2.10. The conditions are clearly necessary: if F is represented by an algebraic space $\tilde{F}$, then (a) is equivalent to $\tilde{F}$ being locally of finite-type over X; if $\tilde{F}$ is étale and locally separated over X, then $\tilde{F} \times_X \operatorname{spec} \hat{A}$ is a direct sum of copies of spec $\hat{A}$, which implies (b).

The sufficiency of the conditions is a representability criterion; (2.9) essentially reduces the proof of the local constructibility of $R^i\pi_*F$ in (2.1) to the proof of (2.7) with S the spectrum of a *complete* local (not merely Henselian) ring. The passage from complete ring to the Henselian ring is one of the difficult points in the original proof that was resolved by Artin's approximation theorem.

Step 1. If F is finite and constant, π_*F is locally constructible.

This may be reduced to the assertion that for Y proper over spec $\hat{A}$, the map $Y_0 \hookrightarrow Y$ of the closed fiber into Y induces a bijection on the sets of connected components, $\pi_0(Y_0) \xrightarrow{\approx} \pi_0(Y)$ (compare II.2.18a). This follows from the Zariski connectedness theorem (compare II.3.8).

Step 2. If F is finite and constant, π_*F is constructible.

Let $Y \xrightarrow{\pi_1} Y' \xrightarrow{\pi_2} X$ be the Stein factorization of π (Hartshorne [2, III.11.5]). Since the fibers of π_1 are geometrically connected and $\pi_{1*}F$ is locally constructible, $\pi_{1*}F$ is the constant sheaf $M_{Y'}$ where $M = \Gamma(Y, F)$. Thus we may assume that $\pi = \pi_2$ is finite. There exists a surjective family $X_i \to X$ of morphisms of finite-type such that $Y \times X_i$ is a disjoint union of copies of X_i. Clearly $\pi_*(F|Y \times X_i)$ is constructible. As π_*F is locally constructible,

$$\pi_*(F|Y \times X_i) = (\pi_*F)|Y \times X_i.$$

That π_*F is constructible follows from (V.1.8d).

Step 3. If F is constructible, $\pi_* F$ is constructible.

There is a finite family of finite maps $f_i: Y_i \to Y$ and finite constant sheaves F_i on Y_i such that F is a subsheaf of $F' = \prod f_{i*}F$ (Artin [9, VII.5.1]). On repeating this construction for F'/F, we obtain an exact sequence $0 \to F \to F' \to F''$. According to Step 2, $\pi_* F'$ and $\pi_* F''$ are constructible, and (V.1.9c) shows that $\pi_* F = \ker(\pi_* F' \to \pi_* F'')$ is also.

Step 4. The theorem is true for a finite morphism $\pi: Y \to X$.

This follows from Step 3, as $R^i\pi_* F = 0$ for $i > 0$.

Step 5. If F is constructible and π is of relative dimension ≤ 1, then $R^i\pi_* F$ is locally constructible.

We check the conditions of (2.9). The isomorphism of $R^i\pi_* F(\varinjlim A_j)$ and $\varinjlim R^i\pi_* F(A_j)$ is a consequence of the commutativity of cohomology with inverse limits of schemes (III.1.16, 3.17). For (2.9b) we must show that $\phi^i: H^i(Y, F) \to H^i(Y_0, F_0)$ is an isomorphism for $i \geq 0$, where Y is proper, of relative dimension one, over a complete local ring $\hat{A}$ and Y_0 is the closed fiber of $Y/\operatorname{spec} \hat{A}$. For $i > 0$, $H^i(Y, F)$ is effaceable in the category of constructible sheaves on Y in the sense that, for any $\gamma \in H^i(Y, F)$, there is an injection $F \to F_\gamma$ with F_γ constructible such that $\gamma \mapsto 0$ under $H^i(Y, F) \to H^i(Y, F_\gamma)$ (Artin [9, VII.4.6]). From this it follows that ϕ^i only has to be shown to be surjective when $i > 0$. We know that ϕ^0 is an isomorphism (Step 3) and that $H^i(Y_0, F_0) = 0$ for $i > 2$ (1.1), and so it only remains to show that ϕ^1 and ϕ^2 are surjective. With the method of Step 3, one reduces this to the case $F = \mathbb{Z}/(l)$. We know that

$$H^1(Y_0, \mathbb{Z}/(l)) = \pi^1(Y_0, \mathbb{Z}/(l))$$

and so, for the surjectivity of ϕ^1, we have to show that any Galois covering of Y_0 with Galois group $\mathbb{Z}/(l)$ lifts to a covering of Y. This follows from Grothendieck's existence theorem (Artin [9, VII.11.7]). If $l = \operatorname{char}(k)$, then ϕ^2 is surjective because $H^2(Y_0, \mathbb{Z}/(l))$ is then zero. Otherwise

$$H^2(Y_0, \mathbb{Z}/(l)) \approx \operatorname{Pic}(Y_0)/l \operatorname{Pic}(Y_0),$$

and we must lift an invertible sheaf from Y_0 to Y. That this can be done again follows from Grothendieck's existence theorem (Artin [9, VII.11.10]).

Step 6. The theorem is true if π is of relative dimension ≤ 1.

One shows that it is only necessary to consider the following situation: X is reduced and $\pi: Y \to X$ is smooth with geometrically connected fibers of dimension one. Then $R^1\pi_*\mathcal{O}_Y$ is locally free of finite rank on X, and its formation commutes with base change. Thus the genus g_x of the fiber Y_x is a locally constant function on X. Again it is only necessary to consider the case that $F = \mathbb{Z}/(l)$. If $l \neq \operatorname{char}(k)$, then $H^i(Y_x, \mathbb{Z}/(l))$ has order $l, l^{2g_x}, l, 1$ respectively for $i = 0, 1, 2, i \geq 3$. Thus $R^i\pi_* F$ is constructible by Step 5 and (V.1.8d). If $l = \operatorname{char}(k)$, then $R^0\pi_*\mathbb{Z}/(l) = \mathbb{Z}/(l)$ and $R^i\pi_*\mathbb{Z}/(l) = 0$ for $i > 1$. Since $R^1\pi_*(\mathbb{Z}/(l))$ is the kernel of an étale map $R^1\pi_*\mathcal{O}_Y \to R^1\pi_*\mathcal{O}_Y$ (where $R^1\pi_*\mathcal{O}_Y$ is regarded as a vector group on X), it is represented by an étale group scheme over X. In particular, it is constructible. (For more details, see Artin [9, VII. Section XII].)

The final step, the reduction to the case of relative dimension ≤ 1, may be found in Artin [9, VII. Section XIII].

§3. Higher Direct Images with Compact Support

Given a complete variety X, an open subscheme $j: U \hookrightarrow X$, and a torsion sheaf F on U_{et}, we defined (III.1.29) the cohomology groups of F with compact support $H^i_c(U, F)$ to be the groups $H^i(X, j_! F)$. Using the proper base change theorem, one may show that these groups are independent of the complete variety in which U is embedded. More generally it is possible to define a good notion of *higher direct images with compact support* for any torsion sheaf F relative to any *compactifiable* morphism $\pi: U \to S$, that is, a morphism that may be embedded in a commutative diagram

$$\begin{array}{ccc} U & \xhookrightarrow{\;j\;} & X \\ & {\scriptstyle\pi}\searrow \quad \swarrow{\scriptstyle\bar{\pi}} & \\ & S & \end{array}$$

with j an open immersion and $\bar{\pi}$ proper; we simply set $R^i_c\pi_* F = R^i\bar{\pi}(j_! F)$.

PROPOSITION 3.1. *The sheaves $R^i_c\pi_* F$ are independent of the embedding $j: U \hookrightarrow X$.*

Proof. Suppose that we have two such S-embeddings, $j: U \hookrightarrow X$ and $j': U \hookrightarrow X'$. After replacing X' by the closure of the image of U in $X \times_S X'$, we may assume that there exists a proper S-morphism $g: X' \to X$ such that $gj' = j$ and $\bar{\pi}' = \bar{\pi}g$. Thus there is a spectral sequence

$$(R^i\bar{\pi}_*)(R^j g_*)(j'_! F) \Rightarrow (R^{i+j}\bar{\pi}'_*)(j'_! F).$$

According to the proper base change theorem (especially (2.5)), the stalks of $R^j g_*(j'_! F)$ may be computed on the fibers of X'/X. Since the fiber X'_x consists of one point if $x \in U$ and $j'_! F | X'_x = 0$ otherwise, it is clear that $g_* j'_! F \approx j_! F$ and $R^i g_*(j'_! F) = 0$ for $i > 0$.

THEOREM 3.2. (a) *The $R_c^i\pi_*$ form a δ-functor.*

(b) *Assume that S is quasi-compact; if π is quasi-finite and separated, then $R_c^i\pi_* = 0$ for $i > 0$; if, moreover, π is étale, then $R_c^0\pi_* = \pi_! =$ "extension by zero"* (II.3.18).

(c) *If in $U' \xrightarrow{\pi'} U \xrightarrow{\pi} S$, π and $\pi \circ \pi'$ are compactifiable, then there is a spectral sequence*

$$(R_c^i\pi_*)(R_c^j\pi'_*)F \Rightarrow R_c^{i+j}(\pi\pi')_*F$$

for all torsion sheaves F on U'.

(d) *If F is constructible, so also is $R_c^i\pi_*F$.*

(e) *The $R_c^i\pi_*F$ commute with base change.*

Proof. (a) The functor $j_!$ is exact.

(b) According to Zariski's main theorem (I.1.8), we may take the compactification X to be finite over S, and then the first assertion follows from the fact that $\bar{\pi}_*$ is exact (II.3.6). The second assertion is proved in (V.1.13).

(c) Note: (i) If

$$\begin{array}{ccc} U' & \overset{j'}{\hookrightarrow} & X' \\ \downarrow{\scriptstyle \pi} & & \downarrow{\scriptstyle \bar{\pi}} \\ U & \overset{j}{\hookrightarrow} & X \end{array}$$

is Cartesian, π and $\bar{\pi}$ are proper, and j and j' are open immersions, then the proper base change theorem shows that $(R^i\bar{\pi}_*)(j'_!F) = j_!(R^i\pi_*F)$ for any torsion sheaf F on U'.

(ii) If in (c) π' is a closed immersion, we may construct a diagram in which X is a compactification of π and X' is the closure of $j\pi'(U')$ in X:

$$\begin{array}{ccc} U' & \overset{j'}{\hookrightarrow} & X' \\ \downarrow{\scriptstyle \pi'} & & \downarrow{\scriptstyle \bar{\pi}'} \\ U & \overset{j}{\hookrightarrow} & X \\ \downarrow{\scriptstyle \pi} & \swarrow{\scriptstyle \bar{\pi}} & \\ S & & \end{array}$$

Then

$$\begin{aligned} (R_c^i(\pi\pi')_*)(F) &= (R^i(\bar{\pi}\bar{\pi}')_*)(j'_!F) && \text{(definition)} \\ &= (R^i\bar{\pi}_*)(\bar{\pi}'_*j'_!F) && \text{(exactness of } \bar{\pi}'_*) \\ &= (R^i\bar{\pi}_*)(j_!\pi'_*F) && \text{(i)} \\ &= (R_c^i\pi_*)(\pi'_*F) && \text{(definition).} \end{aligned}$$

(iii) In the general case we may construct a diagram,

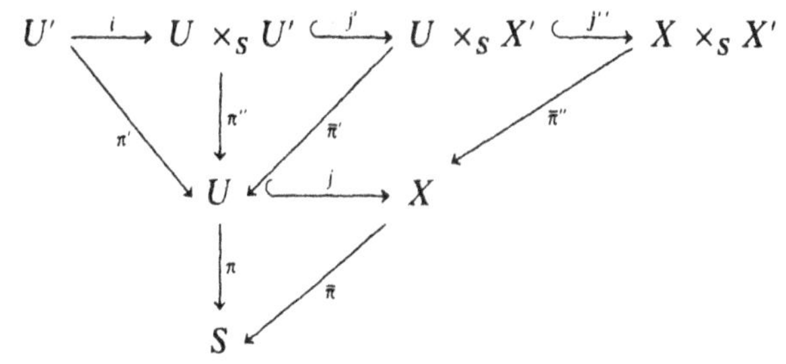

in which X and X' are compactifications of U and U' over S. Thus j, j', j'' are open immersions, $\bar{\pi}$, $\bar{\pi}'$, $\bar{\pi}''$ are proper, and i is a closed immersion. Then

$$\begin{aligned}(R^i_c\pi_*)(R^j_c\pi'_*)(F) &= (R^i_c\pi_*)(R^j_c\pi''_*)(i_*F) && \text{(ii)}\\ &= (R^i\bar{\pi}_*)j_! \circ (R^j\bar{\pi}'_*)(j'_!i_*F) && \text{(definition)}\\ &= (R^i\bar{\pi}_*)(R^j\bar{\pi}''_*)(j''j')_!i_*F && \text{(i)}\\ &\Rightarrow (R^{i+j}(\bar{\pi}\bar{\pi}'')_*)(j''j')_!i_*F\\ &= R^{i+j}_c(\pi\pi')_*F && (\text{(iii), (definition)}).\end{aligned}$$

(d) Clearly $j_!F$ is constructible, and so we may apply (2.1).

(e) This follows from the proper base change theorem (2.3).

Remarks 3.3. (a) One usually writes $R^i\pi_!$ for $R^i_c\pi_*$. However, $R^i\pi_!$ is *not* the ith right derived functor of $R^0\pi_!$ (see the discussion preceeding (III.1.29)).

(b) All statements of (III.1.29) generalize (together with their proofs).

(c) For a scheme U over a separably closed field we write $H^i_c(U, F)$ for $R^i_c\pi_*F$. If F is constructible, then $H^i_c(U, F)$ is finite.

(d) If U is separated of finite-type over $\mathbb{C}$ and is compactifiable (in the above sense), then $H^i_c(U, F) \approx H^i_c(U(\mathbb{C}), F)$ for any torsion sheaf F on U, where the right hand term means cohomology with compact support (in the usual sense) of F on the topological space $U(\mathbb{C})$ [SGA. 4, XVII.5.3.5].

(e) In Nagata [2] it is shown that any separated morphism of finite-type between Noetherian schemes is compactifiable; in particular, a variety is compactifiable if and only if it is separated.

(f) Given a morphism $\pi: Y \to X$, there do not usually exist restriction maps $H^i_c(X, F) \to H^i_c(Y, F|Y)$ unless π is proper.

Exercise 3.4. Let $\pi: X' \to X$ be a proper morphism of compactifiable schemes; let F and F' be sheaves on X and X', and let φ be a map $F \to \pi_*F'$. Assume that there exist open subschemes $U \subset X$ and $U' = \pi^{-1}(U) \subset X'$ such that $\pi|U'$ is an isomorphism $U' \xrightarrow{\approx} U$ and $\varphi: F|U \xrightarrow{\approx} \pi_*F'|U$. Show that there exists an exact sequence

$$\cdots \to H_c^i(X, F) \to H_c^i(X', F') \oplus H_c^i(X - U, F) \to$$
$$H_c^i(X' - U', F') \to H_c^{i+1}(X, F) \to \cdots.$$

(Hint: compare the sequences in (III.1.30) for F', X', U' and F, X, U.)

§4. The Smooth Base Change Theorem

For any scheme X we write *char* (X) for the set of primes (or zero) that occur as char $(k(x))$ for some $x \in X$; equivalently char (X) is the image of the canonical map $X \to \operatorname{spec} \mathbb{Z}$. The torsion of a sheaf F is *prime to* char (X) if $p: F \to F$ is injective for all $p \neq 0$ in char (X).

THEOREM 4.1. (Smooth base change theorem). *Let*

$$\begin{array}{ccc} Y & \xleftarrow{\;g'\;} & Y' \\ \downarrow{\scriptstyle \pi} & & \downarrow{\scriptstyle \pi'} \\ X & \xleftarrow{\;g\;} & X' \end{array}$$

be a Cartesian square with g smooth and π quasi-compact. If F is a torsion sheaf on Y_{et} whose torsion is prime to char (X), *then the base change morphism $g^*(R^i\pi_*F) \to R^i\pi'_*(g'^*F)$ is an isomorphism for all i.*

Before considering the proof of (4.1), we list some of its consequences.

COROLLARY 4.2. (Smooth specialization of cohomology groups). *Let $\pi: Y \to X$ be proper and smooth, and let F be a constructible locally constant sheaf on Y_{et} whose torsion is prime to* char (X). *Then for any $i \geq 0$, $R^i\pi_*F$ is constructible and locally constant; in particular, if X is connected, the groups $H^i(Y_{\bar{x}}, F|Y_{\bar{x}})$ $(=(R^i\pi_*F)_{\bar{x}})$ are isomorphic for all geometric points $\bar{x}$ of X.*

Proof. Let $x_1 \in X$ and let x_0 specialize x_1, that is, $x_0 \in \overline{\{x_1\}}$; choose geometric points $\bar{x}_1$ and $\bar{x}_0$. Any étale neighborhood of $\bar{x}_0$ may be regarded also as an étale neighborhood of $\bar{x}_1$, but possibly in several different ways. However, if we choose a map $\mathcal{O}_{X,\bar{x}_0} \to \mathcal{O}_{X,\bar{x}_1}$, then any étale connected neighborhood of $\bar{x}_0$ will have a unique structure as an étale neighborhood of $\bar{x}_1$ that is compatible with this map. From this we obtain a map $(R^i\pi_*F)_{\bar{x}_0} \to (R^i\pi_*F)_{\bar{x}_1}$ since the direct system whose limit is the first group may be regarded as subsystem of the second direct system.

If $R^i\pi_*F$ is locally constant, both direct systems become constant, and the map will be an isomorphism. Conversely, since we know $R^i\pi_*F$ to be constructible (2.1), if the maps $(R^i\pi_*F)_{\bar{x}_0} \to (R^i\pi_*F)_{\bar{x}_1}$ are isomorphisms for all pairs x_0, x_1, then $R^i\pi_*F$ is locally constant (V.1.10).

Fix a pair x_0, x_1. There exists a strictly local, normal ring A with separably closed field of fractions, and a morphism spec $A \to X$ that sends the generic point of spec A to x_1 and the closed point to x_0; for example, consider

$$\operatorname{spec} A \to \operatorname{spec} A_1 \to \operatorname{spec} A_2 \to X$$

with $A_2 = \mathcal{O}_{X,\bar{x}_0}$, $A_1 = A_2/\mathfrak{p}$ where $\mathfrak{p}$ is the point of spec A_2 mapping to x_1, and A is the integral closure of A_1 in the separable closure of its field of fractions. Since π is proper, $R^i\pi_* F$ commutes with base change, and we may replace X by spec A (and henceforth write $X = \operatorname{spec} A$). If we write Y_1 for the generic fiber of Y/X, then $(R^i\pi_*F)_{\bar{x}_1} = H^i(Y_1, F)$ (according to (2.5)). As $(R^i\pi_*F)_{\bar{x}_0} = H^i(Y, F)$, we only have to show that the restriction map $H^i(Y, F) \to H^i(Y_1, F)$ is an isomorphism.

Consider the diagram:

$$\begin{array}{ccc} Y & \xleftarrow{g'} & Y_1 \\ \downarrow{\scriptstyle \pi} & & \downarrow{\scriptstyle \pi_1} \\ X & \xleftarrow{g} & x_1 \end{array}$$

If we can show that $g'_*(F|Y_1) \approx F$ and $R^i g'_*(F|Y_1) = 0$ for $i > 0$, then the Leray spectral sequence for g' will show that the restriction maps $H^i(Y, F) \to H^i(Y_1, F|Y_1)$ are isomorphisms. Since these conditions are local on Y_{et}, we may assume that F is constant, say $F = M_Y$ (and lose the properness of π!). Let U be an open affine in X and consider

$$\begin{array}{ccc} Y & \xleftarrow{j'} & Y \times U \\ \downarrow{\scriptstyle \pi} & & \downarrow{\scriptstyle \pi'} \\ X & \xleftarrow{j} & U \end{array}$$

According to (4.1),

$$\pi^*(R^i j_* M_U) \approx R^i j'_*(\pi'^* M_U) = R^i j'_*(F|Y \times U).$$

Since higher direct images commute with certain inverse limits of schemes (III.1.16), on passing to the limit over all U, we obtain an isomorphism $\pi^*(R^i g_* M_{k(x_1)}) \approx R^i g'_*(F|Y_1)$. As X is normal, $g_* M_{k(x_1)} = M_X$ (II.3.7) and, as $k(x_1)$ is separably closed, $R^i g_* M_{k(x_1)} = 0$ for $i > 0$. Thus the left hand term in the above isomorphism is M_Y for $i = 0$ and is zero for $i > 0$. This completes the proof of (4.2).

COROLLARY 4.3. *Let $k \subset K$ be separably closed fields, and let X be a scheme over k. For any torsion sheaf F on X whose torsion is prime to* char (k), $H^i(X, F) \xrightarrow{\approx} H^i(X_K, F|X_K)$, $i \geq 0$.

Proof. From (II.3.17) we see that we may replace k and K by their algebraic closures. Now $K = \varinjlim A_i$ where the A_i are smooth k-algebras, and we may apply (4.1).

Remark 4.4. The last corollary should be compared with (2.6). Note that, for $p = \text{char}(k)$, $H^1(\mathbb{A}_k^1, \mathbb{Z}/(p)) = k[T]/(F - 1)k[T]$ (III.4.12) is *not* invariant under change of base field, for example, $aT \notin (F - 1)k[T]$ for any $a \in k$. It follows that (4.3), and hence (4.1), is false without the condition that the torsion of F is prime to char (X).

COROLLARY 4.5. *Let $\pi: X \to S$ be a morphism of finite-type of schemes that are locally of finite-type over a field k, and let F be a constructible sheaf on X. Then $R^i\pi_*F$ is constructible if* char $(k) = 0$ *or* dim $(X) \leq 2$ *and the torsion of F is prime to* char (k).

Proof. [SGA. 4, XVI.5.1]. (The conditions ensure that singularities may be resolved. See also (5.5) and (5.7) below.)

Before proving (4.1), we give an outline. First one defines the notion of a (locally) acyclic morphism $g: Y \to X$; roughly, this means that a sheaf F on X and its restriction g^*F to Y have the same cohomology (locally). Since we know intuitively that any smooth morphism is locally a projection $X \times$ (open ball) $\to X$ and an open ball is contractible, a smooth morphism should be locally acyclic. In the two main steps in the proof of (4.1), we show (4.10) that for any universally locally acyclic morphism $g: Y \to X$ the base change morphism (as in (4.1)) is an isomorphism and (4.15) that any smooth morphism is universally locally acyclic. Each step requires preliminaries; before proving (4.10), we show (4.8) that local acyclicity is equivalent to the base change morphism being an isomorphism for direct images with respect to quasi-finite morphisms; before proving (4.15), we develop a criterion (4.12) that says that acyclicity is implied by a local condition (local acyclicity) plus a global condition on the geometric fibers (acyclicity).

A morphism $g: Y \to X$ is *n-acyclic*, $n \geq -1$, if for all X'/X étale of finite-type and all torsion sheaves F on X' whose torsion is prime to char (X), $H^i(X', F) \to H^i(Y', F \mid Y')$ is bijective for $0 \leq 1 \leq n$ and injective for $i = n + 1$ (where $Y' = Y \times_X X'$). From the Leray spectral sequence for $g' = g_{(X')}: Y' \to X'$ and (III.1.13, 2.11a) one sees easily that this is equivalent to: $F \xrightarrow{\approx} g'_*g'^*F$ and $R^ig'_*(g'^*F) = 0$, $1 \leq i \leq n$, if $n \geq 0$, or $F \to g'_*g'^*F$ is injective if $n = -1$. The morphism g is *acyclic* if it is n-acyclic for all n and is *universally (n-)acyclic* if $g_{(X')}$ is (n-)acyclic for all X-schemes X'. A morphism $g: Y \to X$ is (*universally*) *locally* (*n-*)*acyclic* if for every geometric point $\bar{y}$ of Y, the map

$$\tilde{g}: \tilde{Y} = \text{spec } \mathcal{O}_{Y,\bar{y}} \to \tilde{X} = \text{spec } \mathcal{O}_{X,\bar{y}}$$

is (universally) (n-)acyclic. (The case $n = -1$ of the definitions is needed to start some induction arguments. Since any surjective map is obviously (-1)-acyclic and any quasi-compact (-1)-acyclic map is surjective, the reader may instead assume the maps to be surjective.)

LEMMA 4.6. *A morphism $g: Y \to X$ is n-acyclic if and only if for any Cartesian square,*

$$\begin{array}{ccc} Y & \xleftarrow{f'} & Y' \\ \downarrow{\scriptstyle g} & & \downarrow{\scriptstyle g'} \\ X & \xleftarrow{f} & X' \end{array}$$

with f quasi-finite, and any torsion sheaf F on X' whose torsion is prime to char(X), *$H^i(X', F) \to H^i(Y', F|Y')$ is bijective for $i \leq n$ and injective for $i = n + 1$.*

Proof. Since any étale morphism of finite-type is quasi-finite, the condition is clearly sufficient. For the converse, we must show that $F \xrightarrow{\approx} g'_*g'^*F$ and $(R^i g'_*)g'^*F = 0$ for $1 \leq i \leq n$. (We leave to the reader the modifications required for the case $n = -1$.) Since this is local for the étale topology on X', the next result allows us to assume that f is finite.

SUBLEMMA 4.7. *If $f: X' \to X$ is quasi-finite, then there is a family of commutative squares,*

$$\begin{array}{ccc} X_i & \xleftarrow{f_i} & X'_i \\ \downarrow{\scriptstyle h_i} & & \downarrow{\scriptstyle h'_i} \\ X & \xleftarrow{f} & X' \end{array}$$

in which the f_i are finite, the h_i are étale, and $(h_i: X'_i \to X')$ is a covering of X'_{et}; that is, f is locally, for the étale topology on X', finite.

Proof. Fix an $x' \in X'$; let $x = f(x')$; let $\tilde{X} = \operatorname{spec}(\mathcal{O}^h_{X,x})$, and let $\tilde{X}' = \tilde{X} \times_X X'$. Then (I.4.2c) $\tilde{X}'$ contains an open subscheme $\tilde{U}$ whose image in X' contains x' and that is finite over $\tilde{X}$. As $\tilde{X} = \varprojlim U_\nu$, with the U_ν étale over X, one sees easily [EGA. IV.9] that for some ν, there is an open subscheme $U_{x'}$ of $U_\nu \times X'$ such that $U_{x'}$ is finite over U_ν and $U = X \times_{U_\nu} U_{x'}$. On varying x' in

$$\begin{array}{ccc} U_\nu & \longleftarrow & U_{x'} \\ \downarrow & & \downarrow \\ X & \longleftarrow & X' \end{array}$$

we obtain the required family. (Compare [EGA. IV.18.12.1].)

We now prove (4.6) with f finite. According to the proper base change theorem (2.3), $g^*f_*F \xrightarrow{\approx} f'_*g'^*F$, and so

$$\begin{aligned} H^0(X', F) = H^0(X, f_*F) \to H^0(Y, g^*f_*F) &\approx H^0(Y, f'_*g'^*F) \\ &= H^0(Y', g'^*F) \end{aligned}$$

is bijective if g is 0-acyclic. If $n \geq 1$, then from the exactness of f_* and f'_* (II.3.6), $f_*(R^ig'_*)g'^*F = R^i(fg')_*g'^*F = R^i(gf')_*g'^*F = (R^ig_*)f'_*g'^*F \approx (R^ig_*)g^*f_*F$, which is zero for $1 \leq i \leq n$. It follows that $(R^ig'_*)g'^*F = 0$ for $1 \leq i \leq n$.

LEMMA 4.8. *A morphism $g: Y \to X$ is locally n-acyclic if and only if for any diagram of Cartesian squares,*

$$\begin{array}{ccccc} Y & \xleftarrow{j'} & Y' & \xleftarrow{f'} & Y'' \\ \downarrow{\scriptstyle g} & & \downarrow{\scriptstyle g'} & & \downarrow{\scriptstyle g''} \\ X & \xleftarrow{j} & X' & \xleftarrow{f} & X'' \end{array}$$

with j étale of finite-type and f quasi-finite, and any torsion sheaf F on X'' whose torsion is prime to char (X), *the base change morphism*

$$g'^*(R^if_*F) \to R^if'_*(g''^*F)$$

is bijective for $i \leq n$ and injective for $i = n + 1$.

Proof. $\Rightarrow$: It suffices to show that the maps on stalks

$$(g'^*(R^if_*F))_{\bar{y}'} \to (R^if'_*(g''^*F))_{\bar{y}'}$$

are injective or bijective for the appropriate i and all geometric points $\bar{y}'$ of Y'. Let $x' = g'(y')$ and $\bar{x}' = \bar{y}'$, and consider the diagram,

$$\begin{array}{ccc} \tilde{Y}' & \xleftarrow{\tilde{f}'} & \tilde{Y}'' \\ \downarrow{\scriptstyle \tilde{g}'} & & \downarrow{\scriptstyle \tilde{g}''} \\ \tilde{X}' & \xleftarrow{\tilde{f}} & \tilde{X}'' \end{array}$$

in which $\tilde{X}' = \operatorname{spec}(\mathcal{O}_{X',\bar{x}'})$, $\tilde{Y}' = \operatorname{spec}(\mathcal{O}_{Y',\bar{y}'})$, $\tilde{X}'' = X'' \times_{X'} \tilde{X}'$, and $\tilde{Y}'' = Y'' \times_{Y'} \tilde{Y}' = \tilde{Y}' \times_{\tilde{X}'} \tilde{X}''$. Then

$$(g'^*(R^if_*F))_{\bar{y}'} = (R^if_*F)_{\bar{x}'} = H^i(\tilde{X}'', F|\tilde{X}'')$$

and

$$(R^if'_*(g''^*F))_{\bar{y}'} = H^i(\tilde{Y}'', F|\tilde{Y}'') \qquad \text{(III.1.15).}$$

Since $\tilde{g}'$ is n-acyclic and $\tilde{f}$ is quasi-finite, the map

$$H^i(\tilde{X}'', F|\tilde{X}'') \to H^i(\tilde{Y}'', F|\tilde{Y}'')$$

is bijective for $i \leq n$ and injective for $i = n + 1$ according to (4.6).

$\Leftarrow$: Clearly we may assume X to be affine. Let $\tilde{g}: \tilde{Y} \to \tilde{X}$ be the map arising from a geometric point $\bar{y}$ of Y. We must show that for $\tilde{X}'/\tilde{X}$ étale of finite-type, and F an appropriately torsion sheaf on $\tilde{X}'$, $H^i(\tilde{X}', F) \to H^i(\tilde{Y}', F)$ is bijective for $i \leq n$ and injective for $i = n + 1$. Write $\tilde{X} = \varprojlim X_\nu$ with the X_ν affine and étale over X. For some ν, say $\nu = 0$, there will exist an étale map $X'_0 \to X_0$ such that $\tilde{X}' \to \tilde{X} = (X'_0 \to X_0) \times_{X_0} \tilde{X}$. Then $\tilde{X}' = \varprojlim X'_\nu$ where $X'_\nu = X_\nu \times_{X_0} X'_0$ ($\nu \geq 0$), and we write $h_\nu: \tilde{X}' \to \tilde{X}'_\nu$ for the canonical map. We know F is a direct limit of constructible sheaves and any constructible sheaf on $\tilde{X}'$ is obviously the inverse image of a sheaf on X'_ν for some ν. Since cohomology commutes with direct limits of sheaves (III.3.6d), we may assume that F is of the form $h_0^* F_0$ for an F_0 on X'_0. On applying the hypothesis to the diagram

$$\begin{array}{ccccc}
Y & \longleftarrow & Y_0 & \xleftarrow{f'} & Y'_0 \\
\downarrow{\scriptstyle g} & & \downarrow{\scriptstyle g_0} & & \downarrow{\scriptstyle g'_0} \\
X & \longleftarrow & X_0 & \xleftarrow{f} & X'_0,
\end{array}$$

we find that $g_0^*(R^i f_* F_0) \to R^i f'_*(g_0'^* F_0)$ is bijective for $i \leq n$ and injective for $i = n + 1$. But $\bar{y}$ can be regarded (in a canonical way) as a geometric point of Y_0 and of X_0, and

$$g_0^*(R^i f_* F_0)_{\bar{y}} = (R^i f_* F_0)_{\bar{y}} = H^i(\tilde{X}', F)$$

whereas $(R^i f'_*(g_0'^* F_0))_{\bar{y}} = H^i(\tilde{Y}', F)$, which completes the proof.

COROLLARY 4.9. *A morphism $g: Y \to X$ that is locally of finite-type is locally n-acyclic if $\tilde{g}: \tilde{Y} = \operatorname{spec} \mathcal{O}_{Y,\bar{y}} \to \tilde{X} = \operatorname{spec} \mathcal{O}_{X,\bar{y}}$ is n-acyclic for all $\bar{y}$ such that y is closed in its fiber of Y/X.*

Proof. In the proof of the necessity in (4.8) it was only necessary to consider points $\bar{y}'$ such that y' is closed in its fiber (II.2.17b). The sufficiency part of (4.8) now shows that g is locally n-acyclic.

PROPOSITION 4.10. *The conclusion of* (4.1) *holds if the condition that g is smooth is replaced by the condition that it is universally locally acyclic.*

Proof. We first prove the proposition under the assumption that π is compactifiable, for example, affine of finite-type [EGA. II.5.3.4]. In this case the diagram may be embedded in a diagram,

$$\begin{array}{ccc} Y & \xleftarrow{g'} & Y' \\ \downarrow{\scriptstyle j} & & \downarrow{\scriptstyle j'} \\ \bar{Y} & \xleftarrow{\bar{g}} & \bar{Y}' \\ \downarrow{\scriptstyle \bar{\pi}} & & \downarrow{\scriptstyle \bar{\pi}'} \\ X & \xleftarrow{g} & X' \end{array}$$

in which $\bar{\pi} j = \pi$, $\bar{\pi}' j' = \pi'$, $\bar{\pi}$ is proper, j is an open immersion, and both squares are Cartesian. Then,

$$\begin{aligned} g^* R^i \bar{\pi}_*(R^j j_* F) &\overset{\approx}{\to} (R^i \bar{\pi}'_*) \bar{g}^*(R^j j_* F) && \text{according to (2.3)} \\ &\overset{\approx}{\to} (R^i \bar{\pi}'_*)(R^j j'_*)(g'^* F) && \text{according to (4.8)} \end{aligned}$$

and the local acyclicity of $\bar{g}$. This proves the lemma in this case because

$$\begin{array}{ccc} g^* R^i \bar{\pi}_*(R^j j_* F) & \xrightarrow{\approx} & (R^i \bar{\pi}'_*)(R^j j'_*)(g'^* F) \\ \Downarrow & & \Downarrow \\ g^* R^{i+j} \pi_* F & \longrightarrow & R^{i+j} \pi'_*(g'^* F) \end{array}$$

commutes (compare [SGA. 4, XII.4.4]).

Since the proposition is local on X we may assume, in proving the general case, that X is affine. If V is an open affine of Y (or an open subscheme of an open affine), then $V = \varprojlim V_i$ where each V_i is affine and of finite-type over X (or is an open subscheme of such a scheme). Since such a V_i is compactifiable, the proposition holds for each V_i, and the commutativity of higher direct images with certain inverse limits implies that it holds for V. The proposition now follows by an easy induction argument from the next lemma.

LEMMA 4.11. *Let*

$$\begin{array}{ccc} Y & \xleftarrow{g'} & Y' \\ \downarrow{\scriptstyle \pi} & & \downarrow{\scriptstyle \pi'} \\ X & \xleftarrow{g} & X' \end{array}$$

be Cartesian; let $Y = Y_1 \cup Y_2$ *where* Y_1 *and* Y_2 *are Zariski open subsets of* Y, *let* $Y_{12} = Y_1 \cap Y_2$, *and write* π_1 *for* $\pi | Y_1$, π_{12} *for* $\pi | Y_{12}$, Y'_1 *for*

$Y_1 \times_X X'$, etc. *If the base change morphism is an isomorphism for* π_1, π_2, *and* π_{12}, *then it is an isomorphism for* π.

Proof. Starting with the Mayer-Vietoris sequence (III.2.24), we can construct an exact commutative diagram

$$\begin{array}{ccccccc}
\cdots \to g^*R^{i-1}\pi_{12*}F \mid Y_{12} & \to & g^*R^i\pi_*F & \to & g^*R^i\pi_{1*}F \mid Y_1 \oplus g^*R^i\pi_{2*}F \mid Y_2 & \to \cdots \\
\downarrow \approx & & \downarrow & & \downarrow \approx & \\
\cdots \to R^{i-1}\pi'_{12*}F \mid Y'_{12} & \to & R^i\pi'_*F \mid Y' & \to & R^i\pi'_{1*}F \mid Y'_1 \oplus R^i\pi'_{2*}F \mid Y'_2 & \to \cdots
\end{array}$$

to which a five-lemma argument may be applied.

We next prove a criterion for acyclicity.

PROPOSITION 4.12. *If* $g: Y \to X$ *is quasi-compact, locally* $(n-1)$*-acyclic, and all geometric fibers* $Y_{\bar{x}} \to \bar{x}$ *with* $k(\bar{x})$ *algebraic over* $k(x)$ *are* n*-acyclic, then* g *is* n*-acyclic.*

Proof. Since the conclusion is local on X (see the second definition of acyclic), we may assume X to be quasi-compact. For any $X' \to X$ étale of finite-type, the map $g' = g_{(X')}: Y' \to X'$ satisfies the same hypotheses as g. If $u: \bar{x} \to X'$ is a geometric point of X' with $k(\bar{x})$ algebraic over $k(x)$, then $\bar{x} = \varprojlim U_i$ with each U_i quasi-finite over X'. From (4.8) and the commutativity of higher direct images with inverse limits of schemes, we see that the base change morphism for,

$$\begin{array}{ccc}
Y' & \xleftarrow{u'} & Y'_{\bar{x}} \\
\downarrow g' & & \downarrow g'' \\
X' & \xleftarrow{u} & \bar{x},
\end{array}$$

$g'^*(R^iu_*F_0) \to R^iu'_*(g''^*F_0)$, is bijective for $i \leq n-1$ and injective for $i = n$, where F_0 is any appropriately torsion sheaf on $\bar{x}$. Since $k(\bar{x})$ is separably closed, this implies that $g'^*u_*F_0 \to u'_*g''^*F_0$ is injective $(n \geq 0)$ or that it is bijective and $R^iu'_*(g''^*F_0) = 0$ for $1 \leq i \leq n-1$ $(n > 0)$.

Consider $F = u_*F_0$. To prove the proposition for F we have to show that $F \to g'_*g'^*F$ is injective $(n = -1)$ or that $F \to g'_*g'^*F$ is bijective and $H^i(Y', F \mid Y') \approx H^i(X', F) = 0$ for $1 \leq i \leq n$ $(n \geq 0)$. Consider,

$$F \to g'_*g'^*F \xrightarrow{\phi} g'_*u'_*g''^*F_0 = u_*g''_*g''^*F_0$$

where $\phi = g'_*$(base change morphism for above diagram). The composite is $u_*(F_0 \to g''_*g''^*F_0)$, which is injective because g'' is at least (-1)-acyclic; it follows that $F \hookrightarrow g'_*g'^*F$. If $n \geq 0$, then ϕ is injective (as $g'^*u_*F_0 \hookrightarrow u'_*g''^*F_0$), and the composite is the isomorphism $u_*(F_0 \xrightarrow{\approx} g''_*g''^*F_0)$; it follows that $F \xrightarrow{\approx} g'_* g'^* F$. If $n > 0$, the Leray spectral sequence for u' shows that

$$H^i(Y', F|Y') \hookrightarrow H^i(Y'_{\bar{x}}, F_0|Y'_{\bar{x}}) = 0$$

for $1 \leq i \leq n$ as we have shown that $R^j u'_*(F_0|Y'_{\bar{x}}) = 0$ for $1 \leq j \leq n-1$.

To extend the proof to an arbitrary F we need to use two facts:

(a) any torsion sheaf on X' is a direct limit of constructible sheaves (V.1.10b);

(b) any constructible sheaf F on X' can be embedded $F \hookrightarrow \bigoplus F_i$ in a finite direct sum of sheaves F_i of the type $u_* F_0$ studied in the last paragraph (this follows easily from (V.1.8c); recall that X, and hence X', is quasi-compact).

Consider $\phi^i(F): H^i(X', F) \to H^i(Y', F)$ where F is an appropriately torsion sheaf on X'. Assume that $\phi^{i-1}(F)$ is bijective for all F and some i, $0 \leq i \leq n+1$. Using (b), one can show that $\phi^i(F)$ is injective for all constructible F; then using (a) that $\phi^i(F)$ is injective for all F. Now assume that $i \leq n$; using (b) again, one shows that $\phi^i(F)$ is bijective for F constructible, and (according to (a)) for any F. Since $\phi^{-1} = 0$ is trivially bijective for all F, this induction argument completes the proof.

COROLLARY 4.13. *A morphism $g: Y \to X$ is locally acyclic if for any geometric point $\bar{y}$ of Y with corresponding map*

$$\tilde{g}: \tilde{Y} = \operatorname{spec} \mathcal{O}_{Y,\bar{y}} \to \tilde{X} = \operatorname{spec} \mathcal{O}_{X,\bar{y}},$$

and any geometric point $\bar{x}$ of $\tilde{X}$ such that $k(\bar{x})$ is algebraic over $k(x)$, the geometric fiber $\tilde{Y}_{\bar{x}} \to \bar{x}$ is acyclic.

Proof. Assume that g is locally $(n-1)$-acyclic. (4.12) applied to $\tilde{g}$ shows that it is n-acyclic, which implies that g is locally n-acyclic. The corollary now follows by induction.

Remark 4.14. If g is locally of finite-type, then (4.9) shows that, in the above corollary, only geometric points $\bar{y}$ such that y is closed in its fiber need to be considered.

THEOREM 4.15. *Any smooth morphism $g: Y \to X$ is locally acyclic (and hence universally locally acyclic).*

Proof. Since a composite of locally acyclic morphisms is clearly locally acyclic and (I.3.24b) any smooth morphism factors locally into

$$Y \xrightarrow{g_0} \mathbb{A}^n_X \xrightarrow{g_1} \mathbb{A}^{n-1}_X \to \cdots \mathbb{A}^1_X \xrightarrow{g_n} X$$

with g_0 étale, we need only prove the theorem for each g_i. But as g_0 is obviously locally acyclic (every $\tilde{g}_0$ is an isomorphism) and each g_i for $i \geq 1$ is of the form $\mathbb{A}^1_Z \to Z$, we need only consider the case that Y is the affine line over X. Let y be a point of Y that is closed in its fiber, let

$x = g(y)$, and let $\bar{y}$ and $\bar{x}$ be geometric points over y and x; for simplicity we assume that $k(\bar{y}) = k(y)_{\text{sep}}$ and $k(\bar{x})$ is the separable closure of $k(x)$ in $k(\bar{y})$. We must show that for any geometric point $\bar{z}$ of $\tilde{X} = \operatorname{spec} \mathcal{O}_{X,\bar{x}}$ with $k(\bar{z})$ algebraic over $k(z)$, the geometric fiber $\tilde{Y}_{\bar{z}} \to \bar{z}$ of $\tilde{g}: \tilde{Y} = \operatorname{spec} \mathcal{O}_{Y,\bar{y}} \to \tilde{X}$ is acyclic (4.14). Clearly we may replace X by $\tilde{X}$, that is, we may assume that $X = \operatorname{spec} A$ with A strictly local (and $\bar{x} = x$). Then $Y = \operatorname{spec} A[T]$ and y is a closed point of the closed fiber $\mathbb{A}^1_{k(x)}$ of Y/X. We shall show that y may be taken to be the origin in $\mathbb{A}^1_{k(x)}$; note that in general $k(y)$ is a finite purely inseparable extension of $k(x)$.

LEMMA 4.16. *Let $A \to B$ be a local homomorphism of Henselian local rings such that the residue field extension $k(B)/k(A)$ is purely inseparable. For any finite local A-algebra A', $B \otimes_A A'$ is a local Henselian ring.*

Proof. Any maximal ideal of $B \otimes_A A'$ lies over the unique maximal ideal $\mathfrak{m}_B$ of B, and $(B/\mathfrak{m}_B) \otimes_B (B \otimes_A A') \approx k(B) \otimes_{k(A)} (A'/\mathfrak{m}_A)$ has only one prime ideal. Thus $B \otimes_A A'$ is local, and (I.4.3) shows it to be Henselian.

Let $A \to A'$ be a finite local homomorphism; I claim that, in the proof of the theorem, $X = \operatorname{spec} A$ may be replaced by $X' = \operatorname{spec} A'$. Let $g' = g_{(X')}: Y' \to X'$, and choose a geometric point $\bar{y}'$ of Y' lying over $\bar{y}$. The lemma shows that $\tilde{Y}' \stackrel{df}{=} \operatorname{spec} \mathcal{O}_{Y',\bar{y}'} \approx \tilde{Y} \times_X X'$. Thus the geometric fibers of $\tilde{Y}'/X'$ are the same as those of $\tilde{Y}/X$ apart possibly from a purely inseparable change of base field. According to (II.3.17), such a change of base field alters neither the étale site nor its cohomology, and the claim follows.

We may therefore assume that $\mathcal{O}_{Y,\bar{y}}$ and $\mathcal{O}_{X,\bar{x}} (= A)$ have the same residue field, that is, that $\bar{y}$ corresponds to a rational point a_0 of the closed fiber $\mathbb{A}^1_{k(\bar{x})}$ of $\tilde{Y}/X$. After making a translation $T \mapsto T - a$ with respect to a point of $Y = \mathbb{A}^1_A$ lying over a_0, we may assume that $a_0 = 0$, and so that $\tilde{Y} = \operatorname{spec} A\{T\}$ where $A\{T\}$ is the Henselization of $A[T]$ at the ideal generated by the maximal ideal of A and T. Thus it suffices to show:

LEMMA 4.17. *If A is a Henselian local ring, then any geometric fiber of* spec $A\{T\}$ *over* spec A *is acyclic.*

Proof. Note that $A\{T\} = \varinjlim B$, where each spec B is a smooth affine curve over A. Thus any geometric fiber $\bar{Z}$ of spec $A\{T\}$ over spec A is an inverse limit of smooth affine curves over a separably closed field, from which it follows that $H^i(\bar{Z}, F) = 0$, $i > 1$, for any finite group F whose torsion is prime to char(A) (V.2.4). It remains to show that $H^0(\bar{Z}, F) = F$ and $H^1(\bar{Z}, F) = 0$ for any such F. One sees easily that $A = \varinjlim A_i$ where each A_i is the Henselization of a finitely generated $\mathbb{Z}$-algebra and therefore is excellent (I.1.2). Moreover, $A\{T\} = \varinjlim A_i\{T\}$ and any geometric fiber of spec $A\{T\}$ over spec A is an inverse limit of geometric fibers of spec $A_i\{T\}$ over spec A_i. Thus in proving these statements we may assume A is excellent. The statement for H^0 follows from the next lemma.

LEMMA 4.18. *Any geometric fiber $\bar{Z}$ of* spec $A\{T\}$ *over* spec A *is nonempty and connected.*

Proof. Since spec $A\{T\}$ has the section "$T = 0$" over spec A, it is clear that $\bar{Z}$ is nonempty. We shall need to use the following corollary of (4.16): for any A' finite and local over A, $A'\{T\} \approx A' \otimes_A A\{T\}$. Using this to replace A by a quotient, we may assume that A is an integral domain and $\bar{Z}$ is a geometric fiber above the generic point of spec A. Thus we must show that for any finite separable extension K' of the field of fractions K of A, $Z_{K'} \stackrel{df}{=} \operatorname{spec} A\{T\} \otimes_A K'$ is connected. Let A' be the integral closure of A in K', and let Z' be the generic fiber of spec $A'\{T\}$ over spec A'. As A' is finite over A (I.1.2), $A'\{T\} \approx A' \otimes_A A\{T\}$, which shows $Z' = Z_{K'}$. We may replace A by A', that is, we may assume A to be normal. But then $A[T]$ is normal, which implies that $A\{T\}$ is normal (I.4.10a). Thus $A\{T\} \otimes_A K$ is normal, which certainly implies that $Z = \operatorname{spec}(A\{T\} \otimes_A K)$ is connected.

The statement for H^1 follows from:

LEMMA 4.19. *If A is strictly Henselian and excellent, and $\bar{Z}$ is a geometric fiber of* spec $A\{T\}$/spec A, *then any Galois covering of $\bar{Z}$ with finite abelian Galois group G splits provided G has order prime to the residue characteristic of A.*

Proof. For the proof of this purely algebraic result, we refer the reader to [SGA. 4, XV.3, p. 197–202]. See also [SGA. $4\frac{1}{2}$, Arcata, V.2.3].

Theorem (4.15) has an important corollary, which we shall need later.

COROLLARY 4.20. *For any scheme X, $g:\mathbb{A}^m_X \to X$ is acyclic.*

Proof. Since a composite of acyclic morphisms is obviously acyclic, we need only consider the case $m = 1$. As (4.15) shows that g is locally acyclic, (4.12) allows us to assume that X is the spectrum of a separably closed field k. Then g is obviously 0-acyclic, and to show that it is acyclic we have to show that $H^i(\mathbb{A}^1_k, F) = 0$, $i > 0$, for all torsion constant sheaves F whose torsion is prime to char (k). But $H^1(\mathbb{A}^1_k, F) = 0$ according to (I.5.2f), and $H^i(\mathbb{A}^1_k, F) = 0$ for $i \geq 2$ according to (V.2.4).

Remark 4.21. Let $S = \operatorname{spec} A$ where A is a discrete valuation ring of mixed characteristic whose residue field k and field of fractions K are both separably closed, and let $\pi: X \to S$ be a smooth proper morphism. Write X_0 and X_1 for the closed and generic fibers of π. Thus X_1 can be regarded as a lifting of the variety X_0 from characteristic p to characteristic zero. For any locally constant constructible sheaf F on X whose torsion is prime to char (k), there are isomorphisms,

$$\begin{aligned} H^i(X_0, F|X_0) &\approx H^i(X_1, F|X_1) && (4.2)\\ &\approx H^i(X_{\mathbb{C}}, F|X_{\mathbb{C}}) && (4.3),\\ &\approx H^i(X(\mathbb{C}), F|X(\mathbb{C})) && (\text{III}.3.12),\\ & && [\text{SGA. 4, XVI.4}]. \end{aligned}$$

§5. Purity

Let S be a scheme. A *smooth S-pair* (Z, X) is a closed immersion $i:Z \hookrightarrow X$ of smooth S-schemes. If we write U for $X - Z$, then there is a commutative diagram

$$\begin{array}{ccccc} Z & \xhookrightarrow{i} & X & \xhookleftarrow{j} & U \\ & {\scriptstyle h}\searrow & \downarrow{\scriptstyle f} & \swarrow{\scriptstyle g} & \\ & & S & & \end{array}$$

with i a closed immersion, j an open immersion, and f, g, and h smooth. We say (Z, X) has *codimension* c if, for all $s \in S$, the fiber Z_s has pure codimension c in X_s. For any such pair (Z, X) and any $z \in Z$, there is an open neighborhood X' of $i(z)$ and an étale morphism

$$X' \to \mathbb{A}_S^m = \operatorname{spec} \mathcal{O}_S[T_1, \ldots, T_m]$$

such that $Z \cap X'$ is the inverse image of the closed subscheme defined by the equations $T_{m-c+1} = \cdots = T_m = 0$ (compare (I.3.24), [SGA. 1, II.4.10]), that is, locally for the étale topology, any smooth pair (Z, X) of codimension c is isomorphic to a standard pair $(\mathbb{A}_S^{m-c}, \mathbb{A}_S^m)$. A morphism $\phi:(Z', X') \to (Z, X)$ of smooth S-pairs is an S-morphism $\phi: X' \to X$ such that $Z' = Z \times_X X'$.

For any sheaf F on X_{et}, we write $\underline{H}_Z^i(X, F) = (R^i i^!)F$ (compare III, paragraph preceding 1.25). Thus $\underline{H}_Z^i$ is a sheaf on Z, and there is a spectral sequence $H^i(Z, \underline{H}_Z^j(X, F)) \Rightarrow H_Z^{i+j}(X, F)$ since $i^!$ preserves injectives and $H_Z^0(X, F) = H^0(Z, i^! F)$. Note that $i_* \underline{H}_Z^i(X, F)$ is the sheaf associated with $V \mapsto H_{Z \times V}^i(V, F|V)$.

THEOREM 5.1. (Cohomological purity). *Let (Z, X) be a smooth S-pair of codimension c, and let F be a locally constant torsion sheaf on X_{et} whose torsion is prime to* char (X). *Then $\underline{H}_Z^i(X, F) = 0$ for $i \neq 2c$, and $\underline{H}_Z^{2c}(X, F)$ is locally isomorphic to i^*F on Z_{et}. Equivalently, $F \xrightarrow{\approx} j_* j^* F$, $(R^i j_*) j^* F = 0$ for $i \neq 0, 2c - 1$, and $i^*((R^{2c-1} j_*) j^* F)$ is locally isomorphic to i^*F.*

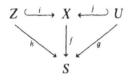

Proof. The equivalence of the two assertions follows from the exact sequence (compare (III.1.25)),

$$\cdots \to \underline{\mathrm{Ext}}^i_X(i_*i^*\mathbb{Z}, F) \to \underline{\mathrm{Ext}}^i_X(\mathbb{Z}, F) \to \underline{\mathrm{Ext}}^i_X(j_!j^*\mathbb{Z}, F) \to \cdots$$

and the isomorphisms,

$$\underline{\mathrm{Ext}}^i_X(i_*i^*\mathbb{Z}, F) \approx i_* \underline{\mathrm{Ext}}^i_Z(i^*\mathbb{Z}, i^!F) = i_*\underline{H}^i_Z(X, F),$$
$$\underline{\mathrm{Hom}}_X(\mathbb{Z}, F) = F, \underline{\mathrm{Ext}}^i_X(\mathbb{Z}, F) = 0, i \neq 0,$$
$$\underline{\mathrm{Ext}}^i_X(j_!j^*\mathbb{Z}, F) \approx j_* \underline{\mathrm{Ext}}^i_U(j^*\mathbb{Z}, j^*F) \approx R^ij_*(j^*F).$$

As the statement of the theorem is local for the étale topology, we may assume that F is the constant sheaf $\mathbb{Z}/(n)$ and that (Z, X) is a standard pair $(\mathbb{A}^{m-c}_S, \mathbb{A}^m_S)$. We first consider the case that $c = 1$ and prove the second form of the theorem. On replacing S by $\mathbb{A}^{m-1}_S$, we may assume that $X = \mathbb{A}^1_S = \mathrm{spec}\, \mathcal{O}_S[T]$ and Z is defined by the equation $T = 0$. The excision theorem (III.1.27) shows that nothing essential is changed if we take X to be $\mathbb{P}^1_S$, Z to be the "point at infinity", and $j: U \hookrightarrow X$ to be the standard embedding $\mathbb{A}^1 \hookrightarrow \mathbb{P}^1$. We use the same notations as in the definition of smooth S-pair.

Consider the Leray spectral sequence,

$$(R^if_*)(R^jj_*)\mathbb{Z}/(n) \Rightarrow (R^{i+j}g_*)\mathbb{Z}/(n).$$

According to (4.20), $g_*\mathbb{Z}/(n) = \mathbb{Z}/(n)$ and $R^{i+j}g_*(\mathbb{Z}/(n)) = 0$ for $i + j > 0$. I claim that $R^ij_*\mathbb{Z}/(n)$ has support on Z for $i > 0$ and that the canonical map $\mathbb{Z}/(n) \to j_*\mathbb{Z}/(n)(=j_*j^*\mathbb{Z}/(n))$ is an isomorphism. These statements may be checked on stalks, so choose a geometric point $\bar{x}$ of $X = \mathbb{P}^1_S$; let

$$\tilde{g}: \tilde{X} = \mathrm{spec}\, \mathcal{O}_{X,\bar{x}} \to \tilde{S} = \mathrm{spec}\, \mathcal{O}_{S,\bar{x}}$$

be the corresponding map on strictly local schemes, and let $\tilde{U} = \tilde{X} \times_X U$. Then $(R^ij_*\mathbb{Z}/(n))_{\bar{x}} = H^i(\tilde{U}, \mathbb{Z}/(n))$. If $x \in U$, then $\tilde{U} = \tilde{X}$, which proves the first part of the claim. For the second part, it suffices to show that $\tilde{U}$ is 0-acyclic over $\tilde{S}$, for then

$$(j_*\mathbb{Z}/(n))_{\bar{x}} = H^0(\tilde{U}, \mathbb{Z}/(n)) = H^0(\tilde{U}, g^*\mathbb{Z}/(n)) \approx H^0(\tilde{S}, \mathbb{Z}/(n)) = \mathbb{Z}/(n).$$

For this it suffices (4.12) to show that the geometric fibers of $\tilde{U}/\tilde{S}$ are 0-acyclic, that is, that they are nonempty and connected. But (4.18) we know this is true of the fibers of $\tilde{X}/\tilde{S}$, and the fibers of $\tilde{U}/\tilde{S}$ are open and dense in these fibers. This proves the claim.

Thus

$$(R^if_*)(R^jj_*)\mathbb{Z}/(n) = R^if_*\mathbb{Z}/(n)$$

if $j = 0$ and is zero if both $i > 0$ and $j > 0$ as f maps Z isomorphically onto S. The spectral sequence therefore gives isomorphisms,

$$f_* j_* \mathbb{Z}/(n) \approx \mathbb{Z}/(n)$$
$$f_*(R^j j_*)\mathbb{Z}/(n) \approx (R^{j+1} f_*)\mathbb{Z}/(n), \quad j > 0.$$

According to (4.2), $(R^{j+1}f_*)\mathbb{Z}/(n)$ is a locally constant sheaf, each of whose stalks is isomorphic to $H^{j+1}(\mathbb{P}^1_k, \mathbb{Z}/(n))$, where k is separably closed. This we know (V.2) to be zero for $j \neq -1, 1$ and to be isomorphic to $\mathrm{Hom}(\mu_n, \mathbb{Z}/(n))$ for $j = 1$. Thus $f_* j_* \mathbb{Z}/(n) \approx \mathbb{Z}/(n)$, $f_*(R^1 j_*)\mathbb{Z}/(n) \approx \mathbb{Z}/(n)$ locally, and $f_*(R^j j_*)\mathbb{Z}/(n) = 0$ for $j > 1$; it follows that $R^1 j_* \mathbb{Z}/(n)|Z \approx \mathbb{Z}/(n)$ (locally) and $R^j j_* \mathbb{Z}/(n) = 0$ for $j > 1$ (as f induces an isomorphism $Z \approx S$). This completes the proof in the case that $c = 1$.

The general case follows by induction since, locally for the étale topology, any smooth S-pair (Z, X) of codimension $c > 1$ can be embedded in a commutative diagram

$$\begin{array}{ccc} Z & \xrightarrow{u} & Y \\ & {\scriptstyle i}\searrow \quad \swarrow{\scriptstyle v} & \\ & X & \end{array}$$

with (Z, Y) and (Y, X) smooth S-pairs of codimension $c - 1$ and 1 respectively, and there is a spectral sequence $(R^i u^!)(R^j v^!)F \Rightarrow (R^{i+j} i^!)F$.

Remark 5.2. Assume, in the situation of the theorem, that F is killed by n (where n is prime to char(X)). For any V étale over X and any $s \in \Gamma(V, F)$, the map $1 \mapsto s: \mathbb{Z}/(n) \to F|V$ defines a map

$$H^{2c}_{Z \times V}(V, \mathbb{Z}/(n)) \to H^{2c}_{Z \times V}(V, F|V).$$

On letting V vary, we obtain a map

$$F \to \underline{\mathrm{Hom}}_X(\underline{H}^{2c}_Z(X, \mathbb{Z}/(n)), \underline{H}^{2c}_Z(X, F))$$

where we have identified $\underline{H}^{2c}_Z$ with $i_* \underline{H}^{2c}_Z$. By following through the method of proof of the theorem, one shows (easily) that this is an isomorphism. Since

$$T_{Z/X} \overset{df}{=} \underline{H}^{2c}_Z(X, \mathbb{Z}/(n))$$

is locally isomorphic to $\mathbb{Z}/(n)$, it is locally free, and the isomorphism may also be written

$$i^*F \otimes T_{Z/X} \xrightarrow{\approx} \underline{H}^{2c}_Z(X, F).$$

COROLLARY 5.3. (Gysin sequence). *Let (Z, X) be a smooth S-pair of codimension c, and let F be a locally constant sheaf on X_{et} that is killed by* *n, where n is prime to* char (X). *With the above notations,*

$$R^jf_*F \xrightarrow{\approx} R^jg_*(F|U), \qquad 0 \leq j \leq 2c-2,$$

and there is an exact sequence

$$0 \to R^{2c-1}f_*F \to R^{2c-1}g_*(F|U) \to h_*(i^*F \otimes T_{Z/X}) \to R^{2c}f_*F \to \cdots$$
$$\cdots \to R^{j-1}g_*(F|U) \to R^{j-2c}h_*(i^*F \otimes T_{Z/X}) \to R^jf_*F \to \cdots.$$

Proof. The canonical isomorphism $i^*F \otimes T_{Z/X} \xrightarrow{\approx} \underline{H}_Z^{2c}(X, F)$ induces isomorphisms

$$H^j(Z, i^*F \otimes T_{Z/X}) \xrightarrow{\approx} H^j(Z, \underline{H}_Z^{2c}(X, F)).$$

As $\underline{H}_Z^j(X, F) = 0$ for $j \neq 2c$, $H^j(Z, \underline{H}_Z^{2c}(X, F)) \approx H_Z^{j+2c}(X, F)$. On using this isomorphism to replace the groups $H_Z^j(X, F)$ in the sequence

$$\cdots \to H_Z^j(X, F) \to H^j(X, F) \to H^j(U, F) \to \cdots,$$

we obtain a sequence

$$\cdots \to H^{j-2c}(Z, i^*F \otimes T_{Z/X}) \to H^j(X, F) \to H^j(U, F) \to \cdots.$$

If we replace S by S' étale over S, and X, Z and U by $X \times S'$, $Z \times S'$, and $U \times S'$, then we obtain a similar sequence. On letting S' vary and sheafifying, we obtain the required sequence.

Remark 5.4. (a) In the situation of the corollary, (2.3) now implies that the formation of $R^jg_*(F|U)$ commutes with base change, and (4.2) implies that if F is constructible, then $R^jg_*(F|U)$ is constructible and locally constant, when f is proper.

(b) In the next section we show that, when S is the spectrum of a separably closed field, then $T_{Z/X}$ is canonically isomorphic to $(\mathbb{Z}/(n))(-c)$ and so $\underline{H}_Z^{2c}(X, F)$ is canonically isomorphic to $(i^*F)(-c)$. Thus, in this case, the Gysin sequence may be written

$$H^j(X, F) \xrightarrow{\approx} H^j(U, F|U), \qquad 0 \leq j \leq 2c-2,$$

and

$$0 \to H^{2c-1}(X, F) \to H^{2c-1}(U, F|U) \to H^0(Z, i^*F(-c)) \xrightarrow{i_*} H^{2c}(X, F) \to \cdots$$
$$\cdots \to H^{2(m-c)}(Z, i^*F(-c)) \xrightarrow{i_*} H^{2m}(X, F) \to H^{2m}(U, F) \to 0,$$

where $m = \dim(X)$. The map $i_*: H^j(Z, i^*F(-c)) \to H^{j+2c}(X, F)$ is the *Gysin map*.

COROLLARY 5.5. *Let X be a smooth variety over a field k, and let F be a finite, locally constant sheaf on X whose torsion is prime to* char (k); *then* $H^i(X, F)$ *is finite for all i.*

Proof. According to (III.2.20) and (II.3.17), we may replace k by its algebraic closure. We assume inductively that the corollary holds for varieties of dimension less than that of X.

Suppose first that F is constant and that X is an Artin neighborhood. Then there is an elementary fibration $g: X \to S$ with S also an Artin neighborhood; in particular S is smooth and $\dim(S) = \dim(X) - 1$. Using (5.4a), one sees that there is an exact sequence

$$\cdots \to H^i(S, g_*F) \to H^i(X, F) \to H^{i-1}(S, R^1g_*F) \to \cdots$$

and that g_*F and R^1g_*F are finite and locally constant. Thus this case follows from the induction assumption.

We next suppose that there is a nonempty open subscheme U of X such that the groups $H^i(U, F|U)$ are finite. There are smooth subschemes X_i of X such that $X = U \cup X_1 \cup \cdots \cup X_r$ (disjoint union), $\dim(X) > \dim(X_1) > \cdots > \dim(X_r)$, and X_j is closed in $U \cup X_1 \cup \cdots \cup X_j$. On applying (5.3) successively to the smooth pairs $(X_j, U \cup X_1 \cup \cdots \cup X_j)$, we find that the groups $H^i(X, F)$ are finite.

We now prove the corollary by induction on i. According to (III.3.13), there exists an étale map $\pi: U' \to X$ such that U' is an Artin neighborhood and $F|U'$ is constant; let $U = \pi(U')$. We may assume that $\pi: U' \to U$ is finite. The cohomology sequence of $0 \to F|U \to \pi_*(F|U') \to F' \to 0$ is

$$\cdots \to H^i(U, F|U) \to H^i(U, \pi_*(F|U')) \to H^i(U, F') \to \cdots.$$

As the groups $H^i(U, \pi_*(F|U')) \approx H^i(U', F|U')$ are finite and F' satisfies the hypotheses of the corollary, if we assume the corollary holds for $i \leq i_0$, then we may deduce it for $i_0 + 1$; this completes the proof.

Example 5.6. Let k be a separably closed field, and let $\Lambda = \mathbb{Z}/(n)$ where n is prime to char (k). As $\mathbb{P}_k^m = \mathbb{A}_k^m \cup \mathbb{P}_k^{m-1}$ (disjoint union) and $H^i(\mathbb{A}_k^m, \Lambda) = 0$ for $i \neq 0$ (4.20), the Gysin sequence shows $H^0(\mathbb{P}_k^m, \Lambda) \approx H^0(\mathbb{A}_k^m, \Lambda) \approx \Lambda$, $H^1(\mathbb{P}_k^m, \Lambda) = 0$, and $H^i(\mathbb{P}_k^{m-1}, \Lambda(-1)) \xrightarrow{\approx} H^{i+2}(\mathbb{P}_k^m, \Lambda)$ for $i \geq 0$. Since $H^i(\mathbb{P}_k^0, \Lambda) = 0$ for $i > 0$, an easy induction argument shows that

$$\begin{aligned} H^i(\mathbb{P}_k^m, \Lambda) &\approx \Lambda(-i/2),\ i \text{ even},\ 0 \leq i \leq 2m \\ &\approx 0 \text{ otherwise.} \end{aligned}$$

In (7.2) below we show that $H^i(U, \Lambda) = 0$ for $i > \dim(U)$ when U is an affine variety over k. If this is assumed, then the Gysin sequence shows that for a smooth hyperplane section X_{m-1} of a smooth projective variety X of dimension m,

$$\begin{aligned} H^{2m-3}(X_{m-1}, \Lambda(-1)) &\xrightarrow{\approx} H^{2m-1}(X, \Lambda),\ m \geq 3, \\ H^{2m-2}(X_{m-1}, \Lambda(-1)) &\xrightarrow{\approx} H^{2m}(X, \Lambda),\ m \geq 2. \end{aligned}$$

There always exists a sequence $X = X_m \supset X_{m-1} \supset \cdots \supset X_1$ of smooth projective varieties such that each X_{i-1} is a hyperplane section of X_i, according to Bertini's theorem (Hartshorne [2, III.7.9.1]). Thus

$$\Lambda = H^2(X_1, \Lambda(1)) \xrightarrow{\approx} \cdots \xrightarrow{\approx} H^{2m-2}(X_{m-1}, \Lambda(m-1)) \xrightarrow{\approx} H^{2m}(X, \Lambda(m))$$
$$H^1(X_1, \Lambda(1)) \to H^3(X_2, \Lambda(2)) \xrightarrow{\approx} \cdots \xrightarrow{\approx} H^{2m-1}(X, \Lambda(m)).$$

From the Poincaré duality theorem (11.2) it follows that $H^1(X, \Lambda(1)) \hookrightarrow H^1(X_1, \Lambda(1))$, that is, for a generic curve Y on X, the restriction map $H^1(X, \Lambda(1)) \to H^1(Y, \Lambda(1))$ is injective. The restatement of this in terms of points on the Jacobian of Y and the Picard variety of X was known classically. (Note also the similarity to (V.3.18).)

Arguments similar to the above show that if X is a smooth complete intersection of dimension m in $\mathbb{P}_k^r$, then

$$H^i(X, \Lambda) \approx \Lambda(-i/2), \ i \text{ even}, \ m < i \leq 2m,$$
$$= 0, \ i \text{ odd}, \ i > m.$$

By the Poincaré duality theorem, these statements extend to $i < m$.

Since complete intersections can be lifted to characteristic zero, we could have deduced this from the classical results over $\mathbb{C}$ (4.21). The usual formulas hold for $\beta_m = \dim H^m(X, \mathbb{Q}_l)$, X a smooth complete intersection of dimension m in $\mathbb{P}^r$ (Hirzebruch [1] and [SGA. 7, XI]). For example, if X is a hypersurface of degree δ, then

$$\beta_m = \delta^{-1}((\delta - 1)^{m+2} + (-1)^m(\delta - 1)) + \varepsilon,$$

where ε is 0 or 1 according to whether m is odd or even.

Remark 5.7. If resolution of singularities is assumed, then it is possible to obtain more complete purity results [SGA. 4, XIX]. In particular, X need not be smooth in (5.5). However, without using resolution of singularities, Deligne [SGA. 4½, Th. finitude] has proved the following theorem: let S be a regular quasi-compact scheme of dimension one or zero, and let $\Lambda = \mathbb{Z}/(n)$ where n is prime to char (S); for any morphism $\pi: Y \to X$ of S-schemes of finite-type and any constructible sheaf F of Λ-modules on Y, the sheaves $R^i\pi_*F$ are constructible.

Moreover, for any quasi-compact scheme S, morphism $\pi: Y \to X$ of S-schemes of finite-type, and constructible sheaf F of Λ-modules (Λ as before), there is an open dense subscheme U of S such that:

(a) above U, the $R^i\pi_*F$ are constructible, and zero except for a finite number of i;

(b) the formation of the $R^i\pi_*F$ is compatible with all base changes $T \to U \subset S$. For example, if S is the spectrum of a field, then $U = S$.

The proof uses the Poincaré duality theorem for a smooth morphism.

Exercise 5.8. Let $\pi: X' \to X$ be a map of smooth varieties over a separably closed field, let F and F' be locally constant sheaves on X and X' killed by n prime to char (k), and let φ be a map $F \to \pi_* F'$. Assume there exist open subvarieties $U \subset X$ and $U' = \pi^{-1}(U) \subset X'$ such that $\pi | U'$ is an isomorphism $U' \xrightarrow{\approx} U$ and $\varphi: F|U \xrightarrow{\approx} \pi_* F'|U$. Show that if $X - U$ and $X' - U'$ are smooth and have codimensions c and c', then there is an exact sequence

$$\cdots \to H^{i-2c}(Z, F(-c)) \to H^i(X, F) \oplus H^{i-2c'}(Z', F'(-c'))$$
$$\to H^i(X', F') \to \cdots.$$

§6. The Fundamental Class

In this section, we fix a separably closed field k and an integer n prime to char (k). All schemes will be smooth varieties over $S = \operatorname{spec} k$, and all sheaves will be sheaves of $\mathbb{Z}/(n)$-modules. We write $\Lambda = \mathbb{Z}/(n)$, and so $\Lambda(r) = \mu_n \otimes \cdots \otimes \mu_n$ (r copies). We use the same notations for smooth S-pairs as in Section 5. If (Z, X) is such a pair, of codimension c, then the spectral sequence $H^i(Z, \underline{H}^j_Z(X, F)) \Rightarrow H^{i+j}_Z(X, F)$ gives canonical isomorphisms

$$H^i(Z, \underline{H}^{2c}_Z(X, F)) \xrightarrow{\approx} H^{2c+i}_Z(X, F)$$

for any locally free sheaf F of finite rank on X. In particular,

$$\Gamma(Z, \underline{H}^{2c}_Z(X, F)) \xrightarrow{\approx} H^{2c}_Z(X, F).$$

In this section we define a canonical class $s_{Z/X} \in H^{2c}_Z(X, \Lambda(c))$ that generates $\underline{H}^{2c}_Z(X, \Lambda(c))$ in the sense that the map $\Lambda \to \underline{H}^{2c}_Z(X, \Lambda(c))$ that sends 1 to the restrictions of $s_{Z/X}$ is an isomorphism. (Recall that we know $\underline{H}^{2c}_Z(X, \Lambda(c))$ is locally isomorphic to $\Lambda(c)$ on Z, and $\Lambda(c)$ is noncanonically isomorphic to Λ). The element $s_{Z/X}$, or its image in $H^{2c}(X, \Lambda(c))$, will be called the *fundamental class* of Z in X.

First consider the case that $c = 1$, that is, that Z is a smooth divisor on X, and assume that Z has only one component. On interpreting everything in terms of Weil divisors (III.2.22), one obtains a canonical isomorphism between the following two exact sequences,

$$\begin{array}{ccccccc}
H^0(U, \mathbb{G}_m) & \longrightarrow & H^1_Z(X, \mathbb{G}_m) & \longrightarrow & H^1(X, \mathbb{G}_m) & \longrightarrow & H^1(U, \mathbb{G}_m) \\
\downarrow \approx & & \downarrow \approx & & \downarrow \approx & & \downarrow \approx \\
\Gamma(U, \mathcal{O}^*_U) & \xrightarrow{\operatorname{ord}_Z} & \mathbb{Z} & \longrightarrow & \operatorname{Pic}(X) & \longrightarrow & \operatorname{Pic}(U)
\end{array}$$

From the Kummer sequence, we obtain an exact sequence

$$\begin{array}{ccccc} H_Z^1(X, \mathbb{G}_m) & \xrightarrow{n} & H_Z^1(X, \mathbb{G}_m) & \longrightarrow & H_Z^2(X, \mu_n) = H_Z^2(X, \Lambda(1)), \\ \| & & \| & & \\ \mathbb{Z} & \xrightarrow{n} & \mathbb{Z} & & \end{array}$$

and we may define $s_{Z/X}$ to be the image of 1 under the boundary map. Clearly $s_{Z/X}$ has order n, and as the same is true of $s_{Z'/X'}$ $(=s_{Z/X}|X')$ for any X' étale over X, it follows that $s_{Z/X}$ generates $\underline{H}_Z^2(X, \Lambda(1))$.

THEOREM 6.1. *There is a unique function* $(Z, X) \mapsto s_{Z/X}$ *associating with any smooth S-pair* (Z, X) *of codimension* c *a fundamental class* $s_{Z/X} \in H_Z^{2c}(X, \Lambda(c))$ *satisfying the following:*

(a) $s_{Z/X}$ *has order* n;

(b) *if* $c = 1$ *and* Z *is connected, then* $s_{Z/X}$ *is the class defined above;*

(c) *if* $\phi:(Z', X') \to (Z, X)$ *is a morphism of smooth S-pairs of codimension* c, *then* $\phi^*(s_{Z/X}) = s_{Z'/X'}$;

(d) *if*

$$\begin{array}{ccc} Z & \xhookrightarrow{v} & Y \\ & {\scriptstyle i}\searrow \quad \swarrow{\scriptstyle u} & \\ & X & \end{array}$$

is a commutative diagram in which (Z, Y), (Y, X), *and* (Z, X) *are smooth pairs of codimensions* a, b, *and* c *respectively, then* $s_{Z/Y} \otimes s_{Y/X} = s_{Z/X}$ *once the identifications induced by the following canonical isomorphisms have been made:*

(i)

$$H_Z^{2a}(Y, \underline{H}_Y^{2b}(X, \Lambda(c))) \xrightarrow{\approx} H_Z^{2c}(X, \Lambda(c))$$

induced by the spectral sequence

$$H_Z^i(Y, \underline{H}_Y^j(X, \Lambda(c))) \Rightarrow H_Z^{i+j}(X, \Lambda(c));$$

(ii)

$$H_Z^{2a}(Y, \underline{H}_Y^{2b}(X, \Lambda(c))) \xrightarrow{\approx} H_Z^{2a}(Y, \Lambda(a)) \otimes H_Y^{2b}(X, \Lambda(b))$$

which uses the fact that

$$H_Z^i(Y, F) \approx H_Z^i(Y, \Lambda) \otimes H^0(Y, F)$$

for any constant, free sheaf F.

Remark 6.2. Conditions (a) and (c) imply that $s_{Z/X}$ generates $\underline{H}_Z^{2c}(X, \Lambda(c))$; this implies that, in (d), $\underline{H}_Y^{2b}(X, \Lambda(b))$ is constant, and hence $s_{Z/Y} \otimes s_{Y/X}$ makes sense as an element of $H_Z^{2c}(X, \Lambda(c))$ if it is known that the theorem holds for smooth S-pairs of codimension $<c$.

Proof of (6.1). Let (Z, X) be a pair with $c = 1$. Then Z can be written as a finite disjoint union, $Z = \bigcup Z_r$, of its irreducible components, and

$$H_Z^{2c}(X, \Lambda(c)) \approx \bigoplus H_{Z_r}^{2c}(X, \Lambda(c))$$

(by excision). According to (c), we must define

$$s_{Z/X} = \sum s_{Z_r/X}$$

where each $s_{Z_r/X}$ is uniquely determined by (b). This definition of $s_{Z/X}$ obviously satisfies condition (a) and is easily seen to satisfy (c), while (d) is empty. Assume now that $c > 1$ and that both the existence and uniqueness are known for smooth pairs of codimension $<c$. As

$$H_Z^{2c}(X, \Lambda(c)) \approx \Gamma(Z, \underline{H}_Z^{2c}(X, \Lambda(c)) = \Gamma(X, i_* \underline{H}_Z^{2c}(X, \Lambda(c)),$$

the uniqueness question is local on X; but every smooth pair (Z, X) can locally be embedded in a smooth triple (Z, Y, X) with codim $(Y, X) = 1$, and so the uniqueness follows from (d). The existence follows from the next lemma, because we can use (d) to define classes locally, and the lemma implies that they then patch.

LEMMA 6.3. *Let* (Z, Y_0, X) *and* (Z, Y_1, X) *be smooth triples, as in* (d), *with* $b = 1$. *Assume that* (6.1) *holds for smooth pairs of codimension less than* $c =$ codim (Z, X). *If we make the same identifications as in* (d), *then*

$$s_{Z/Y_0} \otimes s_{Y_0/X} = s_{Z/Y_1} \otimes s_{Y_1/X}.$$

Proof. For simplicity we assume Z to be connected. Again the problem is local on X, so we may assume that Y_0 and Y_1 are defined by single equations $f_0 = 0$ and $f_1 = 0$. Define $\bar{X} = \mathbb{A}_X^1 = \mathbf{spec}\, \mathcal{O}_X[T]$, $\bar{Z} = \mathbb{A}_Z^1 \subset \bar{X}$, and $\bar{Y}$ to be the subscheme of $\bar{X}$ defined by the equation $(f_1 - f_0)T + f_0 = 0$. There is a projection map $\bar{X} \to \mathbb{A}_k^1$, and if we write $\bar{X}_t(\approx X)$ for the fiber over $t \in \mathbb{A}_k^1$, then $\bar{Y} \cap \bar{X}_0 \approx Y_0$ and $\bar{Y} \cap \bar{X}_1 \approx Y_1$. Let C be the closed subset of $\bar{Y}$ of points where $\bar{Y}$ is not smooth (C is disjoint from Y_0 and Y_1), and replace $\bar{X}$, $\bar{Y}$, $\bar{Z}$ by $\bar{X} - C$, $\bar{Y} - C$, and $\bar{Z} - C \cap \bar{Z}$ so that $(\bar{X}, \bar{Y}, \bar{Z})$ is now a smooth triple. The maps $X \rightrightarrows \bar{X}$ that identify X with $\bar{X}_0$ and $\bar{X}_1$ respectively define morphisms of triples

$$\varepsilon_i:(X, Y_i, Z) \to (\bar{X}, \bar{Y}, \bar{Z}),\ i = 0, 1.$$

Consider the maps

$$H_Z^{2c}(X, \Lambda(c)) \xrightarrow{\varepsilon^*} H_{\bar{Z}}^{2c}(\bar{X}, \Lambda(c)) \underset{\varepsilon_1^*}{\overset{\varepsilon_0^*}{\rightrightarrows}} H_Z^{2c}(X, \Lambda(c))$$

where ε is the projection map $(\bar{X}, \bar{Z}) \to (X, Z)$.

The induction assumption shows that $\underline{H}_Z^{2c}(X, \Lambda(c))$ is generated by $s_{Z/Y_0} \otimes s_{Y_0/X}$ and hence is isomorphic to Λ. On interpreting $\underline{H}_Z^{2c}(X, \Lambda(c))$

as $i^*(R^{2c-1}j_*)j^*F$ where j is the open immersion $X - Z \hookrightarrow X$, we see that the smooth base change theorem implies that

$$\varepsilon^*: \underline{H}_Z^{2c}(X, \Lambda(c)) \xrightarrow{\approx} \underline{H}_{\bar{Z}}^{2c}(\bar{X}, \Lambda(c)).$$

The map $\bar{Z} \to Z$ obviously induces an isomorphism $\pi_0(\bar{Z}) \to \pi_0(Z)$ on the sets of connected components; thus $H^0(Z, \Lambda) \xrightarrow{\approx} H^0(\bar{Z}, \Lambda)$ and

$$\varepsilon^*: H_Z^{2c}(X, \Lambda(c)) \xrightarrow{\approx} H_{\bar{Z}}^{2c}(\bar{X}, \Lambda(c)).$$

Since $\varepsilon_0^*\varepsilon^* = id^* = \varepsilon_1^*\varepsilon^*$, this implies that $\varepsilon_0^* = \varepsilon_1^*$. But (c) of (6.1) shows that

$$\varepsilon_i^*(s_{\bar{Z}/\bar{Y}} \otimes s_{\bar{Y}/\bar{X}}) = s_{Z/Y_i} \otimes s_{Y_i/X}$$

for $i = 0, 1$, which completes the proof.

COROLLARY 6.4. *Let (Z, X) be a smooth pair of codimension c. Then $T_{Z/X} \overset{\mathrm{df}}{=} \underline{H}_Z^{2c}(X, \Lambda)$ is canonically isomorphic to $\Lambda(-c)$.*

Proof. One only has to tensor the isomorphism of the theorem, $\underline{H}_Z^{2c}(X, \Lambda(c)) \approx \Lambda$, with $\Lambda(-c)$.

PROPOSITION 6.5. *S is the spectrum of a separably closed field.*

(a) *Let $i: Z \hookrightarrow X$ be a smooth S-pair of codimension c; let $i_*: H^r(Z, \Lambda) \to H^{r+2c}(X, \Lambda(c))$ be the Gysin map, and let 1_Z denote the identity element in $H^0(Z, \Lambda) = \Lambda$. Then*

$$i_*(1_Z) = \textit{image of } s_{Z/X} \textit{ under } H_Z^{2c}(X, \Lambda(c)) \to H^{2c}(X, \Lambda(c));$$
$$i_*i^*(x) = x \cup i_*(1_Z),\ x \in H^r(X, \Lambda);$$
$$i_*(i^*(x) \cup z) = x \cup i_*z,\ x \in H^{r+2c}(X, \Lambda),\ z \in H^s(Z, \Lambda).$$

(b) *Let $Z \overset{i_1}{\hookrightarrow} Y \overset{i_2}{\hookrightarrow} X$ be a smooth S-triple; then*

$$H^r(Z, \Lambda) \xrightarrow{i_{1*}} H^{r+2a}(Y, \Lambda(a)) \xrightarrow{i_{2*}} H^{r+2a+2b}(X, \Lambda(a+b))$$

is the Gysin map for $Z \xrightarrow{i_2 i_1} X$, that is, $(i_2 i_1)_* = i_{2*}i_{1*}$.

Proof. (a) The formula for $i_*(1_Z)$ follows directly from the definitions, and the formula for $i_*i^*(x)$ follows from the third formula. This last says that the diagram

$$\begin{array}{lccccc}
\text{cup-product:} & H^r(Z, \Lambda) & \times & H^s(Z, \Lambda) & \longrightarrow & H^{r+s}(Z, \Lambda) \\
& \uparrow{\scriptstyle i^*} & & \downarrow{\scriptstyle i_*} & & \downarrow{\scriptstyle i_*} \\
\text{cup-product:} & H^r(X, \Lambda) & \times & H^{s+2c}(X, \Lambda(c)) & \longrightarrow & H^{r+s+2c}(X, \Lambda(c))
\end{array}$$

commutes.

Write Λ_Z and Λ_X for the constant sheaves on Z and X respectively defined by Λ, and let $\Lambda_Z \to \Lambda_Z^{\cdot}$ and $\Lambda_X \to \Lambda_X^{\cdot}$ be injective resolutions. Then $i^!\Lambda_X^{\cdot}$ is a complex of injectives on Z with $H^{2c}(i^!\Lambda_X^{\cdot}) \approx \Lambda_Z(-c)$ and $H^s(i^!\Lambda_X^{\cdot}) = 0$, $s \neq 2c$ (5.1). From this it is easy to construct a map of degree $-2c$, $i^!\Lambda_X^{\cdot} \to \Lambda_Z^{\cdot}(-c)$, that is a quasi-isomorphism and induces the given isomorphism $H^{2c}(i^!\Lambda_X^{\cdot}) \xrightarrow{\approx} H^0(\Lambda_Z^{\cdot}(-c)) = \Lambda(-c)$. On tensoring with $\Lambda(c)$, we get a map $i^!\Lambda_X^{\cdot}(c) \to \Lambda_Z^{\cdot}[-2c]$, which induces the two top vertical arrows in the following diagram:

$$
\begin{array}{ccccc}
\mathrm{Hom}_Z(\Lambda_Z, \Lambda_Z^{\cdot}[r]) & \times & \mathrm{Hom}_Z(\Lambda_Z^{\cdot}, \Lambda_Z^{\cdot}[s]) & \longrightarrow & \mathrm{Hom}_Z(\Lambda_Z, \Lambda_Z^{\cdot}[r+s]) \\
\| & & \uparrow & & \uparrow \\
\mathrm{Hom}_Z(\Lambda_Z, \Lambda_Z^{\cdot}[r]) & \times & \mathrm{Hom}_Z(\Lambda_Z^{\cdot}, i^!\Lambda_X^{\cdot}(c)[2c+s]) & \longrightarrow & \mathrm{Hom}_Z(\Lambda_Z, i^!\Lambda_X^{\cdot}(c)[r+s+2c]) \\
\uparrow\approx & & \wr & & \downarrow \\
\mathrm{Hom}_X(\Lambda_X, i_*\Lambda_Z^{\cdot}[r]) & \times & \mathrm{Hom}_X(i_*\Lambda_Z^{\cdot}, \Lambda_X^{\cdot}(c)[2c+s]) & \longrightarrow & \mathrm{Hom}_X(\Lambda_X, \Lambda_X^{\cdot}(c)[r+s+2c]) \\
\uparrow & & \downarrow & & \| \\
\mathrm{Hom}_X(\Lambda_X, \Lambda_X^{\cdot}[r]) & \times & \mathrm{Hom}_X(\Lambda_X^{\cdot}, \Lambda_X^{\cdot}(c)[2c+s]) & \longrightarrow & \mathrm{Hom}_X(\Lambda_X, \Lambda_X^{\cdot}(c)[r+s+2c]).
\end{array}
$$

The middle vertical arrows are defined by the adjointness of the pairs $(i_*, i^!)$ and (i^*, i_*), the isomorphism $i^*\Lambda_X \approx \Lambda_Z$, and the canonical map $\Lambda_X \to i_*i^*\Lambda_X = i_*\Lambda_Z$. The bottom maps are both induced by this last map. All pairings are composition. The diagram is commutative and defines a commutative diagram of complexes. We pass to homotopy classes of maps and take cohomology. Since $i^!\Lambda_X^{\cdot}(c) \to \Lambda_Z^{\cdot}[-2c]$ is a quasi-isomorphism, the top two vertical arrows become isomorphisms (Appendix C). Thus we may remove the middle two rows of the diagram and obtain the required diagram.

(b) This follows easily from the definitions and (6.1d).

Remark 6.6. The map $i^!\Lambda_X^{\cdot}(c) \xrightarrow{\approx} \Lambda_Z^{\cdot}[-2c]$ defined in the above proof is unique up to homotopy. We leave it as an exercise to the reader to list its functorial properties and behavior under composition.

Remark 6.7. At one technical point in the proof of the Poincaré duality theorem we need a generalization of the diagram in the proof of (6.5).

(a) Let $i: Z \hookrightarrow X$ be a smooth pair as in (6.5), and let F be a sheaf on Z. We may replace $\Lambda_Z^{\cdot}$ in the first three rows of the large diagram by an injective resolution $F^{\cdot}$ of F. On passing to homotopy classes of maps and taking cohomology, we get a diagram

$$\begin{array}{ccccc}
H^r(Z, F) & \times & \mathrm{Ext}^s_Z(F, \Lambda_Z) & \longrightarrow & H^{r+s}(Z, \Lambda_Z) \\
\approx\uparrow i^* & & \approx\downarrow i_* & & \downarrow i_* \\
H^r(X, i_*F) & \times & \mathrm{Ext}^{s+2c}_X(i_*F, \Lambda(c)) & \longrightarrow & H^{r+s+2c}(X, \Lambda(c)).
\end{array}$$

(b) Next let $i: Z \hookrightarrow X$ be a closed immersion with Z reduced but not necessarily smooth. Assume that $(R^r i^!)\Lambda(c) = 0$ for $r < 2c$ and there is a canonical map $\Lambda_Z \to (R^{2c} i^!)(\Lambda(c))$ (this is always so, see (9.1)). Then we have canonical maps

$$\Lambda_Z(-c) \to \sigma'_{\geq 2c}(i^! \Lambda^{\cdot}_X) \xleftarrow{\approx} i^! \Lambda^{\cdot}_X$$

where $\sigma'_{\geq 2c}(i^!\Lambda^{\cdot}_X) = (\cdots \to 0 \to i^!\Lambda^{2c}/\mathrm{im}(d^{2c-1}) \to i^!\Lambda^{2c+1} \to \cdots)$. Since $i^!$ is right adjoint to i_* and preserves injectives, we have maps

$$\begin{array}{ccc}
\mathrm{Ext}^r_X(i_*F, \Lambda_X) = \mathbb{E}\mathrm{xt}^r_X(i_*F, \Lambda^{\cdot}_X) & \xrightarrow{\approx} & \mathbb{E}\mathrm{xt}^r_Z(F, i^!\Lambda^{\cdot}_X) \\
 & & \downarrow\approx \\
\mathrm{Ext}^r_Z(F, \Lambda_Z(-c)) & \longrightarrow & \mathbb{E}\mathrm{xt}^r_Z(F, \sigma'_{\geq 2c}(i^!\Lambda^{\cdot}_X))
\end{array}$$

for any sheaf F on Z. This gives us a commutative diagram

$$\begin{array}{ccccc}
H^r(Z, F) & \times & \mathrm{Ext}^s_Z(F, \Lambda_Z) & \longrightarrow & H^{r+s}(Z, \Lambda_Z) \\
\approx\uparrow i^* & & \downarrow i_* & & \downarrow i_* \\
H^r(X, i_*F) & \times & \mathrm{Ext}^{s+2c}_X(i_*F, \Lambda(c)) & \longrightarrow & H^{r+s+2c}(X, \Lambda(c)).
\end{array}$$

(c) If F has support on a smooth open subvariety Z' of Z, so that $F = j_!(F|Z')$ for $j: Z' \hookrightarrow Z$, then

$$\mathrm{Ext}^r_Z(F, \Lambda_Z) = \mathrm{Ext}^r_Z(j_!(F|Z'), \Lambda_Z) \approx \mathrm{Ext}^r_{Z'}(F|Z', \Lambda_{Z'}).$$

Thus, if we write X' for $(X - Z) \cup Z'$, we have a commutative diagram

$$\begin{array}{ccccc}
H^r(Z, F) & \times & \mathrm{Ext}^s_{Z'}(F|Z', \Lambda_{Z'}) & \longrightarrow & H^{r+s}(Z, \Lambda_Z) \\
\approx\uparrow i^* & & \approx\downarrow i_* & & \downarrow i_* \\
H^r(X, i_*F) & \times & \mathrm{Ext}^{s+2c}_{X'}(i_*F|X', \Lambda(c)) & \longrightarrow & H^{r+s+2c}(X, \Lambda(c))
\end{array}$$

where the map i_* on the Exts is the same as that in (a).

§7. The Weak Lefschetz Theorem

It states:

THEOREM 7.1. (Weak Lefschetz). *Let X be a projective variety of dimension m over a separably closed field k, and let $i: Z \to X$ be the inclusion of a hyperplane section.*

(a) *For any locally constant sheaf F of $\mathbb{Z}/(n)$-modules on X with n prime to* char (k), *the map $i_*: H^i(Z, F) \to H^{i+2}(X, F(1))$ of the Gysin sequence* (5.3) *is an isomorphism for $i \geq m$ and a surjection for $i = m - 1$, provided X and Z are smooth.*

(b) *For any torsion sheaf F on X, the canonical map $H^i_Z(X, F) \to H^i(X, F)$ is an isomorphism for $i \geq m + 2$ and a surjection for $i = m + 1$.*

Proof. Since $U = X - Z$ is affine, these follow from the next theorem and the Gysin sequence (statement (a)) or (III.1.25) (statement (b)).

THEOREM 7.2. *Let X be a scheme that is affine and of finite-type over a separably closed field. Then $cd(X) \stackrel{df}{=} \sup_l \mathrm{cd}_l(X) = \dim(X)$.*

It will be necessary to prove a more general result. For any sheaf F on a scheme X, we write $d(F) = \sup\{d(x) \mid F_{\bar{x}} \neq 0\}$, where $d(x) = \dim$ (closure of $\{x\}$), so $d(\text{zero sheaf}) = -\infty$. Note that if F is constructible, then $d(F) \leq d$ if and only if F has support on a closed subscheme of dimension $\leq d$ (V.1.8c).

THEOREM 7.3(d). *Let $\pi: Y \to X$ be an affine morphism of schemes of finite-type over a field k. If F is a torsion sheaf on Y and $d(F) \leq d$, then $d(R^i\pi_*F) \leq d - i$.*

Proof. We first dispose of the case $d \leq 1$. Since every torsion sheaf is a direct limit of its constructible subsheaves (V.1.10b), we may assume F to be constructible. Then F has support on a closed subscheme of Y of dimension $\leq d$, and so, after replacing Y by this subscheme, we may assume that Y itself has dimension $d = d(F)$. Also, since passing from $\pi: Y \to X$ to $\pi \otimes 1: Y \otimes k_s \to X \otimes k_s$ does not change the strictly local ring at a geometric point of X or the stalk of $R^i\pi_*F$, we may assume that k is separably closed. If $d = 0$, then π is finite, and the theorem follows from (II.3.6). If $d = 1$, then clearly $d(\pi_*F) \leq 1$. Moreover, there is a finite closed subscheme Z of $\pi(X)$ such that π is finite over $\pi(X) - Z$ (I.1.12) and $R^1\pi_*F$ then has support on Z, which shows that $d(R^1\pi_*F) \leq 0$. According to (1.1), we know that $R^i\pi_*F = 0$ for $i > 2$, which leaves the case $i = 2$. This case follows from the next lemma.

LEMMA 7.4. *Theorem* (7.2) *holds if X is a curve.*

Proof. This is proved in (V.2.4).

In order to proceed, we have to consider a local version of (7.3(d)). If $\pi: Y \to X$ is of finite-type, then for a point $y \in Y$ with image $x \in X$ we write $\delta(y) = d(x) + \operatorname{tr} \deg_{k(x)} k(y)$, and for a sheaf F on Y we write $\delta(F) = \sup \{\delta(y) \mid F_{\bar{y}} \neq 0\}$.

STATEMENT 7.5(d). *Let $\pi: Y \to X = \operatorname{spec} A$ be affine of finite-type, where A is the strictly local ring at a geometric point of a scheme algebraic over a field. If F is a torsion sheaf on Y such that $\delta(F) \leq d$, then $H^i(Y, F) = 0$ for $i > d$.*

LEMMA 7.6. *The statement (7.5(d)) is equivalent to (7.3(d)).*

Proof. $\Rightarrow$. Let $\pi: Y \to X$ and F on Y be as in (7.3(d)), let $\bar{x}_0$ be a geometric point of X, and consider

$$\pi': Y' = Y \times_X X' \to X' = \operatorname{spec} \mathcal{O}_{X,\bar{x}_0}.$$

Then $(R^i\pi_* F)_{\bar{x}_0} = H^i(Y', F \mid Y')$ and the next lemma shows that $\delta(F \mid Y') = d(F) - d(x_0)$, from which (7.3(d)) follows.

SUBLEMMA 7.7. *With the notations of the last paragraph let $y' \in Y'$, and let y, x', and x be the images of y' in Y, X', and X respectively. Then $\delta(y') = d(y) - d(x_0)$.*

Proof.

$$d(y) = d(x) + \operatorname{tr} \deg_{k(x)} k(y) \qquad \text{Mumford [3, I.7]},$$

$$\delta(y') = d(x') + \operatorname{tr} \deg_{k(x')} k(y') \qquad \text{(definition)}$$

and $\operatorname{tr} \deg_{k(x)} k(y) = \operatorname{tr} \deg_{k(x')} k(y')$ as $k(x')/k(x)$ and $k(y')/k(y)$ are algebraic. Thus we must show that $d(x') = d(x) - d(x_0)$. We may replace X by the closure of $\{x\}$, so that $\{x'\}$ has closure X', and

$$d(x') = \dim \mathcal{O}_{X',\bar{x}} = \dim \mathcal{O}_{X,x} = \dim (X) - d(x_0) = d(x) - d(x_0).$$

Proof of 7.6. (continued). $\Leftarrow$ Let $\pi': Y' \to X' = \operatorname{spec} A$ and F' on Y' be as in (7.5(d)); if π', Y', X', and F' arise, as above, from a π, Y, X, and F as in (7.3(d)), then the above argument may be reversed. But, by assumption, $X' = \varprojlim X_\alpha$ in which each X_α is a scheme algebraic over a field, Y' is of finite-type over X', and F (if we assume it to be constructible, as we may) is represented by an algebraic space of finite-type over Y'. Thus there will exist, for some α, an appropriate triple π_α, Y_α, F_α over X_α giving rise to π', Y', F' by base change.

We next consider a statement that is a special case of (7.5(d)).

STATEMENT 7.8(d). *Let $X = \operatorname{spec} A$, where A is the strictly local ring at a geometric point of a scheme algebraic over a field. For any $f \in A$, $cd(X_f) \leq d = \dim A$, where $X_f = \operatorname{spec} A_f = \operatorname{spec} A[f^{-1}]$ and cd means $\sup_l cd_l$.*

For the remainder of the proof we assume that $d \geq 2$.

LEMMA 7.9. *The statement* (7.8(d)) *holds for all* $d \leq d_0$ *if and only if* (7.5(d)) *holds for all* $d \leq d_0$.

Proof. $\Leftarrow$. Choose $Y = X_f$.

$\Rightarrow$. We first consider the case that $Y = \mathbb{A}^1_X$ and embed it in the projective line,

$$\begin{array}{ccc} \mathbb{A}^1_X & \overset{j}{\hookrightarrow} & \mathbb{P}^1_X \\ & {\scriptstyle \pi}\searrow \quad \swarrow{\scriptstyle g} & \\ & X & \end{array}$$

Write $X_\infty = \mathbb{P}^1_X - j(\mathbb{A}^1_X)$ for the "point at infinity". Consider the spectral sequence, $H^i(\mathbb{P}^1_X, R^j j_* F) \Rightarrow H^{i+j}(\mathbb{A}^1_X, F)$ where F is a torsion sheaf on $\mathbb{A}^1$ such that $\delta(F) \leq d \leq d_0$. To complete the proof in this case we show:

(a) $H^i(\mathbb{P}^1_X, R^j j_* F) = 0$ if both $i > 0$ and $j > 0$;

(b) $H^i(\mathbb{P}^1_X, j_* F) = 0$ if $i > 2$;

(c) $H^0(\mathbb{P}^1_X, R^j j_* F) = 0$ if $j > d$.

The equality (a) follows from the fact that $R^j j_* F$ has support on X_∞ if $j > 0$, and $X_\infty \approx X$ is strictly local. Equality (b) follows from the proper base change theorem (2.3) and (1.1) and (1.5). For (c) we, as usual, may assume that F is constructible. Since $\delta(F) \leq d$, there exists a closed subscheme Z of $\mathbb{P}^1$ of dimension $\leq d$ containing the support of F. Let $\bar{x}$ be the closed point of X_∞ lying over the closed point of X. Then

$$\begin{aligned} H^0(\mathbb{P}^1, R^j j_* F) &= H^0(X_\infty, (R^j j_* F)|X_\infty), j > 0, \\ &= (R^j j_* F)_{\bar{x}} \\ &= (R^j j'_*(F|Z \cap \mathbb{A}^1))_{\bar{x}}, j' : Z \cap \mathbb{A}^1 \hookrightarrow Z, \\ &\doteq H^j(U, F|U) \end{aligned}$$

where $U = \tilde{Z} \times_{\mathbb{P}^1} \mathbb{A}^1$, $\tilde{Z} = \operatorname{spec} \mathcal{O}_{Z,\bar{x}}$. Thus (c) follows from (7.8).

For the general case, note that we may assume that $Y = \mathbb{A}^n_X$ since Y is affine over X. We assume inductively that (7.5) holds for $\mathbb{A}^{n-1}_X$ and construct a diagram

$$\begin{array}{ccc} \mathbb{A}^n_X & \xrightarrow{g} & \mathbb{A}^{n-1}_X \\ & {\scriptstyle \pi}\searrow \quad \swarrow{\scriptstyle \pi'} & \\ & X & \end{array}$$

in which g identifies $\mathbb{A}^n_X$ with the affine line over $\mathbb{A}^{n-1}_X$. Let F be a torsion sheaf on $\mathbb{A}^n_X$ with $\delta(F) \leq d \leq d_0$. From the case $n = 1$ applied to the stalks of $R^j g_* F$ we find that $\delta(R^j g_* F) \leq d - j$. In the spectral sequence

$$H^i(\mathbb{A}^{n-1}_X, R^j g_* F) \Rightarrow H^{i+j}(\mathbb{A}^n_X, F)$$

the left hand term is zero if $i > d - j$, that is, if $i + j > d$, which completes the proof.

The next lemma will complete the proof of (7.1), (7.2), (7.3), and (7.5).

LEMMA 7.10. *If (7.5(d')) holds for all $d' < d$ then (7.8(d')) holds for $d' \leq d$.*

Proof. Let X, A, f be as in the statement 7.8(d), and let $U = X_f$. We may assume that A is the strictly local ring at a closed geometric point of an algebraic scheme X_0 over a separably closed field k (1.3), so that $X = \varprojlim X_\alpha$ with each X_α affine of finite-type over k, and of dimension d. Let F be a sheaf on U, which we may assume to be constructible; we have to show that $H^i(U, F) = 0$ for $i > d$.

Clearly f and F will arise from pairs $f_\alpha \in \Gamma(X_\alpha, \mathcal{O}_{X_\alpha})$ and F_α on $U_\alpha = (X_\alpha)_{f_\alpha}$ for all sufficiently final α, and $H^i(U, F) = \varinjlim H^i(U_\alpha, F_\alpha)$. To complete the proof, we show that for each of these α, $U \to U_\alpha$ factors as $U \to U'_\alpha \to U_\alpha$ with $cd(U'_\alpha) \leq d$. We may regard f_α as a map $X_\alpha \to \mathbb{A}^1_k$. Consider the diagram:

$$\begin{array}{ccccc}
U & \longrightarrow & U'_\alpha & \longrightarrow & \eta \\
\downarrow & & \cap & & \downarrow \\
X & \longrightarrow & X'_\alpha & \longrightarrow & Y \\
& \searrow & \downarrow & & \downarrow \\
& & X_\alpha & \xrightarrow{f_\alpha} & \mathbb{A}^1_k
\end{array}$$

where Y is spec of the strictly local ring at the origin in $\mathbb{A}^1_k$, $X'_\alpha = Y \times_{\mathbb{A}^1} X_\alpha$ ($X \to X_\alpha$ factors through X'_α because X is strictly local), $\eta =$ generic point of Y, and $U'_\alpha = X'_\alpha \times_Y \eta$. Note that $Y = \{\eta\} \cup \{\text{origin of } \mathbb{A}^1_k\}$; as U_α is the set of points of X_α where $f_\alpha \neq 0$, U'_α can be identified with $U_\alpha \times_{X_\alpha} X'_\alpha \subset X'_\alpha$. Thus $U \to U_\alpha$ factors canonically into $U \to U'_\alpha \to U_\alpha$. It remains to prove that $cd(U'_\alpha) \leq d$. Clearly, $\dim(U'_\alpha) \leq d - 1$ and so, by induction (namely, (7.2) for $d - 1$), $\bar{U}'_\alpha = U'_\alpha \otimes_{k(\eta)} k(\eta)_{\text{sep}}$ has cohomological dimension $\leq d - 1$. But $k(\eta)$ has transcendence degree one over a separably closed field, and so has cohomological dimension one (Shatz [2, Thm. 28, p. 119]); now the Hochschild-Serre spectral sequence for $\bar{U}'_\alpha/U'_\alpha$ shows that $cd(U'_\alpha) \leq d$.

§8. The Künneth Formula

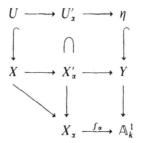

We prove the Künneth formula for cohomology groups with compact support. Both the statement and the proof of the formula become simpler if one works on the level of complexes rather than cohomology groups, and so we begin with some generalities on complexes of sheaves. Λ will denote a finite ring and also the constant sheaf it defines; unless indicated otherwise, all sheaves will be sheaves of Λ-modules and tensor products will be over Λ. Although the reader should assume Λ to be commutative, nothing is altered if it is not, except that the tensor product of two sheaves of Λ-modules will not, in general, then have a Λ-module structure.

If $F^{\cdot}$ and $G^{\cdot}$ are complexes of sheaves on some scheme X, we write $F^{\cdot} \otimes G^{\cdot}$ for the total complex of the double complex $(F^r \otimes G^s)$; more precisely

$$(F^{\cdot} \otimes G^{\cdot})^m = \sum_{r+s=m} F^r \otimes G^s$$

and

$$d^m = \sum d_F^r \otimes 1 + (-1)^r 1 \otimes d_G^s.$$

LEMMA 8.1. *If the maps $g_1, g_2 : G_1^{\cdot} \to G_2^{\cdot}$ are homotopic, then so also are*

$$1 \otimes g_1, 1 \otimes g_2 : F^{\cdot} \otimes G_1^{\cdot} \to F^{\cdot} \otimes G_2^{\cdot}$$

for any complex $F^{\cdot}$.

Proof. This is easy.

LEMMA 8.2. *Let $g : G_1^{\cdot} \to G_2^{\cdot}$ be a quasi-isomorphism, and let $F^{\cdot}$ be a bounded above complex of flat sheaves. Then $1 \otimes g : F^{\cdot} \otimes G_1^{\cdot} \to F^{\cdot} \otimes G_2^{\cdot}$ is a quasi-isomorphism if either:*

(a) *$G_1^{\cdot}$ and $G_2^{\cdot}$ are bounded above, or*

(b) *$F^{\cdot}$ is bounded below.*

Proof. It suffices to show that $F^{\cdot} \otimes C^{\cdot}$ is acyclic, where $C^{\cdot}$ is the mapping cone of g. Note that $C^{\cdot}$ is acyclic, and is bounded above in case (a). The double complex $(F^r \otimes C^s)$ gives rise to a spectral sequence

$$E_1^{r,s} = H^s(F^r \otimes C^{\cdot}) \Rightarrow H^{r+s}(F^{\cdot} \otimes C^{\cdot})$$

(Godement [1, I.4], compare (B. 3)), and the E_1 term is obviously zero. (The conditions (a) and (b) are needed only to ensure that the spectral sequence converges.)

LEMMA 8.3. *If $F^{\cdot}$ is a complex of sheaves such that $H^r(F^{\cdot}) = 0$ for $r \gg 0$, then there exists a quasi-isomorphism $P^{\cdot} \to F^{\cdot}$ with $P^{\cdot}$ a bounded above complex of flat sheaves.*

Proof. It is clear from the second proof of (III.1.1) that for any sheaf F there is a surjection $P(F) \to F$ where $P(F)$ is of the form $\bigoplus \Lambda_U$, $U \xrightarrow{\pi} X$ étale, $\Lambda_U = \pi_!(\Lambda|U)$. Obviously $P(F)$ is flat since its stalks are. Let r_0 be such that $H^r(F^{\cdot}) = 0$ for $r > r_0$; in $F^{\cdot}$ we may replace F^{r_0} with $\ker(d^{r_0})$ and F^r, for $r > r_0$, with zero. Define $P^r = 0$ for $r > r_0$ and P^{r_0} to be some $P(F^{r_0})$. Having defined $P^{r_0}, \ldots, P^{r_0-s}$, we choose $P^{r_0-(s+1)}$ to be

$$P(F^{r_0-(s+1)} \times_{F^{r_0-s}} \ker(P^{r_0-s} \to P^{r_0-s+1})).$$

It is easily checked that the complex so obtained is quasi-isomorphic to $F^{\cdot}$.

LEMMA 8.4. *Let $f: X \to S$ be a proper morphism with S quasi-compact. For any complex $F^{\cdot}$ of sheaves on X there is a quasi-isomorphism $F^{\cdot} \simeq A^{\cdot}(F^{\cdot})$ with $A^{\cdot}(F^{\cdot})$ a complex of f_*- acyclic sheaves. If $H^r(F^{\cdot}) = 0$ for $r \ll 0$, then $A^{\cdot}(F^{\cdot})$ may be taken to be a bounded below complex of injectives. If $F^{\cdot} \simeq A_1^{\cdot}$ and $F^{\cdot} \simeq A_2^{\cdot}$ with $A_1^{\cdot}$ and $A_2^{\cdot}$ complexes of f_*-acyclics, then $f_*A_1^{\cdot} \simeq f_*A_2^{\cdot}$. If $\alpha: F_1^{\cdot} \to F_2^{\cdot}$ is a map of complexes, it is possible to choose $A^{\cdot}(F_1^{\cdot})$ and $A^{\cdot}(F_2^{\cdot})$ such that there is a commutative diagram*

$$\begin{array}{ccc} F_1^{\cdot} & \xrightarrow{\alpha} & F_2^{\cdot} \\ \downarrow \wr & & \downarrow \wr \\ A^{\cdot}(F_1^{\cdot}) & \xrightarrow{\beta} & A^{\cdot}(F_2^{\cdot}). \end{array}$$

If α is a quasi-isomorphism, so also is $f_*(\beta)$.

Proof. See the last part of Appendix C, which applies because of (2.5).

Let S be a quasi-compact scheme and let $f: X \to S$ be a compactifiable morphism with compactification $X \overset{j}{\hookrightarrow} \bar{X} \overset{\bar{f}}{\to} S$. For any complex of sheaves $F^{\cdot}$ on X we write $\mathbf{R}_c f_* F^{\cdot}$ for $\bar{f}_* A^{\cdot}(j_! F^{\cdot})$ where $A^{\cdot}(j_! F^{\cdot})$ is any complex of $\bar{f}_*$-acyclic sheaves which is quasi-isomorphic to $j_! F^{\cdot}$. According to the lemma, $\mathbf{R}_c f_* F^{\cdot}$ is well-defined up to a quasi-isomorphism. We have $H^r(\mathbf{R}_c f_* F^{\cdot}) = \mathbb{R}_c^r f_* F^{\cdot}$. In the case of most interest to us, where $F^{\cdot}$ consists of a single sheaf F, $A^{\cdot}(j_! F)$ may be taken to be an injective resolution of F and $H^r(\mathbf{R}_c f_* F) = R_c^r f_* F$. If $H^r(F^{\cdot}) = 0$ for $r \gg 0$, then $H^r(\mathbf{R}_c f_* F^{\cdot}) = 0$ for $r \gg 0$ and (8.3) shows that there is a bounded above complex of flat sheaves $\bar{\mathbf{R}}_c f_* F^{\cdot}$ and a quasi-isomorphism $\bar{\mathbf{R}}_c f_* F^{\cdot} \to \mathbf{R}_c f_* F^{\cdot}$. Also, in this last situation we can, and always shall, choose $A^{\cdot}(j_! F^{\cdot})$, and hence $\mathbf{R}_c f_* F^{\cdot}$, to be bounded above.

THEOREM 8.5. (Künneth formula). *Consider the diagram*

$$\begin{array}{ccccc} & & X \times_S Y & & \\ & \swarrow^{p} & \Big\downarrow h & \searrow^{q} & \\ X & & & & Y \\ & \searrow_{f} & & \swarrow_{g} & \\ & & S & & \end{array}$$

in which S is quasi-compact and f and g are compactifiable. Let F and G be sheaves on X and Y respectively, with F flat, and write $F \boxtimes G$ for the sheaf $p^*F \otimes q^*G$ on $X \times_S Y$. Then there is a canonical quasi-isomorphism

$$\bar{\mathbf{R}}_c f_* F \otimes \mathbf{R}_c g_* G \simeq \mathbf{R}_c h_*(F \boxtimes G).$$

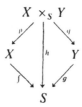

If, moreover, $R_c^r f_* F$ is flat for all r, there are canonical isomorphisms

$$\sum_{r+s=m} R_c^r f_* F \otimes R_c^s g_* G \xrightarrow{\approx} R_c^m h_*(F \boxtimes G).$$

Our first lemma shows that the second statement in the theorem follows from the first.

LEMMA 8.6. *Let $F^\cdot$ and $G^\cdot$ be complexes of sheaves and assume that $F^\cdot$ is flat. Then there is a spectral sequence*

$$E_2^{r,s} = \sum_{i+j=s} \mathrm{Tor}_{-r}^{\Lambda}(H^i(F^\cdot), H^j(G^\cdot)) \Rightarrow H^{r+s}(F^\cdot \otimes G^\cdot).$$

Proof. As Godement [1, I.5.5.1].

We next consider a lemma that is essentially the case $Y = S$ of the theorem.

LEMMA 8.7. *Let S be a quasi-compact scheme and $f: X \to S$ a compactifiable morphism. If F is a flat sheaf on X and $G^\cdot$ is a bounded above complex of sheaves on S, then there is a canonical quasi-isomorphism*

$$\bar{\mathbf{R}}_c f_* F \otimes G^\cdot \simeq \mathbf{R}_c f_*(F \otimes f^* G^\cdot).$$

Proof. After replacing f by $\bar{f}: \bar{X} \to S$ and F by $j_! F$, we may assume f to be proper. Let $P^\cdot(G^\cdot) \xrightarrow{\sim} G^\cdot$ be the quasi-isomorphism whose existence is guaranteed by (8.3). As $\bar{\mathbf{R}}f_* F \otimes P^\cdot(G^\cdot) \xrightarrow{\approx} \bar{\mathbf{R}}f_* F \otimes G^\cdot$ according to (8.2), and

$$\mathbf{R}f_*(F \otimes f^* P^\cdot(G^\cdot)) \xrightarrow{\approx} \mathbf{R}f_*(F \otimes f^* G^\cdot)$$

according to (8.2) and (8.4), we may replace $G^\cdot$ by $P^\cdot(G^\cdot)$, that is, we may assume that $G^\cdot$ is a complex of flat sheaves. Let $I^\cdot(F)$ be an injective resolution of F. As $H^r(f_* I^\cdot(F)) = 0$ for $r \gg 0$, we may truncate $I^\cdot(F)$ to obtain a bounded complex $A^\cdot(F) \simeq F$ of f_*-acyclics. Choose a quasi-isomorphism

$$A^\cdot(F) \otimes f^* G^\cdot \xrightarrow{\sim} A^\cdot(A^\cdot(F) \otimes f^* G^\cdot).$$

On applying f_*, we obtain a map $\mathbf{R}f_* F \otimes f_* f^* G^\cdot \to \mathbf{R}f_*(A^\cdot(F) \otimes f^* G^\cdot)$ which, when composed with the canonical map $\mathbf{R}f_* F \otimes G^\cdot \to \mathbf{R}f_* F \otimes f_* f^* G^\cdot$, gives a map

$$\alpha: \mathbf{R}f_* F \otimes G^\cdot \to \mathbf{R}f_*(A^\cdot(F) \otimes f^* G^\cdot).$$

Since (8.2) and (8.4) show that

$$\mathbf{R}f_*(F \otimes f^* G^\cdot) \simeq \mathbf{R}f_*(A^\cdot(F) \otimes f^* G^\cdot),$$

it remains to show α is a quasi-isomorphism. The next lemma proves this when $G^\cdot$ consists of a single sheaf, and it is obviously true for

$$\tau_{\geq r}(G^\cdot) = (0 \to G^r \to G^{r+1} \to \cdots)$$

for $r \gg 0$ (as $\tau_{\geq r}(G^{\cdot}) = 0$ for $r \gg 0$). Descending induction on r and a five-lemma argument using the exact sequence $0 \to \tau_{\geq r+1}G^{\cdot} \to \tau_{\geq r}G^{\cdot} \to G^{r} \to 0$ now show that α is a quasi-isomorphism for $\tau_{\geq r}G^{\cdot}$ for all r. But, for a fixed r_0,

$$H^{r_0}(\mathbf{R}f_*F \otimes G^{\cdot}) \approx H^{r_0}(\mathbf{R}f_*F \otimes \tau_{\geq r}G^{\cdot})$$

all $r \ll r_0$ and

$$H^{r_0}(\mathbf{R}f_*(A^{\cdot}(F) \otimes f^*G^{\cdot}) \approx H^{r_0}(\mathbf{R}f_*(A^{\cdot}(F) \otimes f^*\tau_{\geq r}G^{\cdot})$$

all $r \ll r_0$, which completes the proof.

LEMMA 8.8. *Let $f: X \to S$ be as in (8.7). For any sheaf F on X and flat sheaf G on S, there are canonical quasi-isomorphisms*

$$\bar{\mathbf{R}}_c f_*F \otimes G \xrightarrow{\sim} \mathbf{R}_c f_*F \otimes G \xrightarrow{\sim} \mathbf{R}_c f_*(F \otimes f^*G).$$

Proof. Again we may assume that f is proper. It is obvious that the first map is a quasi-isomorphism. The second is defined (using α) as in the last lemma. Since G is flat,

$$H^r(\mathbf{R}f_*F \otimes G) = H^r(\mathbf{R}f_*F) \otimes G = R^rf_*F \otimes G.$$

Thus, to prove the lemma it suffices to show that $R^rf_*F \otimes G \xrightarrow{\approx} R^rf_*(F \otimes f^*G)$ for all r. As this may be checked on stalks, the proper base change theorem allows us to assume that S is the spectrum of a separably closed field. In this case we must show that $H^r(X, F) \otimes G \xrightarrow{\approx} H^r(X, F \otimes G)$. If G is a free Λ-module of finite rank, this is obvious but G, being flat, is a direct limit of such modules (Lazard [1]).

Before proving the theorem, we make a slight refinement to the proper base change theorem.

LEMMA 8.9. *Let*

$$\begin{array}{ccc} X & \xleftarrow{p} & X \times_S Y \\ \downarrow{\scriptstyle f} & & \downarrow{\scriptstyle q} \\ S & \xleftarrow{g} & Y \end{array}$$

be Cartesian with g proper, and let G be a sheaf on Y. There is a canonical quasi-isomorphism $f^*\mathbf{R}g_*G \xrightarrow{\sim} \mathbf{R}p_*(q^*G)$.

Proof. Let $G \to I^{\cdot}(G)$ and $q^*G \to I^{\cdot}(q^*G)$ be injective resolutions of G and q^*G. Since $q^*G \to q^*I^{\cdot}(G)$ is still exact, there is a map $q^*I^{\cdot}(G) \to I^{\cdot}(q^*G)$. On applying p_* and composing with the base change map $f^*g_*I^{\cdot}(G) \to p_*q^*I^{\cdot}(G)$, we obtain a map $f^*g_*I^{\cdot}(G) \to p_*I^{\cdot}(q^*G)$, that is, a map $f^*\mathbf{R}g_*G \to \mathbf{R}p_*(q^*G)$. On taking cohomology, we get the base change map $f^*R^rg_*G \to R^rp_*(q^*G)$, which we know to be an isomorphism.

Proof of (8.5). We have

$$
\begin{aligned}
\bar{\mathbf{R}}_c f_* F \otimes \mathbf{R}_c g_* G &\simeq \mathbf{R}_c f_*(F \otimes f^* \mathbf{R}_c g_* G) && (8.7)\\
&\xrightarrow{\sim} \mathbf{R}_c f_*(F \otimes \mathbf{R}_c p_*(q^* G)) && (8.9)\\
&\simeq \mathbf{R}_c f_*(\mathbf{R}_c p_*(F \boxtimes G)) && (8.8)\\
&\xrightarrow{\sim} \mathbf{R}_c h_*(F \boxtimes G) && (3.2c).
\end{aligned}
$$

COROLLARY 8.10. *Let $f: X \to S$ be a compactifiable morphism with S quasi-compact, and let F be a flat sheaf on X. There exists a quasi-isomorphism $\bar{\mathbf{R}}_c f_* F \to P^{\cdot}$ with $P^{\cdot}$ a bounded complex of flat sheaves such that $P^r = 0$ for $r < 0$. If $R^r_c f_* F = 0$ for $r > r_0$, then $P^{\cdot}$ may be chosen so that $P^r = 0$ for $r > r_0$.*

Proof. Write $Q^{\cdot}$ for $\bar{\mathbf{R}}_c f_* F$; from the construction of $\bar{\mathbf{R}}_c f_* F$ we may assume that $Q^r = 0$ if $r > r_0$. Lemma (8.7) shows that there is a quasi-isomorphism $Q^{\cdot} \otimes G \simeq \mathbf{R}_c f_*(F \otimes f^* G)$ for any sheaf G on S, and so

$$H^r(Q^{\cdot} \otimes G) = R^r_c f_*(F \otimes f^* G) = 0$$

for $r < 0$. Thus, if we tensor the diagram

$$
\begin{array}{ccccc}
Q^{-2} & \longrightarrow & Q^{-1} & \longrightarrow & Q^0 \\
 & \searrow^{0} & \downarrow\!\!\!\downarrow & \nearrow & \\
 & & B & &
\end{array}
\qquad B = \operatorname{im}(Q^{-1} \to Q^0)
$$

with G, then the top row remains exact, the surjection remains surjective, and, it follows, $B \otimes G \to Q^0 \otimes G$ remains injective. Thus, for any injection $G' \hookrightarrow G$, the diagram

$$
\begin{array}{ccc}
B \otimes G' & \longrightarrow & B \otimes G \\
\downarrow & & \downarrow \\
Q^0 \otimes G' & \hookrightarrow & Q^0 \otimes G
\end{array}
$$

shows $B \otimes G' \to B \otimes G$ to be injective. Hence B is flat, and, if we define $P^{\cdot}_1$ to be $(\cdots \to 0 \to B \to Q^0 \to Q^1 \to \cdots)$, there is a quasi-isomorphism $Q^{\cdot} \to P^{\cdot}_1$. As B and Q^0 are flat and $B \otimes G \to Q^0 \otimes G$ is injective for any G, Q^0/B^0 is flat, and we may take $P^{\cdot}$ to be $(\cdots \to 0 \to Q^0/B \to Q^1 \to \cdots)$.

COROLLARY 8.11. *In the situation of the theorem, let F_0 be a flat flabby sheaf on a compactification $\bar{X}$ of X and G_0 a flabby sheaf on a compactification $\bar{Y}$ of Y. Then $\bar{p}^* F_0 \otimes \bar{q}^* G_0$ is $\bar{h}_*$-acyclic. If F and G are as in the theorem, then*

$$j_!(F \boxtimes G) \to \bar{p}^* C^{\cdot}(j_! F) \otimes \bar{q}^* C^{\cdot}(j_! G),$$

where $C^{\cdot}(j_! F)$ and $C^{\cdot}(j_! G)$ are the Godement resolutions of $j_! F$ and $j_! G$, is an $\bar{h}_*$-acyclic resolution of $j_!(F \boxtimes G)$.

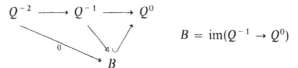

Proof. As $R^r\bar{f}_*F_0 = 0$ for $r > 0$, (8.10) shows that $\bar{f}_*F_0 = P^0$ is flat. Thus (8.5) shows that $R^r\bar{h}_*(F_0 \boxtimes G_0) = \bar{f}_*F_0 \otimes R^r\bar{g}_*G_0 = 0$ for $r > 0$. Since for any sheaf F_0,

$$0 \to F_0 \to C^0(F_0) \to C^0(F_0)/F_0 \to 0$$

is split-exact on stalks (V.1.15), the complex $F_0 \to C^{\cdot}(F_0)$ is homotopically trivial on stalks Godement [1, I.2.4.1a]. In particular this is so of $j_!F \to C^{\cdot}(j_!F)$ and, using (8.2), one sees that

$$\bar{p}^*(j_!F) \otimes \bar{q}^*(j_!G) \to \bar{p}^*C^{\cdot}(j_!F) \otimes \bar{q}^*C^{\cdot}(j_!G)$$

is a resolution. Finally,

$$j_!(F \boxtimes G) \xrightarrow{\sim} \bar{p}^*(j_!F) \otimes \bar{q}^*(j_!G).$$

Remark 8.12. This makes it possible to give an explicit description of the maps in (8.5). To simplify the notation, we take f and g to be proper. Set

$$\begin{aligned} \mathbf{R}f_*F &= f_*C^{\cdot}(F) \\ \mathbf{R}g_*G &= g_*C^{\cdot}(G) \\ \mathbf{R}h_*(F \boxtimes G) &= C^{\cdot}(F) \boxtimes C^{\cdot}(G). \end{aligned}$$

There is a canonical pairing

$$f_*C^{\cdot}(F) \times g_*C^{\cdot}(G) \to h_*(C^{\cdot}(F) \boxtimes C^{\cdot}(G)),$$

and hence

$$\bar{\mathbf{R}}f_*F \times \bar{\mathbf{R}}g_*G \to \mathbf{R}f_*F \times \mathbf{R}g_*G \to \mathbf{R}h_*(F \boxtimes G).$$

This defines the required map $\bar{\mathbf{R}}f_*F \otimes \bar{\mathbf{R}}g_*G \to \mathbf{R}h_*(F \boxtimes G)$.

In particular, the maps $R^rf_*F \otimes R^sg_*G \to R^{r+s}h_*(F \boxtimes G)$ are induced by the restriction maps

$$R^rf_*F \to R^rh_*(p^*F),\ R^sg_*G \to R^sh_*(q^*G)$$

and the cup-product pairings defined by $p^*F \times q^*G \to F \boxtimes G$.

COROLLARY 8.13. *Let X and Y be proper over a separably closed field k, and let F and G be sheaves on X and Y respectively. Assume that F and the groups $H^r(X, F)$ are flat. The maps $H^r(X, F) \to H^r(X \times Y, F|X \times Y)$ and $H^s(Y, G) \to H^s(X \times Y, G|X \times Y)$ and the cup-product pairings*

$$H^r(X \times Y, F|X \times Y) \times H^s(X \times Y, G|X \times Y) \to H^{r+s}(X \times Y, F \boxtimes G)$$

induce isomorphisms

$$\sum_{r+s=m} H^r(X, F) \otimes H^s(Y, G) \to H^m(X \times Y, F \boxtimes G)$$

for all m.

Proof. This is a special case of the theorem.

Remark 8.14. In the language of derived categories, the above results can be expressed a little more smoothly and generally. If $F^{\cdot}$ and $G^{\cdot}$ are any bounded above complexes of sheaves on X and $P^{\cdot}(F^{\cdot}) \xrightarrow{\sim} F^{\cdot}$ is a flat resolution, then (8.5) extends (by the usual descending induction argument) to show

$$\bar{\mathbf{R}}_c f_* F^{\cdot} \otimes \mathbf{R}_c g_* G^{\cdot} \simeq \mathbf{R}_c h_*(P^{\cdot}(F^{\cdot}) \boxtimes G^{\cdot}).$$

In the language of derived categories, this reads as

$$\mathbf{R}_c f_* F^{\cdot} \otimes^{\mathbf{L}} \mathbf{R}_c g_* G^{\cdot} \simeq \mathbf{R}_c h_*(F^{\cdot} \boxtimes^{\mathbf{L}} G^{\cdot}).$$

In particular, in the case $Y = S$ we have

$$\mathbf{R}_c f_* F^{\cdot} \otimes^{\mathbf{L}} G^{\cdot} \simeq \mathbf{R}_c f_*(F^{\cdot} \otimes^{\mathbf{L}} f^* G^{\cdot})$$

for $F^{\cdot}$ and $G^{\cdot}$ bounded above complexes on X and S respectively.

A complex $F^{\cdot}$ of sheaves is of *finite tor-dimension* $< d$ if $H^r(F^{\cdot} \otimes^{\mathbf{L}} G) = 0$ for $r < -d$ and all sheaves G. If $F^{\cdot}$ consists of a single sheaf F, this says that $\mathrm{Tor}_r^{\Lambda}(F, G) = 0$ for $r > d$. If, in the situation of (8.10), $F^{\cdot}$ is a complex of sheaves on X with finite tor-dimension $\leq d$, then the proof of (8.10) shows that $\mathbf{R}_c f_* F^{\cdot}$ has finite tor-dimension $\leq d$; therefore, there exists a bounded complex $P^{\cdot}$ of flat sheaves on S with $P^r = 0$ for $r < -d$ and a quasi-isomorphism $P^{\cdot} \simeq \mathbf{R}_c f_* F^{\cdot}$.

We wish to prove a Künneth formula for sheaves of $\mathbb{Z}_l$-modules and $\mathbb{Q}_l$-vector-spaces, but it will be necessary first to know that their cohomologies can be represented by particularly good complexes.

LEMMA 8.15. *Let X be a separated variety of dimension m over a separably closed field, and let $f : X \to S$ be the structure map. For any flat constructible sheaf F of Λ-modules on X, there is a quasi-isomorphism $P^{\cdot} \xrightarrow{\sim} \mathbf{R}_c f_* F$ where $P^{\cdot}$ is a complex of finitely generated projective Λ-modules with $P^r = 0$ except for $0 \leq r \leq 2m$.*

Proof. As in (8.3) we may construct a quasi-isomorphism $P^{\cdot} \xrightarrow{\sim} \mathbf{R}_c f_* F$ with $P^{\cdot}$ a complex of flat Λ-modules with $P^r = 0$ for $r > 2m$. Since

$$H^r(P^{\cdot}) = H^r(\mathbf{R}_c f_* F) = H_c^r(X, F)$$

is finitely generated (2.8), a slight refinement of the same construction (Mumford [4, II.5, p. 47]) gives a complex of finitely generated flat, and hence projective, modules. As in the proof of (8.10), P^0/B^0, where $B^0 = \mathrm{im}(P^{-1} \to P^0)$, is flat and hence projective. Thus $P^0 = B^0 \oplus P^0/B^0$, and we may replace P^0 by P^0/B^0.

A complex of modules over a ring A is said to be *perfect* if it is a bounded complex of projective, finitely generated modules. We shall write $\mathbf{H}_c^{\cdot}(X, F)$ for any perfect complex of Λ-modules which satisfies the conditions of (8.15).

PROPOSITION 8.16. *Let $f: X \to S$ be as in (8.15), and let $F = (F_n)$ be a flat constructible sheaf of A-modules, where A is the integral closure of $\mathbb{Z}_l$ in a finite field extension of $\mathbb{Q}_l$. (So F_n is a flat constructible sheaf of Λ_n-modules, where $\Lambda_n = A/\mathfrak{m}^n$). Then it is possible to choose the complexes $\mathbf{H}_c^{\cdot}(X, F_n)$ so that, for all n, there are maps $\mathbf{H}_c^{\cdot}(X, F_{n+1}) \to \mathbf{H}_c^{\cdot}(X, F_n)$ inducing isomorphisms $\mathbf{H}_c^{\cdot}(X, F_{n+1}) \otimes \Lambda_n \xrightarrow{\approx} \mathbf{H}_c^{\cdot}(X, F_n)$.*

Proof. As $F_{n+1} \otimes \Lambda_n \xrightarrow{\approx} F_n$, (8.8) shows that

$$\mathbf{H}_c^{\cdot}(X, F_{n+1}) \otimes \Lambda_n \xrightarrow{\sim} \mathbf{R}_c f_* F_n \xleftarrow{\sim} \mathbf{H}_c^{\cdot}(X, F_n).$$

Thus the proposition follows from the next two lemmas.

LEMMA 8.17. *Let A be a (Noetherian) ring, and let $M^{\cdot} \xrightarrow{\phi} L^{\cdot} \xleftarrow{\pi} N^{\cdot}$ be maps of complexes of A-modules with π a quasi-isomorphism. If $M^{\cdot}$ is perfect, there exists a quasi-isomorphism $\psi: M^{\cdot} \to N^{\cdot}$ such that $\pi\psi = \phi$.*

Proof. Let $C^{\cdot}$ be the mapping cone of $L^{\cdot} \xrightarrow{id} L^{\cdot}$. It is an exact complex, and we may replace π by $N^{\cdot} \oplus C^{\cdot} \xrightarrow{\sim} L^{\cdot}$. Thus we may assume π to be surjective. $K^{\cdot} = \ker(N^{\cdot} \xrightarrow{\pi} L^{\cdot})$ is exact, as may be seen from the cohomology sequence of $0 \to K^{\cdot} \to N^{\cdot} \to L^{\cdot} \to 0$. For $r \gg 0$ we define $\psi^r: M^r \to N^r$ to be zero. Suppose that ψ^r has been defined for all $r > r_0$. There exists a map $\psi_1^{r_0}: M^{r_0} \to N^{r_0}$ such that $\pi^{r_0}\psi_1^{r_0} = \phi^{r_0}$. Moreover,

$$d_N^{r_0}\psi_1^{r_0} - \psi^{r_0+1}d_M^{r_0}$$

maps M^{r_0} into $K^{r_0+1} \subset N^{r_0+1}$. As

$$d_K^{r_0+1}(d_N^{r_0}\psi_1^{r_0} - \psi^{r_0+1}d_M^{r_0}) = 0$$

there exists a map $\psi_2: M^{r_0} \to K^{r_0}$ such that

$$d_K^{r_0}\psi_2 = d_N^{r_0}\psi_1^{r_0} - \psi^{r_0+1}d_M^{r_0}.$$

We define ψ^{r_0} to be $\psi_1 - \psi_2$.

LEMMA 8.18. *Let A be a local Artin ring, and let A_0 be some quotient ring of A. We write $M \mapsto M_0$, $\phi \mapsto \phi_0$ for the functor $A_0 \otimes_A -$. Let $M^{\cdot}$ and $N^{\cdot}$ be perfect complexes of A and A_0-modules respectively, and let ψ be a quasi-isomorphism $M_0^{\cdot} \xrightarrow{\sim} N^{\cdot}$. Then there exists a perfect complex $L^{\cdot}$ of A-modules, a quasi-isomorphism $\phi: M^{\cdot} \to L^{\cdot}$, and an isomorphism $L_0^{\cdot} \xrightarrow{\approx} N^{\cdot}$ such that*

$$\begin{array}{ccc} M_0^{\cdot} & \xrightarrow[\sim]{\phi_0} & L_0^{\cdot} \\ {\scriptstyle\psi}\downarrow{\scriptstyle\sim} & \swarrow{\scriptstyle\approx} & \\ N^{\cdot} & & \end{array}$$

commutes.

Proof. Note that any finitely generated, projective A_0-module, being free, is of the form M_0 for some finitely generated, projective A-module M and that any A_0-homomorphism $\psi: M_0 \to M'_0$ of such modules, being represented by a matrix, is of the form ϕ_0 for some A-homomorphism $\phi: M \to M'$.

SUBLEMMA 8.19. *For any exact perfect complex $N^\cdot$ of A_0-modules, there exists an exact perfect complex $M^\cdot$ of A-modules such that $M_0^\cdot \approx N^\cdot$.*

Proof. From $N^\cdot$ we get short exact sequences

$$0 \to Z^{r_1-1} \to N^{r_1-1} \to N^{r_1} \to 0$$
$$0 \to Z^{r_1-2} \to N^{r_1-2} \to Z^{r_1-1} \to 0$$
$$\cdots$$
$$0 \to N^{r_0} \to N^{r_0+1} \to Z^{r_0+2} \to 0.$$

By descending induction, each Z^r is finitely generated and free. We lift $N^{r_1-1} \to N^{r_1}$ to $M^{r_1-1} \to M^{r_1}$, and define $Y^{r_1-1} = \ker(M^{r_1-1} \to M^{r_1})$. Then we lift $N^{r_1-2} \to Z^{r_1-1}$ to $M^{r_1-2} \to Y^{r_1-1}$ and continue.

SUBLEMMA 8.20. *Let $\phi: L^\cdot \to M^\cdot$ be a map of perfect complexes of A-modules. Any map $L_0^\cdot \to M_0^\cdot$ that is homotopic to ϕ_0 is of the form ψ_0, where $\psi: L^\cdot \to M^\cdot$ is homotopic to ϕ.*

Proof. Let $\psi': L_0^\cdot \to M_0^\cdot$ be homotopic to ϕ_0:

$$\phi_0 - \psi' = dk_0 + k_0 d,$$

where we may assume that $k = (k^r)$ is a family of maps $k^r: L^r \to M^{r-1}$. We define $\psi = \phi - (dk + kd)$.

We now prove the lemma. The mapping cone of $N^\cdot \xrightarrow{id} N^\cdot$ is exact and perfect and hence lifts to a complex $C^\cdot$ of A-modules. After replacing $M^\cdot$ by $M^\cdot \oplus C^\cdot$, we may assume that ψ is surjective. Now Ker (ψ) is an exact perfect complex and hence may be lifted to a similar complex $K^\cdot$ of A-modules. The inclusion $K_0^\cdot \hookrightarrow M_0^\cdot$ is homotopic to the zero map (compare 8.22 below) and so, according to (8.20), lifts to a map $K^\cdot \to M^\cdot$. According to (IV.1.11), this map identifies each K^r with a direct summand of M^r. Thus $L^\cdot = M^\cdot/K^\cdot$ is perfect and has the required properties.

Let $f: X \to S$ and $F = (F_n)$ be as in (8.16). The complex $\mathbf{H}_c^\cdot(X, F) = \varprojlim \mathbf{H}_c^\cdot(X, F_n)$ is a perfect complex of A-modules and $\mathbf{H}_c^\cdot(X, F) \otimes \Lambda_n \xrightarrow{\approx} \mathbf{H}_c^\cdot(X, F_n)$. Moreover,

$$H^r(\mathbf{H}_c^\cdot(X, F)) = H^r(\varprojlim \mathbf{H}_c^\cdot(X, F_n)) = \varprojlim H^r(\mathbf{H}_c^\cdot(X, F_n))$$
$$= \varprojlim H_c^r(X, F_n) \stackrel{df}{=} H_c^r(X, F)$$

(compare the proof of (V.1.11)).

THEOREM 8.21. *Consider the diagram*

$$\begin{array}{ccccc} & & X\times_S Y & & \\ & \swarrow^{p} & \Big\downarrow h & \searrow^{q} & \\ X & & & & Y \\ & \searrow_{f} & & \swarrow_{g} & \\ & & S & & \end{array}$$

in which S is the spectrum of a separably closed field and f and g are compactifiable. For any flat constructible sheaves F and G of A-modules (A as in (8.16)) *on X and Y, there is a canonical quasi-isomorphism*

$$\mathbf{H}^{\cdot}_c(X, F) \otimes \mathbf{H}^{\cdot}_c(Y, G) \xrightarrow{\sim} \mathbf{H}^{\cdot}_c(X \times Y, F \boxtimes G).$$

Thus there are exact sequences,

$$0 \to \sum_{r+s=m} H^r_c(X, F) \otimes H^s_c(Y, G) \to H^m_c(X \times Y, F \boxtimes G)$$

$$\to \sum_{r+s=m+1} \operatorname{Tor}^A_1 (H^r_c(X, F), H^s_c(X, G)) \to 0.$$

Proof. As A is a principal ideal domain, $\operatorname{Tor}^A_r = 0$ for $r > 1$. Thus the second part of the theorem follows from the first (and (8.6)).

For the first part, note that we have quasi-isomorphisms

$$\mathbf{H}^{\cdot}_c(X, F_n) \otimes \mathbf{H}^{\cdot}_c(Y, G_n) \xrightarrow[\sim]{\phi_n} \mathbf{R}_c h_*(F_n \boxtimes G_n) \xleftarrow[\sim]{\pi_n} \mathbf{H}^{\cdot}_c(X \times Y, F_n \boxtimes G_n)$$

and (8.17) shows that we get quasi-isomorphisms

$$\psi_n : \mathbf{H}^{\cdot}_c(X, F_n) \otimes \mathbf{H}^{\cdot}_c(Y, G_n) \xrightarrow{\sim} \mathbf{H}^{\cdot}_c(X \times Y, F_n \boxtimes G_n).$$

It remains to show that we can choose these so that $\psi_{n+1} \otimes \Lambda_n = \psi_n$.

By construction,

$$\begin{array}{ccc} \mathbf{H}^{\cdot}_c(X, F_{n+1}) & \xrightarrow{\sim} & \mathbf{R}_c f_* F_{n+1} \\ \Big\downarrow & & \Big\downarrow \text{can} \\ \mathbf{H}^{\cdot}_c(X, F_n) & \xrightarrow{\sim} & \mathbf{R}_c f_* F_n \end{array}$$

commutes, and similarly for G. It follows that

$$\begin{array}{ccc} \mathbf{H}^{\cdot}_c(X, F_{n+1}) \otimes \mathbf{H}^{\cdot}_c(Y, G_{n+1}) & \longrightarrow & \mathbf{R}'_c h_*(F_{n+1} \boxtimes G_{n+1}) \\ \Big\downarrow & & \Big\downarrow \\ \mathbf{H}^{\cdot}_c(X, F_n) \otimes \mathbf{H}^{\cdot}_c(Y, G_n) & \longrightarrow & \mathbf{R}_c h_*(F_n \boxtimes G_n) \end{array}$$

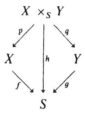

commutes. Thus $\phi_{n+1} \otimes \Lambda_n = \phi_n$, that is, $\pi_n \circ (\psi_{n+1} \otimes \Lambda_n) = \pi_n \circ \psi_n$. The next lemma shows that $\psi_{n+1} \otimes \Lambda_n$ and ψ_n are homotopic, and (8.20) then shows that ψ_{n+1} can be modified to get $\psi_{n+1} \otimes \Lambda_n = \psi_n$.

LEMMA 8.22. *Let ψ_1 and ψ_2 be maps $M^\cdot \to N^\cdot$ such that $\pi\psi_1 = \pi\psi_2$ for some quasi-isomorphism $\pi: N^\cdot \to L^\cdot$. If $N^\cdot$ is a bounded above complex of projectives, then ψ_1 is homotopic to ψ_2.*

Proof. According to (the dual of) Hartshorne [1, I.4.5], π has a homotopy inverse π', so $\pi'\pi \sim 1$. Then $\psi_1 \sim \pi'\pi\psi_1 = \pi'\pi\psi_2 \sim \psi_2$.

COROLLARY 8.23. *In the situation of (8.21), there are canonical isomorphisms*

$$\sum_{r+s=m} H_c^r(X, F \otimes \Omega) \otimes H_c^s(Y, G \otimes \Omega) \xrightarrow{\approx} H_c^m(X \times Y, (F \boxtimes G) \otimes \Omega)$$

where Ω is the field of fractions of A.

Proof. Tensor the exact sequence in (8.21) with Ω.

Remark 8.24. The maps in (8.23) are as described (8.12). If X and Y are proper over S, they are induced by

$$H^r(X, F \otimes \Omega) \xrightarrow{\text{rest}} H^r(X \times Y, F \otimes \Omega)$$

$$H^s(Y, G \otimes \Omega) \xrightarrow{\text{rest}} H^s(X \times Y, G \otimes \Omega)$$

and cup-product

$$H^r(X \times Y, F \otimes \Omega) \times H^s(X \times Y, G \otimes \Omega) \to H^{r+s}(X \times Y, F \boxtimes G \otimes \Omega).$$

If we make

$$H^*(X, \mathbb{Q}_l) \overset{df}{=} \sum_r H^r(X, \mathbb{Q}_l),$$

$H^*(Y, \mathbb{Q}_l)$, and $H^*(X \times Y, \mathbb{Q}_l)$ into graded rings using the cup-product pairing, then (8.23) shows that there is a canonical isomorphism

$$H^*(X, \mathbb{Q}_l) \otimes H^*(Y, \mathbb{Q}_l) \xrightarrow{\approx} H^*(X \times Y, \mathbb{Q}_l)$$

of graded rings. Moreover, if $\phi: X \to X'$ and $\psi: Y \to Y'$ are morphisms of proper S-schemes, then

$$\begin{array}{ccc} H^*(X, \mathbb{Q}_l) \otimes H^*(Y, \mathbb{Q}_l) & \xrightarrow{\approx} & H^*(X \times Y, \mathbb{Q}_l) \\ \uparrow{\scriptstyle \phi^* \otimes \psi^*} & & \uparrow{\scriptstyle (\phi \times \psi)^*} \\ H^*(X', \mathbb{Q}_l) \otimes H^*(Y', \mathbb{Q}_l) & \xrightarrow{\approx} & H^*(X' \times Y', \mathbb{Q}_l) \end{array}$$

commutes.

Remark 8.25. If one uses the base change theorem of Deligne cited in (5.7) instead of the proper base change theorem, then the same arguments as in the proof of (8.5) show the following: let X and Y be schemes of finite-type over a separably closed field k, and let F and G be sheaves of Λ-modules on X and Y respectively; then (in the notation of derived categories),

$$\mathbf{R}\Gamma(X, F) \otimes^{\mathbf{L}} \mathbf{R}\Gamma(Y, G) \xrightarrow{\sim} \mathbf{R}\Gamma(X \times Y, F \,\square^{\mathbf{L}}\, G).$$

§9. The Cycle Map

All varieties will be smooth over a fixed algebraically closed field k, and Λ will be the constant sheaf $\mathbb{Z}/(n)$ with n prime to char (k). We shall associate a cohomology class with each algebraic cycle on such a variety.

Recall that a *prime r-cycle* on a variety X is a closed integral subscheme of codimension r, that an *algebraic r-cycle* is an element of the free abelian group $C^r(X)$ on the set of prime r-cycles, and that an *algebraic cycle* is an element of the graded group $C^*(X) = \sum C^r(X)$. A prime r-cycle W and a prime s-cycle Z *intersect properly* if each irreducible component of $W \cap Z$ has codimension $r + s$; then $W.Z$ *is defined* and belongs to $C^{r+s}(X)$. Two algebraic cycles W and Z intersect properly, and $W.Z$ is defined, if each prime cycle occurring in W intersects properly with each prime cycle occurring in Z. For any map $\pi: Y \to X$ and $Z \in C^*(X)$, we set $\pi^*Z = \Gamma_{\pi}.(Y \times Z)$ when this is defined. Any proper map $\pi: Y \to X$ defines a map $\pi_*: C^*(Y) \to C^*(X)$ such that, for Z a prime cycle on Y, $\pi_*Z = 0$ if $\dim(\pi(Z)) < \dim Z$, and $\pi_*Z = d(\pi(Z))$ where d is the degree of $\pi|Z: Z \to \pi(Z)$ otherwise. These maps are related by the *projection formula*, $\pi_*(\pi^*W.Z) = W.\pi_*Z$, which holds whenever both sides are defined.

We write $H^*(X, \Lambda)$, or simply $H^*(X)$, for $\sum_r H^r(X, \Lambda(r'))$ where r' is the integral part of $r/2$. Cup-product makes $H^*(X)$ into an anticommutative graded ring.

We shall define a homomorphism of graded groups, which doubles degrees, $cl_X: C^*(X) \to H^*(X)$. If Z is a smooth prime r-cycle so Z is a smooth subvariety of X, then we define $cl_X(Z)$ to be the image of the fundamental class $s_{Z/X}$ of Z in X under the map $H_Z^{2r}(X, \Lambda(r)) \to H^{2r}(X, \Lambda(r))$. Equivalently (6.5), $cl_X(Z) = i_*(1_Z)$, where i_* is the Gysin map $H^0(Z, \Lambda) \to H^{2r}(X, \Lambda(r))$ and $1_Z \in H^0(Z, \Lambda)$ is the identity element of the ring $H^*(Z)$. To extend this definition to singular cycles, we need a lemma.

LEMMA 9.1. *For any reduced closed subscheme Z of codimension r in X, $H_Z^s(X, \Lambda) = 0$ for $s < 2r$.*

Proof. When Z is smooth, this follows from (5.1). The general case is proved by descending induction on r. If $r = \dim(X)$, then Z is a finite set of points and therefore is smooth. In general we may choose an open subset U of X such that $U \cap Z$ is smooth, U is dense in every irreducible component of Z of codimension r, and $X \supset U \supset X - Z$. The exact cohomology sequence of this last triple is (III.1.26):

$$\cdots \to H^s_{X-U}(X, \Lambda) \to H^s_Z(X, \Lambda) \to H^s_{U \cap Z}(U, \Lambda) \to \cdots.$$

Now $H^s_{X-U}(X, \Lambda) = 0$ for $s < 2(r + 1)$ by induction as $X - U$ has codimension at least $r + 1$ in X, and $H^s_{U \cap Z}(U, \Lambda) = 0$ for $s < 2r$ as $U \cap Z$ is smooth, from which the lemma follows.

Now let Z be any prime r-cycle on X, and choose an open subset U of X as in the lemma. We have

$$s_{U \cap Z/U} \in H^{2r}_{U \cap Z}(U, \Lambda(r)) \overset{\approx}{\leftarrow} H^{2r}_Z(X, \Lambda(r)) \to H^{2r}(X, \Lambda(r)),$$

where the isomorphism comes from the exact sequence of the triple $X \supset U \supset X - Z$. The image of $s_{U \cap Z/U}$ in $H^{2r}(X, \Lambda(r))$ is defined to be $cl_X(Z)$. According to (6.1c), it is independent of the choice of U. We extend cl_X to the whole of $C^*(X)$ by linearity. Note that in determining $cl_X(Z)$ when Z is an r-cycle, we are allowed to remove any closed subvariety of X of codimension $>r$.

PROPOSITION 9.2. *Let $\pi: Y \to X$ be a map of varieties and Z an algebraic cycle on X. If, for every prime cycle Z' occurring in Z, $Y \times_X Z'$ is integral, then π^*Z is defined and $cl_Y(\pi^*Z) = \pi^*cl_X(Z)$.*

Proof. We may assume that Z is prime. The condition implies that, after X and Y have been replaced by open subvarieties, Z and $Y \times_X Z$ will be smooth, and $\pi^*Z = Y \times_X Z$. Then

$$\begin{aligned} cl_Y(\pi^*Z) &= \text{image}\,(s_{Z \times Y/Y}) && \text{(definition)}\\ &= \pi^*\,\text{image}\,(s_{Z/X}) && \text{(6.1c)}\\ &= \pi^*cl_X(Z) && \text{(definition).}\end{aligned}$$

PROPOSITION 9.3. *Let $i: Z \hookrightarrow X$ be a closed immersion of smooth varieties. For any $W \in C^r(Z)$, $i_*(cl_Z(W)) = cl_X(W)$, where i_* is the Gysin map $H^{2r}(Z, \Lambda(r)) \to H^{2(r+c)}(X, \Lambda(r + c))$.*

Proof. We may assume that W is a smooth prime cycle. Then $cl_Z(W) = i_{1*}(1_W)$ where i_{1*} is the Gysin map for $i_1: W \hookrightarrow Z$, and

$$i_*(cl_Z(W)) = i_*(i_{1*}(1_W)) = (i \circ i_1)_*(1_W)$$

according to (6.5b).

PROPOSITION 9.4. *For any $W \in C^*(X)$ and $Z \in C^*(Y)$,*

$$cl_{X \times Y}(W \times Z) = p^*cl_X(W) \cup q^*cl_Y(Z),$$

where p and q are the projections $X \times Y \to X$ and $X \times Y \to Y$.

Proof. We may assume that W and Z are smooth prime cycles. Let i be the closed immersion $X \times Z \hookrightarrow X \times Y$. Then

$$\begin{aligned} i_* i^* cl_{X \times Y}(W \times Y) &= i_* cl_{X \times Z}(i^*(W \times Y)) && (9.2) \\ &= i_* cl_{X \times Z}(W \times Z) \\ &= cl_{X \times Y}(W \times Z). && (9.3) \end{aligned}$$

On the other hand, (6.5) shows that

$$\begin{aligned} i_* i^* cl_{X \times Y}(W \times Y) &= cl_{X \times Y}(W \times Y) \cup cl_{X \times Y}(X \times Z) \\ &= cl_{X \times Y}(p^* W) \cup cl_{X \times Y}(q^* Z) \\ &= p^* cl_X(W) \cup q^* cl_Y(Z). \end{aligned}$$

PROPOSITION 9.5. *Let W and Z be algebraic cycles on X such that each prime cycle in W intersects each prime cycle in Z transversally. Then*

$$cl_X(W.Z) = cl_X(W) \cup cl_X(Z).$$

Proof. We may assume that W and Z are prime. Then $W \times Z$ intersects the diagonal in $X \times X$ transversally, and so $\Delta: X \to X \times X$ and $W \times Z$ satisfy the hypotheses of (9.2). Thus

$$\begin{aligned} cl_X(W.Z) &= cl_X(\Delta^*(W \times Z)) && (\text{definition of } \Delta^*) \\ &= \Delta^* cl_{X \times X}(W \times Z) && (9.2) \\ &= \Delta^*(p^* cl_X(W) \cup q^* cl_X(Z)) && (9.4) \\ &= cl_X(W) \cup cl_X(Z). \end{aligned}$$

Remark 9.6. The map $cl_X: C^1(X) \to H^2(X, \Lambda(1))$ is the composite of the canonical maps

$$C^1(X) \to \mathrm{Pic}\,(X) \text{ and } \mathrm{Pic}\,(X) \to H^2(X, \Lambda(1)),$$

as follows from (6.1b). In particular, $\mathrm{cl}_X(Z)$, for Z a divisor, depends only on the linear equivalence class of Z.

Example 9.7. We saw in (5.6) that, for any linear subspace L^r of $\mathbb{P}^m$ of codimension r, the Gysin map $\Lambda = H^0(L^r, \Lambda) \xrightarrow{i_*} H^{2r}(\mathbb{P}^m, \Lambda(r))$ is an isomorphism. Thus $H^{2r}(\mathbb{P}^m, \Lambda(r))$ is generated by $cl_{\mathbb{P}^m}(L^r)$. Since $\mathrm{Pic}\,(\mathbb{P}^m) \approx \mathbb{Z}$ is generated by the class of any hyperplane, $cl(L^1)$ is independent of L^1. Also, as L^r is the transverse intersection of r hyperplanes, (9.5) shows that

$$cl(L^r) = cl(L^1) \cup \cdots \cup cl(L^1)$$

is also independent of L^r. It follows that the map

$$\Lambda[T]/(T^{m+1}) \to H^*(\mathbb{P}^m)$$

that sends T^r to $cl_{\mathbb{P}^m}(L^r)$ is an isomorphism of graded rings.

Remark 9.8. A much more general, and complete, treatment of the cycle map can be found in [SGA. $4\frac{1}{2}$, Cycle].

§10. Chern Classes

We fix an algebraically closed field k and consider only varieties X that are quasi-projective and smooth over k; again Λ denotes the constant sheaf $\mathbb{Z}/(n)$ with n prime to char (k). We shall associate cohomology classes, called *Chern classes*, with each vector bundle (or coherent sheaf) on X. In order to apply the method of construction of Grothendieck [2], we have to know the cohomology ring of a projective bundle on X. Recall that the projective bundle of a vector bundle E over X (or locally free, coherent $\mathcal{O}_X$-module) is a variety $\mathbb{P}(E)$ with a morphism $\mathbb{P}(E) \to X$ such that each fiber of $\mathbb{P}(E)$ over X is canonically isomorphic to the projective space of lines in the fiber (or stalk) of E over X (Hartshorne [2, II.7]). As in Section 9, we write $H^*(X)$ for the cohomology ring $\sum_r H^r(X, \Lambda([r/2]))$ of X, where $[r/2]$ is the integral part of $r/2$.

Proposition 10.1. *Let E be a vector bundle of rank m over X, and let $P = \mathbb{P}(E) \xrightarrow{p} X$ be the associated projective bundle. Let $\xi \in H^2(P, \Lambda(1))$ be the image of the canonical line bundle $\mathcal{O}_P(1)$ on P under the map* $\operatorname{Pic}(P) \to H^2(P, \Lambda(1))$ *arising from the Kummer sequence. Then the map $H^*(X)[T]/(T^m) \to H^*(P)$, which is p^* on $H^*(X)$ and sends T to ξ, is an isomorphism of graded $H^*(X)$-modules, that is, $H^*(\mathbb{P}(E))$ is a free $H^*(X)$-module with basis $1, \xi, \xi^2, \ldots, \xi^{m-1}$.*

Proof. To simplify the proof, we choose an isomorphism $\Lambda \xrightarrow{\approx} \Lambda(1)$ and use it to identify each $\Lambda(s)$ with Λ. Note that if $X = \operatorname{spec} k$ the proposition is proved in (9.7). The next lemma implies it in the case that E is the trivial bundle.

Lemma 10.2. *The Künneth formula holds for $X \times \mathbb{P}^{m-1}$ and the sheaf $\Lambda = \Lambda \boxtimes \Lambda$. More precisely, the maps*

$$p^*: H^r(X, \Lambda) \to H^r(X \times \mathbb{P}^{m-1}, \Lambda),$$
$$q^*: H^s(\mathbb{P}^{m-1}, \Lambda) \to H^s(X \times \mathbb{P}^{m-1}, \Lambda)$$

together with the cup-product pairings

$$H^r(X \times \mathbb{P}^{m-1}, \Lambda) \times H^s(X \times \mathbb{P}^{m-1}, \Lambda) \to H^{r+s}(X \times \mathbb{P}^m, \Lambda)$$

induce isomorphisms

$$\sum_{r+s=t} H^r(X, \Lambda) \otimes H^s(\mathbb{P}^{m-1}, \Lambda) \xrightarrow{\approx} H^t(X \times \mathbb{P}^{m-1}, \Lambda).$$

Proof. If X is complete, this is a special case of (8.5) (even (8.13)), and it is possible to modify the proof of (8.5) so that (in this special situation) it applies to a noncomplete X. We use the same notations as in Section 8; in particular, we have a diagram

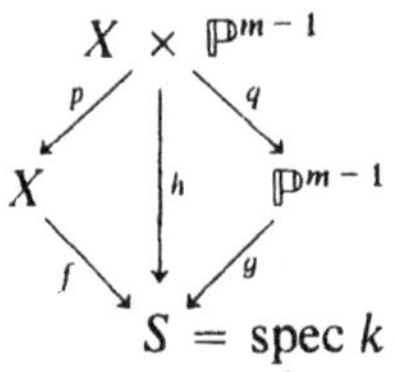

and (8.15) a perfect complex $\mathbf{H}^{\cdot}(\mathbb{P}^{m-1}, \Lambda)$ of Λ-modules such that

$$H^r(\mathbf{H}^{\cdot}(\mathbb{P}^{m-1}, \Lambda)) = H^r(\mathbb{P}^{m-1}, \Lambda).$$

Since $H^{2r}(\mathbb{P}^{m-1}, \Lambda) \approx \Lambda$ and $H^{2r+1}(\mathbb{P}^{m-1}, \Lambda) = 0$, there exist maps of complexes $\Lambda \to \mathbf{H}^{\cdot}(\mathbb{P}^{m-1}, \Lambda)$ of degree $2r$ that induce isomorphisms $\Lambda \xrightarrow{\approx} H^{2r}(\mathbf{H}^{\cdot}(\mathbb{P}^{m-1}, \Lambda))$ on homology. Thus there is a quasi-isomorphism from the complex

$$\sum_{r=0}^{m} \Lambda[-2r] = \Lambda \to 0 \to \Lambda \to \cdots \to 0 \to \Lambda \to \cdots$$

to $\mathbf{H}^{\cdot}(\mathbb{P}^{m-1}, \Lambda)$ and so we may replace $\mathbf{H}^{\cdot}(\mathbb{P}^{m-1}, \Lambda)$ by $\sum \Lambda[-2r]$. Since obviously $H^r(X, \Lambda) \otimes \Lambda \xrightarrow{\approx} H^r(X, \Lambda)$ we have a quasi-isomorphism

$$\mathbf{R}f_*\Lambda \otimes \mathbf{H}^{\cdot}(\mathbb{P}^{m-1}, \Lambda) \xrightarrow{\approx} \mathbf{R}f_*(f^*\mathbf{H}^{\cdot}(\mathbb{P}^{m-1}, \Lambda))$$

Now

$$\begin{aligned} \mathbf{R}f_*(f^*\mathbf{H}^{\cdot}(\mathbb{P}^{m-1}, \Lambda)) &\xrightarrow{\approx} \mathbf{R}f_*(\mathbf{R}p_*(q^*\Lambda)) \qquad (8.9) \\ &\xrightarrow{\approx} \mathbf{R}f_*(\mathbf{R}p_*(\Lambda)) \\ &\xrightarrow{\approx} \mathbf{R}h_*\Lambda, \end{aligned}$$

which proves the lemma.

This shows that we have isomorphisms of graded rings

$$H^*(X)[T]/(T^m) \xrightarrow{\approx} H^*(X) \otimes H^*(\mathbb{P}^{m-1}) \xrightarrow[\approx]{(p^*, q^*)} H^*(X \times \mathbb{P}^{m-1}),$$

which proves the proposition in the case that E is trivial. Since E is locally trivial for the Zariski topology on X, it only remains to show that the proposition holds for X if it holds for open subsets X_0, X_1, $X_0 \cap X_1$, where $X_0 \cup X_1 = X$. This follows from comparing the Mayer-Vietoris sequences (III.2.24),

$$\cdots \to H^r(\mathbb{P}(E_0)) \oplus H^r(\mathbb{P}(E_1)) \to H^r(\mathbb{P}(E_{01})) \to H^{r+1}(\mathbb{P}(E)) \to \cdots$$
$$\Lambda[T]/(T^m) \otimes (\cdots \to H^r(X_0) \oplus H^r(X_1) \to H^r(X_{01}) \to H^{r+1}(X) \to \cdots)$$

where we have written X_{01} for $X_0 \cap X_1$ and E_0, E_1, and E_{01} for the restrictions of E to X_0, X_1, and X_{01}.

To define Chern classes on the category $\mathbf{V}$ of all quasi-projective smooth varieties, we need the following data (Grothendieck [2]):

(a) a contravariant functor from $\mathbf{V}$ to graded, anti-commutative rings; this we take to be $X \mapsto H^*(X)$;

(b) a functorial homomorphism $\operatorname{Pic}(X) \xrightarrow{p_X} H^2(X)$; this we take to be that defined by the Kummer sequence;

(c) for any closed immersion $i: Z \hookrightarrow X$, with Z and X in $\mathbf{V}$, a group homomorphism $i_*: H^*(Y) \to H^*(X)$ that raises degrees by twice the codimension of Y in X; this we take to be the Gysin map. As in Section 9, we write $cl_X(Z)$ for $i_*(1_Z)$.

This data is to satisfy the following hypotheses:

A1. that proved in (10.1);

A2. if Z is a smooth divisor on X, then the image of Z (as an element of $\operatorname{Pic}(X)$) under p_X is $cl_X(Z)$; this is obvious from the various definitions;

A3. if W, Z and X are in $\mathbf{V}$ and $W \overset{i_1}{\hookrightarrow} Z \overset{i_2}{\hookrightarrow} X$ are closed immersions, then $(i_2 i_1)_* = i_{2*} i_{1*}$; this is proved in (6.5);

A4. if Z and X are in $\mathbf{V}$ and $i: Z \hookrightarrow X$ is a closed immersion, then $i_*(i^*(x) \cup z) = x \cup i_* z$ for any $x \in H^*(X)$ and $y \in H^*(Y)$; this is proved in (6.5).

Let E be a vector bundle of rank m over X, and let $\xi \in H^2(\mathbb{P}(E))$ be the element corresponding to the canonical line bundle on $\mathbb{P}(E)$. Proposition (10.1), that is, (A1), shows that there are unique elements $c_r(E) \in H^{2r}(X)$ such that,

$$\begin{cases} \sum_{r=0}^{m} c_r(E)\xi^{m-r} = 0 \\ c_0(E) = 1, \qquad c_r(E) = 0 \text{ for } r > m. \end{cases}$$

Then $c_r(E)$ is called the rth *Chern class* of E, $c(E) = \sum c_r(E)$ the *total Chern class* of E, and $c_t(E) = 1 + c_1(E)t + \cdots + c_m(E)t^m$ the *Chern polynomial* of E.

THEOREM 10.3. *The Chern classes satisfy the following:*

(a) (Functoriality). *if $\pi: Y \to X$ is a morphism with X and Y in $\mathbf{V}$ and E is a vector bundle on X, then $c_r(\pi^{-1}(E)) = \pi^*(c_r(E))$ all r;*

(b) (Normalization). *if* E *is a line bundle on* X, *then* $c_t(E) = 1 + p_X(E)t$ *where, on the right hand side,* E *is regarded as an element of* Pic (X);

(c) (Additivity). *if* $X \in \mathbf{V}$, *and* $0 \to E' \to E \to E'' \to 0$ *is an exact sequence of vector bundles on* X, *then*

$$c_t(E) = c_t(E')c_t(E'').$$

Moreover, these three properties characterize the Chern classes uniquely.

Proof. This is Theorem 1 of Grothendieck [2].

Remark 10.4. If

$$c_t(E) = 1 + c_1(E)t + c_2(E)t^2 + \cdots + c_m(E)t^m$$

is formally put equal to $\prod (1 + \gamma_i(E)t)$, then the $\gamma_i(E)$ are called the *Chern roots* of E. The Chern roots of $\check{E}$ are $-\gamma_i(E)$, those of $E \otimes E'$ are $\gamma_i(E) + \gamma_j(E')$, and those of $\Lambda^p E$ are $\gamma_{i_1} + \cdots + \gamma_{i_p}$, $i_1 < i_2 < \cdots < i_p$.

Condition (c) of (10.3) says that the map c factors through the Grothendieck group $K(X)$ of vector bundles on X. Since X is smooth, $K(X)$ is equal to the Grothendieck group of coherent $\mathcal{O}_X$-modules. Thus a prime cycle Z on X defines an element $\gamma(Z) =$ (class of $\mathcal{O}_Z$) in $K(X)$, and $Z \mapsto \gamma(Z)$ may be extended by linearity to a map $\gamma: C^*(X) \to K(X)$. If $K^r(X)$ denotes the subgroup of $K(X)$ generated by all coherent sheaves supported in codimension $\geq r$, then $\sum_{s \leq r} C^s(X)$ maps onto $K^r(X)$. The groups $K^r(X)$ define a decreasing filtration of $K(X)$ and the associated graded group $GK^*(X)$ becomes a ring under the product law defined by Tor, that is, under $[M][N] = \sum (-1)^r [\underline{\mathrm{Tor}}_r^{\mathcal{O}_X}(M, N)]$ where $[M]$ denotes the class of M in $GK^*(X)$. Serre's description of intersection products in terms of Tor (Serre [9]) shows that the homomorphism of graded groups, $\gamma': C^*(X) \to GK^*(X)$, induced by γ preserves products: $\gamma'(W.Z) = \gamma'(W)\gamma'(Z)$. Since cycles that are rationally equivalent to zero map to zero under γ', this shows that γ' defines a surjective homomorphism of graded rings $\phi: CH^*(X) \to GK^*(X)$, where $CH^*(X)$ denotes the Chow ring of X (Grothendieck [2, p. 150]). (For a recent exposition of the Chow ring, see Douady-Verdier [1, I, II].)

Grothendieck [2, p. 147–151] also defines a completed Chern class map $\tilde{c}: K(X) \to (H^*(X))^\sim$ where $(H^*(X))^\sim$ is a certain *λ-ring* defined by $H^*(X)$. On passing to the associated graded objects, we deduce a map of graded rings $\psi: GK^*(X) \to H^*(X)'$ where, as graded groups, $H^*(X)'$ equals $H^*(X)$, but $H^*(X)'$ has the multiplicative structure

$$x_r * x_s = \frac{-(r + s - 1)!}{(r - 1)!(s - 1)!} x_r x_s, \qquad x_r \in H^r(X), \qquad x_s \in H^s(X);$$

compare [SGA. 6, V.6.5]. The map ψ is functorial in X.

Assume now that $(\dim(X) - 1)!$ is invertible in Λ, that is, that if p is prime and p divides n, then $p \geq \dim(X)$. Then there is an isomorphism

$$x_r \mapsto x_r/(-1)^{r-1}(r-1)!: H^*(X)' \to H^*(X).$$

Define cl'_X to be the composite of the maps

$$C^*(X) \twoheadrightarrow CH^*(X) \xrightarrow{\phi} GK^*(X) \xrightarrow{\psi} H^*(X)' \xrightarrow{\approx} H^*(X).$$

This has the properties:

C1. cl'_X is a homomorphism of graded rings (that doubles degrees);

C2. if Z is rationally equivalent to W, then $cl'_X(Z) = cl'_X(W)$;

C3. cl'_X is functorial in X (provided we restrict to varieties X such that $(\dim(X) - 1)!$ is invertible in Λ).

Of these, (C1) and (C3) hold because cl'_X is a composite of homomorphisms that are functorial in X and (C2) follows directly from the definition.

LEMMA 10.5. *Let E be a vector bundle of rank r on X, and let $s: X \to E$ be a section of E such that $s(X)$ intersects the zero section of E transversally, so that the zero set of s is a smooth subscheme Z of X. Then $cl'_X(Z) = cl_X(Z)$.*

Proof. According to Theorem 2 of Grothendieck [2], which holds in our situation because of (9.2), $cl_X(Z) = c_r(E)$.

Let $\check{E}$ denote the dual of E. Almost by definition of Z, $\mathcal{O}_Z$ is the cokernel of the map $\check{E} \to \mathcal{O}_X$ that is dual to the map $\mathcal{O}_X \to E$ induced by s. The Koszul complex (compare Hartshorne [1, III.7]) gives a resolution

$$0 \to \Lambda^r(\check{E}) \to \Lambda^{r-1}(\check{E}) \to \cdots \to \check{E} \to \mathcal{O}_X \to \mathcal{O}_Z \to 0$$

of $\mathcal{O}_Z$ by locally free sheaves. Thus $[\mathcal{O}_Z] = \sum(-1)^i[\Lambda^i(\check{E})]$ in $K(X)$. The image of $\Lambda^i(\check{E})$ in $H^*(X)'$ is $(-1)^r$ times the image of $\Lambda^i(E)$, and the image of $\sum(-1)^i[\Lambda^i(E)]$ is $-(r-1)!c_r(E)$; compare [SGA. 6, V.6.6.1]. Thus

$$cl'_X(Z) = ((-1)^{r-1}(r-1)!)^{-1}(-1)^r(-(r-1)!)c_r(E) = c_r(E) = cl_X(Z).$$

PROPOSITION 10.6. *The two maps $cl_X, cl'_X: C^*(X) \to H^*(X)$ are equal.*

Proof. It suffices to show that $cl_X(Z) = cl'_X(Z)$ for Z a prime r-cycle on X. We may replace X by an open subvariety U provided the codimension of $X - U$ in X is $>r$, for then (9.1) shows that $H^{2r}(X, \Lambda(r)) \to H^{2r}(U, \Lambda(r))$ is injective. Thus we may assume that Z is smooth and that it is defined (as a scheme) by equations $f_1 = f_2 = \cdots = f_r = 0$ in some neighborhood U of Z. Define E to be the vector bundle on X whose restrictions to U and $V = X - Z$ are trivial and that is such that the patching map $\mathbb{A}^r_{U \cap V} \to \mathbb{A}^r_{U \cap V}$ is $(u, t_1, \ldots, t_r) \mapsto (u, t_1/f_1(u), \ldots, t_r/f_r(u))$.

Then $s: X \to E$ such that $s|U$ is $u \mapsto (u, f_1(u), \ldots, f_r(u))$ and $s|V$ is $u \mapsto (u, 1, \ldots, 1)$ is a section of E over X that intersects the zero section transversally and has Z as its zero set. Thus, according to (10.5), $cl'_X(Z) = cl_X(Z)$.

COROLLARY 10.7. *Assume that* (dim $(X) - 1)!$ *is invertible in* Λ. *Then the map* cl_X *(as defined in Section* 9) *is a homomorphism of graded rings* $CH^*(X) \to H^*(X)$ *that is functorial in* X.

Remark 10.8. If we replace Λ by $\mathbb{Q}_l$ in the above, then we obtain exactly the same results except that now the condition that (dim $(X) - 1)!$ be invertible in $\mathbb{Q}_l$ is vacuous.

§11. The Poincaré Duality Theorem

Throughout Λ denotes the constant sheaf $\mathbb{Z}/(n)$, where n will always be prime to char (X) for any scheme X that occurs.

We first state and prove the Poincaré duality theorem for a smooth separated variety over a separably closed field. (Recall (3.3e) that separated varieties are compactifiable.) Then we show how the theorem implies an important geometric result: the group of numerical equivalence classes of cycles on a smooth projective variety is finitely generated. Finally we state a relative Poincaré duality theorem for a smooth compactifiable morphism.

In mild disagreement with the notations of the last two sections, we shall write $cl_X(P)$ for the image of $s_{P/X}$ under the canonical map $H_P^{2d}(X, \Lambda(d)) \to H_c^{2d}(X, \Lambda(d))$ when P is a closed point on a separated variety.

THEOREM 11.1. (Poincaré duality). *Let X be a smooth separated variety of dimension d over a separably closed field k.*

(a) *There is a unique map* $\eta(X): H_c^{2d}(X, \Lambda(d)) \to \Lambda$ *such that* $cl_X(P) \mapsto 1$ *for any closed point P of X; moreover, $\eta(X)$ is an isomorphism.*

(b) *For any constructible sheaf F of Λ-modules on X, the canonical pairings*

$$H_c^r(X, F) \times \operatorname{Ext}_X^{2d-r}(F, \Lambda(d)) \to H_c^{2d}(X, \Lambda(d)) \xrightarrow[\approx]{\eta(X)} \Lambda$$

are nondegenerate.

As in (V.2), this has the immediate consequence:

COROLLARY 11.2. *For any locally constant, constructible sheaf F of Λ-modules on X, the cup-product pairings*

$$H_c^r(X, F) \times H^{2d-r}(X, \check{F}(d)) \to H_c^{2d}(X, \Lambda(d)) \approx \Lambda$$

are nondegenerate.

Before proving (a) of the theorem, we need two lemmas.

LEMMA 11.3. *For any separated variety X of dimension d over a separably closed field, $H_c^{2d}(X, \Lambda(d)) \approx \Lambda$.*

Proof. We use induction on d. If X_0 is a dense open subvariety of X, then the exact sequence (III.1.30)

$$\cdots \to H_c^r(X_0, \Lambda(d)) \to H_c^r(X, \Lambda(d)) \to H_c^r(X - X_0, \Lambda(d)) \to \cdots$$

and (1.1) show that $H_c^{2d}(X_0, \Lambda(d)) \xrightarrow{\approx} H_c^{2d}(X, \Lambda(d))$. Thus we may replace X by such an X_0 (or X_0 by such an X). Therefore we may assume that there is a smooth variety S and a projective smooth map $\pi: X \to S$ whose fibers are connected of dimension 1 (III.3.13). In

$$\begin{array}{ccccccc} \cdots \to & \underline{\mathrm{Pic}}_{X/S} & \xrightarrow{n} & \underline{\mathrm{Pic}}_{X/S} & \to & R^2\pi_*\Lambda(1) & \to \cdots \qquad \text{(Kummer sequence)} \\ & {\scriptstyle \text{degree}}\big\downarrow & & \big\downarrow{\scriptstyle \text{degree}} & & & \\ \cdots \longrightarrow & \mathbb{Z} & \xrightarrow{n} & \mathbb{Z} & \longrightarrow & \Lambda & \longrightarrow 0 \end{array}$$

the map $\underline{\mathrm{Pic}}_{X/S} \to \Lambda$ factors through $R^2\pi_*\Lambda(1)$ (since it does on each fiber (V.2)), and the resulting map $R^2\pi_*\Lambda(1) \to \Lambda$ is an isomorphism (since it is on each fiber). The spectral sequence (3.2c) gives an isomorphism

$$H_c^{2d}(X, \Lambda(d)) \xrightarrow{\approx} H_c^{2d-2}(S, R^2\pi_*\Lambda(d)),$$

and

$$H_c^{2d-2}(S, R^2\pi_*\Lambda(d)) \xrightarrow{\approx} H_c^{2d-2}(S, \Lambda(d-1)) \xrightarrow{\approx} \Lambda$$

by induction.

LEMMA 11.4. *Let $\pi: Y \to X$ be a separated étale morphism where X is a smooth separated variety of dimension d over a separably closed field. Let P be a closed point of Y and let $Q = \pi(P)$. The map*

$$\pi_*: H_c^{2d}(Y, \Lambda(d)) \to H_c^{2d}(X, \Lambda(d))$$

induced by the map $R_c^0\pi_*\Lambda(d) = \pi_!\pi^*\Lambda(d) \xrightarrow{tr} \Lambda(d)$ defined in (V.1.13) *sends $cl_Y(P)$ to $cl_X(Q)$.*

Proof. Consider a compactification $Y \overset{j}{\hookrightarrow} \bar{Y} \xrightarrow{\bar{\pi}} X$ of π with $\bar{\pi}$ finite. There is a commutative diagram

$$\begin{array}{ccccccc} H^{2d}_{\bar{\pi}^{-1}(Q)}(\bar{Y}, j_!\Lambda(d)) & \to & H^{2d}_P(\bar{Y}, j_!\Lambda(d)) & \to & H^{2d}_c(\bar{Y}, j_!\Lambda(d)) & = & H^{2d}_c(Y, \Lambda(d)) \\ \big\downarrow & & & & \big\downarrow & & \\ H^{2d}_Q(X, \Lambda(d)) & & \longrightarrow & & H^{2d}_c(X, \Lambda(d)) & & \end{array}$$

in which both vertical maps are induced by $\operatorname{tr}:\bar{\pi}_* j_! \Lambda(d) \to \Lambda(d)$. Thus it suffices to show that some inverse image of $s_{P/\bar{Y}}$ in

$$H^{2d}_{\pi^{-1}(Q)}(\bar{Y}, j_! \Lambda(d))$$

maps to $s_{Q/X}$. By excision (III.1.28) we may assume that X is strictly local. But then Y is a disjoint union of copies of open subsets of X, and the assertion is obvious.

We now prove (a) of the theorem. Recall (9.7) that $H^{2d}(\mathbb{P}^d, \Lambda(d))$ is generated by $cl_{\mathbb{P}^d}(P)$ for any closed point P of $\mathbb{P}^d$ and that this class is independent of P; thus we have a unique choice for $\eta(\mathbb{P}^d)$. Let X be as in the theorem, and fix a closed point $P_0 \in X$. According to (I.3.24), there exists an open neighborhood X_0 of P_0 and maps

$$\begin{array}{ccccc} X & \stackrel{j}{\hookleftarrow} & X_0 & & \\ & & \downarrow{\scriptstyle \pi} & & \\ & & U & \stackrel{j}{\hookrightarrow} & \mathbb{P}^d \\ & & \downarrow & \swarrow & \\ & & S = \operatorname{spec} k & & \end{array}$$

in which both maps j are open immersions and π is étale and separated. Consider

$$H^{2d}_c(X, \Lambda(d)) \xleftarrow{\approx} H^{2d}_c(X_0, \Lambda(d)) \xrightarrow{\pi_*} H^{2d}_c(U, \Lambda(d))$$
$$\xrightarrow{\approx} H^{2d}(\mathbb{P}^d, \Lambda(d)) \xrightarrow[\approx]{\eta(\mathbb{P}^d)} \Lambda.$$

For any Q in U, $cl_U(Q)$ is mapped to $1 \in \Lambda$ and, according to (11.4), the same is true of any P in X_0. This shows that $H^{2d}_c(X_0, \Lambda(d)) \to \Lambda$ is surjective, and (11.3) implies that it is an isomorphism. We define $\eta(X)$ to be the composite $H^{2d}_c(X, \Lambda(d)) \xrightarrow{\approx} \Lambda$. Clearly $cl_X(P) \mapsto 1$ for any $P \in X_0$, and this property (for a single P) determines $\eta(X)$ uniquely. Thus, if we begin with a different P_0 and X_0, we obtain the same map, which shows that $cl_X(P) \mapsto 1$ for any closed point $P \in X$.

We now prove (b) of the theorem by induction on the dimension d of X. Note that the case $d = 1$ is proved in (V.2). We write $\phi^r(X, F)$ for the map $\operatorname{Ext}^{2d-r}_X(F, \Lambda(d)) \to H^r_c(X, F)^\vee$ induced by the pairing in the theorem. (The reader is advised to skip the proofs that diagrams commute, at least on a first reading.)

Step 1. Let $\pi: X' \to X$ be a finite étale map, where X' and X are varieties as in (11.1); for a sheaf F on X', $\phi^r(X', F)$ is an isomorphism if and only if $\phi^r(X, \pi_* F)$ is an isomorphism.

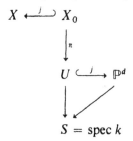

Proof. The degree of π will be constant, and so following (V.1.14), we only have to show that

$$\begin{array}{ccc} H_c^{2d}(X', \Lambda(d)) & \xrightarrow{\eta(X')} & \Lambda \\ \Big\downarrow{\scriptstyle tr} & & \Big\| \\ H_c^{2d}(X, \Lambda(d)) & \xrightarrow{\eta(X)} & \Lambda \end{array}$$

commutes, but this follows from (11.4).

Step 2. $\phi^r(X, F)$ is an isomorphism if F has support on a smooth closed subvariety $Z \neq X$ of X.

Proof. Regard F as a sheaf on Z. We have to show that there exists a commutative diagram

$$(*) \qquad \begin{array}{ccccccc} H_c^r(Z, F) & \times & \mathrm{Ext}_Z^{2a-r}(F, \Lambda(a)) & \longrightarrow & H_c^{2a}(Z, \Lambda(a)) & \xrightarrow{\eta(Z)} & \Lambda \\ \approx\Big\uparrow{\scriptstyle i^*} & & \approx\Big\downarrow{\scriptstyle i_*} & & \approx\Big\downarrow{\scriptstyle i_*} & & \Big\| \\ H_c^r(X, i_*F) & \times & \mathrm{Ext}_X^{2d-r}(i_*F, \Lambda(d)) & \longrightarrow & H_c^{2d}(X, \Lambda(d)) & \xrightarrow{\eta(X)} & \Lambda. \end{array}$$

where a is the dimension of Z.

Let $\bar{X}$ be a compactification of X and let $\bar{Z}$ be the closure of Z in $\bar{X}$. Thus we have a diagram:

$$\begin{array}{ccc} Z & \xhookrightarrow{i} & X \\ \Big\downarrow{\scriptstyle j} & & \Big\downarrow{\scriptstyle j} \\ \bar{Z} & \xhookrightarrow{i} & \bar{X}. \end{array}$$

Note that $j_!i_* = i_*j_!$ and that

$$H_c^{2a}(Z, \Lambda(a)) \xrightarrow{\approx} H^{2a}(\bar{Z}, \Lambda(a))$$

and

$$H_c^{2d}(X, \Lambda(d)) \xrightarrow{\approx} H^{2d}(\bar{X}, \Lambda(d)).$$

Thus we need a diagram

$$\begin{array}{ccccccc} H^r(\bar{Z}, j_!F) & \times & \mathrm{Ext}_Z^{2a-r}(F, \Lambda(a)) & \longrightarrow & H^{2a}(\bar{Z}, \Lambda(a)) & \longrightarrow & \Lambda \\ \approx\Big\uparrow{\scriptstyle i^*} & & \approx\Big\downarrow{\scriptstyle i_*} & & \approx\Big\downarrow{\scriptstyle i_*} & & \Big\| \\ H^r(\bar{X}, i_*j_!F) & \times & \mathrm{Ext}_X^{2d-r}(i_*F, \Lambda(d)) & \longrightarrow & H^{2d}(\bar{X}, \Lambda(d)) & \longrightarrow & \Lambda \end{array}$$

The left hand part of this diagram was constructed in (6.7c), and the right hand square commutes because of (6.5):

$$s_{P/Z} \in H_P^{2a}(Z, \Lambda(a)) \longrightarrow H_c^{2a}(Z, \Lambda(a)) \longrightarrow H^{2a}(\bar{Z}, \Lambda(a))$$

$$\downarrow \qquad\qquad \downarrow^{i_*} \qquad\qquad \downarrow^{i_*} \qquad\qquad \downarrow^{i_*}$$

$$s_{P/X} \in H_P^{2d}(X, \Lambda(d)) \longrightarrow H_c^{2d}(X, \Lambda(d)) \longrightarrow H^{2d}(\bar{X}, \Lambda(d)).$$

Step 3.   Let U be an open subvariety of X. Then $\phi^r(X, F)$ is an isomorphism for all r if and only if $\phi^r(U, F|U)$ is an isomorphism for all r.

Proof.   Construct a sequence $X = X_0 \supset X_1 \supset X_2 \supset \cdots \supset X_s = U$ such that $X_i - X_{i+1}$ is a smooth variety and is closed in X_i. A five-lemma argument, using Step 2, shows that $\phi^r(X_i, F|X_i)$ is an isomorphism for all r if and only if $\phi^r(X_{i+1}, F|X_{i+1})$ is an isomorphism for all r (compare Step 3 of (V.2.1)).

Step 4.   Let X be a variety for which (11.1) holds, and let $F^{\cdot}$ be a complex of sheaves on X that is bounded below and such that $H^r(F^{\cdot})$ is zero for $r \gg 0$ and constructible for all r. Then there are canonical nondegenerate pairings

$$\mathbb{H}_c^r(X, F^{\cdot}) \times \mathbb{E}\mathrm{xt}_X^{2d-r}(F^{\cdot}, \Lambda(d)) \to H_c^{2d}(X, \Lambda(d)) \xrightarrow[\approx]{\eta} \Lambda.$$

Proof.   Let $j_!F^{\cdot} \tilde{\to} I^{\cdot}(j_!F^{\cdot})$ be a quasi-isomorphism of $j_!F^{\cdot}$ into a bounded below complex of injective sheaves, and let $j_!\Lambda(d) \to I^{\cdot}(j_!\Lambda(d))$ be an injective resolution of $j_!\Lambda(d)$. Then $\mathbb{H}_c^r(X, F^{\cdot}) = $ (homotopy classes of maps $\Lambda_{\bar{X}} \to I^{\cdot}(j_!F^{\cdot})[r]$) and $\mathbb{E}\mathrm{xt}_X^{2d-r}(F^{\cdot}, \Lambda(d)) = $ (homotopy classes of maps $j_!F^{\cdot} \to I^{\cdot}(j_!\Lambda(d))[2d - r]$, or $I^{\cdot}(j_!F^{\cdot}) \to I^{\cdot}(j_!\Lambda(d))[2d - r]$), and so the pairing may be defined to be composition of homotopy classes, as before (compare Hartshorne [1, I.§6]). Define:

$$\sigma_{\geq r}(F^{\cdot}) = \cdots \to 0 \to \mathrm{im}\,(d^{r-1}) \hookrightarrow F^r \to F^{r+1} \to \cdots$$

$$\sigma'_{\geq r}(F^{\cdot}) = \cdots \to 0 \to F^r/\mathrm{im}\,(d^{r-1}) \to F^{r+1} \to \cdots .$$

There is a quasi-isomorphism $\sigma_{\geq r}(F^{\cdot}) \tilde{\to} \sigma'_{\geq r}(F^{\cdot})$ and an exact sequence

$$0 \to H^r(F^{\cdot}) \to \sigma'_{\geq r}(F^{\cdot}) \to \sigma_{\geq r+1}(F^{\cdot}) \to 0.$$

For $r_0 \gg 0$, $\sigma_{\geq r_0}(F^{\cdot})$ is exact, and both groups to be paired are zero; thus the pairing is trivially nondegenerate. By applying a five-lemma argument to the long exact sequences arising from the above short exact sequence, we may deduce that the pairing is nondegenerate for $\sigma'_{\geq r_0 - 1}(F^{\cdot})$. The quasi-isomorphism $\sigma_{\geq r_0 - 1}(F^{\cdot}) \tilde{\to} \sigma'_{\geq r_0 - 1}(F^{\cdot})$ now shows that it is nondegenerate for $\sigma_{\geq r_0 - 1}(F^{\cdot})$, and we may continue inductively. For $r \ll 0$ we have $\sigma_{\geq r}(F^{\cdot}) = F^{\cdot}$.

Step 5. Let S be a smooth separated variety of dimension $d - 1$ over a separably closed field k, and let $\pi: X \to S$ be projective and smooth with all fibers of dimension 1. Then $\phi^r(X, F)$ is an isomorphism for any locally constant, constructible sheaf F on X.

Proof. We use the fact that we know the theorem for complexes of sheaves on S (by Step 4 and induction) and for the fibers of π. Since π is projective, it is easy to construct a diagram

$$\begin{array}{ccc} X & \xhookrightarrow{\ j\ } & \bar{X} \\ {\scriptstyle\pi}\big\downarrow & & \big\downarrow{\scriptstyle\bar{\pi}} \\ S & \xhookrightarrow{\ j\ } & \bar{S} \\ \big\downarrow & \swarrow & \\ \operatorname{spec} k & & \end{array}$$

in which $\bar{\pi}$ is projective, $\bar{S}$ is a compactification of S, $\bar{X}$ is the closure of X, and $\bar{\pi}^{-1}(S) = X$. Let $j_!\Lambda(d) \to \Lambda^{\cdot}_! = (j_!\Lambda(d))^{\cdot}$ and $j_!\Lambda_S(d-1) \to \Lambda^{\cdot}_{S!} = (j_!\Lambda_S(d-1))^{\cdot}$ be injective resolutions and consider

$$\begin{array}{ccccc} \mathrm{Hom}_{\bar{X}}(\Lambda, F^{\cdot}_![r]) & \times & \mathrm{Hom}_{\bar{X}}(F^{\cdot}_![r], \Lambda^{\cdot}_![2d]) & \longrightarrow & \mathrm{Hom}_{\bar{X}}(\Lambda, \Lambda^{\cdot}_![2d]) \\ \big\updownarrow{\scriptstyle\approx} & & \big\downarrow & & \big\updownarrow{\scriptstyle\approx} \\ \mathrm{Hom}_{\bar{S}}(\Lambda, \bar{\pi}_*F^{\cdot}_![r]) & \times & \mathrm{Hom}_{\bar{S}}(\bar{\pi}_*F^{\cdot}_![r], \bar{\pi}_*\Lambda^{\cdot}_![2d]) & \longrightarrow & \mathrm{Hom}_{\bar{S}}(\Lambda, \bar{\pi}_*\Lambda^{\cdot}_![2d]) \\ & & \big\downarrow & & \big\downarrow \\ & & \mathrm{Hom}_{\bar{S}}(\bar{\pi}_*F^{\cdot}_![r], \Lambda^{\cdot}_{S!}[2d-2]) & \longrightarrow & \mathrm{Hom}_{\bar{S}}(\Lambda, \Lambda^{\cdot}_{S!}[2d-2]) \end{array}$$

where $j_!F \to F^{\cdot}_! = (j_!F)^{\cdot}$ is an injective resolution of $j_!F$; the isomorphisms come from writing $\Lambda_X = \bar{\pi}^*\Lambda_S$ and using the adjointness of $\bar{\pi}_*$ and $\bar{\pi}^*$, and the lower two vertical arrows come from the map $\bar{\pi}_*(j_!\Lambda(1))^{\cdot} \to (j_!\Lambda_S)^{\cdot}[-2]$ induced by the canonical map $R^2\pi_*\Lambda(1) \to \Lambda$ defined in the proof of (11.3). On passing to homotopy classes of maps and identifying $\mathrm{Ext}^s_{\bar{X}}(j_!F, j_!\Lambda)$ with $\mathrm{Ext}^s_X(F, \Lambda)$, etc., we find the commutative diagram

$$\begin{array}{ccccccc} H^r_c(X, F) & \times & \mathrm{Ext}^{2d-r}_X(F, \Lambda(d)) & \longrightarrow & H^{2d}_c(X, \Lambda(d)) & \xrightarrow[\approx]{\eta(X)} & \Lambda \\ \big\uparrow{\scriptstyle\approx} & & \big\downarrow{\scriptstyle\beta} & & \big\downarrow{\scriptstyle\gamma} & & \\ \mathbb{H}^r_c(S, \mathbf{R}\pi_*F) & \times & \mathbb{E}\mathrm{xt}^{2d-2-r}_S(\mathbf{R}\pi_*F, \Lambda(d-1)) & \longrightarrow & H^{2d-2}_c(S, \Lambda(d-1)) & \xrightarrow[\approx]{\eta(S)} & \Lambda \end{array}$$

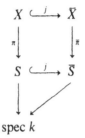

where $\mathbf{R}\pi_* F = \pi_* F^{\cdot}$, $F \to F^{\cdot}$ being an injective resolution of F, and we have made use of the obvious quasi-isomorphism $j_! \pi_* F^{\cdot} \xrightarrow{\approx} \bar{\pi}_*(j_! F)^{\cdot}$. The map γ is an isomorphism because $R^2 \pi_* \Lambda(d) \xrightarrow{\approx} \Lambda(d-1)$, and the bottom pairing is nondegenerate by Step 4. Thus it remains to show that β is an isomorphism. By a similar process to the above we may define a map $\tilde{\beta}: \pi_* \underline{\mathrm{Hom}}_X(F, \Lambda(1)^{\cdot}) \to \underline{\mathrm{Hom}}_S(\mathbf{R}\pi_* F, \Lambda^{\cdot}[-2])$ such that $H^{2d-r}(\Gamma(S, \tilde{\beta})) = \beta$ (apart from a Tate twist). Since both complexes are of flabby sheaves (III.2.13c), it suffices to show that $\tilde{\beta}$ is a quasi-isomorphism (Appendix C), and this may be checked on stalks. Write $\underline{\mathrm{Hom}}_X(F, \Lambda(1)^{\cdot}) = \check{F}(1)^{\cdot}$; it is a flabby resolution of $\check{F}(1)$. For any geometric point $\bar{s}$ of S,

$$H^r(\pi_* \check{F}(1)^{\cdot}_{\bar{s}}) = (H^r(\pi_* \check{F}(1)^{\cdot}))_{\bar{s}} = (R^r \pi_* \check{F}(1))_{\bar{s}} = H^r(X_{\bar{s}}, \check{F}(1) | X_{\bar{s}})$$

according to (2.3) whereas

$$H^r(\underline{\mathrm{Hom}}_S(\mathbf{R}\pi_* F, \Lambda^{\cdot}[-2])_{\bar{s}}) = H^{r-2}(\mathrm{Hom}((\mathbf{R}\pi_* F)_{\bar{s}}, \Lambda^{\cdot}_{\bar{s}}))$$

by a slight generalization of (III.1.31). (Note that $H^r(\mathbf{R}\pi_* F)$ is locally constant and constructible for all r.) Since $\Lambda \to \Lambda^{\cdot}_{\bar{s}}$ is a quasi-isomorphism of injective Λ-modules, $\Lambda^{\cdot}_{\bar{s}}$ may be replaced by Λ, and we see that $\tilde{\beta}_{\bar{s}}$ is a map

$$H^r(X_{\bar{s}}, \check{F}(1) | X_{\bar{s}}) \to \mathrm{Hom}(H^{2-r}(X_{\bar{s}}, F | X_{\bar{s}}), \Lambda).$$

A routine calculation shows that $\tilde{\beta}_{\bar{s}}$ agrees with the map in (V.2.2) and hence is an isomorphism.

Step 6. If F is constant, $\phi^r(X, F)$ is an isomorphism.

Proof. According to (III.3.13), we may find an open dense subvariety X_0 of X and maps

$$\begin{array}{ccccc} X & \hookleftarrow & X_0 & \hookrightarrow & \bar{X}_0 \\ & & & \searrow & \downarrow \pi \\ & & & & S \end{array}$$

such that $\pi: \bar{X}_0 \to S$ satisfies the hypothesis of Step 5. There exists a constant sheaf $\bar{F}$ on $\bar{X}_0$ such that $F | X_0 = \bar{F} | X_0$. According to Step 3, it suffices to prove that $\phi^r(\bar{X}_0, \bar{F})$ is an isomorphism, but Step 5 does this.

Step 7. If F is locally constant, $\phi^r(X, F)$ is an isomorphism.

Proof. According to (V.1.1), we know that there is a finite étale map $\pi: X' \to X$ such that $F | X'$ is constant. Since $\phi^r(X', F | X')$ is an isomorphism for all r, it follows from Step 1 that $\phi^r(X, \pi_*(F | X'))$ is an isomorphism for all r. The step may now be proved by induction on r, using the exact sequence

$$0 \to F \to \pi_*(F | X') \to F_1 \to 0.$$

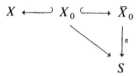

(Note that F_1 is also locally constant, and

$$\mathrm{Ext}^{2d-r}(F, \Lambda(d)) = H^{2d-r}(\check{F}(d)) = 0$$

for $r < 0$.)

Step 3 now completes the proof of the theorem because any constructible sheaf is locally constant over an open subvariety of X (V.1.8).

COROLLARY 11.5. *Let X be a smooth separated variety of dimension d over a separably closed field. If X is not complete, then $H^{2d}(X, F) = 0$ for any locally constant torsion sheaf F on X.*

Proof. We may assume that F is constructible, and then (11.2) shows that $H^{2d}(X, F) = H_c^0(X, F)^\vee = 0$.

Remark 11.6. Let $\pi: Y \to X$ be a proper map of smooth separated varieties over a separably closed field, and let $a = \dim(X)$, $d = \dim(Y)$ and $c = d - a$. From (3.2c) we know there are maps

$$\pi^*: H_c^{2d-r}(X, \Lambda(d)) \to H_c^{2d-r}(Y, \Lambda(d)).$$

By duality (11.2), π^* induces a map $\pi_*: H^r(Y, \Lambda) \to H^{r-2c}(X, \Lambda(-c))$, which is uniquely determined by the equation

$$\eta_X(\pi_*(y) \cup x) = \eta_Y(y \cup \pi^*(x)), \qquad x \in H_c^{2d-r}(X, \Lambda(d)), \qquad y \in H^r(Y, \Lambda),$$

where η_X and η_Y are the trace maps (11.1).

The maps π_* have the following properties.

(a) If π is étale (hence finite), then π_* is the trace map (V.1.12). This is obvious from the proof of Step 1 above.

(b) If π is a closed immersion, then π_* is the Gysin map. This is obvious from the proof of Step 2 above.

(c) If $Z \overset{\pi_1}{\to} Y \overset{\pi_2}{\to} X$ are proper, then $(\pi_2\pi_1)_* = \pi_{2*}\pi_{1*}$; for

$$\begin{aligned}\eta_X(\pi_{2*}\pi_{1*}(z) \cup x) = \eta_Y(\pi_{1*}(z) \cup \pi_2^*(x)) &= \eta_Z(z \cup \pi_1^*\pi_2^*(x)) \\ &= \eta_Z(z \cup (\pi_2\pi_1)^*(x)) \\ &= \eta_Z((\pi_2\pi_1)_* z \cup x).\end{aligned}$$

(d) If Y and X are complete, then $\eta_X(\pi_*(y)) = \eta_Y(y)$ for $y \in H^{2d}(Y, \Lambda(d))$; moreover, $\pi_*(y \cup \pi^*(x)) = \pi_*(y) \cup x$ for $x \in H^r(X)$, $y \in H^s(Y)$.

Indeed,

$$\eta_X(\pi_*(y)) = \eta_X(\pi_*(y) \cup 1_X) = \eta_Y(y \cup \pi^*(1_X)) = \eta_Y(y);$$

moreover,

$$\eta_X(\pi_*(y \cup \pi^*(x)) \cup x') = \eta_Y(y \cup \pi^*(x \cup x')) = \eta_X(\pi_*(y) \cup x \cup x')$$

all $x' \in H^{2d-r-s}(X)$.

(e) If π is a closed immersion and X is complete, then $\eta_X(cl_X(Y) \cup x) = \eta_Y(\pi^*(x))$ for all $x \in H^{-2c}(X, \Lambda(-c))$; this equation characterizes $cl_X(Y)$.

This follows directly from (b) above.

THEOREM 11.7. *Let X be a projective variety of dimension d over a separably closed field; the group N^r of r-cycles on X modulo numerical equivalence is finitely generated.*

Proof. Since X is projective, the intersection product defines a pairing $CH^r(X) \times CH^{d-r}(X) \to \mathbb{Z}$ and, by definition of N^r, this becomes a nondegenerate pairing $N^r \times N^{d-r} \to \mathbb{Z}$ on the quotient groups. Thus N^r is torsion free, and it suffices to show that $N^r \otimes_{\mathbb{Z}} \mathbb{Q}_l$ is finite-dimensional. Since the intersection product on CH^* agrees with the cup-product on $H^*(X, \mathbb{Q}_l)$ (10.8) and the latter is nondegenerate, the kernel of the map

$$cl_X : CH^r(X) \otimes \mathbb{Q}_l \to H^{2r}(X, \mathbb{Q}_l(r))$$

is contained in the kernel of $CH^r(X) \otimes \mathbb{Q}_l \to N^r(X) \otimes \mathbb{Q}_l$. As $H^{2r}(X, \mathbb{Q}_l(r))$ is finite-dimensional (2.8), so also is $N^r(X) \otimes \mathbb{Q}_l$.

We now state the relative Poincaré duality theorem. To avoid notational complications we use the language of derived categories (Hartshorne [1, Chapter I] and [SGA. $4\frac{1}{2}$, C.D.]). In particular $D(X)$ denotes the derived category of the category of sheaves of Λ-modules on X_{et}. Also, to preserve the symmetry of the formulae, we write $\mathbf{R}\pi_!$ for $\mathbf{R}_c\pi_*$. First we define a trace map.

PROPOSITION 11.8. *Let $\pi : X \to S$ be a smooth compactifiable morphism, all of whose fibers have dimension d. There exists a unique map $\eta = \eta(X/S) : R^{2d}\pi_!\Lambda(d) \to \Lambda$ such that, for any geometric point $\bar{s}$ of S and any closed point P of $X_{\bar{s}}$, $\eta_{\bar{s}} : H_c^{2d}(X_{\bar{s}}, \Lambda(d)) \to \Lambda$ maps $cl_X(P)$ to 1. If the fibers of π are connected, then η is an isomorphism.*

Proof. If S is the spectrum of a separably closed field, then X is a disjoint union $X = X_1 \cup \cdots \cup X_r$ of smooth varieties. In this case $\eta(X/S)$ must be defined to be $\sum \eta(X_i)$ with $\eta(X_i)$ as in (11.1a); it therefore exists, is unique, and is an isomorphism if X is connected.

In the general case $\eta(X/S)$ is uniquely determined since it is on stalks. Thus it suffices to define it locally for the étale topology on S. We may therefore assume that S is connected and that X is a disjoint union of smooth S-schemes with connected fibers, and we need only consider the case that X itself has connected fibers. Moreover, we may assume that $\pi : X \to S$ has a section $i : S \to X$ (I.3.26). Define $R_S^{2d}\pi_*\Lambda(d)$ to be the sheaf on S_{et} associated with the presheaf $T \mapsto H_T^{2d}(X_{(T)}, \Lambda(d))$ where we regard T as a subscheme of $X_{(T)}$ through $i_{(T)} : T \to X_{(T)}$. According to (6.1), there is a unique isomorphism $H_T^{2d}(X_{(T)}, \Lambda(d)) \xrightarrow{\approx} \Lambda$ such that $s_{S/X} \mapsto 1$. The maps

$$H^{2d}_T(X_{(T)}, \Lambda(d)) \to H^{2d}_c(X_{(T)}, \Lambda(d))$$

induce a canonical map

$$R^{2d}_S \pi_* \Lambda(d) \to R^{2d}\pi_! \Lambda(d).$$

This is an isomorphism since its stalk at any geometric point $\bar{s}$ of S is

$$H^{2d}_{\bar{s}}(X_{\bar{s}}, \Lambda(d)) \xrightarrow{\approx} H^{2d}_c(X_{\bar{s}}, \Lambda(d)).$$

We define $\eta(X/S)$ to be

$$R^{2d}\pi_!\Lambda(d) \approx R^{2d}_S\pi_*\Lambda(d) \xrightarrow{\approx} \Lambda.$$

Exercise 11.9. Show that $\eta(X/S)$ is compatible with change of base and composition of morphisms, is the canonical morphism coming from the adjunction formula when X/S is étale, and is the map arising from the Kummer sequence when $X \to S$ is proper with fibers of dimension one. (Only compatibility with composites, $\pi = \pi_2\pi_1$, is nontrivial. One may assume that the base scheme is the spectrum of a separably closed field and that π_1 has a section. Then an argument similar to that in the last paragraph works.)

Let $\pi: X \to S$ be as in (11.8). The *twisted inverse image* functor $\pi^!$ maps a complex $G^{\cdot}$ of sheaves on S to $\pi^! G^{\cdot} = (\pi^* G^{\cdot} \otimes \Lambda(d))[2d]$. Clearly $\pi^!$ defines a functor $\pi^!: D(S) \to D(X)$.

For any sheaf G on S there are canonical maps

$$\mathbf{R}\pi_!\pi^!(G) \to R^{2d}\pi_!(\pi^*G \otimes \Lambda(d)) \to \pi^*G \otimes R^{2d}\pi_!\Lambda(d).$$

On composing this with $1 \otimes \eta(X/S)$, we obtain a trace morphism $\mathbf{R}\pi_!\pi^!(G) \to G$. This definition extends to complexes of sheaves $G^{\cdot}$.

As $\mathbf{R}\pi_!$ is a functor, there is, for any $F^{\cdot} \in D^-(X)$ and $G^{\cdot} \in D^+(X)$, a morphism

$$\mathbf{R}\pi_*\mathbf{R}\,\underline{\mathrm{Hom}}\,(F^{\cdot}, G^{\cdot}) \to \mathbf{R}\,\underline{\mathrm{Hom}}\,(\mathbf{R}\pi_!F^{\cdot}, \mathbf{R}\pi_!G^{\cdot}).$$

On replacing $G^{\cdot}$ by $\pi^! G^{\cdot}$ and using the trace map, we find a morphism

$$\Delta^1_{X/S}: \mathbf{R}\pi_*\mathbf{R}\,\underline{\mathrm{Hom}}_X\,(F^{\cdot}, \pi^!G^{\cdot}) \to \mathbf{R}\,\underline{\mathrm{Hom}}_S\,(\mathbf{R}\pi_!F^{\cdot}, G^{\cdot}).$$

On applying $\mathbf{R}\Gamma(S, -)$, we find a morphism

$$\Delta^2_{X/S}: \mathbf{R}\,\mathrm{Hom}_X\,(F^{\cdot}, \pi^!G^{\cdot}) \to \mathbf{R}\,\mathrm{Hom}_S\,(\mathbf{R}\pi_!F^{\cdot}, G^{\cdot})$$

which, on cohomology, is

$$\Delta^3_{X/S}: \mathbb{E}\mathrm{xt}^r_X\,(F^{\cdot}, \pi^!G^{\cdot}) \to \mathbb{E}\mathrm{xt}^r_S\,(\mathbf{R}\pi_!F^{\cdot}, G^{\cdot}).$$

For example, if $F^{\cdot} = F$ and $G^{\cdot} = \Lambda$, Δ^3 is a map

$$\mathrm{Ext}^{r+2d}_X\,(F, \Lambda(d)) \to \mathrm{Ext}^r_S\,(\mathbf{R}\pi_!F, \Lambda).$$

When S is the spectrum of a separably closed field, Λ_S is injective, and so $\mathrm{Ext}_S^r(\mathbf{R}\pi_! F, \Lambda) = \mathrm{Hom}(H_c^{-r}(X, F), \Lambda)$. In fact, Δ^3 is the map

$$\mathrm{Ext}_X^{2d-r}(F, \Lambda(d)) \to \mathrm{Hom}(H_c^r(X, F), \Lambda)$$

defined by the pairing in (11.1).

THEOREM 11.10. *The maps $\Delta^1_{X/S}, \Delta^2_{X/S}, \Delta^3_{X/S}$ are isomorphisms.*

The proof is given in detail in [SGA. 4, XVIII] and sketched in [SGA. $4\frac{1}{2}$, Arcata VI]. See also Verdier [1].

§12. The Rationality of the Zeta Function

For any variety X over a finite field $k = \mathbb{F}_q$, the zeta function of X is defined to be

$$Z(X, t) = \exp\left(\sum_{n>0} \frac{\nu_n(X)t^n}{n}\right)$$

where $\nu_n(X)$ is the number of points of X with coordinates in $\mathbb{F}_{q^n}$. When X is projective and smooth, the Weil conjectures (Weil [1]) state:

W1. (Rationality) $Z(X, t) = \dfrac{P_1(X, t) \cdots P_{2d-1}(X, t)}{P_0(X, t) \cdots P_{2d}(X, t)}$ where $d = \dim(X)$

and each $P_r(X, t)$ is a polynomial with coefficients in a field of characteristic zero;

W2. (Integrality) $P_0(X, t) = 1 - t$, $P_{2d}(X, t) = 1 - q^d t$, and each $P_r(X, t) = \prod (1 - a_{r,i}t)$ where the $a_{r,i}$ are algebraic integers;

W3. (Functional Equation) $Z(1/q^d t) = \pm q^{d\chi/2} t^{\chi} Z(t)$ where $\chi = (\Delta . \Delta)$;

W4. (Riemann Hypothesis) each $a_{r,i}$ and its conjugates have absolute value $q^{r/2}$;

W5. if X is the specialization of a smooth projective variety Y over a number field, then the degree of $P_r(X, t)$ is equal to the rth Betti number of the complex manifold $Y(\mathbb{C})$.

Conjecture (W1) is a formal consequence of a Lefschetz trace formula, itself an easy consequence of results that we have already proved. Conjecture (W3) follows from the Poincaré duality theorem and (W5) from the smooth specialization of cohomology groups (4.2). Conjectures (W2) and (W4) are more difficult and we only prove (in the next section) the more general form of the Lefschetz trace formula that is required for their proof.

In the next two lemmas and theorem, X is a smooth projective variety over an algebraically closed field, and p and q denote the first and second projection maps $X \times X \rightrightarrows X$. We identify $\mathbb{Q}_l$ with $\mathbb{Q}_l(1)$ and write $H^*(X)$

for the cohomology ring $\bigoplus H^r(X, \mathbb{Q}_l)$ where l is prime to the characteristic. Also we identify $H^*(X \times X)$ with $H^*(X) \otimes H^*(X)$ by means of the Künneth formula.

LEMMA 12.1. *Let $\phi: X \to X$ be an endomorphism of X. For any $b \in H^*(X)$,*

$$p_*(cl_{X \times X}(\Gamma_\phi) \cup q^*(b)) = \phi^*(b).$$

Proof. The map $(1, \phi): X \to X \times X$ has image Γ_ϕ, and $p \circ (1, \phi) = id$.

$$\begin{aligned} p_*(cl_{X \times X}(\Gamma_\phi) \cup q^*(b)) &= p_*((1, \phi)_*(1) \cup q^*(b)) && \text{(Section 9)} \\ &= p_*(1, \phi)_*(1 \cup (1, \phi)^* q^*(b)) && \text{(11.6d)} \\ &= (\mathrm{id})_*(1 \cup \phi^*(b)) && \text{(11.6c)} \\ &= \phi^*(b). \end{aligned}$$

LEMMA 12.2. *Let ϕ be as in (12.1); let (e_i) be a basis for $H^*(X)$, and let (e_i') be the dual basis (so that if $e_i \in H^r$, then $e_i' \in H^{2d-r}$). Then*

$$cl_{X \times X}(\Gamma_\phi) = \sum_i \phi^*(e_i) \otimes e_i'.$$

Proof. Let $cl_{X \times X}(\Gamma_\phi) = \sum a_i \otimes e_i'$. Then, according to (12.1),

$$\begin{aligned} \phi^*(e_j) &= p_*\left(\left(\sum_i a_i \otimes e_i'\right)(1 \otimes e_j)\right) \\ &= p_*(a_j \otimes e_{2d}), \end{aligned}$$

where e_{2d} is the canonical generator of $H^{2d}(X)$. But $p_*(a_j \otimes e_{2d}) = a_j$, which completes the proof.

THEOREM 12.3. (Lefschetz trace formula). *Let $\phi: X \to X$ be an endomorphism such that $(\Gamma_\phi \cdot \Delta)$ is defined. Then*

$$(\Gamma_\phi \cdot \Delta) = \sum_{r=0}^{2d} (-1)^r Tr(\phi \mid H^r(X, \mathbb{Q}_l))$$

Proof. (Compare Steenrod [1, p. 26–27]). Let (e_i^r) be a basis for $H^r(X, \mathbb{Q}_l)$, and let (f_i^{2d-r}) be the dual basis for $H^{2d-r}(X, \mathbb{Q}_l)$. Then

$$cl_{X \times X}(\Gamma_\phi) = \sum_{r,i} \phi^*(e_i^r) \otimes f_i^{2d-r}$$

and

$$\begin{aligned} cl_{X \times X}(\Delta) &= \sum_{r,i} e_i^r \otimes f_i^{2d-r} \\ &= \sum_{r,i} (-1)^{r(2d-r)} f_i^{2d-r} \otimes e_i^r \\ &= \sum_{r,i} (-1)^r f_i^{2d-r} \otimes e_i^r. \end{aligned}$$

Thus

$$cl_{X\times X}(\Gamma_\phi\cdot\Delta)=\sum_{r,i}(-1)^r\phi^*(e_i^r)f_i^{2d-r}\otimes e^{2d}\quad(\text{where } e^{2d}\xmapsto{\eta(X)}1)$$

$$=\sum_{r=0}^{2d}(-1)^r Tr(\phi^*)(e^{2d}\otimes e^{2d})$$

as $\phi^*(e_i^r)f_j^{2d-r}$ is the coefficient of e_j^r when $\phi^*(e_i^r)$ is expressed in terms of the basis (e_j^r). On applying $\eta(X\times X)$ to both sides, we obtain the required formula.

THEOREM 12.4. *Let X be a smooth projective variety of dimension d over a finite field $k=\mathbb{F}_q$. Then*

$$Z(X,t)=\frac{P_1(X,t)\cdots P_{2d-1}(X,t)}{(1-t)P_2(X,t)\cdots(1-q^dt)}$$

where

$$P_i(X,t)=\det(1-Ft\,|\,H^r(\bar{X},\mathbb{Q}_l)),$$

F being the Frobenius endomorphism of $\bar{X}=X\otimes k_{\text{al}}$.

Proof. This follows from (12.3) exactly as (V.2.6) follows from (V.2.5).

Remark 12.5. (a) We have shown only that $Z(X,t)\in\mathbb{Q}_l(t)$, but it follows that $Z(X,t)\in\mathbb{Q}(t)$. Indeed,

$$Z(X,t)=\prod_{x\in X^0}\frac{1}{1-t^{\deg(x)}}=1+a_1t+a_2t^2+\cdots\in\mathbb{Z}[[t]].$$

Moreover, $Z(X,t)\in\mathbb{Q}(t)$ if and only if there exists an N such that $\det(a_{i+j+M})_{0\le i,j\le N}=0$ for all $M\gg 0$ (Bourbaki [1, IV.5, Exercise 3]). But since $Z(X,t)\in\mathbb{Q}_l(t)$, the same criterion shows that the determinants are zero as elements of $\mathbb{Q}_l$, and hence as elements of $\mathbb{Q}$.

(b) In the expression $Z(X,t)=P_1(X,t)\cdots/(1-t)\cdots$ of (12.4) we have not shown that the polynomials P_i are independent of l. However, (a) shows that, after removing any common factors, the numerator and denominator will be independent of l and have coefficients in $\mathbb{Q}$. Of course, the Riemann hypothesis shows that there can be no cancellations.

(c) Write $Z(X,t)=P(t)/Q(t)$ where $P(t),Q(t)\in\mathbb{Q}[t]$, have no common factors, and are of the form

$$P(t)=1+b_1t+\cdots,\ Q(t)=1+c_1t+\cdots.$$

Then I claim that $b_i,c_i\in\mathbb{Z}$. Indeed, suppose some $c_i\notin\mathbb{Z}$; then there exists a prime l and an algebraic number β such that l divides some positive power of β and $Q(\beta)=0$. Thus β is a pole of $Z(X,t)$, but $Z(X,\beta)=1+a_1\beta+a_2\beta^2+\cdots$ converges l-adically, which is a contradiction. Since $Z(X,t)^{-1}\in\mathbb{Z}[[t]]$ the same argument shows the $b_i\in\mathbb{Z}$.

THEOREM 12.6. *For any X as in* (12.4),

$$Z(X, 1/q^d t) = \pm q^{d\chi/2} t^{\chi} Z(t)$$

where $\chi = (\Delta \cdot \Delta)$.

Proof. $F:\bar{X} \to \bar{X}$ has degree q^d; this is obvious if $X \supset \mathbb{A}^d$ (for F on $\mathbb{A}^d$ is induced by

$$T_i \mapsto T_i^q : \bar{k}[T_1, \ldots, T_d] \to \bar{k}[T_1, \ldots, T_d])$$

but, since F is purely inseparable and $\bar{X}$ contains an open subvariety which is étale over $\mathbb{A}^d$, the general case follows. For any point P in $\bar{X}$, (10.8) shows that

$$(F^* cl_{\bar{X}}(P) = cl_{\bar{X}}(F^* P) = cl_{\bar{X}}(q^d P) = q^d cl_{\bar{X}}(P).$$

Thus F acts as multiplication by q^d on $H^{2d}(\bar{X}, \mathbb{Q}_l)$. From this, and the Poincaré duality theorem, it follows that if $\alpha_1, \ldots, \alpha_s$ are the eigenvalues of F acting on $H^r(\bar{X}, \mathbb{Q}_l)$, then $q^d/\alpha_1, \ldots, q^d/\alpha_s$ are the eigenvalues of F acting on $H^{2d-r}(\bar{X}, \mathbb{Q}_l)$. An easy calculation now shows that the required formula holds with χ replaced by

$$\sum (-1)^r \dim H^r(\bar{X}, \mathbb{Q}_l),$$

but (12.3) (with $\phi = \mathrm{id}$) shows that this sum is equal to $(\Delta \cdot \Delta)$. (Compare Hartshorne [2, Appendix C.4.3].)

§13. The Rationality of L-Series

For k a field, X a scheme over k, and E a sheaf on X_{et}, we write $\bar{k}$ for the algebraic closure of k, $\bar{X}$ for $X \otimes_k \bar{k}$, and $\bar{E}$ for the restriction of E to $\bar{X}_{\mathrm{et}}$. Also Λ will denote the constant sheaf defined by a finite local ring and Ω the constant sheaf defined by a finite field extension of $\mathbb{Q}_l$. We always assume that l and the order of Λ are prime to the characteristic of any ground field k.

One aim of this section is to extend the main theorem of the last section to arbitrary separated varieties.

THEOREM 13.1. *Let X be a separated variety over a finite field* $k = \mathbb{F}_q$, *and let F be the Frobenius endomorphism of* $\bar{X}$. *Then*

$$Z(X, t) = \prod_r \det(1 - Ft \mid H_c^r(\bar{X}, \mathbb{Q}_l))^{(-1)^{r+1}}$$

Both for its intrinsic interest, and because it is needed for the induction argument that proves (13.1), we shall consider a more general statement in which X is allowed to be any compactifiable scheme of finite-type over k and $\mathbb{Q}_l$ is replaced by a constructible sheaf of Ω-vector spaces. First we must describe how F acts on cohomology groups. Note that if $\phi: X \to X$

is a proper map and E is a sheaf on X, then ϕ defines a map $H_c^r(X, E) \to H_c^r(X, \phi^*E)$; for ϕ to define an endomorphism of $H_c^r(X, E)$ there must be given a map $\phi^*E \to E$.

Fix $k = \mathbb{F}_q$ and consider the category of schemes over k. The Frobenius $F_{X/k}$ of such a scheme is the k-morphism $X \to X$ that is the identity map on the underlying space of X and the qth power map on $\mathcal{O}_X$. It is functorial in the sense that for any k-morphism $\pi: Y \to X$,

$$\begin{array}{ccc} Y & \xleftarrow{F_{Y/k}} & Y \\ \downarrow{\scriptstyle \pi} & & \downarrow{\scriptstyle \pi} \\ X & \xleftarrow{F_{X/k}} & X \end{array}$$

commutes. Thus, from such a π, we may construct a diagram

$$\begin{array}{ccccc} Y & \xleftarrow{(F_{X/k})_{(Y)}} & Y^{(q)} & \xleftarrow{F_{Y/X}} & Y \\ \downarrow{\scriptstyle \pi} & & \downarrow{\scriptstyle \pi^{(q)}} & \swarrow{\scriptstyle \pi} & \\ X & \xleftarrow{F_{X/k}} & X & & \end{array}$$

in which the square is Cartesian and $F_{Y/X}$, the *relative Frobenius* of Y/X, is the unique X-morphism such that $(F_{X/k})_{(Y)} \circ F_{Y/X} = F_{Y/k}$.

LEMMA 13.2. *If π is étale, then $F_{Y/X}$ is an isomorphism; moreover, the composite of*

$$s \mapsto s \times 1 : \operatorname{Hom}_X(X, Y) \to \operatorname{Hom}_X(X, Y^{(q)})$$

and

$$t \mapsto F_{Y/X}^{-1} \circ t : \operatorname{Hom}_X(X, Y^{(q)}) \to \operatorname{Hom}_X(X, Y)$$

is the identify map.

Proof. For any k-scheme, $F_{/k}$ is separated, bijective, and radicial. Since $F_{Y/k} = (F_{X/k})_{(Y)} \circ F_{Y/X}$, it follows that $F_{Y/X}$ is also separated, bijective, and radicial. As it is also étale (I.3.6), it is an isomorphism (I.3.11).

For the second statement, let $s \mapsto s_1$ under the mapping of the lemma, so that s_1 is the unique X-morphism such that $(F_{X/k})_{(Y)} \circ F_{Y/X} \circ s_1 = s \circ F_{X/k}$. From the functoriality of $F_{/k}$, we know that

$$s \circ F_{X/k} = F_{Y/k} \circ s = (F_{X/k})_{(Y)} \circ F_{Y/X} \circ s,$$

which proves $s_1 = s$.

For any k-schemes X and Y, $F_{X \times Y/k} = F_{X/k} \times F_{Y/k}$. In particular, since $\bar{X} = X \otimes_k \bar{k}$,

$$F_{\bar{X}/k} = F_{X/k} \otimes F_{\bar{k}/k} = (1 \otimes F_{\bar{k}/k}) \circ (F_{X/k} \otimes 1).$$

Thus $F_{X/k} \otimes 1_{\bar{k}} = F_{\bar{X}/\bar{k}}$ and, as in the case that X is a variety over k, will be referred to simply as the *Frobenius endomorphism* $F = F_{(q)}$ of $\bar{X}$. Note that it depends on $k = \mathbb{F}_q$; if k is replaced by $\mathbb{F}_{q^n}$ and X by $X \otimes \mathbb{F}_{q^n}$, then $\bar{X}$ is unchanged but F is replaced by $F_{(q^n)} = F^n_{(q)}$.

Let E be a sheaf on X_{et}. According to (V.1), we may regard E as an étale algebraic space over X, and, as the above discussion also applies to algebraic spaces, we have an isomorphism $F_{E/X}: E \xrightarrow{\approx} E^{(q)} = F^*_{X/k}E$. On tensoring this with $id_{\bar{k}}$, we get an isomorphism $\phi_E: \bar{E} \to F^*\bar{E}$, which may also be described as the unique map making the following diagram (in which the square is Cartesian) commute:

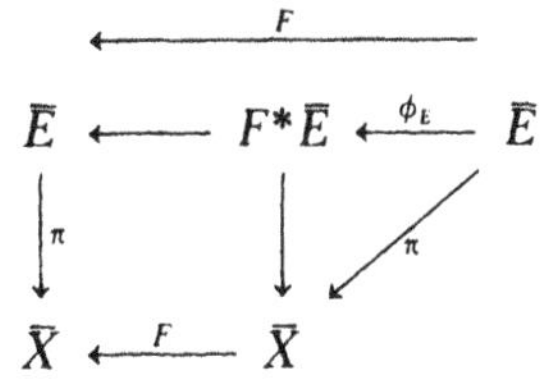

We write F^* for $\phi_E^{-1}: F^*\bar{E} \to \bar{E}$.

Since $F: \bar{X} \to \bar{X}$ is finite, and hence proper, the pair (F, F^*) defines an endomorphism of $H^r_c(\bar{X}, \bar{E})$ which is again denoted by F.

Let

$$\bar{X}^{F^n} = \{\bar{x} \in \bar{X}^0 \mid F^n(\bar{x}) = \bar{x}\} \approx X(\mathbb{F}_{q^n}).$$

Any closed point $\bar{x}$ of $\bar{X}$ can be regarded as a geometric point of X, and hence E has a stalk at such a point. The map F^{*n} defines a map $E_{F^n(\bar{x})} \to E_{\bar{x}}$, which, when $\bar{x} \in \bar{X}^{F^n}$, is an endomorphism $F^n_{\bar{x}}$ of $E_{\bar{x}}$. Suppose now that E is a constructible sheaf of Ω-vector-spaces on X; let x be a closed point of X, let $\bar{x}$ be a point of $\bar{X}$ mapping to x, and let F_x be the endomorphism $F^{\deg(x)}_{\bar{x}}$ of $E_{\bar{x}}$ where $\deg(x) = [k(x):k]$. The pair $(E_{\bar{x}}, F_x)$ is, up to isomorphism, independent of the choice of $\bar{x}$. Thus we may define

$$Z(X, E, t) = \prod_{x \in X^0} \det(1 - F_x t^{\deg(x)} \mid E_{\bar{x}})^{-1} \in \Omega[[t]].$$

THEOREM 13.3. *If X is separated and of finite-type over $k = \mathbb{F}_q$, then*

$$Z(X, E, t) = \prod_r \det(1 - Ft \mid H^r_c(\bar{X}, \bar{E}))^{(-1)^{r+1}}.$$

On applying (V.2.7) to each side of this equation, we find that

$$\begin{aligned} \log(Z(X, E, t)) &= \sum_{x \in X^0} \sum_m Tr(F^m_x \mid E_{\bar{x}})(t^{m \deg x}/m) \\ &= \sum_n \sum_{\bar{x} \in \bar{X}^{F^n}} Tr(F^n_{\bar{x}} \mid E_{\bar{x}}) t^n / n \end{aligned}$$

whereas

$$\log\left(\prod_r \det(1 - Ft|H_c^r(\bar{X}))^{(-1)^{r+1}}\right) = \sum_n \sum_r (-1)^r Tr(F^n|\ H_c^r(\bar{X}, \bar{E})).$$

Thus (13.3) is equivalent to the following trace formula.

THEOREM 13.4. $\sum_{\bar{x}\in\bar{X}^{F^n}} Tr(F_{\bar{x}}^n|E_{\bar{x}}) = \sum_r (-1)^r Tr(F^n|H_c^r(\bar{X}, \bar{E}))$.

Since replacing X by $X \otimes \mathbb{F}_{q^m}$ in (13.4) corresponds to replacing F by F^m, we need only prove (13.4) for $n = 1$.

Remark 13.5. Write $f = F_{\bar{k}/k} = (a \mapsto a^q)$ for the canonical generator of $\mathrm{Gal}(\bar{k}/k)$. It defines an automorphism $1 \otimes f$ of $X \otimes \bar{k} = \bar{X}$ and, as $(1 \otimes f)^*\bar{E} \approx \bar{E}$ (canonically) for any sheaf E on X, it also defines an automorphism $(1 \otimes f)^*$ of $H_c^r(\bar{X}, \bar{E})$. In fact $(1 \otimes f)^*$ and F are inverse elements of $\mathrm{Aut}(H_c^r(\bar{X}, \bar{E}))$. Since $(1 \otimes f) \circ F = F_{\bar{X}/k}$, to prove this we must show that the pair

$$(F_{\bar{X}/k}: \bar{X} \to \bar{X},\ F_{\bar{E}/\bar{X}}^{-1}: F_{\bar{X}/k}^{*}\bar{E} \to \bar{E})$$

defines the identity map $H^r(\bar{X}, \bar{E}) \to H^r(\bar{X}, \bar{E})$ for any sheaf $\bar{E}$ on $\bar{X}$. Using the uniqueness properties of derived functors, one finds it suffices to do this for $r = 0$, but this case is proved in (13.2) (take X to be $\bar{X}$ and Y to be the algebraic space defined by $\bar{E}$).

This partially justifies the following notations. The elements $f = (a \mapsto a^q) \in \mathrm{Gal}(\bar{k}/k)$ and

$$f_x = (a \mapsto a^{q\ \deg(x)}) \in \mathrm{Gal}(k(\bar{x})/k(x)),$$

where $\bar{x} \mapsto x$, are called the *arithmetic Frobenius elements*, and $F = f^{-1} \in \mathrm{Gal}(\bar{k}/k)$ and $F_x = f_x^{-1} \in \mathrm{Gal}(k(\bar{x})/k(x))$ are the *geometric Frobenius elements*. The two definitions of F and F_x agree when these elements are regarded as elements of $\mathrm{End}(H_c^r(\bar{X}, \bar{E}))$ and $\mathrm{End}(E_{\bar{x}})$.

Example 13.6. Assume X is a separated variety over a finite field.

(a) Let $\rho: \mathrm{Gal}(\overline{k(X)}/k(X)) \to \mathrm{Aut}_\Omega(V)$ be a representation on a finite-dimensional vector space V, and assume that ρ is unramified at almost all divisors of X, that is, that it factors through $\pi_1(U, \bar{\eta})$ for some open immersion $U \overset{j}{\hookrightarrow} X$. Then ρ defines a twisted-constant constructible sheaf E' on U such that $E'_{\bar{\eta}} = V$ (as $\pi_1(U, \bar{\eta})$-modules) and a sheaf $E = j_*E'$ on X. Moreover,

$$\begin{aligned} L(X, \rho, t) &\overset{df}{=} \prod_{x \in X^0} \det(1 - \rho(f_x^{-1})t|V^{I_x}) \\ &= Z(X, E, t) \end{aligned}$$

and hence, according to (13.3), is a rational function.

(b) Let Y be Galois over X, with finite Galois group G, and let ρ be a representation of G on a finite-dimensional complex vector space V. We may assume that ρ is defined over a finite extension of $\mathbb{Q}$ and hence over a finite extension Ω of $\mathbb{Q}_l$. Then,

$$L^{\text{Artin}}(Y, \rho, t) \stackrel{df}{=} \prod_{x \in X^0} \frac{1}{\det(1 - t^{\deg(x)}\rho(f_y)|V^{I_y})}$$
$$= Z(X, E, t)$$

where E is the sheaf of Ω-vector spaces on X defined by the contragredient of ρ (compare V.2.21h).

PROPOSITION 13.7. *It suffices to prove* (13.3) *in the case that X is a smooth separated curve over $k = \mathbb{F}_q$ and E is a twisted-constant constructible sheaf of Ω-vector spaces.*

Proof. We write $\det(1 - Ft|H_c^*(\bar{X}, \bar{E}))$ for

$$\prod_r \det(1 - Ft|H_c^r(\bar{X}, \bar{E}))^{(-1)^{r+1}}.$$

LEMMA 13.8. (a) (Multiplicativity in E) *For any exact sequence*

$$0 \to E' \to E \to E'' \to 0,$$
$$Z(X, E, t) = Z(X, E', t)Z(X, E'', t)$$

and

$$\det(1 - Ft|H_c^*(\bar{X}, \bar{E}))$$
$$= \det(1 - Ft|H_c^*(\bar{X}, \bar{E}'))\det(1 - Ft|H_c^*(\bar{X}, \bar{E}'')).$$

(b) (Multiplicativity in X) *If U is an open subscheme of X and $Y = X - U$, then*

$$Z(X, E, t) = Z(U, E|U, t)Z(Y, E|Y, t)$$

and

$$\det(1 - Ft|H_c^*(\bar{X}, \bar{E}))$$
$$= \det(1 - Ft|H_c^*(\bar{U}, \bar{E}|\bar{U}))\det(1 - Ft|H_c^*(\bar{Y}, \bar{E}|\bar{Y})).$$

(c) (Multiplicativity under fibration) *If $\pi: X \to S$ is a k-morphism, then*

$$Z(X, E, t) = \prod_{s \in S^0} Z(X_s, E|X_s, t)$$

and

$$\det(1 - Ft|H_c^*(\bar{X}, \bar{E})) = \prod_j \det(1 - Ft|H_c^*(\bar{S}, \overline{R_c^j\pi_*(E)})^{(-1)^j}.$$

Proof. The proofs are all easy. For example, the first part of (a) follows from considering the exact sequences

$$0 \to E'_{\bar{x}} \to E_{\bar{x}} \to E''_{\bar{x}} \to 0$$

and the second part of (b) from considering the exact sequence (III.1.30)

$$\cdots \to H_c^r(U, E) \to H_c^r(X, E) \to H_c^r(Y, E) \to \cdots.$$

The second part of (c) follows from the spectral sequence

$$H_c^r(\bar{S}, \overline{R_c^s\pi_*E}) \Rightarrow H_c^{r+s}(\bar{X}, \bar{E})$$

and the general fact that if $V_2^{r,s} \Rightarrow V^n$ is a spectral sequence of finite-dimensional vector spaces in which $V_2^{r,s} = 0$ for $r \gg 0$, $s \gg 0$, and $\phi = (\phi_2^{r,s}, \phi^n)$ is a compatible family of endomorphisms of the $V_2^{r,s}$ and V^n such that the $\phi_2^{r,s}$ commute with the differentials, then

$$\begin{aligned}\prod_{r,s} \det(\phi | V_2^{r,s})^{(-1)^{r+s+1}} &= \prod \det(\phi | V_3^{r,s})^{(-1)^{r+s+1}} \\ &\cdots \\ &= \prod \det(\phi | V_\infty^{r,s})^{(-1)^{r+s+1}} \\ &= \prod \det(\phi | V^n)^{(-1)^{n+1}}.\end{aligned}$$

We now assume that (13.3) holds for schemes of dimension ≤ 1, and prove by induction on $d = \dim(X)$ that it holds in general. According to (13.8b) we may replace X by an open subscheme, and hence assume that there is a k-morphism $\pi: X \to S$ with S and all fibers of π having dimension $\leq d - 1$. According to (13.8c),

$$Z(X, E, t) = \prod_{s \in S^0} Z(X_s, E | X_s, t)$$

whereas

$$\begin{aligned}\det(1 - Ft | H^*(\bar{X}, \bar{E})) &= \prod_j \det(1 - Ft | H_c^*(\bar{S}, \overline{R_c^j\pi_*E}))^{(-1)^j} \\ &= \prod_j Z(S, R_c^j\pi_*E, t)^{(-1)^j} \qquad \text{(by induction)} \\ &= \prod_j \prod_{s \in S^0} \det(1 - F_s t^{\deg(s)} | (R_c^j\pi_*E)_{\bar{s}})^{(-1)^{j+1}} \\ &= \prod_{s \in S^0} \det(1 - F_s t^{\deg(s)} | H_c^*(\bar{X}_s, \overline{E | X_s})) \\ &= \prod_{s \in S^0} Z(X_s, E | X_s, t) \qquad \text{(induction).}\end{aligned}$$

Thus we may assume X has dimension ≤ 1. Also (X, E) may be replaced by $(X_{\text{red}}, E | X_{\text{red}})$ since this changes neither side of the equation (II.3.17). When $\dim(X) = 0$, (13.4) is easy and in fact we have the following more general statement. (To deduce (13.4) from it, take $A = \Omega$, $X = \bar{X}$, $\phi = F$, $\phi^* = F^*$.)

LEMMA 13.9. *Let X be a finite discrete topological space, let $\phi: X \to X$ be a map, let A be a ring, let E be a sheaf of A-modules on X whose stalks E_x are free and finitely generated, and let ϕ^* be a map $\phi^*: \phi^*E \to E$. Then*

$$\sum_{x \in X^\phi} Tr(\phi_x^* | E_x) = Tr(\phi | H^0(X, E)).$$

Proof. Explicitly, $H^0(X, E) = \bigoplus_{x \in X} E_x$, $(\phi^*E)_x = E_{\phi(x)}$, ϕ^* is a family of maps $\phi_x^*: E_{\phi(x)} \to E_x$, and $\phi: H^0(X, E) \to H^0(X, E)$ is the map $\bigoplus E_x \to \bigoplus E_x$, $(e_x)_{x \in X} \mapsto (f_x)_{x \in X}$ where $f_x = \phi_x^* e_{\phi(x)}$. On writing out the matrix of ϕ in blocks, one sees easily that

$$Tr(\phi | H^0(X, E)) = \sum_{\phi(x) = x} Tr(\phi_x^* | E_x).$$

It follows that we may assume X to have dimension exactly one and that we may replace it by any dense open subset. Thus we may assume that it is smooth and E is twisted-constant. Since X is a disjoint union of its irreducible components, we may also assume that it is irreducible. Then X is a smooth curve over the algebraic closure k' of k in $k(X)$, and, to complete the proof of (13.7), it remains to show that

$$Z(X/k, E, t) = Z(X/k', E, t^n)$$

and

$$\det(1 - Ft | H^*(X \otimes_k \bar{k}, \bar{E})) = \det(1 - Ft^n | H^*(X \otimes_{k'} \bar{k}, \bar{E}))$$

where $n = [k':k]$. The first of these is obvious and, for the second, we note that $X \otimes_k \bar{k}$ is a disjoint union of n copies of $X \otimes_{k'} \bar{k}$ and that $(H_c^r(X \otimes_k \bar{k}, \bar{E}), F)$ is isomorphic to (V^n, F_n') where $V = H_c^r(X \otimes_{k'} \bar{k}, \bar{E})$, $F_n'(v_1, \ldots, v_n) = (F'v_n, v_1, \ldots, v_{n-1})$ and F' is the Frobenius endomorphism of $X \otimes_{k'} \bar{k}$. It remains to show that

$$\det(1 - F_n' t | V^n) = \det(1 - Ft^n | V).$$

If V is one-dimensional, this is easy, and we may essentially reduce the problem to that case by choosing a basis $\{e_1, \ldots, e_m\}$ of V for which the matrix of F is upper triangular and then choosing $\{(e_1, 0, \ldots, 0), (0, e_1, 0, \ldots, 0), \ldots, (e_2, 0, \ldots, 0), \ldots\}$ as basis for V^n.

From now on we may assume that X is a smooth connected curve over $k = \mathbb{F}_q$ and E is twisted-constant. We shall prove (13.4) in this case by reducing it to the case of a constant sheaf and applying (V.2.20). However, we cannot do this directly because there need not be a finite map $Y \to X$ such that $E|Y$ is constant. Instead we write $E = (E_n)_n \otimes_A \Omega$ where A is the ring of integers in Ω and each E_n is a sheaf over $\Lambda_n = A/\mathfrak{m}^n$, prove an analogue of (13.4) for E_n, and deduce (13.4) for E by passing to the inverse limit and tensoring with Ω. One additional difficulty arises in stating the theorem for E_n, namely, that the groups $H_c^r(\bar{X}, \bar{E}_n)$ are not usually free and so $Tr(F, H_c^r(\bar{X}, \bar{E}_n))$ is not defined. To avoid this problem we make use of the complexes $\mathbf{H}_c^{\cdot}(\bar{X}, \bar{E}_n)$ of (8.16).

LEMMA 13.10. *Let A be a local Noetherian ring, $M^{\cdot}$ a complex of A-modules, and $\alpha: M^{\cdot} \to M^{\cdot}$ an endomorphism of $M^{\cdot}$. For any quasi-isomorphism $\gamma: P^{\cdot} \to M^{\cdot}$ with $P^{\cdot}$ a perfect complex of A-modules, there exists an endomorphism β of $P^{\cdot}$ such that*

$$\begin{array}{ccc} P^{\cdot} & \xrightarrow{\beta} & P^{\cdot} \\ \downarrow{\scriptstyle\gamma} & & \downarrow{\scriptstyle\gamma} \\ M^{\cdot} & \xrightarrow{\alpha} & M^{\cdot} \end{array}$$

commutes; β is unique up to homotopy. Moreover,

$$Tr(\beta | P^{\cdot}) \overset{df}{=} \sum (-1)^r Tr(\beta | P^r)$$

is independent of both $P^{\cdot}$ and β; if $H^r(M^{\cdot})$ is free for each r, then

$$Tr(\beta | P^{\cdot}) = \sum (-1)^r Tr(\beta | H^r(M^{\cdot}));$$

if A is an integral domain with field of fractions Ω, then

$$Tr(\beta | P^{\cdot}) = \sum (-1)^r Tr(\beta \otimes 1 | H^r(M^{\cdot}) \otimes_A \Omega).$$

Proof. The existence of β follows from (8.17). Let β_1, β_2 be two maps such that $\gamma\beta_1 = \alpha\gamma = \gamma\beta_2$. We may construct a commutative diagram

$$\begin{array}{ccccccccc} & & & & P^{\cdot} & & & & \\ & & & \swarrow & \downarrow{\scriptstyle\binom{\beta_1 - \beta_2}{0}} & \searrow{\scriptstyle 0} & & & \\ 0 & \to & K^{\cdot} & \to & P^{\cdot} \oplus C^{\cdot} & \xrightarrow[\sim]{(\gamma,\text{can.})} & M^{\cdot} & \to & 0 \end{array}$$

in which $C^{\cdot}$ is the mapping cone of $M^{\cdot} \overset{id}{\to} M^{\cdot}$ and $K^{\cdot} = \ker(P^{\cdot} \oplus C^{\cdot} \to M^{\cdot})$. Since $C^{\cdot}$ is exact, $P^{\cdot} \oplus C^{\cdot} \xrightarrow{\sim} M^{\cdot}$ and $K^{\cdot}$ is exact. It follows that the map $P^{\cdot} \to K^{\cdot}$ is homotopic to zero (dual of Hartshorne [1, I.4.4]) which implies $\beta_1 - \beta_2$ is homotopic to zero.

Let (k^r) be a homotopy relating β_1 and β_2, so that $k^r: P^r \to P^{r-1}$ and $\beta_1^r - \beta_2^r = d^{r-1}k^r + k^{r+1}d^r$. Then

$$\begin{aligned} Tr(\beta_1 - \beta_2 | P^{\cdot}) &= \sum (-1)^r Tr(d^{r-1}k^r + k^{r+1}d^r) \\ &= \sum (-1)^r Tr(d^{r-1}k^r - k^r d^{r-1}) \\ &= 0 \end{aligned}$$

because $Tr(AB) = Tr(BA)$ if A and B are any two rectangular matrices such that AB and BA are both defined (and hence square).

Suppose we are given two pairs $(P_1^{\cdot} \xrightarrow{\gamma_1} M^{\cdot}, \beta_1)$ and $(P_2^{\cdot} \xrightarrow{\gamma_2} M^{\cdot}, \beta_2)$. According to (8.17), there exists a map $\varepsilon: P_2^{\cdot} \to P_1^{\cdot}$ such that $\gamma_1\varepsilon = \gamma_2$. Let $C_1^{\cdot} = M^{\cdot} \oplus M^{\cdot}[1]$ be the mapping cone of $\mathrm{id}: M^{\cdot} \to M^{\cdot}$, and let $C_\varepsilon^{\cdot} = P_1^{\cdot} \oplus P_2^{\cdot}[1]$ be the mapping cone of ε. Then

$$\gamma = \begin{pmatrix} \gamma_1 & 0 \\ 0 & \gamma_2[1] \end{pmatrix} : C_\varepsilon^{\cdot} \to C_1^{\cdot}$$

is a quasi-isomorphism and

$$\gamma\beta = \begin{pmatrix} \alpha & 0 \\ 0 & \alpha[1] \end{pmatrix} \quad \text{where} \quad \beta = \begin{pmatrix} \beta_1 & 0 \\ 0 & \beta_2[1] \end{pmatrix}.$$

Since ε is a quasi-isomorphism, $C_\varepsilon^{\cdot}$ is exact, and the identity map is homotopic to zero, $1 \sim 0$. Thus $\beta = \beta 1 \sim 0$ and, according to the above argument, $Tr(\beta | C_\varepsilon^{\cdot}) = 0$; but clearly

$$Tr(\beta | C_\varepsilon^{\cdot}) = Tr(\beta_1 | P_1^{\cdot}) - Tr(\beta_2 | P_2^{\cdot}).$$

Suppose that the modules $H^r(M^{\cdot})$ are free. Let r_1 be such that $P^r = 0$ for $r > r_1$, and consider

$$0 \to B^{r_1} \to P^{r_1} \to H^{r_1}(P^{\cdot}) \to 0$$
$$0 \to Z^{r_1-1} \to P^{r_1-1} \to B^{r_1} \to 0$$
$$0 \to B^{r_1-1} \to Z^{r_1-1} \to H^{r_1-1} \to 0.$$
$$\cdots$$

Since $H^r(P^{\cdot}) \approx H^r(M^{\cdot})$ is free, descending induction shows that $B^{r_1}, Z^{r_1-1}, B^{r_1-1}, \ldots$ are flat and hence free. Our assertion now follows from the additivity of traces.

If A is an integral domain, then

$$\begin{aligned} Tr(\beta | P^{\cdot}) = Tr(\beta | P^{\cdot} \otimes \Omega) &= \sum (-1)^r Tr(\beta | H^r(P^{\cdot} \otimes \Omega)) \\ &= \sum (-1)^r Tr(\beta | H^r(P^{\cdot}) \otimes \Omega) \\ &= \sum (-1)^r Tr(\beta | H^r(M^{\cdot}) \otimes \Omega). \end{aligned}$$

Let X be a separated variety over $k = \mathbb{F}_q$, let E be a constructible sheaf of Λ-modules on X, and let $f: \bar{X} \to \operatorname{spec} \bar{k}$ be the structure map. We put $\mathbf{R}_c f_* \bar{E} = \pi_* C^{\cdot}(j_! \bar{E})$ where $C^{\cdot}(j_! \bar{E})$ is the Godement resolution of $j_! \bar{E}$ and we write $\mathbf{H}_c^{\cdot}(\bar{X}, \bar{E})$ for any perfect complex of Λ-modules for which there is given a quasi-isomorphism $\mathbf{H}_c^{\cdot}(\bar{X}, \bar{E}) \to \mathbf{R}_c f_* \bar{E}$ (see 8.15). By a slight refinement of the construction at the start of this section one sees that the Frobenius maps $F: H_c^r(\bar{X}, \bar{E}) \to H_c^r(\bar{X}, \bar{E})$ are induced by a map of complexes $F:\mathbf{R}_c f_* \bar{E} \to \mathbf{R}_c f_* \bar{E}$. Now (13.10) shows that F lifts to a map $F:\mathbf{H}_c^{\cdot}(\bar{X}, \bar{E}) \to \mathbf{H}_c^{\cdot}(\bar{X}, \bar{E})$ and that $Tr(F|\mathbf{H}_c^{\cdot}(\bar{X}, \bar{E}))$ is well defined. If also E is a constructible sheaf of flat Λ-modules then $E_{\bar{x}}$ is flat and hence free. Thus all terms in the following theorem are defined.

THEOREM 13.11. *Let X be a smooth separated curve over $k = \mathbb{F}_q$, and let E be a locally constant, constructible sheaf of flat Λ-modules on X. Then*

$$\sum_{\bar{x} \in \bar{X}^F} Tr(F_{\bar{x}}|E_{\bar{x}}) = Tr(F|\mathbf{H}_c^{\cdot}(\bar{X}, \bar{E})).$$

LEMMA 13.12. *Theorem* (13.11) *implies* (13.4) (*hence also* (13.3) *and* (13.1)).

Proof. After (13.7) (and the last assertion of (13.10)) it suffices to prove the analogue of (13.11) in which E is a twisted-constant constructible sheaf of Ω-vector-spaces. Let E be such a sheaf; it can be written $E = E_\infty \otimes_A \Omega$, $E_\infty = (E_n)_n$, where A is the ring of integers in Ω and each E_n is a locally constant constructible sheaf of flat $\Lambda_n(=A/\mathfrak{m}^n)$-modules. According to (8.16) (and following discussion), there exists a perfect complex of A-modules $\mathbf{H}_c^{\cdot}(\bar{X}, \bar{E}_\infty)$ such that $H^r(\mathbf{H}_c^{\cdot}(\bar{X}, \bar{E}_\infty) \otimes \Omega) = H_c^r(\bar{X}, \bar{E})$ and $\mathbf{H}_c^{\cdot}(\bar{X}, \bar{E}_\infty) \otimes \Lambda_n = \mathbf{H}_c^{\cdot}(\bar{X}, \bar{E}_n)$. Thus

$$Tr(F|\mathbf{H}_c^{\cdot}(\bar{X}, \bar{E}_n)) = Tr(F|\mathbf{H}_c^{\cdot}(\bar{X}, \bar{E}_\infty)) \qquad (\text{mod } \mathfrak{m}^n)$$

whereas (obviously)

$$Tr(F_{\bar{x}}|\bar{E}_{n\bar{x}}) = Tr(F_{\bar{x}}|\bar{E}_{\infty\bar{x}}) \qquad (\text{mod } \mathfrak{m}^n).$$

Hence

$$\sum_{\bar{x} \in \bar{X}^F} Tr(F_{\bar{x}}|E_{\bar{x}}) = Tr(F|\mathbf{H}_c^{\cdot}(\bar{X}, \bar{E}))$$

since they are both in A and agree mod $\mathfrak{m}^n$ for all n.

We begin the proof of (13.11). Let $Y \to X$ be a Galois map such that $E|Y$ is constant, and write π for $\bar{Y} \to \bar{X}$ (where $\bar{Y} = Y \otimes_k \bar{k}$ may not be connected). Let G be the Galois group of Y/X (hence also of $\bar{Y}/\bar{X}$), and let R be the group ring $\Lambda[G]$. Recall that a perfect complex of R-modules is a bounded complex of finitely generated projective modules. Let Λ_0 be the image of $\mathbb{Z}$ in Λ, so that $\Lambda_0 = \mathbb{Z}/(l^n)$ for some n, and let $R_0 = \Lambda_0[G]$.

LEMMA 13.13. *The complex $\mathbf{H}_c^{\cdot}(\bar{Y}, \Lambda_0)$ may be chosen to be a perfect complex of R_0-modules; $\mathbf{H}_c^{\cdot}(\bar{X}, \bar{E}) \xrightarrow{\sim} \mathbf{H}_c^{\cdot}(\bar{Y}, \Lambda_0) \otimes_{R_0} M$ where $M = E(\bar{Y})$. (G acts on $\mathbf{H}_c^{\cdot}(\bar{Y}, \Lambda_0)$ and M through its action on $\bar{Y}$.)*

Remark 13.14. If N and M are G-modules, then $N \otimes_{\Lambda_0} M$ is a G-module with the action $\sigma(n \otimes m) = \sigma n \otimes \sigma m$. In particular, if N is an R_0-module and M is an R-module, then $N \otimes_{\Lambda_0} M$ is an R-module. Let

M' denote M regarded as a G-module with trivial action; then $\sigma \otimes m \mapsto \sigma \otimes \sigma m$ determines an R-isomorphism $R_0 \otimes_{\Lambda_0} M' \to R_0 \otimes_{\Lambda_0} M$. Thus if M is free as a Λ-module, then $R_0 \otimes_{\Lambda_0} M$ is projective, and it follows easily that the same is true of $N \otimes_{\Lambda_0} M$ for any projective R_0-module N.

Proof of (13.13). Since both Λ_0 and G act on $\pi_*\Lambda_0$, it is an R_0-module, and as such is locally free (of rank one). Thus (8.15) shows that there is a perfect complex of R_0-modules $\mathbf{H}^{\cdot}_c(\bar{X}, \pi_*\Lambda_0)$ and a quasi-isomorphism $\mathbf{H}^{\cdot}_c(\bar{X}, \pi_*\Lambda_0) \xrightarrow{\sim} \mathbf{R}_c f_*(\pi_*\Lambda_0)$, where f is the structure map of $\bar{X}$. Since $\mathbf{R}_c(f\pi)_*\Lambda_0 \xrightarrow{\sim} \mathbf{R}_c f_*(\pi_*\Lambda_0)$, π being exact, we may choose $\mathbf{H}^{\cdot}_c(\bar{Y}, \Lambda_0)$ to be $\mathbf{H}^{\cdot}_c(\bar{X}, \pi_*\Lambda_0)$, which proves the first part of (13.13).

The trace $\pi_*\pi^*\bar{E} \to \bar{E}$ induces an isomorphism $(\pi_*\pi^*\bar{E})_G \to \bar{E}$, and

$$(\pi_*\pi^*\bar{E})_G = (\pi_*\pi^*\bar{E}) \otimes_R \Lambda = (\pi_*\Lambda_0 \otimes_{\Lambda_0} f^*M) \otimes_R \Lambda$$

(regarding M as a constant sheaf on spec $\bar{k}$). According to (8.7),

$$\mathbf{H}^{\cdot}_c(\bar{X}, \pi_*\Lambda_0) \otimes_{\Lambda_0} M \xrightarrow{\sim} \mathbf{H}^{\cdot}_c(\bar{X}, \pi_*\Lambda_0 \otimes_{\Lambda_0} f^*M)$$

since $\pi_*\Lambda_0$ is a free Λ_0-module. By assumption, M is a projective Λ-module, and so the above remark shows that $\mathbf{H}^{\cdot}_c(\bar{X}, \pi_*\Lambda_0) \otimes_{\Lambda_0} M$ is a complex of projective R-modules. Moreover, $\pi_*\Lambda_0 \otimes_{\Lambda_0} f^*M$ is a sheaf of projective R-modules, and so $\mathbf{H}^{\cdot}_c(\bar{X}, \pi_*\Lambda_0 \otimes_{\Lambda_0} f^*M)$ may be chosen to be a perfect complex of R-modules. Thus tensoring over R preserves the quasi-isomorphism

$$\mathbf{H}^{\cdot}_c(\bar{X}, \pi_*\Lambda_0) \otimes_{\Lambda_0} M \otimes_R \Lambda \xrightarrow{\sim} \mathbf{H}^{\cdot}_c(\bar{X}, \pi_*\Lambda_0 \otimes_{\Lambda_0} f^*M) \otimes_R \Lambda.$$

The left hand side is $\mathbf{H}^{\cdot}_c(\bar{Y}, \Lambda_0) \otimes_{R_0} M$ whereas an application of (8.7) shows

$$\mathbf{H}^{\cdot}_c(\bar{X}, \pi_*\Lambda_0 \otimes_{\Lambda_0} f^*M) \otimes_R \Lambda \xrightarrow{\sim} \mathbf{H}^{\cdot}_c(\bar{X}, \pi_*\Lambda_0 \otimes_{\Lambda_0} f^*M \otimes_R \Lambda) = \mathbf{H}^{\cdot}_c(\bar{X}, \bar{E});$$

this completes the proof.

Consider

$$1 \to G \to \mathrm{Gal}\,(\bar{k}(\bar{Y})/k(X)) \to \mathrm{Gal}\,(\bar{k}(\bar{X})/k(X)) \to 1,$$

and note that

$$\mathrm{Gal}\,(\bar{k}(\bar{X})/k(X)) \xrightarrow{\approx} \mathrm{Gal}\,(\bar{k}/k) \xrightarrow{\approx} \hat{\mathbb{Z}}$$

where the first isomorphism is restriction and the second sends $f = F_{\bar{k}/k}$ to 1 (compare 13.5). The *degree*, deg (σ), of $\sigma \in \mathrm{Gal}\,(\bar{k}(\bar{Y})/k(X))$ is its image in $\hat{\mathbb{Z}}$; thus σ acts on $\bar{k} \subset \bar{k}(\bar{Y})$ as $f^{\deg(\sigma)}$. Let

$$W^- = \{\sigma \in \mathrm{Gal}\,(\bar{k}(\bar{Y})/k(X)) \mid \deg(\sigma) \in \mathbb{Z}, \deg(\sigma) \le 0\},$$

so that there is an exact sequence of monoids,

$$1 \to G \to W^- \to \mathbb{Z}^- \to 0$$

where $\mathbb{Z}^- = \{n \in \mathbb{Z} \mid n \leq 0\}$. Any $\sigma \in \mathrm{Gal}(\bar{k}(\bar{Y})/k(X))$ defines an X-morphism of $\bar{Y}$, and $\sigma \mapsto \tilde{\sigma} = \sigma \circ (F_{\bar{Y}/k})^{-\deg(\sigma)}$ identifies W^- with the set of $\bar{k}$-morphisms of $\bar{Y}$ lying over some power of F, that is, such that $\pi\tilde{\sigma} = F^n\pi$ for some $n \geq 0$ (because

$$\sigma \circ F_{\bar{Y}/k}^{-\deg(\sigma)} \big| \bar{X} = (1 \otimes f^{\deg(\sigma)}) \circ F_{\bar{X}/k}^{-\deg(\sigma)}$$

and $(1 \otimes f) \circ F = F_{\bar{X}/k}$).

Write $G_{-1} = \{\sigma \in W^- \mid \deg(\sigma) = -1\}$; $\alpha \mapsto \tilde{\alpha}$ identifies G_{-1} with $\{\bar{k}\text{-morphisms } \alpha : \bar{Y} \to \bar{Y} \mid \pi\alpha = F\pi\} = \{F\sigma \mid \sigma \in G\}$. The group G acts on G_{-1} by conjugation, $\alpha \mapsto \sigma^{-1}\alpha\sigma$, and we let S be the set of conjugacy classes. Let $Z(\alpha)$ be the centralizer of α in G, $Z(\alpha) = \{\sigma \in G \mid \alpha\sigma = \sigma\alpha\}$, and let $z(\alpha)$ be its order $[Z(\alpha)]$. Clearly $z(\alpha)$, $Tr(\tilde{\alpha} \mid M)$, and $Tr(\tilde{\alpha} \mid \mathbf{H}_c^\cdot(\bar{Y}, \Lambda_0))$ depend only on the class of α in S.

LEMMA 13.15.

$$Tr(F \mid \mathbf{H}_c^\cdot(\bar{X}, \bar{E})) = Tr(F \mid \mathbf{H}_c^\cdot(\bar{Y}, \Lambda_0) \otimes_{R_0} M)$$
$$= \sum_{\alpha \in S} \frac{Tr(\tilde{\alpha} \mid \mathbf{H}_c^\cdot(\bar{Y}, \Lambda_0))}{z(\alpha)} \cdot Tr(\tilde{\alpha} \mid M).$$

Before proving (13.15), we show that it implies (13.11) (hence also (13.4), (13.3), ...). Let $\bar{Y} \hookrightarrow \bar{Y}^c$ be the embedding of $\bar{Y}$ into its smooth projective closure. Any $\alpha \in G_{-1}$ defines an endomorphism $\tilde{\alpha}$ of $\bar{Y}^c$ and the differential $d\tilde{\alpha} = dF \circ d\alpha = 0$, and so the fixed points of $\tilde{\alpha}$ in $\bar{Y}^c$ all have multiplicity one. According to (13.8b) and (13.9), we know that it suffices to prove (13.11) for a nonempty open subset of X. Thus we may assume that $\bar{X}^F$ is empty, which implies that the $\tilde{\alpha}$ have no fixed points in $\bar{Y}$ since they would lie over fixed points of F. Thus (V.2.20) shows that $Tr(\tilde{\alpha} \mid \mathbf{H}_c^\cdot(\bar{Y}, \mathbb{Z}/(l^N))) = 0$ for all N. Since

$$Tr(\tilde{\alpha} \mid \mathbf{H}_c^\cdot(\bar{Y}, \Lambda_0)) = Tr(\tilde{\alpha} \mid \mathbf{H}_c^\cdot(\bar{Y}, \mathbb{Z}/(l^N))) \pmod{l^n}$$

for all $N \geq n$, we see that

$$Tr(\tilde{\alpha} \mid \mathbf{H}_c^\cdot(\bar{Y}, \Lambda_0)) \,.\, z(\alpha)^{-1} = 0.$$

Thus $Tr(F \mid \mathbf{H}_c^\cdot(\bar{X}, \bar{E})) = 0$ and (13.11) is true, since both sides of the equation are zero.

The first equality in (13.15) follows immediately from (13.13) while the second is an exercise in noncommutative traces. Let R be a noncommutative ring and let $R^\natural$ be the quotient of the additive group of R by the subgroup generated by elements of the form *ab-ba*. If ϕ is an endomorphism of a finitely generated free R-module M and has matrix (a_{ij}) with respect to some basis, then $Tr(\phi) = Tr_R(\phi|M)$ is defined to be the image of $\sum a_{ii}$ in $R^{\natural}$; it is independent of the basis. If ϕ is an endomorphism of a finitely generated projective R-module M, then we write M as a direct summand of a free R-module, $N = M \oplus M'$, and define $Tr(\phi)$ to be $Tr(\phi \oplus 0|M \oplus M')$; it is independent of the choice of N and M'.

Return to the situation that R is a group ring, $R = \Lambda[G]$. The map $\sum a_\sigma \sigma \mapsto a_e: R \to \Lambda$ induces a map $\varepsilon: R^{\natural} \to \Lambda$, and for any endomorphism ϕ of a finitely generated projective R-module M, $Tr_\Lambda^G(\phi) = Tr_\Lambda^G(\phi|M)$ is defined to be $\varepsilon(Tr_R(\phi|M))$.

LEMMA 13.16. $Tr_\Lambda(\phi|M) = [G]Tr_\Lambda^G(\phi|M)$.

Proof. We may assume that M is free and, as the traces depend only on diagonal terms of the matrix of ϕ, that $M = R$. Then ϕ is right multiplication by some element $\sum a_\sigma \sigma$, and $Tr_\Lambda(\phi) = [G]a_e$, $Tr_\Lambda^G(\phi) = a_e$.

Let Λ_0 be a subring of Λ and let $R_0 = \Lambda_0[G]$.

LEMMA 13.17. *Let ϕ be an endomorphism of a finitely projective R_0-module N, and let ψ be an endomorphism of a finitely generated R-module M that is free as a Λ-module. Then*

$$Tr_\Lambda^G(\phi \otimes \psi|N \otimes_{\Lambda_0} M) = Tr_{\Lambda_0}^G(\phi|N)Tr_\Lambda(\psi).$$

Proof. Note that, according to (13.14), $N \otimes_{\Lambda_0} M$ is a projective R-module and so all terms are defined. We need only consider the case $N = R_0$. Then ϕ is multiplication on the right by some element $\sum a_\sigma \sigma$, and the isomorphism $R_0 \otimes_{\Lambda_0} M' \approx R_0 \otimes_{\Lambda_0} M$ of (13.14) transforms $\phi \otimes \psi$ into the endomorphism

$$n \otimes m \mapsto \sum_\sigma a_\sigma n\sigma \otimes \sigma^{-1}\psi(m) \text{ of } R_0 \otimes_{\Lambda_0} M'.$$

One calculates easily, using (13.16), that

$$Tr_\Lambda^G(n \otimes m \mapsto a_\sigma n\sigma \otimes \sigma^{-1}\psi(m)) = \begin{cases} a_e Tr_\Lambda(\psi), & \sigma = e \\ 0, & \text{otherwise,} \end{cases}$$

which completes the proof.

Suppose again that G fits into an exact sequence of monoids

$$1 \to G \to W^- \to \mathbb{Z}^- \to 0$$

and define G_{-1}, $Z(\alpha)$, $z(\alpha)$, and S as before. Let P be a Λ-module on which W^- acts Λ-linearly and which is projective when regarded as an R-module. Then P_G is a projective Λ-module, and any $\sigma \in W^-$ defines an endomorphism of P_G that depends only on the degree of σ.

LEMMA 13.18.

$$Tr\left(\sum_{\beta\in G_{-1}}\beta|P\right) = Tr\left(\sum_{\beta\in G_{-1}}\beta|P^G\right) = Tr\left(\sum_{\beta\in G_{-1}}\beta|P_G\right).$$

Proof. Fix an element $\beta_0 \in G_{-1}$ and let $\nu = \sum_{\sigma\in G}\sigma$. Then $P^G \hookrightarrow P \xrightarrow{\nu} P^G$ is multiplication by $[G]$. Thus

$$Tr(\sum \beta|P) = Tr(\beta_0\nu|P) = Tr([G]\beta_0|P^G) = Tr(\sum \beta|P^G).$$

Multiplication by ν defines an isomorphism $P_G \to P^G$ (Serre [7, VIII.1, Prop. 1], and so P^G may be replaced by P_G.

PROPOSITION 13.19. *For any* $\beta_0 \in G_{-1}$,

$$Tr_\Lambda(\beta_0|P_G) = \sum_{\alpha\in S} Tr_\Lambda^{Z(\alpha)}(\alpha|P).$$

Proof. After multiplication by $[G]$, the formula may be proved directly:

$$\begin{aligned}[G]Tr_\Lambda(\beta_0|P_G) &= Tr_\Lambda\left(\sum_{\beta\in G_{-1}}\beta|P_G\right)\\ &= Tr_\Lambda\left(\sum_{\beta\in G_{-1}}\beta|P\right) && (13.18)\\ &= \sum_{\beta\in S}\frac{[G]}{[Z(\beta)]}Tr(\beta|P)\\ &= \sum_{\beta\in S}[G]Tr_\Lambda^{Z(\beta)}(\beta|P) && (13.16)\end{aligned}$$

Unfortunately this does not prove the proposition as $[G]$ may be a zero-divisor in Λ. Instead we argue as follows. Let $\phi: R \to R$ be the map induced by $(\sigma \mapsto \beta_0\sigma\beta_0^{-1}): G \to G$. To give a $\Lambda[W^-]$-module is the same as to give an R-module P plus a ϕ-linear map $\gamma: P \to P$, the correspondence being described by $(\beta_0^n\sigma)p = \gamma^n(\sigma p)$ for $\beta_0^n\sigma \in W^-$, $p \in P$. Clearly we may replace the given pair (P, γ) by a pair in which P is free as an R-module, and we may choose $P = R$. Then γ will be a map of the form $x \mapsto \phi(x)a$, where $a = \sum a_\sigma\sigma \in R$. We now replace Λ by the polynomial ring $\mathbb{Z}[a_\sigma]_{\sigma\in G}$, P by the group ring over $\mathbb{Z}[a_\sigma]$, and define γ by the same formula as before but with the new meaning for a_σ. Since $\mathbb{Z}[a_\sigma]$ is torsion free, the above argument shows that the required identity holds over $\mathbb{Z}[a_\sigma]$, which implies that it holds over Λ.

We now complete the proof of (13.15). Let $\alpha \in W^-$ act on $\mathbf{H}_c^r(\bar{Y}, \Lambda_0)$ and M as $\tilde{\alpha}$. Note that $\beta_0 = (1\otimes f)^{-1} \in G_{-1}$ acts as $\tilde{\beta}_0 = (1\otimes f)^{-1} \circ F_{\bar{Y}/k} = F$.

$$\begin{aligned}
&Tr_\Lambda(F\,|\,\mathbf{H}_c^r(\bar{Y}, \Lambda_0) \otimes_{R_0} M) \\
&\quad = Tr_\Lambda(F\,|\,(\mathbf{H}_c^r(\bar{Y}, \Lambda_0) \otimes_{\Lambda_0} M)_G) \\
&\quad = \sum_{\alpha \in S} Tr_\Lambda^{Z(\alpha)}(\tilde{\alpha}\,|\,\mathbf{H}_c^r(\bar{Y}, \Lambda_0) \otimes_{\Lambda_0} M) \qquad \text{according to (13.19)} \\
&\quad = \sum_{\alpha \in S} Tr_{\Lambda_0}^{Z(\alpha)}(\tilde{\alpha}\,|\,\mathbf{H}_c^r(\bar{Y}, \Lambda_0))Tr_\Lambda(\tilde{\alpha}\,|\,M) \qquad \text{according to (13.17)} \\
&\quad = \sum_{\alpha \in S} \frac{Tr_{\Lambda_0}(\tilde{\alpha}\,|\,\mathbf{H}_c^r(\bar{Y}, \Lambda_0))}{z(\alpha)} \cdot Tr_\Lambda(\tilde{\alpha}\,|\,M) \qquad \text{according to (13.16).}
\end{aligned}$$

On forming the alternating sum over r, we obtain the second equality in (13.15).

Remark 13.20. (*a*) The above proof of (13.3) is due to Grothendieck. Our exposition of the proof of (13.11) follows notes of a course given by N. Katz at Princeton in 1973–1974 (in turn based on a letter of L. Illusie) and that by Deligne in [SGA. $4\frac{1}{2}$, Rapport].

(b) The L-series associated with representations of the Galois group (or, more generally, the Weil group) of $k(X)$ satisfy functional equations. For this, see Deligne [1].

(c) In [SGA. $4\frac{1}{2}$, Fonction L mod l^n] Deligne proves a multiplicative trace formula for the Frobenius map that is both stronger and more general than (13.11) above.

(d) The proof of the very general Lefschetz trace formula, sketched by Verdier in [2], has been completed by Illusie, and appears in [SGA. 5].

APPENDIX A

Limits

We need a slight generalization of the usual notion of limit. The proofs are not difficult and will be omitted (see Artin [1, I.1]).

Let **I** and **C** be categories and consider covariant functors $\mathbf{I} \to \mathbf{C}$. Each object M in **C** defines such a functor, namely, the constant functor h_M with $h_M(i) = M$ for all i in **I**, and $h_M(i \to j) = id_M$ for all maps $i \to j$. An object M of **C** is the *direct limit* of a functor $F{:}\mathbf{I} \to \mathbf{C}$, and we write $M = \varinjlim F$, if there is given a morphism of functors $\phi{:}F \to h_M$ such that any other morphism of functors $\phi'{:}F \to h_N$ factors uniquely into $\phi' = \psi\phi$. Explicitly, this means that there are given maps $F_i \to M$ for all i in **I** such that the diagrams

$$\begin{array}{ccc} F_i & \longrightarrow & F_j \\ & \searrow \quad \swarrow & \\ & M & \end{array} \qquad i \to j$$

commute and that M is universal with respect to these properties.

For example, if I is a partially ordered set that is made into a category by requiring that Hom (i, j) contain one element if $i \leq j$ and none otherwise, then this definition agrees with the usual definition.

Inverse limits are defined similarly.

The category **I** is *pseudofiltered* if:

(f_1) any diagram of the form

$$i \begin{array}{l} \nearrow j \\ \searrow j' \end{array}$$

can be inserted into a commutative diagram

$$i \begin{array}{c} \nearrow j \searrow \\ \searrow j' \nearrow \end{array} k;$$

(f_2) any diagram of the form

$$i \underset{v}{\overset{u}{\rightrightarrows}} j$$

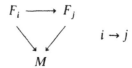

can be inserted into a diagram

$$i \overset{u}{\underset{v}{\rightrightarrows}} j \overset{w}{\rightarrow} k$$

with $wu = wv$. **I** is *connected* if for any two objects i and j of I, there exist morphisms $i \to i_1 \leftarrow j_1 \to i_2 \leftarrow j_2 \cdots \leftarrow j$, and **I** is *filtered* if it is both connected and pseudofiltered. For example, the category arising from a partially ordered set is filtered if and only if the set is directed.

If **I** is a disjoint union of categories, $\mathbf{I} = \coprod \mathbf{I}_i$, then $\varinjlim_{\mathbf{I}} F = \coprod_i \varinjlim_{\mathbf{I}_i} F_i$ where $\coprod_i$ denotes direct sum in the category **C**. In particular, if **I** is discrete, that is, the only morphisms are the identity maps, then $\varinjlim F = \coprod_i F_i$.

If the dual category $\mathbf{I}^\circ$ to **I** is filtered, then **I** is said to be *cofiltered*.

PROPOSITION 1. *Let **I** be a filtered (sufficiently small) category; let $i \mapsto X_i$ be a functor **I** → **Sets**, and let X be the disjoint union of the X_i. Define $x_i \sim x_j$, for $x_i \in X_i$ and $x_j \in X_j$, to mean that there exist morphisms $i \to k, j \to k$ such that x_i and x_j have equal images under the corresponding morphisms $X_i \to X_k$, $X_j \to X_k$.*

(a) *$\sim$ is an equivalence relation on X, and $X/\sim\, = \varinjlim X_i$.*

(b) *For any pair $x_i \in X_i$, $x_j \in X_j$ there exists a k such that x_i and x_j are equivalent to elements of X_k.*

(c) *If x_i, $x_i' \in X_i$, then $x_i \sim x_i'$ if and only if there exists a morphism $i \to j$ such that x_i and x_i' have equal images under the corresponding morphism $X_i \to X_j$.*

COROLLARY 2. *Let **I** be a filtered category, and let F be a functor **I** → **C** where **C** is the category of all groups, or all abelian groups, or all rings,* etc. *Let f be the forgetful functor $f:\mathbf{C} \to$ **Sets**. Then $\varinjlim F$ exists and $f(\varinjlim F) = \varinjlim (fF)$.*

As usual, direct limits commute with direct limits and inverse limits with inverse limits, but not generally the one with the other.

PROPOSITION 3. *If **I** is pseudofiltered, then direct limits in **Ab** commute with finite inverse limits.*

Let $\mathbf{C} \overset{u}{\underset{v}{\rightleftarrows}} \mathbf{C}'$ be functors such that u is the left adjoint of v, that is, $\mathrm{Hom}_{\mathbf{C}'}(u(X), X') \approx \mathrm{Hom}_{\mathbf{C}}(X, v(X'))$ as bifunctors.

PROPOSITION 4. *The functor u commutes with all direct limits and the functor v with all inverse limits.*

PROPOSITION 5. *Let I be a set, and let **J** be the category whose set of objects is $I \cup I \times I$, and whose morphisms are inclusions of the form*

$i \hookrightarrow (i,j)$ or $j \hookrightarrow (i,j)$. For any functor $F\colon \mathbf{J} \to \mathbf{C}$,

$$\varprojlim_{\mathbf{J}} F \to \prod_{i \in I} F_i \rightrightarrows \prod_{i,j} F_{(i,j)}$$

is exact.

PROPOSITION 6. *Let **C** be a category. Arbitrary (respectively, finite) inverse limits exist in **C** if and only if arbitrary (respectively, finite) products exist and difference kernels of pairs of morphisms exist.*

A full subcategory $\mathbf{J} \subset \mathbf{I}$ is said to be *cofinal* if for all i in $\mathbf{I}$ there exists a map $i \to j$ in $\mathbf{I}$ with j in $\mathbf{J}$.

PROPOSITION 7. *If **I** is pseudofiltered or filtered, so also is **J**. If **I** satisfies (f_1), then for any functor $F\colon \mathbf{I} \to \mathbf{C}$ the canonical map*

$$\varinjlim_{\mathbf{I}} F \to \varinjlim_{\mathbf{J}} F|\mathbf{J}$$

is a bijection.

Recall (Bucur-Deleanu [1, 5.8]) that an abelian category **C** is said to satisfy (AB3) if any family of objects has a direct sum, to satisfy (AB4) if it satisfies (AB3) and any direct sum of exact sequences is exact, and to satisfy (AB5) if it satisfies (AB3) and any filtered direct limit of exact sequences is exact. It is said to satisfy (AB3*), (AB4*), or (AB5*) if the dual category satisfies (AB3), (AB4), or (AB5) respectively. For example, **Ab** satisfies (AB5) and (AB4*).

APPENDIX B

Spectral Sequences

Let **A**, **B**, and **C** be abelian categories such that **A** and **B** have enough injectives, and let $f:\mathbf{A} \rightarrow \mathbf{B}$ and $g:\mathbf{B} \rightarrow \mathbf{C}$ be left exact functors. If f takes injectives in **A** to g-acyclic objects in **B**, that is, if I injective implies that $R^p g(I) = 0$ for all $p > 0$, then there are relations between the objects $(R^p g)(R^q f)(A)$ and the objects $R^n(gf)(A)$ for any A in **A**. For example, by applying f and g successively to an injective resolution $0 \rightarrow A \rightarrow I^0 \rightarrow I^1 \rightarrow \cdots$ of A, it is easy to derive an exact sequence

$$0 \rightarrow (R^1 g)(f(A)) \rightarrow R^1(gf)(A) \rightarrow g(R^1 f(A)).$$

Also, if f happens to be exact, then there are canonical isomorphisms $R^p g(fA) \xrightarrow{\approx} R^p(gf)(A)$. All such relations are summarized by the giving of a spectral sequence $R^p g R^q f(A) \Rightarrow R^{p+q}(gf)(A)$.

We shall give only a very brief description of spectral sequences; for more details, see Shatz [2, II.4], or almost any book on homological algebra.

A *spectral sequence* consists of the following.

(a) A family $(E_r^{p,q})$ of objects of an abelian category, where p, q, r are integers and $p, q \geq 0$, $r \geq 1$ (or $r \geq 2$).

(b) Morphisms as follows.

$E_1^{p,q}$

$d_1^{p,q}: E_1^{p,q} \rightarrow E_1^{p+1,q}$

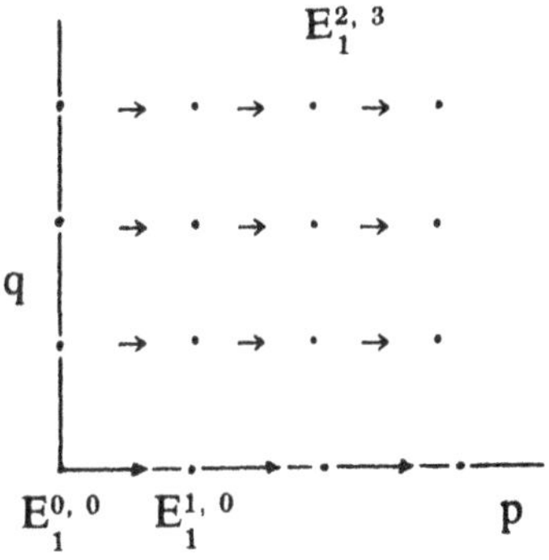

$E_2^{p,q}$

$d_2^{p,q}: E_2^{p,q} \to E_2^{p+2,q-1}$

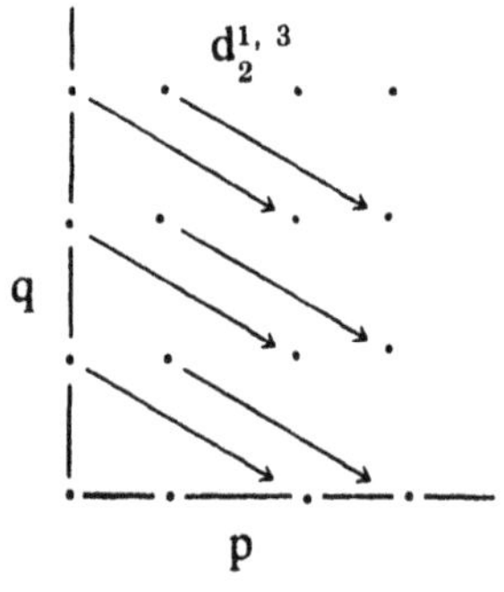

In general, morphisms $d_r^{p,q}: E_r^{p,q} \to E_r^{p+r,q-r+1}$. These morphisms satisfy: $d_r^{p+r,q-r+1} d_r^{p,q} = 0$. Of course, if any one of p, q, $p + r$, or $q - r + 1$ is <0, then $d_r^{p,q}$ is taken to be the zero map.

(c) The objects $E_{r+1}^{p,q}$ on the $(r + 1)^{st}$ level are derived from those on the rth level as follows:

$$E_{r+1}^{p,q} = \frac{\ker (d_r^{p,q})}{\operatorname{im} (d_r^{p-r,q+r-1})}.$$

For example, we have $E_2^{2,0} \twoheadrightarrow E_3^{2,0} = E_4^{2,0} = \cdots \overset{df}{=} E_\infty^{2,0}$, and

$$E_2^{p,0} \twoheadrightarrow E_3^{p,0} \twoheadrightarrow \cdots E_{p+1}^{p,0} = E_{p+2}^{p,0} = \cdots \overset{df}{=} E_\infty^{p,0}.$$

In general, for each (p, q), there is an r_0 depending on (p, q) such that for all $r \geq r_0$, $d_r^{p,q} = 0 = d_r^{p-r,q+r-1}$. Then

$$E_{r_0}^{p,q} = E_{r_0+1}^{p,q} = \cdots \overset{df}{=} E_\infty^{p,q}.$$

Note that there are injective maps $E_\infty^{0,q} \to E_2^{0,q}$.

(d) A family of objects (E^n), $n \geq 0$, and for each E^n a filtration,

$$E^n = E_0^n \supset E_1^n \supset E_2^n \supset \cdots \supset E_n^n \supset 0,$$

such that

$$E_p^n / E_{p+1}^n = E_\infty^{p,n-p}.$$

Pictorially:

$E_\infty^{p,q}$

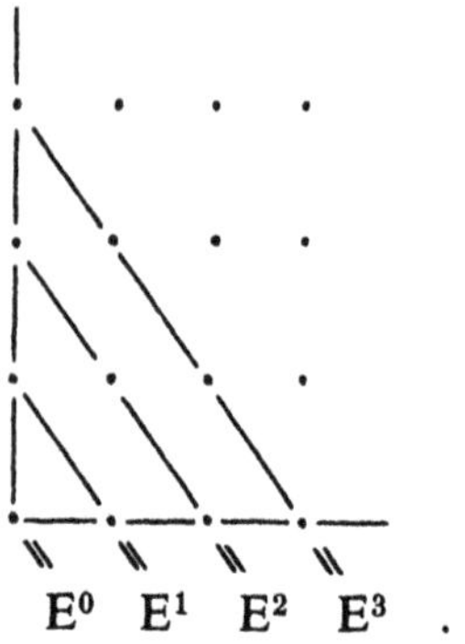

Such a spectral sequence is written: $E_1^{p,q} \Rightarrow E^n$ (or $E_2^{p,q} \Rightarrow E^n$).

The relation between E^n and the $E_2^{p,q}$ can be made explicit for small n. First note that

$$E_2^{0,0} = E_\infty^{0,0} = E^0.$$

Also, $E^1 \supset E_1^1 \supset 0$ where $E_1^1 = E_\infty^{1,0} = E_2^{1,0}$ and $E^1/E_1^1 = E_\infty^{0,1} = \ker(d_2^{0,1})$. Thus $0 \to E_1^1 \to E^1 \to E^1/E_1^1 \to 0$ gives rise to an exact sequence,

$$0 \to E_2^{1,0} \to E^1 \to E_2^{0,1} \xrightarrow{d_2^{0,1}} E_2^{2,0}.$$

With some effort, this can be extended to an exact sequence,

$$0 \to E_2^{1,0} \to E^1 \to E_2^{0,1} \xrightarrow{d_2^{0,1}} E_2^{2,0} \to E_1^2 \to E_2^{1,1} \to E_2^{3,0}$$

where $E_1^2 = \ker(E^2 \to E_2^{0,2})$.

The most important existence theorem for spectral sequences is the following.

THEOREM 1. *Let* **A**, **B**, *and* **C** *be abelian categories; assume that* **A** *and* **B** *have enough injectives, and let* $f:\mathbf{A} \to \mathbf{B}$ *and* $g:\mathbf{B} \to \mathbf{C}$ *be left exact functors. If* f *takes injectives to* g*-acyclics, then there is a spectral sequence*

$$(R^p g)(R^q f)(A) \Rightarrow R^n(gf)(A)$$

for any object A *of* **A**. *In particular, there is an exact sequence*

$$0 \to R^1 g(fA) \to R^1(gf)A \to g(R^1 f)A \to R^2 g(fA) \to \cdots.$$

Proof. See Hilton-Stammbach [1].

COROLLARY 2. *If, in the situation of the theorem,* f *is exact,* $R^p(gf)(A) \approx (R^p g)fA$.

Example 3. Let $A^{\cdot\cdot}$ be a *double complex*, that is, a family of objects $A^{p,q}$, where $p, q \in \mathbb{Z}$, with differentials $d_I^{p,q}: A^{p,q} \to A^{p+1,q}$ and $d_{II}^{p,q}: A^{p,q} \to A^{p,q+1}$ such that $d_I d_I = 0 = d_{II} d_{II}$ and $d_I d_{II} + d_{II} d_I = 0$. Let $A^\cdot$ be the *total complex* $\mathrm{tot}(A^{\cdot\cdot})$ of $A^{\cdot\cdot}$, that is, $A^n = \bigoplus_{p+q=n} A^{p,q}$ with $d^n = \bigoplus(d_I^{p,q} + d_{II}^{p,q})$, and assume that $A^{p,q} = 0$ if $p < 0$ or $q < 0$. Then there are spectral sequences

$$\begin{cases} {}_I E_1^{p,q} = {}_{II}H^q(A^{p,\cdot}) \Rightarrow H^{p+q}(A^\cdot) \\ {}_{II}E_1^{p,q} = {}_I H^q(A^{\cdot,p}) \Rightarrow H^{p+q}(A^\cdot) \end{cases}$$

and

$$\begin{cases} {}_I E_2^{p,q} = {}_I H^p({}_{II}H^q(A^{\cdot\cdot})) \Rightarrow H^{p+q}(A^\cdot) \\ {}_{II}E_2^{p,q} = {}_{II}H^p({}_I H^q(A^{\cdot\cdot})) \Rightarrow H^{p+q}(A^\cdot), \end{cases}$$

where ${}_I H$ and ${}_{II}H$ mean take cohomology with respect to d_I and d_{II} respectively.

Remark 4. Most of the spectral sequences in this book are of the type defined above. More generally one can allow $E_r^{p,q}$ to be nonzero when p or q is negative; it is then not automatic that $E_r^{p,q}$ is constant for large r.

APPENDIX C

Hypercohomology

Let **A** be an abelian category and $C(\mathbf{A})$ the category of complexes in **A**. It is again an abelian category. Let $C^+(\mathbf{A})$, $C^-(\mathbf{A})$ and $C^b(\mathbf{A})$ denote the subcategories of $C(\mathbf{A})$ comprising those complexes $A^\cdot$ that are *bounded below* (that is, such that $A^r = 0$ for $r \ll 0$), *bounded above*, and *bounded* (that is, bounded above and below). A map $A_1^\cdot \to A_2^\cdot$ of complexes is a *quasi-isomorphism*, written $A_1^\cdot \xrightarrow{\approx} A_2^\cdot$, if the induced maps on cohomology, $H^r(A_1^\cdot) \to H^r(A_2^\cdot)$, are isomorphisms for all r.

Let $f:\mathbf{A} \to \mathbf{B}$ be a left exact functor from **A** into a second abelian category, and assume that **A** has enough injectives. Any $A^\cdot \in C^+(\mathbf{A})$ admits a quasi-isomorphism, $A^\cdot \xrightarrow{\approx} I^\cdot$, into a complex $I^\cdot \in C^+(\mathbf{A})$ whose objects are injective. (Easy, Hartshorne [1, I.4.6i].) The functors $A^\cdot \mapsto H^r(fI^\cdot): C^+(\mathbf{A}) \to \mathbf{B}$ are the *right hyper-derived functors* $\mathbb{R}^r f$ of f; they are (essentially) independent of the choices of the quasi-isomorphisms $A^\cdot \xrightarrow{\approx} I^\cdot$. They have the following properties.

(a) For any exact sequence $0 \to A_1^\cdot \to A_2^\cdot \to A_3^\cdot \to 0$ of complexes in $C^+(\mathbf{A})$, there is a long exact sequence

$$\cdots \to \mathbb{R}^r fA_3^\cdot \to \mathbb{R}^{r+1} fA_1^\cdot \to \mathbb{R}^{r+1} fA_2^\cdot \to \mathbb{R}^{r+1} fA_3^\cdot \to \cdots.$$

(b) If an object A of **A** is regarded as an element of $C^+(\mathbf{A})$, with $A^0 = A$ and $A^r = 0$ for $r \neq 0$, then $\mathbb{R}^r fA = R^r fA$.

(c) Let $C_0(\mathbf{A})$ be the subcategory of $C^+(\mathbf{A})$ of complexes $A^\cdot$ such that $A^r = 0$ for $r < 0$. The functors $\mathbb{R}^r f$ when restricted to $C_0(\mathbf{A})$, are the right derived functors of $(A^\cdot \mapsto \mathbb{R}^0 f(A^\cdot) = \ker(A^0 \to A^1)): C_0(\mathbf{A}) \to \mathbf{B}$.

(d) Any quasi-isomorphism $A_1^\cdot \xrightarrow{\approx} A_2^\cdot$ induces isomorphisms $\mathbb{R}^r fA_1^\cdot \to \mathbb{R}^r fA_2^\cdot$.

(e) If $A^\cdot$ is exact, $\mathbb{R}^r fA^\cdot = 0$ for all r.

(f) Let $A^\cdot$ be a complex whose objects are acyclic for f, that is, $R^r fA^s = 0$ for all $r > 0$. Then $\mathbb{R}^r fA^\cdot = H^r(fA^\cdot)$.

(g) For any complex $A^\cdot \in C^+(\mathbf{A})$, there are spectral sequences

$$E_1^{r,s} = R^s fA^r \Rightarrow \mathbb{R}^{r+s} fA^\cdot$$
$$E_2^{r,s} = R^r f(H^s(A^\cdot)) \Rightarrow \mathbb{R}^{r+s} fA^\cdot.$$

Of these, (a) and (b) hold by definition, (d), (e) and (f) follow from (g), (c) follows from the fact that the functors $\mathbb{R}^r f$ satisfy the definition for the right derived functors of $A^{\cdot} \mapsto \mathbb{R}^0 f(A^{\cdot})$, and (g) may be proved by taking an injective resolution $A^{\cdot} \to I^{\cdot\cdot}$ of $A^{\cdot}$, observing that $A^{\cdot} \xrightarrow{\approx} \operatorname{tot}(I^{\cdot\cdot})$ and writing out the two spectral sequences arising from the double complex $fI^{\cdot\cdot}$ (Appendix B.3).

Note that (e) and (f) imply that if $\alpha: A_1^{\cdot} \to A_2^{\cdot}$ is a quasi-isomorphism of bounded below complexes whose objects are acyclic for f, then $f\alpha: fA_1^{\cdot} \to fA_2^{\cdot}$ is also a quasi-isomorphism. For the mapping cone $C_\alpha^{\cdot}$ of α is exact, and so $\mathbb{R}^r f C_\alpha^{\cdot} = 0$ for all r. Thus the exact sequence $0 \to A_2^{\cdot} \to C_\alpha^{\cdot} \to A_1^{\cdot}[1] \to 0$ gives isomorphisms

$$\mathbb{R}^{r-1} f A_1^{\cdot}[1] = \mathbb{R}^r f A_1^{\cdot} \xrightarrow{\approx} \mathbb{R}^r f A_2^{\cdot}.$$

According to (f), these may be written $H^r(fA_1^{\cdot}) \xrightarrow{\approx} H^r(fA_2^{\cdot})$, which shows that $fA_1^{\cdot} \xrightarrow{\approx} fA_2^{\cdot}$.

It is possible to extend some of these definitions and results to unbounded complexes if there exists an n such that $R^r f A = 0$ for all A and all $r > n$. Note that this condition implies that if $A^0 \to A^1 \to \cdots \to A^n \to 0$ is an exact sequence with $A^0, \ldots, A^{n-1}$ f-acyclic then A^n is also f-acyclic, for

$$R^r f A^n = \mathbb{R}^{r+n} f A^{\cdot} = R^{r+n} f(\ker(A^0 \to A^1)) = 0$$

for $r > 0$. Using this, we may construct for any complex $C^{\cdot}$ a quasi-isomorphism $C^{\cdot} \xrightarrow{\approx} A^{\cdot}(C^{\cdot})$ with the objects of $A^{\cdot}(C^{\cdot})$ f-acyclic. Indeed, fix an r_0. There exists a quasi-isomorphism $(\cdots 0 \to C^{r_0} \to C^{r_0+1} \to \cdots) \xrightarrow{\approx} (\cdots 0 \to I^{r_0} \to I^{r_0+1} \to \cdots)$ with each object I^r and each map $C^r \to I^r$ injective. We define $A_1^{\cdot}$ by

$$\begin{cases} A_1^r = I^r, & r \geq r_0 \\ A_1^r = C^r, & r < r_0. \end{cases}$$

Clearly there is a quasi-isomorphism $C^{\cdot} \xrightarrow{\approx} A_1^{\cdot}$. Now fix an $r_1 < r_0$ and construct another quasi-isomorphism $A_1^{\cdot} \hookrightarrow B^{\cdot}$ with B^r injective for $r \geq r_1$ and $B^r = A_1^r$ for $r < r_1$. Let $D^{\cdot} = \operatorname{coker}(A_1^{\cdot} \to B^{\cdot})$. It is exact and D^r is injective for $r \geq r_0$. Thus

$$\operatorname{im}(D_2^{r_0+n-1} \to D_2^{r_0+n})$$

is f-acyclic, and the same is true of

$$A_2^{r_0+n} \stackrel{df}{=} \operatorname{im}(B^{r_0+n-1} \to B^{r_0+n}) \oplus_{A_1^{r_0+n-1}} A_1^{r_0+n}$$

because there is an exact sequence,

$$0 \to A_1^{r_0+n} \to A_2^{r_0+n} \to \operatorname{im}(D^{r_0+n-1} \to D^{r_0+n}) \to 0.$$

Thus, if we define $A_2^{\cdot}$ by

$$A_2^r = \begin{cases} B^r, & r < r_0 + n \\ A_2^{r_0+n}, & r = r_0 + n \\ A_1^r, & r > r_0 + n, \end{cases}$$

then there is a quasi-isomorphism $A_1^{\cdot} \xrightarrow{\sim} A_2^{\cdot}$ and A_2^r is f-acyclic for $r \geq r_1$. We may continue in this way, $A_1^{\cdot} \xrightarrow{\sim} A_2^{\cdot} \xrightarrow{\sim} A_3^{\cdot} \xrightarrow{\sim} \cdots$, and define $A^{\cdot}(C^{\cdot}) = \varinjlim A_i^{\cdot}$, that is, for a fixed r, $A^r(C^{\cdot}) = A_i^r$ all $i \gg 0$.

This suggests that we define $\mathbb{R}^r f C^{\cdot} = H^r(A^{\cdot}(C^{\cdot}))$, but it is technically more convenient to proceed as follows. We say that two complexes $C_1^{\cdot}$ and $C_2^{\cdot}$ are *quasi-isomorphic*, $C_1^{\cdot} \simeq C_2^{\cdot}$, if there is a finite sequence of quasi-isomorphisms $C_1^{\cdot} \xleftarrow{\sim} D_1^{\cdot} \xrightarrow{\sim} \cdots \xleftarrow{\sim} D_s^{\cdot} \xrightarrow{\sim} C_2^{\cdot}$; note that then we are given isomorphisms $H^r(C_1^{\cdot}) \xrightarrow{\approx} H^r(C_2^{\cdot})$. We define $\mathbb{R}^r f C^{\cdot} = H^r(fA^{\cdot})$ if $A^{\cdot}$ is any complex of f-acyclic objects that is quasi-isomorphic to $C^{\cdot}$. For this to make sense we have to show that if $A_1^{\cdot} \simeq A_2^{\cdot}$ and both are complexes of f-acyclic objects, then $fA_1^{\cdot} \simeq fA_2^{\cdot}$. Note first that a map $\alpha_0 : C^{\cdot} \to C'^{\cdot}$ extends step by step to maps,

$$\begin{array}{ccccccc}
C^{\cdot} & \longrightarrow & A_1^{\cdot} & \longrightarrow & A_2^{\cdot} & \longrightarrow \cdots & A^{\cdot}(C^{\cdot}) \\
\downarrow{\scriptstyle \alpha_0} & & \downarrow{\scriptstyle \alpha_1} & & \downarrow{\scriptstyle \alpha_2} & & \downarrow{\scriptstyle \alpha_\infty} \\
C'^{\cdot} & \longrightarrow & A_1'^{\cdot} & \longrightarrow & A_2'^{\cdot} & \longrightarrow \cdots & A^{\cdot}(C'^{\cdot})
\end{array}$$

(To prove this, use the fact that for any maps $I^{\cdot} \xleftarrow{\alpha} C^{\cdot} \underset{\sim}{\xrightarrow{\beta}} D^{\cdot}$ of bounded below complexes with $I^{\cdot}$ a complex of injectives, there exists a map $\gamma : D^{\cdot} \to I^{\cdot}$ such that $\gamma\beta = \alpha$; we prove a dual statement in (VI.8.17).) Thus given $A_1^{\cdot} \xrightarrow{\sim} A_2^{\cdot}$ we may construct a diagram

$$\begin{array}{ccccccccc}
A_1^{\cdot} & \xleftarrow{\sim} & D_1^{\cdot} & \xrightarrow{\sim} & \cdots & \xrightarrow{\sim} & D_s^{\cdot} & \xrightarrow{\sim} & A_2^{\cdot} \\
\downarrow{\scriptstyle \sim} & & \downarrow{\scriptstyle \sim} & & & & \downarrow{\scriptstyle \sim} & & \downarrow{\scriptstyle \sim} \\
A^{\cdot}(A_1^{\cdot}) & \xleftarrow{\sim} & A^{\cdot}(D_1^{\cdot}) & \xrightarrow{\sim} & \cdots & \xrightarrow{\sim} & A^{\cdot}(D_s^{\cdot}) & \xrightarrow{\sim} & A^{\cdot}(A_2^{\cdot}).
\end{array}$$

Thus we may assume that there is a map $A_1^{\cdot} \xrightarrow{\sim} A_2^{\cdot}$. Write

$$\tau_{>r}(A_1^{\cdot}) = (\cdots 0 \to A_1^{r+1} \to A_1^{r+2} \to \cdots).$$

Then

$$H^{r_0}(fA_1^{\cdot}) = H^{r_0}(f(\tau_{>r}A_1^{\cdot})) \approx H^{r_0}(f(\tau_{>r}A_2^{\cdot})) = H^{r_0}(fA_2^{\cdot})$$

for $r < r_0 - n$.

Bibliography

Altman, A., Hoobler, R., and Kleiman, S.
[1] A note on the base change map for cohomology. *Compositio Math. 27*, (1973), 25–38.

Altman, A. and Kleiman, S.
[1] *Introduction to Grothendieck Duality Theory*. Lecture Notes in Math. 146, Springer, Heidelberg, 1970.

Amitsur, S.
[1] Simple algebras and cohomology groups of arbitrary fields. *Trans. Amer. Math. Soc. 90* (1959), 73–112.

Artin, E. and Tate, J.
[1] *Class Field Theory*. W. A. Benjamin, New York, 1968.

Artin, M.
[1] *Grothendieck Topologies*. Lecture Notes, Harvard University Math. Dept, Cambridge, Mass 1962.
[2] Etale coverings of schemes over Hensel rings. *Amer. J. Math. 88* (1966), 915–934.
[3] The etale topology of schemes. *Proc. of Internat. Congr. of Math.* (Moscow 1966). edited by I. G. Petrovsky. Izdat. "Mir", Moscow, 1968, pp. 44–56.
[4] The implicit function theorem in algebraic geometry. *Algebraic Geometry* (Internat. Colloq. Tata Inst. Fund. Res. Bombay, 1968). Oxford University Press, London, 1969, pp. 13–34.
[5] Algebraic approximation of structures over complete local rings. *Inst. Hautes Etudes Sci. Publ. Math. 36* (1969), 23–58.
[6] Algebraization of formal moluli. I. *Global Analysis: Papers in Honor of K. Kodaira*, edited by D. C. Spencer and S. E. Iyanaga. University of Tokyo Press, Tokyo, 1969, pp. 21–71.
[7] On the joins of Hensel rings. *Advances in Math. 7* (1971), 282–296.
[8] *Algebraic Spaces*. Yale Mathematical Monographs 3, Yale University Press, New Haven, 1971.
[9] *Théorèmes de Représentabilité pour les Espace Algébriques*. Presses de l'Université de Montréal, Montréal, 1973.

Artin, M. and Mazur, B.
[1] *Etale Homotopy*. Lecture Notes in Math. 100, Springer Heidelberg, 1969.

Artin, M. and Milne, J.
[1] Duality in the flat cohomology of curves. *Invent. math. 35* (1976), 111–129.

Atiyah, M. and Macdonald, I.
[1] *Introduction to Commutative Algebra*. Addison-Wesley, Reading, Mass., 1969.
Auslander, M. and Goldman, O.
[1] The Brauer group of a commutative ring. *Trans. Amer. Math. Soc. 97* (1960), 367–409.
Azumaya, G.
[1] On maximally central algebras. *Nagoya Math. J. 2* (1951), 119–150.
Bašmakov, M.
[1] The cohomology of Abelian varieties over a number field. *Russ. Math. Surveys 27*, no. 6 (1972), 25–71.
Bass, H.
[1] *Algebraic K-theory*. W. A. Benjamin, New York, 1968.
Beauville, A.
[1] *Surfaces Algébriques Complexes*. Asterisque 54, Soc. Math. de France, 1978.
Blanchard, A.
[1] *Les Corps Non Commutatifs*. Presses Univ. France 1972.
Bombieri, E. and Mumford, D.
[1] Enriques' classification of surfaces in char (p), II, *Complex Analysis and Algebraic Geometry: A collection of papers dedicated to K. Kodaira*, edited by W. Baily and T. Shioda. Cambridge University Press, Cambridge, 1977, pp. 23–42.
Borel, A. and Serre, J.-P.
[1] Le théorème de Riemann-Roch. *Bull. Soc. Math. de France, 86* (1958) 97–136.
Bourbaki, N.
[1] *Algèbre*. Eléments de Math. 4, 6, 7, 11, 14, 23, 24, Hermann, Paris, 1947–59.
[2] *Algèbre Commutative*. Eléments de Math. 27, 28, 30, 31, Hermann, Paris, 1961–65.
Boutot, J.-F.
[1] *Frobenius et cohomologie locale*. (Séminaire Bourbaki 1974/75. no. 453). Lecture Notes in Math. 514, Springer, Heidelberg, 1976.
Bredon, G.
[1] *Sheaf Theory*. McGraw-Hill, New York, 1967.
Bucur, I. and Deleanu, A.
[1] *Introduction to the Theory of Categories and Functors*. Wiley, London, 1968.
Cartier, P.
[1] *Les groupes $Ext^s(A, B)$*. (Séminaire Grothendieck (1957), Exposé 3). Secrétariat Mathematique, Paris, 1958.
Cassels, J. and Fröhlich, A., ed.
[1] *Algebraic Number Theory*. (Proc. Instructional Conf., Brighton, 1965). Thompson, Washington D.C., 1967.
Chase, S. and Rosenberg, A.
[1] Amitsur cohomology and the Brauer group. *Mem. Amer. Math. Soc. 52* (1965), 34–79.
Chow., W.-L. and van der Waerden, B.
[1] Zur Algebraischen Geometrie, IX. *Math. Ann. 113* (1937), 692–794.

Deligne, P.

[1] Les constantes des équations fonctionnelles des fonctions L. *Modular Functions of One Variable, II* (Proc. Internat. Summer School, Univ. Antwerp, 1972). Lecture Notes in Math. 349, Springer, Heidelberg, 1973, pp. 501–597.

[2] La conjecture de Weil, I. *Inst. Hautes Etudes Sci. Publ. Math. 43* (1974), 273–307.

[3] La conjecture de Weil, II ibid. (to appear).

Demazure, M. and Gabriel, P.

[1] *Groupes Algébriques.* vol. 1, *Géométrie algébriques, généralités, groupes commutatifs.* Masson, Paris 1970.

DeMeyer, F. and Ingraham, E.

[1] *Separable Algebras over Commutative Rings.* Lecture Notes in Math. 181, Springer, Heidelberg, 1971.

Douady, A.

[1] Cohomologie des groups compacts totalement discontinus. (Séminaire Bourbaki 1959/60, No. 189). Secrétariat Mathematique, Paris.

Douady, A. and Verdier, J-L., ed.

[1] *Séminaire de Géométrie Analytique.* Astérisque 36–37, Soc. Math. de France, 1976.

Giraud, J.

[1] Analysis situs. (Séminaire Bourbaki 1962/63, no. 256). Reprinted in *Dix Exposés sur la Cohomologie des Schémas,* North-Holland, Amsterdam, 1968, pp. 1–11.

[2] *Cohomologie Non Abélienne.* Springer, Heidelberg, 1971.

Godement, R.

[1] *Topologie Algébriques et Théorie des Faisceaux.* Hermann, Paris, 1958.

Grauert, H. and Remmert, R.

[1] Komplex Räume. *Math. Ann. 136* (1958), 245–318.

Grothendieck, A.

[1] Sur quelques points d'algèbre homologique. *Tôhoku Math. J. 9* (1957), 119–221.

[2] La théorie des classes de Chern. *Bull. Soc. Math. de France 86* (1958), 137–154.

[3] *Fondements de la Géométrie Algébrique.* (Séminaire Bourbaki 1957–62). Secrétariat Mathematique, Paris, 1962.

[4] Le groupe de Brauer. I. Algèbres d'Azumaya et interprétations diverses, II. Théorie cohomologique, III. Exemples et compléments. In *Dix Exposés sur la Cohomologie des Schémas,* North-Holland, Amsterdam, 1968, pp. 46–188.

Grothendieck, A. and Dieudonné, J.

Eléments de Géométrie Algébrique

[EGA. I] *Le langage des schémas.* Springer, Heidelberg 1971.

[EGA. II] *Etude globale élémentaire de quelques classes de morphismes.* Inst. Hautes Etudes Sci. Publ. Math. 8 (1961).

[EGA. III] *Etude cohomologique des faisceaux cohérents. Ibid 11* (1961), *17* (1963).

[EGA. IV] *Etude locale des schémas et des morphismes de schémas. Ibid 20* (1964), *24* (1965), *28* (1966), *32* (1967).

Grothendieck, A. et al.
Séminaire de Géométrie Algébrique
[SGA. 1] *Revêtements étale et groupe fondamental* (1960–61). Lecture Notes in Math. 224, Springer, Heidelberg, 1971.
[SGA. 4] (with Artin, M. and Verdier, J.-L.) *Théorie des topos et cohomologie étale des schémas* (1963–64). Lecture Notes in Math. 269, 270, 305, Springer, Heidelberg, 1972–73.
[SGA. $4\frac{1}{2}$] (by Deligne, P. with Boutot, J.-F., Illusie, L., and Verdier, J.-L.) *Cohomologie étale.* Lecture Notes in Math. 569. Springer, Heidelberg, 1977.
[SGA. 5] *Cohomologie l-adique et fonctions L* (1965–66). Lecture Notes in Math. 589, Springer, Heidelberg, 1977.
[SGA. 6] (with Berthelot, P. and Illusie, L.) *Théorie des intersections et théorème de Riemann-Roch* (1966–67). Lecture Notes in Math. 589, Springer, Heidelberg, 1977.
[SGA. 7] (with Deligne, P. and Katz, N.) *Groupes de monodromie en géométrie algébriques* (1967–68). Lecture Notes in Math. 288, 340. Springer, Heidelberg, 1972–73.

Grothendieck, A. and Murre, J.
[1] *The Tame Fundamental Group of a Formal Neighbourhood of a Divisor with Normal Crossings on a Scheme.* Lecture Notes in Math. 208, Springer, Heidelberg, 1971.

Hartshorne, R.
[1] *Residues and Duality.* Lecture Notes in Math. 20, Springer, Heidelberg, 1966.
[2] *Algebraic Geometry.* Springer, Heidelberg, 1977.

Herstein, I.
[1] *Noncommutative Rings.* Carus Math. Monograph 15, Math. Assn. of America, 1968.

Hilton, P. and Stammbach, U.
[1] *A Course in Homological Algebra.* Springer, Heidelberg, 1970.

Hirzebruch, F.
[1] *Topological Methods in Algebraic Geometry.* 3rd ed., Springer, Heidelberg, 1966.

Hoobler, R.
[1] Brauer groups of abelian schemes. *Ann. Sci. Ecole Norm. Sup. 5* (1972), 45–70.

Igusa, J-i
[1] Fibre systems of Jacobian varieties. *Amer. J. Math. 78* (1956), 171–199.
[2] Fibre systems of Jacobian varieties. II. Local monodromy groups of fibre systems. *Amer. J. Math. 78* (1956), 745–760.
[3] Abstract vanishing cycle theory. *Proc. Japan Acad. 34* (1958), 589–593.
[4] Betti and Picard numbers of abstract algebraic surfaces. *Proc. Nat. Acad. Sci. U.S.A. 46* (1960), 724–726.

Iverson, B.
[1] Critical points of an algebraic function. *Inventiones math. 12* (1971), 210–224.
[2] *Generic Local Structure in Commutative Algebra.* Lecture Notes in Math. 310, Springer, Heidelberg, 1973.

Katz, N.

[1] Algebraic solutions of differential equations (p-curvature and the Hodge filtration). *Inventiones math. 18* (1972), 1–118.

[2] An overview of Deligne's proof of the Riemann hypothesis for varieties over finite fields (Hilbert's problem 8). *Mathematical Developments Arising from Hilbert Problems*, edited by S. Browder, Amer. Math. Soc., Proc. Symp. Pure Math 28, 1976. pp. 275–306.

Kleiman, S.

[1] Algebraic cycles and the Weil conjectures. *Dix Exposés sur la Cohomologie des Schémas*, North-Holland, Amsterdam, 1968, pp. 359–386.

Knus, M. A. and Ojanguren, M.

[1] *Théorie de la Descente et Algèbres d' Azumaya.* Lecture Notes in Math. 389, Springer, Heidelberg, 1974.

Knutson, D.

[1] *Algebraic Spaces.* Lecture Notes in Math. 203, Springer, Heidelberg, 1971.

Kurke, H.

[1] Castelnuovo's criterion of rationality. *Mat. Zametki 11* (1972), 27–32.

Kurke, H., Pfister, G., and Roczen, M.

[1] *Henselche Ring und Algebraische Geometrie.* VEB Deutscher, Berlin, 1975.

Lang, S.

[1] Unramified class field theory over function fields in several variables. *Ann. of Math. 64* (1956), 285–325.

[2] *Abelian Varieties.* Interscience, New York, 1959.

[3] *Diophantine Geometry.* Interscience, New York, 1962.

Lazard, D.

[1] Sur les modules plat. *C.R. Acad. Sci. Paris, 258* (1964), 6313–6316.

Lefschetz, S.

[1] *L'Analysis Situs et la Géométrie Algébrique.* Gauthier-Villar, Paris, 1924.

Manin, Yu.

[1] *Cubic Forms: Algebra, Geometry, Arithmetic.* North-Holland, Amsterdam, 1974.

Matsumura, H.

[1] *Commutative Algebra.* W. A. Benjamin, New York, 1970.

Mazur, B.

[1] Rational points of Abelian varieties with values in towers of number fields. *Inventiones math. 18* (1972), 183–266.

[2] Notes on étale cohomology of number fields. *Ann. Sci. Ecole Norm. Sup. 6* (1973), 521–556.

Messing, W.

[1] Short sketch of Deligne's proof of the hard Lefschetz theorem. *Algebraic Geometry*, (Arcata 1974), edited by R. Hartshorne. Amer. Math. Soc. Proc. Symp. Pure Math. 29, 1975 pp. 563–580.

Mitchell, B.

[1] *Theory of Categories.* Academic Press, New York, 1965.

Miyanishi, M.

[1] *Introduction à la Théorie des Sites et Son Application à la Construction des Préschémas Quotients*. Presses de l'Université de Montréal, Montréal, 1971.

Mumford, D.

[1] *Geometric Invariant Theory*. Springer, Heidelberg, 1965.

[2] *Lectures on Curves on an Algebraic Surface*. Annals of Math. Studies 59, Princeton University Press, Princeton, 1966.

[3] *Introduction to Algebraic Geometry*. Lecture Notes, Harvard University Math. Dept., Cambridge, Mass., 1967.

[4] *Abelian Varieties*. Oxford University Press, Oxford, 1970.

Murre, J.

[1] *Lectures on an Introduction to Grothendieck's Theory of the Fundamental Group*. Lecture Notes, Tata Institute of Fundamental Research, Bombay, 1967.

Nagata, M.

[1] *Local Rings*. Interscience, New York, 1962.

[2] A generalization of the imbedding problem of an abstract variety in a complete variety. *J. Math. Kyoto Univ. 3* (1963), 89–102.

Nori, M.

[1] On the representations of the fundamental group. *Compositio Mathematica 33* (1976), 29–41.

Ogg, A.

[1] Cohomology of Abelian varieties over function fields. *Ann of Math. 76* (1962), 185–212.

[2] Elliptic curves and wild ramification. *Amer. J. Math. 89* (1967), 1–21.

Orzech, M. and Small, C.

[1] *The Brauer Group of Commutative Rings*. Lecture Notes in Pure and Applied Math. 11, Dekker, New York, 1975.

Raynaud, M.

[1] Caractéristique d'Euler-Poincaré d'un faisceau et cohomologie des variétés abeliennes. (Séminaire Bourbaki 1964/65, no. 286). In *Dix Exposés sur la Cohomologie des Schémas*, North-Holland, Amsterdam, 1968. pp. 12–30.

[2] *Faisceaux Amples sur les Schémas en Groupes et les Espaces Homogènes*. Lecture Notes in Math. 119, Springer, Heidelberg, 1970.

[3] *Anneaux Locaux Henséliens*. Lecture Notes in Math. 169, Springer, Heidelberg, 1970.

Reiner, I.

[1] *Maximal Orders*. Academic Press, New York, 1975.

Roquette, P.

[1] On the Galois cohomology of the projective linear group and its applications to the construction of generic splitting fields of algebras. *Math. Ann. 150* (1962), 411–439.

Šafarevič, I.

[1] Principal homogeneous spaces defined over a function field. *Trudy Mat. Inst. Steklov 64* (1961), 316–346 (Translated in *Amer. Math. Soc. Translations 37* (1964), 85–114).

[2] *Lectures on Minimal Models and Birational Transformations of Two Dimensional Schemes*. Lecture Notes, Tata Institute of Fundamental Research, Bombay, 1966.

[3] *Basic Algebraic Geometry*. Springer, Heidelberg, 1974.

Šafarevič, I. et al.

[1] *Algebraic Surfaces*. Proc. Steklov Inst. Math. 75 (1965) (Translated by Amer. Math. Soc. 1967).

Serre, J.-P.

[1] Faisceaux algébriques cohérents. *Ann of Math. 61* (1955), 197–278.

[2] Géométrie algébrique et géométrie analytique. *Ann. Inst. Fourier 6* (1956), 1–42.

[3] Critère de rationalité pour les surfaces algébriques. (Séminaire Bourbaki 1956/57 no. 146). Secrétariat Mathematique, Paris.

[4] Sur la topologie des variétés algébriques en caractéristique p. *Symposium internacional de topología algebraica* Universidad Nacional Autónoma de México and UNESCO, Mexico City, 1958, 24–53.

[5] Espaces fibrés algébriques. (Séminaire Chevalley 2 (1958), Exposé 1). Secrétariat Mathématique, Paris.

[6] *Groupes Algébriques et Corps de Classes*. Hermann, Paris, 1959.

[7] *Corps Locaux*. Hermann, Paris, 1962.

[8] *Cohomologie Galoisienne*. Lecture Notes in Math. 5, Springer, Heidelberg, 1964.

[9] *Algèbre Locale-Multiplicités*. Lecture Notes in Math. 11, Springer, Heidelberg, 1965.

[10] *Cours d' Arithmetique*. Presses Universitaires de France, Paris, 1970.

[11] *Représentations Linéaires des Groupes Finis*. 2nd ed. Herman, Paris, 1971.

Seshadri, C.

[1] L'opération de Cartier. Applications. (Séminaire Chevalley 3 (1958/59) Exposé 6). Secrétariat Mathématique, Paris.

Shatz, S.

[1] The cohomological dimension of certain Grothendieck topologies. *Ann. of Math. 83* (1966), 572–595.

[2] *Profinite Groups, Arithmetic, and Geometry*. Annals of Math. Studies 67, Princeton University Press, Princeton, 1972.

Steenrod, N.

[1] The work and influence of Professor S. Lefschetz in algebraic topology. *Algebraic Geometry and Topology: A. Symposium in Honor of S. Lefschetz*, edited by R. H. Fox, D. C. Spencer, Princeton University Press, Princeton, 1957 pp. 24–43.

Tate, J.

[1] On the conjectures of Birch and Swinnerton-Dyer and a geometric analog. (Séminaire Bourbaki 1965/66, no. 306). W. A. Benjamin, New York, 1966.

Verdier, J.-L.

[1] A duality theorem in the étale cohomology of schemes. *Proc. Conf. Local Fields* (Driebergen 1966), edited by T. Springer, Springer, Heidelberg, 1967, pp. 184–198.

[2] The Lefschetz fixed point formula in étale cohomology. *Proc. Conf. Local Fields* (Driebergen 1966), edited by T. Springer. Springer, Heidelberg, 1967 p. 199–214.

Waterhouse, W.

[1] Profinite groups are Galois groups. *Proc. Amer. Math. Soc. 42* (1973), 639–640.

[2] Basically bounded functors and flat sheaves. *Pac. J. Math. 57* (1975), 597–610.

Weil, A.

[1] Number of solutions of equations over finite fields. *Bull. Amer. Math. Soc. 55* (1949), 497–508.

Yuan, S.

[1] On the Brauer groups of local fields. *Ann. of Math. 82* (1965), 434–444.

Zariski, O.

[1] *Algbraic Surfaces*, (Springer, Heidelberg, 1935), 2nd supplemented ed. with appendixes by S. Abhyankar, J. Lipman, and D. Mumford. Springer, Heidelberg, 1971.

[2] Foundations of a general theory of birational correspondences. *Trans. Amer. Math. Soc. 53* (1943), 490–542.

[3] The theorem of Bertini on the variable singular points of a linear system of varieties. *Trans. Amer. Math. Soc. 56* (1944), 130–140.

[4] *Introduction to the Problem of Minimal Models in the Theory of Algebraic Surfaces.* Publ. Math. Soc. Japan, no. 4, 1958.

[5] The problem of minimal models in the theory of algebraic surfaces. *Amer. J. Math. 80* (1958), 146–184.

[6] On Castelnouvo's criterion of rationality $p_a = P_2 = 0$ of an algebraic surface. *Illinois J. Math.*, *2* (1958), 303–315.

Index